Dementia Care

Mala Kapur Shankardass
Editor

Dementia Care

Issues, Responses and International Perspectives

Springer

Editor
Mala Kapur Shankardass
Sociologist, Gerontologist and Health Social Scientist
Independent Freelance International/National Consultant and Writer
New Delhi, Delhi, India

ISBN 978-981-16-3863-3 ISBN 978-981-16-3864-0 (eBook)
https://doi.org/10.1007/978-981-16-3864-0

This Springer imprint is published by the registered company Springer Nature Singapore Pte Ltd.
The registered company address is: 152 Beach Road, #21-01/04 Gateway East, Singapore 189721, Singapore

Acknowledgements

This book would not have seen the light of the day without the cooperation and support of a whole lot of experts from different countries including my own. They stood with me, encouraged me to go ahead with this project to throw light on dementia issues not from a typical purely neuroscience or pharmacological perspective but bringing in other dimensions associated with dementia care. In particular, I am grateful to my friend Avinash De Sousa, a well-recognized expert in the field of mental health who kept on prompting me to reach out to other known professionals to contribute and enrich this publication. I am grateful to each one of them, without naming them individually here as their names appear in the chapters, for their scholarly contributions to this book. Without their belief in me, I could not have put this volume together, which for me is special as it touches themes that require not only attention but also visibility for taking forward concerns and responses to dementia care. Efforts of each of the authors are much appreciated for highlighting very significant areas of research, providing pertinent evidence and sharing knowledge from the different disciplinary background.

I am very grateful to Satvinder Kaur, Editor—Humanities and Social Sciences, Springer Nature India, for her patience, guidance and consideration for this book Dementia Care: Issues, Responses and International Perspectives. Her help as well of all her team members at many stages of the publication of this volume has been very valuable.

I would like to take this opportunity to thank my husband Suman Shankardass who has always stood beside me in my professional endeavours and helping me in many small and big ways to accomplish my intellectual goals. His enduring partnership is remarkable and desired. I want to also thank my son Varun, daughter-in-law Trisha and 4-year-old grandson Kirat for being with me, providing me necessary distractions from time to time to ease out the pressures of work in compiling this interesting volume.

Finally I am in gratitude of my friends and colleagues from across the world who in their own ways believed in my expertise and handling of this exciting project.

Gurgoan, India

Mala Kapur Shankardass

Contents

About the Editor

Mala Kapur Shankardass is a sociologist, gerontologist, health, and development social scientist. She is a teacher, writer, researcher, consultant, and an activist. She has been involved with aging studies since 1992 when she started postdoctoral work on aging issues. She has done prestigious assignments with United Nations agencies and international as well as national institutions. She has participated in panel discussions, delivered key note addresses, and chaired sessions on aging concerns with reputed organizations both in India and abroad. She has been recognized for her work and contributions in gerontology across the world.

Chapter 1
Introduction Perspectives on Dementia Care—Issues and Responses

Mala Kapur Shankardass

Abstract With ageing of populations in all countries, the number of people living with dementia is also rising. Families getting affected by care of dementia patients face many problems. There is no cure for dementia, yet various therapies are emerging to give relief to patients or delay the progression and manage the disease. Many responses are emerging which equip the families to take care of dementia-affected people. Perspectives from social sciences, legal profession, health sciences, social work, education and human rights are making inroads in different countries along with various care models which are supporting families and patients affected by dementia. This introductory chapter of the book provides some understanding of some measures arising from sciences, alternative therapies and community initiatives to emphasize that many developments are taking place on dementia care aspects internationally. The chapter is based on the author's theoretical and empirical understanding from a gerontologist's perspective about dementia care.

Keywords Dementia care · Concerns · Responses · Social science and health perspective · Legal safeguards · Educational tools · Alternative therapies · Community initiatives

With ageing of the populations, rates of dementia in societies is growing. With increases in dementia cases, care-related issues are receiving attention from multiple dimensions. There is national interest in many countries to deal with concerns related to dementia care, to have more research on dementia matters and to understand the impact on society from different perspectives, social, economic and health (ADI, 2013). This book with focus on dementia care is non-conservative in approach as it refers to various options to deal with the disease with impact on individuals and families but not to all that are available in different societies, nor from only a medical pharmacological viewpoint or either from a totally physiological perspective. This volume, unlike some recent editions on dementia which take into account pathologies and clinical aspects (Martin & Preedy, 2020a, 2020b) or essentially a neuroscience approach on dementia (Martin & Preedy, 2020a, 2020b) reviews different policy

M. K. Shankardass (✉)
Independent Freelance International/National Consultant and Writer, New Delhi, India

M. K. Shankardass (ed.), *Dementia Care*,
https://doi.org/10.1007/978-981-16-3864-0_1

responses in terms of certain national initiatives, discusses present status of care options from a social, psychological, social work, alternative medicine/therapies and diagnostic point of view which are being adopted internationally in various cultural settings. It brings attention to different concerns of families caring for those living with dementia from a sociocultural specific viewpoint but nonetheless relevant globally including emphasis on specific legal matters such as enduring power of attorney as well as focus on certain ethical issues. The book also reflects on what the future might hold for those affected by dementia as individuals or at the family and community level. This book with chapters contributed by experts from different disciplines and various countries goes through old and new developments in the field of dementia care by discussing aspects such as role of education in developing competent workforce for conducting research and innovative approaches to dementia treatment, country-specific policy options and narratives, supporting families with dementia; understanding the emerging issues and challenges for family caregivers, reviewing psychological pressures and concerns affecting caregivers; discussing role of technology; looking at rehabilitation processes, bringing attention to language impairments in dementia and the need for dementia friendly communities, focus on long-term issues; end of life care, different ethical issues, growing incidences of elder abuse among patients with cognitive impairments and also discusses occurrence of suicide among people with dementia. There are chapters highlighting use of music and art therapy, artistic collaborations for diagnosis of dementia, and there is also reference to dementia happening among younger age groups. Implicit in all the chapters of the book is emphasis on rights-based approach to enhance and improve the understanding for better life related to care of those with a diagnosis of dementia and their caregivers. There is much to share about dementia care, and this book helps us move along the road to an increasingly evidence-based knowledge about different selected dimensions.

This century has seen much activity in getting to know about dementia both on prevention and on research front. It is a vast field with many aspects outlining care, some which we know of, others which we are getting to know and lots still left to be known. Countries are bringing out guidelines which emphasize timely diagnosis; living well with dementia and delaying functional decline; managing symptoms through training staff in how to provide person-centred care and using non-pharmacological approaches in the first instance; and training and supporting families and carers to provide care. Many institutions/organizations, government owned as well as those in private sector concerned with brain health, cognitive impairments and memory-related matters emphasize on promotion of issues related with dementia care. For many concerned about this field and working on dementia care issues, it is frustrating not to have a cure yet for managing the disease. Dementia is leading to breakdown among individuals affected. The difficult circumstances which people with dementia go through can be assessed in many ways in terms of rising risk of suicide among them. As Isha Dhingra in her chapter in this book points out the neuro-biomarkers, clinical predictors, prevention, management and phenomenology of suicidal thoughts and behaviours in people with dementia have many implications for public health and policy responses which in contemporary times cannot be ignored.

The implacability of dementia disease is taking all those concerned with it towards many interventions and risk-reduction strategies which can bring relief to those affected by dementia. Diana Kerr for this book outlines the power of music on the lives of people living with dementia. There is growing body of research and practice which highlights for instance how music can be used in different ways to engage people living with dementia and have positive results on their quality of life, while research on finding drugs to treat dementia is continuing unabatedly dealing with those affected is taking recourse in many different ways. There is emphasis in many countries to develop policies and strategies to improve the lives of people affected with dementia. For instance in UK as elsewhere too we see policy being developed at the national level, though inconsistently at the clinical level, and implemented at the local and clinical levels for dementia risk reduction and prevention, especially through the primary healthcare settings. In Australia, it is observed that the quality of clinical practice in dementia care is variable (Laver et al., 2016). The availability of high-quality services to support workforce training, diagnosis and on-going care, advance care planning and support for families to provide care is seen to be inconsistent. But there are guidelines which provide recommendations for an agreed standard of practice for the diagnosis and management of people with dementia. The guidelines address care of people with dementia in community, residential care and hospital settings and are relevant to medical practitioners, nurses, aged care workers and allied health professionals. They are also useful for researchers, educators, policymakers and decision-makers. In this century, certain countries have taken concrete steps to develop healthcare provisions for dementia, long-term services and supports and hospices. But the rising monetary costs of dementia are a matter of concern which is taking an economic toll on nations, especially in the developed countries where cost analysis on dementia is being done. It is estimated that the economic burden on nations due to dementia is greater than heart disease or cancer. Data from America is indicating that in the next three decades the number of people requiring health care for dementia, long-term care services and hospices is going to double and costs of dealing with dementia will rise by 500% (Shih et al., 2014). In many countries with increased life expectancies, the number of older people developing dementia is on the rise, and it is observed that there are also two generations within one family to be living with dementia at the same time. In addition, in these countries with more single-parent and single-child households, the fact that caregiving is most frequently provided by family caregivers raises the question of who will take care of the growing number of older adults who may develop dementia (ibid). Further, as Christopher Johnson and David Dietrich in their chapter in this book bring attention to the disturbing fact of rising commodification of long-term care in societies as it impacts dementia care, the situation in future looks grave. They give example of the American society where using the concept of "McDonalization of Long Term Care" there is growing concern about dementia care in nursing homes, assisted living facilities and in memory care clinics. They voice their apprehensions about staff burnout, high turnover rates, poor quality of care, all having negative consequences on dementia care.

In many countries, dementia prevention is addressed through national policy but has limited scope since it varies in terms of implementation at the regional and local levels. Generally, it is seen that dementia care with regard to various modifiable risk factors is associated with non-communicable diseases by bringing emphasis on reducing obesity, changing lifestyle patterns, stopping smoking, taking care of diabetes, focusing on intake of certain food products and nutritional items, etc. There is less emphasis placed on dementia-specific concerns such as finding a cure. In recent times, many recommendations are being put in public domain on brain health exercises and improving cognitive faculties. These are primarily based on preventive strategies and being delivered through primary healthcare services in UK and other countries where there is an available platform for increasing knowledge and understanding on dementia. It is observed in many regions of the world that knowledge related to dementia is limited even among health professionals, who need specific training programmes for this in order to manage and treat the disease among the general population. Thus, enhancing the role of education, especially at postgraduate and higher levels, to provide a solid base through which lives of those affected by dementia can be improved is much needed as Louise McCabe in his chapter in this book has outlined. The need for adopting multidisciplinary perspectives in dealing with dementia cannot be undermined. Recent professional opinion across many developed and developing countries is for taking into consideration at the time of diagnosis of dementia and at regular intervals too assessment for medical comorbidities and key psychiatric features associated with dementia to ensure optimal management of coexiting conditions. Certain countries, for instance Australia, have recommendations for health and aged care professionals to adopt comprehensive role—appropriate dementia-specific training which can facilitate improving the quality of life for the person with dementia. These recommendations also assist the medical community to understand a person with dementia, to read their body language, behaviour and respond appropriately. A welcome practice in many developed and ageing countries is to have training models for dementia-specific specialists that focus on understanding symptoms and behaviours as well as provide person-centred care. Experts know that people with dementia cannot express their needs through communication but can do so through their actions and behaviour. Thus, comprehensive assessment and analysis of behaviour of the person with dementia become crucial for which adequate tools are needed for monitoring the type and patterns of behaviour and having an objective measurement of behavioural and psychological symptoms. Many international experts emphasize on patient-centred care of persons with dementia through building a dementia-friendly community. This approach is reflected in the chapter by *D*aria Smirnova and **T**atiana Smirnova in relation to treatment of language impairments in dementia with effect on various quality of life concerns and the need for neuro-psychological and psycholinguistic approaches in the choice for treatment and interventions. While reviewing available interventions for treatment, the authors also discuss de-stigmatization activities related to dementia care.

Internationally, the medical community, especially those involved with treating dementia-affected people, are bringing greater emphasis on promoting and maintaining independence of such persons through activities of daily living, continuing exercises, pursuing meaningful tasks that are of interest. In the absence of cure for dementia as a long-term treatment plan and management strategy, focus is also on appropriate nutritional diet, oral health and general fitness techniques. It is interesting that across the world provisions for dementia care are bringing greater emphasis on non-pharmacological interventions. At times, such measures are being adopted even before considering use of medications. It is significant that epidemiological and health data from across the world indicate the presence of dementia in all countries and further that such cases are incurable but characterized by progressive deterioration. Health professionals while pessimistic at present about being able to find a cure in the immediate near future are enthusiastic about progress being made in psychosocial interventions to improve the quality of life of persons with dementia and their caregivers. Very effectively, S. Shaji in his chapter in this book brings out the advantages of psychosocial rehabilitation of persons with dementia by outlining the focus on stabilizing the condition of the patients with control of certain factors contributing towards deterioration. It is now well recognized that dementia care involves cognitive rehabilitation, management of behavioural problems, stimulation- and emotion-oriented interventions.

As many studies from various countries are indicating, empowering families to look after people with dementia is much preferred way of treatment. No doubt there are challenges associated with family caregiving some of which are highlighted by Mathew Varghese, Upasana Baruah and Santosh Loganathan in their chapter on India in this volume. The emerging issues refer to the need to have culturally sensitive programmes for caregiver education and training tailored to the domiciliary and socioeconomic status of the family. It is important to develop models of care for low- and middle-income countries where dementia care is primarily home based. On similar lines, Beata Bugajska and Rafal Iwanski in their chapter in the book emphasize the emerging need for developing local strategies for strengthening the caregiving potential of families in Poland, rightly expressed by them in many countries, the main care force for dementia treatment is family caregivers, yet caregiving potential of families is under estimated. Be it Europe or Asia or any other continent, majority of caregivers are women aged 55–64 years. They are ageing, having specific health and social needs, but their requirements are ignored. Limited attempts are made in most societies to explore their concerns and issues and develop appropriate suitable initiatives for their well-being. Limited programmes exist in all countries to support them, harness their abilities for participation in family-based caring process for dementia-affected patients.

Women-oriented care initiatives and inputs in dementia care practices indicate it being homeward and evidence from different nation's show that increasingly majority of people with dementia are being looked after by their relatives in their homes be it spouse, sibling or adult child. K. S. Shaji and Nimmy Chandran in their chapter bring attention to an important aspect related to dementia in societies and particularly in low- and middle-income countries that of care being home based and not hospital or institutional based. As rightly discussed by them, the challenges associated with providing home-based care are plenty, and these need to be faced as the advantages of such provisions are many, especially when there are limited institutional care facilities. Rehabilitation of carers has emerged as an important intervention in dementia care. Along with medical management and non-pharmacological interventions, providing carers with support and educating them on the disease reduce carers burden and help them achieve the quality of life for themselves and for people with dementia. Saadiya Hurzuk and Anand Ambali in their chapter portray the need for holistic approach with person-centred care for dementia. It is essential that family carers and professional care providers who generally have high risk of being affected by various illnesses as certain studies do indicate should have programmes for reducing their stress, depression, anxiety, habit of indulging in substance abuse and aggravation of various kinds which can be underlying causes of their illnesses. As many studies are showing, universally the impact of providing dementia care by family member has various psychological after effects on the caregiver, and it is seen to be associated with increased burden and strain felt by them. Ipsita Basu and Susmita Mukhopadhyay's chapter focuses on this important aspect with suggestions for need to develop coping strategies by the caregivers.

Within the gambit of non-pharmacological interventions is also the need to work with carers and families which help in managing symptoms and bring out effective communications essential for treatment. The challenge of supporting families living with dementia demands specialized knowledge for understanding the wider disciplinary and cultural settings through which solutions can be found. Interestingly, many countries are reviewing taking into account a relational approach to dementia policy that addresses the needs of all those involved in supporting people with dementia, whether in an unpaid or paid capacity. Marvin Formosa keeping this focus in his chapter describes the Maltese government's *National Dementia Strategy* which aims at enhancing the quality of life of persons with dementia and family carers. There is emerging universal agreement among experts dealing with dementia on including carers and families in the planning, decision-making, managing and caring for people with dementia. However, evidence from many countries indicates that carers are often not provided with enough support or adequate training to effectively provide care to person living with dementia. Yet recent research from different countries is pointing towards the fact that multifaceted programmes involving both the carer and the person with dementia can improve quality of life for both. There is increasing focus on the principles of dignity in care and in recognizing the dementia person as an individual engaged in various activities which need to be respected. These concerns are particularly noteworthy when it comes to judging the testamentary capacity of the patient and certifying for their health. Legal aspects related to dementia care have many dimensions. Taking example of Australia Eileen Webb brings attention to a serious legal concern related to dementia care in terms of appointment of an enduring power of attorney and certain consequences which it has in increasing risk of elder abuse, facing difficulties in undertaking various transactions related to important matters and its various implications for people involved. The growing incidence of elder abuse in dementia care is a matter of grave concern which requires attention in developing various combating strategies. Sara Alon in her chapter with reference to Israel discusses the vulnerability of older adults with dementia to abuse. Through her study and with reference to other research work, she reiterates that dementia, dependency and stress experienced by the caregiver are predictors to the occurrence of elder abuse. Giving a broad range from 5.4 to 62.3% prevalence rate for abusive behaviour towards older adults in community settings, she restates the requirement for detection and methods of intervention in this field.

A human right approach and ethics involved in treating the patient, deciding on the course of treatment and delivery of care has far reaching consequences. In recent times, much emphasis is being laid on meeting the global challenge of ethical issues related to dementia care. Avinash De Sousa and Russell D'Souza bring out the concern for ethics-oriented approach, especially in clinical practice and research eloquently in their chapter. Ethical issues become relevant for early detection and diagnosis, for use of medications in mild cognitive impairments, for genetic testing, in matters assessing the role of neuroscience to enhance cognition in dementia subjects. Ethical dilemmas relate to diagnostic disclosure, appropriate informed consent processes, conflict of interests in research as well with research design issues in dementia care trials, issues with the use of placebos and the use of biological treatments in these subjects. The importance of ethical issues is also pivotal in the later stages of disease as concern about appropriate therapeutic goals and end-of-life caregains significance. Similarly, when it comes to admitting dementia patient in an institution ethics of treatment plays a crucial role as much as it revolves around taking a decision for home care. In some countries, the state's apparent withdrawal of minimum service and delegating most of the task to home care could have ethical implications. Safeguards to quality care most commonly seen in wards or hospitals are missing in home care. This might do more harm for the patient than good. However, medical and nursing professionals have to weigh if choosing to provide care at home would be more beneficial for the patient or otherwise. Research from advanced countries indicates the utility of having team-oriented approach for community care of patients with dementia with nursing and medical staff working dedicatedly with community workers. This implies an increase in workforce requirements for community care and services.

The extent to which technology may be able to support people with dementia and their caregivers is of interest to the patients, caregivers, policy-makers and governments. Avinash De Souza in his chapter throws light on the role of technology in dementia care, treatment and support and its role in helping with preventing disease progression. Globally, technologies identified in dementia help in memory support, treatment, safety and security, training, care delivery, social interaction and with many other aspects of care in different settings including technology-aided reminiscence or therapeutic mechanisms. An illustrative chapter in the book is by Sadie Harmon, Erin Partridge and Bob Savage which exemplifies the role of video conferencing technology as well as the broader implications of such technology for people living with dementia. It puts forth the importance of community art and art therapy practice in dementia care through the equitable collaboration between people with lived experience and professionals as experienced in America. On similar grounds,

Michelle Olson elucidates the use of creative arts therapy, poetry and simple mindful techniques that are grounded in nature to empower people living with dementia in the USA. She provides evidence for opportunities through artistic and verbal expressions, by making meaningful connections with self and others to deal with dementia and improve quality of life of those involved. Professionally, it is believed that if effective communications exist during treatment then there can be reduction in symptoms.

Significantly, in many parts of the world through international and national government and non-government channels, programmes are being outlined that include educational information regarding dementia which benefits all those affected with dementia as carers or as individuals living with dementia. In Australia, Denmark, Germany, India, the Netherlands, UK, to name a few countries, there are since last few years programmes tailored for providing individualized care management strategies, especially to overcome specific problems related with dementia, supporting coping mechanisms, learning effective communication skills to deal with person with dementia, along with programs for providing respite care, stress management, and maintaining personal well-being. Some of the chapters in this volume discuss these strategies and I am sure will make interesting reading. Over the years, since the start of this century, many organizations have come up in various countries which provide support to individuals and families affected by dementia as also referred in some of the chapters in this book. These practices listed in the different chapter texts contribute towards better communication with help of tools for effective interactions and understanding as per dementia person's care plan. End of life care as well as palliative care programmes have been developed and are practiced in many parts of the world, especially for those with advanced dementia. Colin Brewer in his chapter raises a critical issue related to dementia care that of giving choices and preferences for medical assistance in dying (MAID). Based on his research, Colin shares the fact that dementia is now feared more than cancer. Various surveys show that when dementia becomes severe, then those affected do not want to continue living. His opinion is that in Britain and Canada, most dementia patients would like the option of an advance request for MAID in the form of euthanasia when dementia reaches a certain degree of severity. But even when it is legal as in the Benelux countries where doctors are sympathetic to MAID, the process is seen to be difficult. Important differences exist between religious groups in their attitudes to MAID in all diagnostic contexts. The chapter referring to some case histories of patients from Switzerland discusses some of the ethical, philosophical, religious and diagnostic issues that confront Western societies but are also salient in other cultural contexts.

The responses to dementia care seen in different parts of the world are meaningful as every year more and more people are getting affected by it and the number is expected to rise. No doubt different national health ministries are developing and implementing policies and programmes to systematically address the needs of individuals and families affected by dementia, but the progress is seen more in the European region than in the America or the Asia Pacific countries (Sun et al., 2020). Of course the challenge to governments to respond to the growing numbers of people living with dementia are substantial, but as WHO (2015) recommends a broad public health approach can help towards improving the care and quality of life of people living with dementia and family caregivers. WHO (2017) has advised its member countries to initiate policies or develop strategies on dementia by tuning it with specific country healthcare policy or old-age policies and plans or even have a standalone dementia strategy with a public health perspective. Such attempts are also meaningful, especially when dementia is being witnessed to occur among younger populations. Vidya Giri Shankar in his paper based on UK brings attention to early onset of dementia, specifically among those below the age of 65 years, which makes for challenging clinical situation. It presents different symptomatology, aetiology, diagnostic strategy and psychosocial concerns.

Many countries are working on the recommendations of WHO, and some have accomplished framing action plans. So far, Australia, Belgium, Costa Rica, Cuba, Denmark, Finland, France, Ireland, Israel, Italy, Japan, Luxembourg, Mexico, Netherlands, Norway, Republic of Korea, Switzerland, United Kingdom of Great Britain and Northern Ireland (separate plans for England, Northern Ireland, Scotland and Wales) and the USA have developed national dementia policy or plan. Besides the push by WHO to promote a dementia strategy in India Alzheimer's and Related Disorders Society of India is making attempts for rapid action by the government which is working on this. What is significant is that the various national policies and plans have several common key issues and emphasize on effective sustainability to achieve the results. They bring together relevant stakeholders, identify priority areas for action and focus on implementing strategy by committing on resources, suggesting intersectoral collaborations. It is important to have time frame for action, build monitoring and evaluation mechanisms which the different country plans advocate. The various national policies already in place and those in the making stress on raising awareness on dementia care, the significance of timely diagnosis, being committed towards quality continuing care and services, supporting caregivers, voicing the need for workforce training and prominently give importance to prevention and research. Having rights-based approach towards people living with dementia and their caregivers is pivotal in contemporary times; however, this needs to be translated through legislative and regulatory processes to be effective in reducing discriminatory practices. The different existent country policies on dementia adopted in the last couple of years or so strategically provide formal recognition to these issues on dementia care. Imperative in the context of upholding the rights of people with dementia as recommended by WHO is the recognition of protecting for people affected with dementia the right to appropriate autonomy and self-determination. It is pertinent for ageing societies to focus on such concerns which are ethical and appropriate. In many societies, through education, attempts are being made to enhance the capacities of all stakeholders—policymakers, professionals and families to retain rights-based perspective in dementia care.

Burden of care these days is being related to family caregivers of those affected by dementia. A huge impact on the lives of the family members who are primary carers takes place when caring for a person with dementia. The informal caregivers require special attention, respite care provisions, psychological support and mechanisms which can reduce the strain of caregiving and stress levels. A stated above, few chapters in the book discuss these dimensions and provide insights into programmes being adopted to assist caregivers at the community level. Certain studies indicate the beneficial effects of caregiver interventions in decreasing the institutionalization of the care recipient. In developing countries, many practices related to home-based support for caregivers are being planned which impact positively on the caregivers mental health status and minimize the burden of care. However, due to stigma attached to dementia, there are barriers in some countries to avail of home-based dementia care services. In addition, the lack of awareness of available services, cultural and financial constraints restrain proper utilization of such measures existent in societies. To overcome these factors, proactive, culturally relevant and context specific actions are required including advocacy, awareness generation, developing and implementing dementia policies and plans, health system strengthening, capacity building, supporting caregivers and research. Much of these initiatives are viable with strengthening multispecialist teams by integrating healthcare and social care professionals as well as being associated with legal functionaries. Dementia care is a global public health challenge requiring interventions from different perspectives. Being a chronic and complex condition dealing with dementia requires sustained action and coordination between multiple stakeholders. It must be prioritized in policies, plans, laws and services and involve people living with dementia and their caregivers who are affected for many years after the onset of symptoms of dementia. It is pertinent for the society that programmes related to dementia care keep the goal of providing quality of life to patients, their family and care givers. With appropriate national-level support countries can turn into dementia-friendly societies globally. All nations can widen research on dementia care, improve services, increase diagnostic facilities, and have greater reach out at the local levels.

References

ADI. (2013). Policy Brief for G8 Heads of Government. The Global Impact of Dementia 2013–2050. Alzheimer's Disease International.

Laver, K., Cumming, R. G., Dyer, S. M., Agar, M. R., Anstey, K. J., Beattie, E., Brodaty, H., Broe, T., Clemson, L., Crotty, M., Dietz, M., Draper, B. M., Flicker, L., Friel, M., Heuzenroeder, L. M., Koch, S., Kurrle, S., Rhonda Nay, C., Pond, D., … Yates, M. W. (2016). National policy actions on dementia in the Americas and Asia-Pacific: Consensus and challenges. Clinical practice guidelines for dementia in Australia. *Medical Journal of Australia, 204*(5), 191–193.

Martin, C. R., & Preedy, V. R. (Eds.) (2020a). *Diagnosis and management in dementia: The neuroscience of dementia* (Vol !). Academic Press: An imprint of Elsevier.

Martin, C. R., & Preedy, V. R. (Eds.) (2020b). *Genetics, neurology, behaviour, and diet in dementia: The neuroscience of dementia* (Vol. 2). Academic Press: An imprint of Elsevier.

Shih, R. A., Concannon, T. W., Liu, J. L., & Friedman, E. M. (2014). Improving dementia long-term care book subtitle: A policy blueprint. RAND Corporation.

Sun, F., Chima, E, Wharton, T., & Iyengar, V. (2020). National policy actions on dementia on the Americas and Asia—Pacific Consensus and challenges. *Rev Panam Salud Publica.* 44: e2., Published online January 22. https://doi.org/10.26633/RPSP.2020.2. PMCID: PMC6971847.

WHO. (2015). *Dementia a public health priority.* World Health Organization.

WHO. (2017). *Global action plan on the public health response to dementia 2017–2025.* World Health Organization.

Chapter 2
Suicide in Dementia

Isha Dhingra

Phenomenology

> "I started noticing the changes myself few months ago...My family was worried for over a year...I was good with language, reading and writing... but now I struggle even in conversation to get the right words. It's scary...I worry about the future."- Mr K., Chennai (Kumar et al., 2020).

In the penultimate episode of the final season of Mad Men, set in the 1960s, Betty Francis, who has recently found out that she has terminal lung cancer, has this heart-wrenchingly plain conversation with her daughter, Sally.

Sally: He [referring to Betty's husband, Henry] doesn't know you won't get treatment because you love the tragedy.

Betty: Sally, I've learned to believe people when they tell you it's over. They don't want to say it, so it's usually the truth.

Sally: Henry says you could have a year.

Betty: And what would that year be like?

Sally: I'll be with you. I won't let you give up!

Betty: I know that. But I watched my mother die. I won't do that to you. And I don't want you to think I'm a quitter. I've fought for plenty in my life. That's how I know when it's over. It's not a weakness. It's been a gift to me. To know when to move on.

Less straightforward are Emeritus Professor Norman Cantor's deliberations on when and how, not whether, he would end his life were he to receive a diagnosis of Alzheimer's disease. He explains, "My prime object is to avoid the precipitous mental deterioration accompanying advanced Alzheimer's or similar dementia... My

I. Dhingra (✉)
Yale University, New Haven, USA
e-mail: isha.dhingra@yale.edu

M. K. Shankardass (ed.), *Dementia Care*,
https://doi.org/10.1007/978-981-16-3864-0_2

aversion is grounded in my abhorrence of reduced mental function to a degree I deem intolerably demeaning. Such a status is unacceptable to me *whether or not I would experience distress in a future demented state* [italics author's]. Keep in mind that I spent my work career as an academic. My personal satisfaction and self-image have flowed largely from intellectual functions like observation, reflection, and analysis. Inability to understand and process information is, for me, an intolerably undignified status".

Both stories reflect an urgency to preserve control, dignity, autonomy, and fidelity to one's self-image. When control over one's own life feels in waning, there can still be control over its continuity. Betty refuses palliative therapy and writes Sally a note essentially detailing her desire to preserve how she was defined in life, which was by her beauty and grace, in death. She is to be buried in her favourite blue chiffon, with her hair the way she likes to wear it. Her lipstick can be found in her purse. Norman, the academic, asserts, "I plan to engineer my self-deliverance", then visits a myriad of methods to do so with in-depth analysis and debate. Just like Betty, he wants to be known for what he built during his life in death, which in his case is intellectual prowess, not decay.

Professor of Ethics in Medicine Rebecca Dresser argues that "even well-informed individuals can make misguided choices because dementia is a "transformative experience" that affects people in ways that they cannot anticipate. As a result, they can only guess at what they might want and need as a person with dementia". But in conclusion to his argument for suicide in the early stages of his hypothetical dementia, Norman maintains his resolve. "As an adult, I have nurtured and developed a character and a life image. I have earned the prerogative of avoiding an intolerably undignified existence even if my future persona would not be palpably suffering in that condition. My life narrative belongs to me, not my hypothetical future persona" (Cantor, 2020).

Annie Zwijnenberg from the Netherlands, where euthanasia was legalized 2002, was filmed in a documentary entitled "Before It's Too Late", which captures her journey through advancing Alzheimer's disease and her death by euthanasia. Her children say she knew exactly what she wanted the moment she was diagnosed. Annie describes her unacceptably altered quality of life: she was once an avid walker, skier and hiker but was unable to go anyplace once her illness set in because "I get confused all the time". Her daughter, Anneke, says, "There is no right or wrong decision. It's hard to decide you want to die but it's as hard to decide, I think, that you want to live. She [referring to Annie] hated it when someone said: 'It's so brave that you made this decision.' She said choosing to live with dementia is just as brave". A key requirement for euthanasia is that the patient must have the capacity to understand their decision to die, a potentially inaccessible luxury in late stage dementia. Annie chose euthanasia in the relatively early stages of Alzheimer's because she was worried that she would not be afforded the privilege were she to wait longer. This fear is common in people with early stage dementia who are considering euthanasia. "Five to midnight", referring to Cinderella leaving the ball just in time, is an increasingly common euphemism that captures the tension between choosing to die too soon and waiting until it is too late (Bomford & Doyle, 2019). This dilemma has also been labelled the "Loss of Capacity Fear" or "Dementia Fear" (Terman, 2016).

Annie's and Norman's stories bring to light the important place that the popular medicalization of ageing and dementia holds in shaping personal narratives. Both their decisions are based at least in part on informed fears of what their lives might look like should the disease take over completely.

To this end, Elizabeth Herskovits observes that the popular conception of Alzheimer's disease has in fact irreversibly tied a "loss of self" to the illness, and "consequently, the subjective experience of being and becoming old has become increasingly distressing" (Herskovits, 1995).

The evolution of the idea of deteriorating cognitive function from simply being a part of ageing to the forbidding biomedical conceptualization of Alzheimer's disease took place in the 1970s and 80s carrying with it the notion that a person with deteriorating cognitive function, potentially Alzheimer's disease, would become progressively dependent on others for their basic needs (Milne & Phillips, 2020). A counter perspective is offered in a quote from Herskovits's paper:

> [T]he key psychological task in dementia care is that of keeping the sufferer's personhood in being. This requires us to see personhood in social rather than individual terms ... in a context of psychological abundance, where interdependence is openly acknowledged... dementia would not turn out to be such a tragedy, and dementia care not so great a burden .... Rather, it may become an exemplary model of interpersonal life, an epitome of how to be human. (Herskovits, 1995; Kitwood & Bredin, 1992)

Similarly, neuroscientific advances in the early diagnosis of dementia may preempt an individual's actual experience of the condition and predispose to suicide based on the popular medicalized understanding of the illness. On the other hand, accurate biomarkers of advancing illness and scientific ways to know what that advancement could mean for a particular individual could help solve the dilemma of when to act on suicidal intent for patients with dementia who still wish to die and prevent death by suicide for others.

In interviews with The Guardian, people diagnosed with early onset dementia describe relating to various extents to the depiction of the illness in a popular film based on Lisa Genova's novel, Still Alice. They describe personal experiences of guilt, from the idea of burdening or someday not recognizing their loved ones; of being misunderstood and stereotyped by people they thought they could rely on; of being more at the mercy of their moods than their cognition, and sometimes functioning as "emotional" rather than "intellectual" beings, and how that can be a good thing; of processing incoming information just as they normally would but not being able to recall it later; and of having many "foggy" and many "sunny" days. Julianne Moore, who researched her lead role in Still Alice via conversations with Alzheimer's activists, women diagnosed with early onset Alzheimer's, and clinicians, says, "I think an idea still stands that Alzheimer's is all about memory. One of the things I found is that people often simply feel lost. Alzheimer's is more akin to an ongoing panic attack where suddenly nothing has any reference. It's like having to cut through fog every day". In the film, Alice, once diagnosed, sets up an intricate system to end her life before it is too late, which eventually does not succeed (Davis, 2014; Seymour, 2015).

A quite specific phenomenology has been recorded in an American Vietnam war veteran who went on to develop semantic dementia in his mid-to-late 50s due to an underlying anterior temporal lobe atrophy. He attempted to end his life on multiple occasions because he was unable to picture himself doing activities that had in the past defined him, or in other words, he was unable to see himself as "capable" in his own future. For instance, while he had once been a bartender and had an episodic recall of this, albeit with patchy semantic detail, he was now unable to place himself in that role or even picture himself standing behind a bar. As a result of this loss of a sense of his future self, he felt empty, hopeless, depressed and suicidal. The disease "robbed him of his brain". "What can I do without a brain? What am I going to do for the rest of my life? How would I live without a mind?", he said, when describing wanting to end his life during interviews. See (Hsiao et al., 2013) for the complete case report.

Another documented dementia syndrome-specific phenomenology is a sense of alienation and expendability, which has emerged as a consistent theme in patients with HIV. People with HIV dementia are also prone to be of younger age, have coexisting psychiatric morbidities such as depression or addictions and develop disinhibition and impulsivity, so exacerbating their risk for suicide (Alfonso & Cohen, 1994).

Clinical Medicine and Suicide in Dementia

Merriam Webster calls dementia a progressive condition marked by the development of multiple cognitive deficits such as memory impairment, aphasia and the inability to plan and initiate complex behaviour. The DSM-5 says that "major neurocognitive disorder", a substitute term for dementia that attempts to divorce it from any connotations of insanity, involves a significant cognitive decline from baseline substantiated by the individual's or a close informant's concern or by clinical testing, which interferes with independence in everyday activities. This is distinguished from mild neurocognitive disorder, or a noticeable but modest cognitive decline that does not interfere with independence in everyday activities (APA, 2013).

An estimated 131 million or 1 in 85 people will suffer dementia globally by 2050, with a recent meta-analysis showing a striking rise in the incidence and prevalence of the Alzheimer's type in developing nations (Brookmeyer et al., 2007; Gao et al., 2019). In India, the reported prevalence of dementia varies between 1 and 10% owing to methodological variations in the literature and socio-economic and cultural factors that impact reporting like inadequate awareness and the stigmatization of mental illness (Kumar et al., 2019). The Alzheimer's and Related Disorders Society of India, in their report "Dementia in India 2020" (Dementia in India, 2020), note that approximately 5 million people in the country live with dementia with many more at high risk and no cure in sight. They also point out that dementia robs a person of not only their memories and personality, but often of their dignity, a critical perspective that the clinical definitions of the illness appear to overlook. It can be argued that in ignoring the depth of personal loss, particularly the loss of control over one's personal narrative, the clinical definitions of dementia partially defeat their own purpose, for diagnosing something is only as good as the potential to ameliorate the suffering caused by it.

Older adults are known to be at an increased risk for suicide as well as for involving lethal means to suicide, incurring 3–9 times the general suicide risk, with higher rates in developing nations (Conejero et al., 2018a; Upadhyaya et al., 1999). For people with dementia, stages of acute distress and therefore of higher risk of suicide include the time of diagnosis and the phase of illness with perceptible deterioration and retained insight (Cipriani et al., 2013). Other identified stressors are a younger age at diagnosis, distress over potential loss of autonomy or being a burden on caregivers and perceived increasing difficulties with activities of daily living. Conversely, the later stages of dementia with impaired insight and executive function are protective against suicide, although executive deficits may independently predispose to suicide as well (Cipriani et al., 2013; Diehl-Schmid et al., 2017; Richard-Devantoy et al., 2012, 2016; Schneider et al., 2001). This paradoxical drop in psychic distress in the later stages of dementia may be one of the reasons why the reported rates of suicide in people with dementia are lower than those of the general population.

At this point, it is imperative to acknowledge two things: first, that old age itself can bring on an existential debate over whether continuing to live is worth the risk of becoming debilitated or a burden on others; and second, that there exists an array of potential comorbidities that can independently or in concert precipitate suicidal thoughts and acts in older people and in those with dementia. These include chronic medical ailments such as hypertension and other cardiac diseases, diabetes, orthopaedic issues and pain; psychiatric comorbidities such as depression, anxiety, adjustment disorders, bipolar disorder and psychosis; and the social collaterals of ageing such as the loss of loved ones, loss of routines due to retirement from work and, increasingly, the loss of home to nursing home. In this context, a German study interviewed 141 older patients (not specifically suffering from dementia) admitted over nine years (1973–82) for abortive suicide, i.e. people who came close to attempting suicide but did not. They found depression and loneliness to be the predominant cause. Other causes included severe physical complaints (also associated with depression) and paranoid psychoses (Bohm, 1988). Notably, abortive suicide has a similar level of intent and a strong association with attempted suicide, and people who make aborted attempts are twice as likely to attempt suicide as those who have not (Barber et al., 1998).

Determining the prevalence of depression, and, by extension, suicidal thoughts and behaviours in dementia has been a consistently troubling issue. Numbers vary drastically depending on the population studied, the diagnostic tools and other research methodology used, the type of dementia, and the clinical setting. Studies included in a recent meta-analysis across types of dementia (Alzheimer's, vascular, fronto-temporal and dementia with Lewy bodies) reported prevalences of depression between 0 and 93% (Kuring et al., 2018).

Curiously, one study found the prevalence of major depression to vary significantly between patients attending geriatric psychiatry services (~29%) and patients attending a memory clinic (8%), despite there being no significant difference in the severity of cognitive impairment between the two settings (the patients from the memory clinic were included in the interest of putting together a representative sample of people with dementia, as they were expected to have milder cognitive impairments). Demographic variables did not vary between the two settings, and the memory clinic ironically had a higher proportion of non-Alzheimer's type dementias, known to have greater co-occurrence of depression than the Alzheimer's type (Bennett & Thomas, 2014; Kuring et al., 2018). While this discrepancy in the prevalence of depression between the two settings may be attributed to differential referral, with patients meeting the criteria for major depression being referred to psychiatric services, the effect of the clinical setting itself was not considered (Ballard et al., 1996). So, the presumption of the researchers in this particular study was that the memory clinic would yield patients with milder cognitive impairments than geriatric psychiatry services, which did not hold true. Yet, we can assume this to be the presumption of the people attending the clinic themselves, that they are not as terribly off as others who have to attend a psychiatric service. Could this possibly have been a mitigating factor against externally mediated depression in the memory clinic population?

Even in general (i.e. irrespective of the presence of comorbid mood disorders), methodologically robust research into suicide in dementia is lacking, and data specific to vascular, frontotemporal, Lewy body and HIV dementia, in which suicidal thoughts and behaviours may be more common, are sparse (Haw et al., 2009). Available data are derived from varying methodology and are often contradictory, with some studies suggesting an association between dementia and suicide and others negating any (Conejero et al., 2018b).

A cohort study following people diagnosed with dementia during hospitalization over 11 years found a significant association between the diagnosis of dementia and subsequent suicidal deaths independent of comorbid mood disorders (Erlangsen et al., 2008). Notably, suicides in older adults may be "silent" or underreported by as many as 40% or more, including cases due to self-starvation, dehydration, medication overdoses or "accidents" (Alphs et al., 2016).

Yet there is some consensus over there being an increased risk of suicidal thoughts and behaviours in mild dementia (possibly by way of preserved insight and executive functioning), soon after diagnosis, in Huntington's disease, in those with comorbid depression, and in those for whom antidementia medication has not been efficacious (Haw et al., 2009). As mentioned previously, late stage dementia appears to be protective against suicide, potentially by way of impaired executive functioning. However, in Alzheimer's disease, suicide has been found to occur even many years after diagnosis, supporting the case for regular and continued clinical assessment of suicidal thoughts and actions in people diagnosed with dementia (Serafini et al., 2016).

No specific guidelines exist for the clinical assessment and management of suicidal thoughts and behaviours in people with dementia. A future protocol for the clinical assessment for suicidality in patients with dementia may benefit from considering the following:

First, a keen awareness of which patients bear a greater risk for suicidal thoughts and behaviours. As mentioned, these include patients with a history of any kind of self-harm, those who have been recently diagnosed, those with palpably deteriorating cognitive abilities but relatively intact executive functioning, those with impaired executive functioning who as a result are prone to impulsivity, those with certain dementia syndromes such as HIV dementia (Alfonso & Cohen, 1994), Huntington's disease (Fiedorowicz et al., 2011) and frontotemporal disease (Sabodash et al., 2013), those with comorbid psychiatric syndromes or symptoms, especially of mood disorders, substance use disorders (Bellini et al., 1998) or psychoses.

General risk factors for suicide such as a family history and poor social support cannot be ignored either. It has been found that apart from suicidal ideation, or thoughts self-harm with the goal of ending one's own life, "death ideation", or passive death wishes, defined as thoughts that life is not worth living or the desire for one's own death, is prevalent in older adults and associated with an increased risk of suicide (Scocco & Leo, 2002).

Demographic risk factors include age (those diagnosed young as well as those over the age of 70), male sex, and white race, but it is unclear as to how much importance should be given to these factors, especially given that true diversity in research is hard to come by.

Second, a habit of routinely and compassionately assessing mood, emotion, social determinants of health and suicidal thoughts and behaviours in a socioculturally sensitive manner, which has yet to fully develop even in general psychiatry, let alone in neurology, general medicine, geriatrics, urology, or orthopaedics, all of whom may treat patients with dementia.

The ideal assessment would include a narrative component to capture each individual's experience of their dementia syndrome and suicidal motivations, apart from the use of standardized scales for dementia and suicide. A review of scales used to assess suicide in older adults can be found here (Alphs et al., 2016). Scales used for the assessment of dementia in India can be found here (Wadia & Khadilkar, 2015) and here (Swati et al., 2015). The ideal assessment would build towards an individualized treatment plan to include available drug therapies and counselling appropriate to the desires and needs of that particular patient. The inclusion of supportive loved ones in the treatment plan could be a critical component, especially in the Indian context. And of course, follow up, follow up, follow up.

Third is an updated awareness of country or region-specific laws and ethical principles surrounding both suicide and dementia. For instance, section 309 of the Indian Penal Code (IPC) contains—"Whoever attempts to commit suicide and does any act towards the commission of such offence, shall be punished with simple imprisonment for a term which may extend to one year or with fine, or with both" (Ranjan et al., 2014).

As Vadlamani and Gowda (2019) describe, while this law remains in place, the Mental Healthcare Act (MHCA) of 2017 in its section 115 supersedes IPC

Section 309, decriminalizing suicide and enabling access to free treatment and rehabilitation for victims, though details regarding funding for these treatments remain unclear. However, abetment of suicide remains punishable. Further, while the MHCA allows for advance directives wherein a patient may refuse treatment for their mental illness, including attempted suicide, these directives can be revoked by a review board.

Finally, "active" euthanasia is a crime in India while "passive" euthanasia, or the withdrawal of life support for those persistently dependent on it or the refusal of treatment by the terminally ill, has now been legalized. However, accessing passive euthanasia remains impractical owing to the bureaucratic hurdles in place (Srivastava, 2018). Voluntarily stopping eating and drinking (VSED) as a means to a literal end remains a grey area (Terman, 2016).

Last, Indian clinical culture is still heavily paternalistic. It may help in building trust between physician and patient were the physician to let go of the presumption that they "know better" than the patient. Even with a complete lack of formal systemic education, the patient or their caregivers will have far greater insight into their lived experiences than the physician ever could, irrespective of the number of patients that physician has seen or their or their own or a loved ones' lived experiences. Listening to patient experiences could provide key knowledge that may separate successful from unsuccessful therapies.

Biomarkers for Suicide in Dementia

There is some evidence that increased brain amyloid may predispose to suicide in dementia, especially post-diagnosis, via its association with depression and impaired decision-making (Conejero et al., 2018b), alerting us once again to the complexity of the relationship between impaired executive function and suicide. In Alzheimer's disease, the accumulation of β-amyloid further impairs neural serotonergic pathways (Gonzalo-Ruiz et al., 2003) that could already be altered by Alzheimer's disease itself (Ramirez et al., 2014; Verdurand & Zimmer, 2017). Amyloid β may also impair the stress response (Catania et al., 2009; Morgese et al., 2017) and dysregulate the hypothalamic pituitary adrenal (HPA) axis (Jokinen & Nordstrom, 2008; Jokinen et al., 2010; Morgese et al., 2014), enhancing vulnerability to suicidal thoughts and acts.

Biomarkers for Suicide

Low peripheral serotonin activity has consistently been linked with suicidal behaviour, though studies reporting this result have varyingly measured serotonin activity in the platelets, plasma or serum. Platelet or plasma serotonin levels have been shown to be inversely correlated with the severity of suicide attempts.

People failing a dexamethasone suppression test (DST), indicating sustained higher cortisol secretion with impaired feedback, are found to be significantly more likely to commit suicide than those with a normal dexamethasone suppression response. However, the association between the DST and attempted suicide is less consistent.

Other potential biomarkers for suicidality include peripheral levels of noradrenaline and its metabolite MHPG, BDNF and lymphocytic function (Pandey & Dwivedi, 2012).

Resting state fMRI has emerged as a promising tool for identifying potential neurobiomarkers of suicidal behaviour (Stange et al., 2019). Whole brain networks can be imaged and analysed with minimal additional stress to the patient and with potential for clinical translation of findings (Fox & Greicius, 2010; Maknojia et al., 2019). Abnormalities have been reported in brain networks involved in attention, emotion and reward processing in people with depression and suicidal ideation (Du et al., 2017).

To date, the predictive utility of structural (white matter lesions, brain volume, hippocampal volume) imaging alone for all-cause dementia remains limited (Stephan et al., 2015).

Future imaging biomarkers for suicidality in dementia would have to be carefully considered. Early dementia frequently masquerades as depression, indicating the involvement of common brain pathways such as the reward system (Porcelli et al., 2019).

Ethics and Policy

The debates around the ethics of suicide, dementia and suicide in people with dementia are evolving. Three pertinent foci of ethics and policy are touched upon below, the goal being to open up a discussion or line of thought, rather than to expound on any one focus. These are, in order, ethics and policy issues surrounding the development of biomarkers for dementia, the role of society in debilitating illness, and euthanasia.

Advances in our knowledge and clinical use of biomarkers for dementia entail earlier diagnosis and, for those whose views align with Norman's or Annie's or indeed Alice's, may increase the risk of pre-emptive suicide, especially since diagnosis at an early stage of the illness means executive functioning will likely be preserved. On the other hand, these advances may be welcome so that those who wish to do so may carry out their wishes while still motivated and able (Davis, 2014). Yet, from a public health standpoint it seems important to emphasize that test results can only convey risk, not certainty, and are prone to false positives and false negatives.

It has been argued that rather than focusing on the potentially increased risk of suicide due to the prospective development of state-of-the-art predictive biomarkers, suicide in dementia should be taken as a stark and sad reflection of societal inadequacies, just like suicide due to economic hardship. As one author states, "The choice of pre-emptive suicide reduces pressure on society to offer high-quality and humane services for patients with dementia and their families" (Dresser, 2014). In the bigger picture, this implies a societal responsibility to prioritize the care and quality of life not only for those who are able to provide capital gain within currently prevalent socio-economic structures. We must question why people feel "useless' or "burdensome" rather than just accepting these reactions despite the fact that it is driving them towards suicide. It seems relevant to once again quote Elizabeth Herskovits—"…dementia would not turn out to be such a tragedy, and dementia care not so great a burden …. Rather, it may become an exemplary model of interpersonal life, an epitome of how to be human".

Suicide surfaces in both positive (altruistic or heroic) and negative (condemned) lights in Hindu (Kolhatkar, 1994; Young, 1989) and Islamic (Rosenthal, 1946) religious texts and literature. Even prehistorically, it was viewed as a violation of the established order or as an act directed against the welfare of the soul. It is speculated that the presence of stronger religious beliefs precludes discussion and debate about suicide. This is supported by the surge of interest in the topic of suicide during the period of enlightenment in the eighteenth century when religion was losing favour (Rosenthal, 1946).

Like most countries, India has historically held any form of assisted death unlawful. Suicide itself was looked upon as a crime. This situation is gradually evolving, with suicide now being decriminalized and withdrawal of life support legalized, all within only the last five years.

In India, a vocal few advocate for the legalization of euthanasia on the grounds of it being a human right. For the sake of perspective, in the Netherlands and Belgium, euthanasia was legalized in 2002.

The following is synthesized from Subhash Chandra Singh's comprehensive account of some challenging ethical arguments surrounding euthanasia (Singh, 2012).

First, its apparent contradiction to the Hippocratic oath. This author personally believes that oath taking, white coats, Hippocrates and the saviourism pervasive in medicine all need to be reassessed, but that perhaps merits a separate chapter.

Second, whether or not assisting someone with death is moral.

A third facet revolves around a cluster of complex but interrelated dilemmas. These are generally matters of systemic power vs. individual autonomy and include self-determination, an often constitutionally defined role of the state in "protecting" its citizens, and the question of "capacity" or soundness of mind. Some key conundrums in this area are—Who gets to decide, and for whom, whether life is worth living? Should an assumed intrinsic value of human life take precedence over people's own evaluation? Who gets to decide what is a "good" or "bad" way to die? Where do we draw the line between "protection" and tyranny?

Fourth is the question of legal grey areas such as making distinctions between euthanasia and withdrawal of treatment, pain relief that may unintentionally result in death, abetting suicide, and suicide itself.

Fifth is the issue of a reconciliation of various religious beliefs with euthanasia.

And, finally, the slippery slope argument, which revolves around a requisite sustained stringency of euthanasia laws and practices to avoid their abuse. Cited is a situation like that during World War II when Hitler ordered a "mercy killing" of the "sick and disabled".

Conclusion

In conclusion, the prevailing conception of dementia born from medical discourse, popular representation, and also the characteristic language used when talking about the illness may paint a certain picture that could prove to be far from the actual lived experience of at least some patients with dementia, whose accounts are lesser known. The significant risk of pre-emptive suicide in patients with dementia coupled with the lack of insight into the drivers of this phenomenon highlights a dire need for further qualitative research in this area. With this in mind, the importance of a phenomenological approach, the incorporation of narrative interviews and the gradual and patient development of an individual-centric treatment plan while truly sharing decision-making power with the patient and their family (biological or chosen) in the clinic is worth acknowledging. Society's role in providing a higher standard of living for older adults and those with chronic, debilitating illness needs to be examined. The development of predictive biomarkers may provide a much needed respite for those who unequivocally choose death over living with dementia. However, India does not as yet offer a safe or accessible path to this chosen death. It may be years or even decades before we develop into a society that can fearlessly respect an individual's wish to die, or, for that matter, their right to live comfortably should they not be able to conventionally contribute to capital gain.

References

Alfonso, C. A., & Cohen, M. A. (1994). HIV-dementia and suicide. *General Hospital Psychiatry, 16*(1), 45–46.

Alphs, L., Brashear, H. R., Chappell, P., Conwell, Y., Dubrava, S., Khin, N. A., et al. (2016). Considerations for the assessment of suicidal ideation and behavior in older adults with cognitive decline and dementia. *Alzheimers Dement (n Y)., 2*(1), 48–59.

APA. (2013). *Diagnostic and statistical manual of mental disorders* (5th ed.). American Psychiatric Publishing.

Ballard, C., Bannister, C., Solis, M., Oyebode, F., & Wilcock, G. (1996). The prevalence, associations and symptoms of depression amongst dementia sufferers. *Journal of Affective Disorders, 36*(3–4), 135–144.

Barber, M. E., Marzuk, P. M., Leon, A. C., & Portera, L. (1998). Aborted suicide attempts: A new classification of suicidal behavior. *American Journal of Psychiatry, 155*(3), 385–389.

Bellini, M., De Ronchi, D., Forti, P., Maioli, F., & Ravaglia, G. (1998). Suicidal thoughts and dementia. *Archives of Gerontology and Geriatrics*, 33–38.

Bennett, S., & Thomas, A. J. (2014). Depression and dementia: Cause, consequence or coincidence? *Maturitas, 79*(2), 184–190.

Bohm, M. (1988). Phenomenology and psychodynamics of suicidal acts by the elderly. *Zeitschrift Für Alternsforschung, 43*(1), 19–24.

Bomford, A., & Doyle, E. (2019). Wanting to die at 'five to midnight'—before dementia takes over. *BBC News Stories.*

Brookmeyer, R., Johnson, E., Ziegler-Graham, K., & Arrighi, H. M. (2007). Forecasting the global burden of Alzheimer's disease. *Alzheimer's & Dementia, 3*(3), 186–191.

Cantor, N. L. (2020). Bill of Health [Internet]: Harvard Law 2015. [2020/05/23]. Available from: https://blog.petrieflom.law.harvard.edu/2015/04/16/my-plan-to-avoid-the-ravages-of-extreme-dementia/-more-15260

Catania, C., Sotiropoulos, I., Silva, R., Onofri, C., Breen, K. C., Sousa, N., et al. (2009). The amyloidogenic potential and behavioral correlates of stress. *Molecular Psychiatry, 14*(1), 95–105.

Cipriani, G., Vedovello, M., Lucetti, C., Di Fiorino, A., & Nuti, A. (2013). Dementia and suicidal behavior. *Aggression and Violent Behavior, 18*(6).

Conejero, I., Olie, E., Courtet, P., & Calati, R. (2018a). Suicide in older adults: Current perspectives. *Clinical Interventions in Aging, 13*, 691–699.

Conejero, I., Navucet, S., Keller, J., Olie, E., Courtet, P., & Gabelle, A. (2018b). A complex relationship between suicide, dementia, and amyloid: A narrative review. *Frontiers in Neuroscience, 12*, 371.

Davis, D. S. (2014). Alzheimer disease and pre-emptive suicide. *Journal of Medical Ethics: Journal of the Institute of Medical Ethics, 40*(8).

Diehl-Schmid, J., Jox, R., Gauthier, S, Belleville, S., Racine, E., Schule, C., et al. (2017). Suicide and assisted dying in dementia: What we know and what we need to know. A narrative literature review. *International Psychogeriatrics, 29*(8), 1247–1259.

Dresser, R. (2014). Pre-emptive suicide, precedent autonomy and preclinical Alzheimer disease. *Journal of Medical Ethics: Journal of the Institute of Medical Ethics, 40*(8).

Du, L., Zeng, J., Liu, H., Tang, D., Meng, H., Li, Y., et al. (2017). Fronto-limbic disconnection in depressed patients with suicidal ideation: A resting-state functional connectivity study. *Journal of Affective Disorders, 215*, 213–217.

Erlangsen, A., Zarit, S. H., & Conwell, Y. (2008). Hospital-diagnosed dementia and suicide: A longitudinal study using prospective, nationwide register data. *The American Journal of Geriatric Psychiatry, 16*(3), 220–228.

Fiedorowicz, J. G., Mills, J. A., Ruggle, A., Langbehn, D., Paulsen, J. S., & Group P-HIotHS. (2011). Suicidal behavior in prodromal Huntington disease. *Neurodegener Diseases, 8*(6), 483–490.

Fox, M. D., & Greicius, M. (2010). Clinical applications of resting state functional connectivity. *Frontiers in Systems Neuroscience, 4*, 19.

Gao, S., Burney, H. N., Callahan, C. M., Purnell, C. E., & Hendrie, H. C. (2019). Incidence of dementia and Alzheimer disease over time: A meta-analysis. *Journal of the American Geriatrics Society, 67*(7), 1361–1369.

Gonzalo-Ruiz, A., Gonzalez, I., & Sanz-Anquela, J. M. (2003). Effects of beta-amyloid protein on serotoninergic, noradrenergic, and cholinergic markers in neurons of the pontomesencephalic tegmentum in the rat. *Journal of Chemical Neuroanatomy, 26*(3), 153–169.

Haw, C., Harwood, D., & Hawton, K. (2009). Dementia and suicidal behavior: A review of the literature. *International Psychogeriatrics, 21*(3), 440–453.

Herskovits, E. (1995). Struggling over subjectivity: Debates about the "self" and Alzheimer's disease. *Medical Anthropology Quarterly, 9*(2), 146–164.

Hsiao, J. J., Kaiser, N., Fong, S. S., & Mendez, M. F. (2013). Suicidal behavior and loss of the future self in semantic dementia. *Cognitive and Behavioral Neurology, 26*(2), 85–92.

Jokinen, J., & Nordstrom, P. (2008). HPA axis hyperactivity as suicide predictor in elderly mood disorder inpatients. *Psychoneuroendocrinology, 33*(10).

Jokinen, J., Ouda, J., & Nordstrom, P. (2010). Noradrenergic function and HPA axis dysregulation in suicidal behaviour. *Psychoneuroendocrinology, 35*(10), 1536–1542.

Kitwood, T., & Bredin, K. (1992). Towards a theory of dementia care: Personhood and well-being. *Ageing & Society, 12*, 269–287.

Kolhatkar, M. (1994). Euthanasia in Vedic literature. *Bulletin of the Deccan College Post-Graduate and Research Institute., 54*(55), 167–170.

Kumar, C. S., George, S., & Kallivayalil, R. A. (2019). Towards a dementia-friendly India. *Indian Journal of Psychological Medicine, 41*(5), 476–481.

Kumar, C. T. S., Shaji, K. S., Varghese, M., & Nair, M. K.C. (Eds.) (2020). Dementia in India 2020.

Kuring, J. K., Mathias, J. L., & Ward, L. (2018). Prevalence of depression, anxiety and PTSD in people with dementia: A systematic review and meta-analysis. *Neuropsychology Review, 28*(4), 393–416.

Maknojia, S., Churchill, N. W., Schweizer, T. A., & Graham, S. J. (2019). Resting state fMRI: Going through the motions. *Frontiers in Neuroscience, 13*, 825.

Milne, A., & Phillips, J. (2020). Conceptualising dementia. In *Mental health in later life: Taking a life course approach* (pp. 165–98). Bristol University Press.

Morgese, M. G., Tucci, P., Colaianna, M., Zotti, M., Cuomo, V., Schiavone, S., et al. (2014). Modulatory activity of soluble beta amyloid on HPA axis function in rats. *Current Pharmaceutical Design, 20*(15), 2539–2546.

Morgese, M. G., Schiavone, S., & Trabace, L. (2017). Emerging role of amyloid beta in stress response: Implication for depression and diabetes. *European Journal of Pharmacology, 817*, 22–29.

Pandey, G. N., & Dwivedi, Y. (2012). Peripheral biomarkers for suicide. In Y. Dwivedi (Ed.), *The neurobiological basis of suicide. Frontiers in neuroscience.* Boca Raton (FL).

Porcelli, S., Van Der Wee, N., van der Werff, S., Aghajani, M., Glennon, J. C., van Heukelum, S., et al. (2019). Social brain, social dysfunction and social withdrawal. *Neuroscience and Biobehavioral Reviews, 97*, 10–33.

Ramirez, M. J., Lai, M. K., Tordera, R. M., & Francis, P. T. (2014). Serotonergic therapies for cognitive symptoms in Alzheimer's disease: Rationale and current status. *Drugs, 74*(7), 729–736.

Ranjan, R., Kumar, S., Pattanayak, R. D., Dhawan, A., & Sagar, R. (2014). (De-) criminalization of attempted suicide in India: A review. *Industrial Psychiatry Journal, 23*(1), 4–9.

Richard-Devantoy, S., Jollant, F., Kefi, Z., Turecki, G., Olie, J. P., Annweiler, C., et al. (2012). Deficit of cognitive inhibition in depressed elderly: A neurocognitive marker of suicidal risk. *Journal of Affective Disorders, 140*(2), 193–199.

Richard-Devantoy, S., Turecki, G., & Jollant, F. (2016). Neurobiology of elderly suicide. *Archives of Suicide Research, 20*(3), 291–313.

Rosenthal, F. (1946). On suicide in Islam. *Journal of the American Oriental Society., 66*(3), 239–259.

Sabodash, V., Mendez, M. F., Fong, S., & Hsiao, J. J. (2013). Suicidal behavior in dementia: A special risk in semantic dementia. *American Journal of Alzheimer's Disease and Other Dementias, 28*(6), 592–599.

Schneider, B., Maurer, K., & Frolich, L. (2001). Dementia and suicide. *Fortschritte Der Neurologie-Psychiatrie, 69*(4), 164–169.

Scocco, P., & De Leo, D. (2002). One-year prevalence of death thoughts, suicide ideation and behaviours in an elderly population. *International Journal of Geriatric Psychiatry, 17*(9), 842–846.

Serafini, G., Calcagno, P., Lester, D., Girardi, P., Amore, M., & Pompili, M. (2016). Suicide risk in Alzheimer's Disease: A systematic review. *Current Alzheimer Research, 13*(10), 1083–1099.

Seymour, T. (2015). Still Alice is 'shockingly accurate'—people living with dementia give their verdict. *The Guardian.*

Singh, S. C. (2012). Euthanasia and assisted suicide: Revisiting the sanctity of life principle. *Journal of the Indian Law Institute., 54*(2), 196–231.

Srivastava, D. (2018). OxHRH Blog [Internet] 2018. [cited 2020]. Available from: http://ohrh.law.ox.ac.uk/right-to-die-with-dignity-a-fundamental-right-indian-supreme-court-allows-passive-euthanasia-and-living-wills

Stange, J. P., Jenkins, L. M., Pocius, S., Kreutzer, K., Bessette, K. L., DelDonno, S. R., et al. (2019). Using resting-state intrinsic network connectivity to identify suicide risk in mood disorders. *Psychological Medicine*, 1–11.

Stephan, B. C., Tzourio, C., Auriacombe, S., Amieva, H., Dufouil, C., Alperovitch, A., et al. (2015). Usefulness of data from magnetic resonance imaging to improve prediction of dementia: Population based cohort study. *BMJ, 350*, h2863.

Swati, B., Sreenivas, V., Manjari, T., & Ashima, N. (2015). Dementia Assessment by Rapid Test (DART): An Indian screening tool for dementia. *5*(198).

Terman, S. A. (2016). To live long enough to warm the hearts of others: Reflections on informing my patient about a peaceful way to die. *Narrative Inquiry in Bioethics, 6*(2), 102–105.

Upadhyaya, A. K., Conwell, Y., Duberstein, P. R., Denning, D., & Cox, C. (1999). Attempted suicide in older depressed patients: Effect of cognitive functioning. *The American Journal of Geriatric Psychiatry, 7*(4), 317–320.

Vadlamani, L. N., & Gowda, M. (2019). Practical implications of Mental Healthcare Act 2017: Suicide and suicide attempt. *Indian Journal of Psychiatry, 61*(Suppl 4), S750–S755.

Verdurand, M., & Zimmer, L. (2017). Hippocampal 5-HT1A receptor expression changes in prodromal stages of Alzheimer's disease: Beneficial or deleterious? *Neuropharmacology, 123*, 446–454.

Wadia, N. H., & Khadilkar, S. V. (2015). *Neurological practice: An Indian perspective*. Elsevier.

Young, K. K. (1989). *Euthanasia: Traditional Hindu views and the contemporary debate*. Sri Satguru Publications.

Chapter 3
The Power of Music in the Lives of People Living with Dementia

Diana Kerr

Abstract There is a substantial and growing body of research and practice showing the extraordinary impact of music and particularly group singing on the lives of people living with dementia. This chapter will focus on the ways in which people with dementia are affected by music. Research evidence and practical advice will be given to inform and to enable motivated individuals, carers, volunteers and professionals to use music in different ways to engage people living with dementia. It is not about music therapy, which requires highly skilled and trained professionals. This chapter is about using music for the benefit of almost everyone living with dementia. Most people living with dementia do not require therapy but they will enjoy and be enriched by access to music in a variety of settings such as the community, residential homes, hospitals and family homes. Different ways of delivering and enabling people to experience music will be covered. Consideration will be given to the use of music as a means of calming or stimulating, as a vehicle to aid reminiscence, as a way of providing individuals with access to their own playlist, as a stimulus and way to promote dancing and activity. Finally, the powerful and extraordinary benefits that can be gained from enabling people to sing with others will be explored. Issues about organization, content, membership and focus of singing groups for people with dementia as well as their carers will be addressed.

Keywords Dementia · Music · Singing · Dancing · Singing groups

All the interventions described in this chapter are for the benefit and consideration of everyone supporting and caring for people living with dementia, professionals, volunteers, formal and informal carers and families.

This is not a chapter about music therapy. Music therapists provide a highly skilled intervention with people living with dementia who require such therapy. Most people with dementia do not require therapy they require an experience that is common to us all and which while often being therapeutic, it makes people feel better, is a normal activity. It is important not to professionalize or medicalize an activity or pathologies the participants unless there is a need for specialist intervention.

D. Kerr (✉)
Independent Consultant, Edinburgh, Scotland

M. K. Shankardass (ed.), *Dementia Care*,
https://doi.org/10.1007/978-981-16-3864-0_3

There will be an exploration of the importance of music to us as a species, its role and power. This will be followed by a consideration of the power of music in the world of people living with dementia.

The various ways of accessing and using music will be described. The benefits and possible obstacles of different approaches and interventions will be explored.

The Power of Music

A question I am frequently asked is "Why is music so good for people with dementia?" To answer this question, it is helpful to start off by asking "why is music so important to all of us?"

Everyone responds to music, even people who say they are tone deaf or not musical will move to a rhythm and those who claim not to be musical will sing to themselves and be moved by particular music. All humans respond to music. Humans are a musical species as Sacks (2007) states "All of us (with very few exceptions) can perceive music, perceive tones, timbre, pitch intervals, melodic contours, harmony and (perhaps most elementally) rhythm".

Levitin (2008) writes about the ubiquity and antiquity of music stating "No known human culture now or at any time in the recorded past lacked music. Some of the oldest physical artefacts found in human and protohuman excavation sites are musical instruments: bone flutes and animal skins stretched over tree stumps to make drums".

Music exists in every culture. It varies from one culture to another but in a way that shows how central it is. It reflects something about each culture and you can often hear the culture in the music. Think of Spanish flamenco music, Chinese songs and Welsh choirs, Hindi songs and English folk songs how different and reflective of their culture and time they are. No matter what the culture, the music is reflecting and moving its audience.

The fact that music is somehow an integral part of being human is perhaps best illustrated by the response of babies and their mothers to music. New mothers, unprompted, will hum and sing to their infant knowing that that is likely to soothe and calm them. The baby's response to music maybe, partly, because they have already heard this in the womb; the sound of the mother's heart beat (a basic and well used rhythm) will already be familiar as a foundation for music appreciation and response. Very early on babies start to spontaneously vocalize, the sounds that they make really do seem like early attempts to sing and soon they will begin to incorporate sounds around them and to extend their range. Lamont (2001) found that at age one, infants recognized and preferred music they were exposed to in the womb. By the time they are two, children will show a preference for the music of their own culture. During our teens, we begin to identify personal musical preferences, and as adults, the music that we seem to be nostalgic for and to remember best is the music we responded to between the ages of 15 and 24 years of age. This is probably partly because adolescence is a time of heightened emotion and we tend to remember things that have a heightened emotional content. Since music during our teens is often expressing emotions that are hard to articulate, it is not surprising that these are the memories that stick. Also, as Levitin (2008) writes, "it is around fourteen that the wiring of our musical brains is approaching adult like levels of completion". Whatever the reason, the music of those years is the music we remember well.

Our response to music can, of course, run the whole gamut of our emotions. We will often choose music to reflect our mood. If we are feeling down we will listen to music that reinforces the mood and in a way gives us even more strength to wallow and sigh; think of the lovesick youth. Conversely, we will use music to lift our spirits and express our joy.

It is almost as if we are hard wired to use and respond to music The power of music to evoke emotions and influence our feelings about things is understood well by advertisers who will employ music to heighten our response to, and so liking or even craving for, inanimate objects such as cars, clothes, food and drink and even the latest electronic device.

An important element of music is that it has a pulse, remember the mother's heartbeat that is the first pulse the baby hears. Levitin (2008) states that most music is foot-tapping. He states "we listen to music that has a pulse, something you can tap your foot to, or at least tap the foot in your mind to". I am not sure if much foot-tapping goes on when listening to a Beethoven sonata or a Bach Cantata but the point he is making is that listening to music involves our bodies and something physical happens. Nietzsche (1888) maintained that we listen to music with our muscles. He was right. We resonate to the music; it gets into our bodies as well as our minds. Our bodies respond to music. We keep time, perhaps by simply tapping our feet, or maybe by moving to the beat when we dance or March. We can even respond unwittingly. People doing workouts in the gym can change their rate of running, jumping or whatever activity it is they are doing if the music they are listening to is changed. The person exercising may not be conscious of the change but their body is.

So music gets to us emotionally and physically. The very few people who do not respond to music are likely to be suffering from a rare condition. Music gets underneath our skin.

What is So Good About Music for People with Dementia?

Sacks (2007) writes "music is no luxury to them (people with dementia) but a necessity and can have a power beyond anything else to restore them to themselves, and to others, at least for a while".

The rest of this chapter will demonstrate the truth of this.

The previous section made clear why and how music is profoundly important and a core element in all our lives. It was also shown how our experience of music starts before we are born and very early on in our development becomes embedded in our memory. This helps to explain the phenomenon that when much else fails as dementia progresses, the music stays on. It is embedded in our longest memories.

As other experiences become confusing and communication becomes difficult, the role and experience of music becomes even more important. Music stays with us long after speech and other skills have gone. "Musical perception, sensibility, emotion and musical memory can survive long after other forms of memory have disappeared" (Sacks 2007).

Anyone who has worked with people with dementia will have witnessed the remarkable fact that people who have lost the ability to speak coherently or even at all will sing an entire song perfectly. It is not only the words but the musical memory that stays so people will hum or whistle a tune even when the words to the song have gone.

One of the important aspects of supporting people with dementia is to minimize the impact of their losses and to play to their strengths. As music and musical memory remain throughout the relentless progression of the condition which robs people of most of their skills and memories, then clearly we should be using music as an integral part of our support for people living with dementia, (Kerr, 2015). Even at the end stage of the condition when people are close to death, music will reach them. It is important not to assume that the person lying inert and apparently not responding is oblivious to the sound of music. Play or sing to people at this end stage and you will see changes in their facial expression and even vocal activity and physical movement. Music can provide one last way to reach the person and enable them to respond at an emotional level.

Our understanding of the role of music in the lives of people living with dementia has been informed by a considerable body of research carried out over the last 25 years. From some key studies, we know that music is effective in reducing a range of challenging behaviour. Playing calming music will reduce agitation (Denny, 1997). Music can also reduce aggressive behaviour, "wandering," repetitive vocalization and irritability. (Casby & Holm, 1994; Groene, 1993; Ragneskog et al, 1996) This is perhaps not surprising if we recognize the calming effect that music can have. Ragneskog et al. (1996) and Denny (1997) found that playing soothing and quiet background music at meal times reduced restlessness and irritability; people also showed less fear-panic and depressed mood.

Playing the right music at meal times means people will be more relaxed, will sit longer at the meal and will eat more (Ragneskog et al., 1996). Given that people with dementia have problems with eating, this seems to be an opportunity not to be missed.

Clark et al. (1998) suggested that if the right background music is played during bath times, people with Alzheimer's disease showed reduced aggression and hitting behaviours. Rader (2002) reported that a nurse who sang at bath time to someone who had previously refused for ten years to have a bath or shower was able to wash that person completely without incident.

Gina

Gina, a dental nurse, had experienced some distressing times with people with dementia when they were having dental care. Often patient's treatment had to be discontinued because of the level of distress caused. The level of agitation would often begin before the patient entered the room.

Gina had heard about music and how it could be calming. She would sing with the patient as they approached the dentist's room. The singing would continue throughout the treatment. Gina sang childhood songs such as lullabies. Of course, the singing did not calm everyone all the time but many people were able to tolerate the treatment and some even found it a good experience.

We know that if caregivers sing to people with dementia when carrying out intimate tasks that the incidence of challenging behaviour can be significantly reduced. This maybe the result of a number of factors. For the carer, the mere act of singing reduces stress in them and this will be transmitted to the person with dementia. Also, the sound of the person singing maybe calming because it is reminiscent of the mother singing to the child.

Using music appropriately can lead to an improvement in reality orientation, memory recall and social behaviour (Ashida, 2000; Gregory, 2002; Lord & Garner, 1993). Music will often trigger speech (Brontons & Koger, 2000) but it can also allow the person with dementia to sing something that reflects their mood or articulates something they want to say but can only sing it. I was in a residential home when a lady I know came up to me and sang "Show me the way to go home"; she was communicating very well through her singing!

Different Ways to Use Music

There are, of course, many ways in which people living with dementia can engage with and benefit from music. Different approaches will affect people in different ways at different times. What works with one person may not work with another or indeed what works on a Tuesday with one person may not work with them on

Wednesday and morning and afternoon responses may vary considerably. There is not a "one size fits all" solution.

The rest of this chapter will describe some key, well tried and researched ways that music can be used with people living with dementia.

1. As background or individually targeted music to alter mood and interaction by stimulating or calming.
2. As a reminiscence tool used as a way to trigger memories.
3. As a stimulation to encourage and enable dancing and physical activity.
4. As a personal playlist individually tailored to each person enhancing their sense of self and individual choice.
5. As a way to give people an opportunity to sing alongside others.
6. As an entertainment from musicians.

As Background or Individually Targeted Music to Alter Mood and Interaction by Stimulating or Calming

To an extent, the following statement sounds banal "music affects and reflects our mood".

Everyone reading this chapter will know how they like and use different mood music at different times and under different circumstances, upbeat music when you are feeling buoyant and full of the joys and quieter music when feeling contemplative or sad or just relaxing. It is no surprise, therefore, that people with dementia are also affected by the mood and interaction that music can induce.

Clair (1996) describes two very broad musical categories, stimulating and relaxing or sedating. If you are with someone whose mood and/or interaction you feel needs to be changed, then the idea of these two broad categories is helpful.

It is, however, important to use the music in a considered and informed way. Crowe (2004) refers to the "iso principle"; this principle states that when wanting to use music to alter or enhance someone's mood, you should start with music that reflects their present mood and disposition. Think how annoying it would be if you were feeling a bit withdrawn and sad and someone came up to you and started playing a loud jolly tune; if someone is low in mood, then a song like "Danny boy" or some soulful classical music might be a good place to start; you need to know what their preferences would be.

If someone is high and in need of calming, a good start might be something like "Alexander's rag time band" or some dance music or some good old rock. Then gradually reduce the tempo and volume.

It might be useful sometimes to have a small drum so that you can beat a rhythm and then increase or slow it down maybe using the same song more than once but just at a different beat.

It is also useful to use music as background sound to help calm a group of people. The use of calming music in this way should be restricted to 20 min sessions in every hour or as Clair (1996) suggests for some people one hour of music followed by one hour of quiet time. The timings will need to be gauged by watching and listening to the person/people with dementia. Remember that one day 20 min maybe enough and another day a longer time may work.

If the music is not turned off, it just adds to the noise clutter and is at the least irritating and at the worst agitating and intrusive.

I had been doing some training with managers from a particular group of care homes. I had told them about the power of calming music but forgot to mention the need to turn it off after about 20 min. A fortnight later I was training the staff from the same homes. Before I could even start the training, they asked me if I had really said to keep the calming music on all the time. The staff, and presumably the residents, were as they said, 'climbing the walls with the constant noise', melodies they had liked they now loathed!

It is also important to make sure that other noises are not present when using music in this way. People with dementia often find it hard to differentiate between noises. What should they be listening to and what is extraneous?

Also, of course, many people with dementia will have hearing problems and be using hearing aids that exacerbate the confusion of noises. So, turn off the TV and radio and ask others to have their conversations elsewhere. The music is all that needs to be heard.

As a Reminiscence Tool Used as a Way to Trigger Memories

Music can awaken our connection to the world and our history and this is true for people living with dementia.

We all have attachments to some songs more than others. Maybe because they are good pieces of music but, often as not, because they remind us and transport us to an event, a person and a period in our lives, usually with an emotional element. Music that touches our emotions will be well embedded in our memory.

People with dementia will have the same experience of music stirring emotions and memories.

As stated earlier, the songs that are best embedded in our consciousness and which have usually the most emotional appeal are the songs that we experienced during our teens and young adulthood.

Playing music and songs from the past will transport people with dementia to that period. They may even in their reality experience the time in the past as their present.

Songs will evoke memories and provide an opportunity to reminisce. This means that the provider of the music is playing to the strengths of the person with dementia. Their long-term memory is going to be clearer and they will have a better vocabulary and also a motivation to talk about past events. If music can trigger this response, then use it. This stimulation of past memories may need to be on a one to one basis. It can, of course, be equally powerful as a group activity as the following case study illustrates well.

Margaret, Janet and Pauline

Margaret, Janet and Pauline sat in the sitting room every day but said very little to each other, or indeed anyone else. Janet was particularly withdrawn and Margaret was constantly insisting on going home.

They also showed an element of annoyance even jealousy if anyone spoke to one of them and not the others. They did, however, seem to get some comfort from sitting close to each other.

One day, we played some Glenn Miller. The first piece was "In the Mood", then Moonlight Serenade followed by Tuxedo Junction by the time we were doing Chattanooga choo choo all three women were singing along and smiling. Then Margaret started to tell a story about going to the dances during the war. She then reminisced about her time as an ambulance driver. This prompted Pauline to reminisce about her experience of dances and the Blitz in London. Then it seemed like the flood gates had opened. Talk of dances, boyfriends and making clothes out of odds and ends ensued.

It was a history lesson and they were in their element. The memories were vivid, and although there were some sad stories, mostly there was a pride in what they had been through and a great deal of laughter.

Rio (2009) comments that "Enjoying music from these young adult years maybe the most successful way of having a musical memory help someone feel a connection with who she was and she has been".

If, however, we do not know someone's life story and history, we may not be able to determine, except in a general sense, what music might trigger reminiscence. This chapter is not about life story work but it is worth underlining the importance of not making assumptions about people's musical triggers.

Polly

Polly, a woman with Down's Syndrome, was given a diagnosis of Alzheimer's disease. Her father decided to compile a life story book and box for her. She knew how important this would be as her mother had had dementia and Polly had seen how important it was to have things from the past.

Quite unexpectedly and unprompted by anyone else, Polly said that she wanted all her favourite songs to be recorded so that they could be played to her when she had forgotten lots of other things. Six years later when her condition had much deteriorated, her father played these songs to her and she began, as she had anticipated, responding quite dramatically and began to reminisce about the things that they reminded her of. Polly was the same age as me so I found it a bit bewildering and interesting that the songs she listened to and reminisced about were of a generation before me. Polly, because she had Down's Syndrome, had been living with her parents and listening to their music at an age when I was out at parties and clubs listening to the music of my generation.

Reminiscing can help us make sense of our lives and also remind us of good times.

There is a need, of course, to be wary of stirring up negative memories and emotions but there is often no way to know beforehand, just avoid in future.

As a Stimulation to Encourage and Enable Dancing and Physical Activity

People living with dementia tend to experience a decrease in their activity levels. This is the consequence of increased social isolation, of apraxia that will make purposeful movements difficult and lack of suitable stimulation.

The importance of movement for health and emotional well-being means that we should be seeking ways to counteract the increased immobility and sedentary life styles. Music is a wonderful vehicle to achieve this.

People with dementia who have problems with walking will sway rhythmically to a beat; this will still be in place even after people have lost the ability to speak or sing. It is a great joy to see people, apparently disconnected from their environment and feeling lost, move and dance with gusto to music.

Many people now in their old age will have been regular dancers at Saturday night hops; many will have learned ballroom dancing and certainly they will have danced at parties.

Playing dance tunes from the past will stimulate not only old dancing memories but will trigger the bodies to respond. Dance steps and routines will often return.

Memories of dance events from their past maybe resurrected and help stimulate social contact again.

The following two case studies are good example of this.

Julia

Julia attends our singing group with her son. Julia rarely speaks and equally rarely sings the songs but the moment the first note is played she is up and dancing with her son. She is transformed from a fragile elderly woman into a lithe, energetic and smiling woman. She is capable of dancing for the best part of an hour with short breaks.

Mary

Mary, who was a gymnast in her youth, will trip lightly and graciously across the floor in perfect time with the music. Often she is irritated by the lack of involvement of others and tries to get them up dancing as well. She once said to me "That man needs to ask that girl to dance".

The hall we use for our singing group is very much like the halls that people used to go dancing in on a Saturday night. I think that she wonders why the men are not asking the women to dance, that was what used to happen.

Both these women were able, with a little support; to dance, others will need firmer supports. It may be necessary to have a caregiver on each side of the person; what is interesting is how people who cannot easily walk will sway and move their feet from left to right in time with the music and in unison with the two people either side who can act as supports and partners.

Nystrom (2002) noted that people with dementia seemed to participate (in dancing)…when concurrently singing.

This has been the experience of our singing groups. It is a rare person who does not get up and dance or move rhythmically when we sing. The songs that they enjoy most seem to be "Beautiful Sunday" and "I Could have Danced all Night" the latter outstrips any other dancing songs that we use…I cannot recommend it enough. To see people pause for the beat at the end between the last "danced" and "All night" is a joy to behold as it shows the level of involvement and of being in tune with the beat that demonstrates the power of the music to get into their brains and their bodies.

The other wonderful aspect of encouraging people to dance is that often it is an opportunity for life partners to share a joyful physical activity with the person with dementia. Such opportunities can be rare.

The following, taken from a singing group session, is a perfect illustration of this.

Peter

"Peter was very confused and distant. We played a 'walking' song that must have meant something special to him and his wife, they danced together. They stayed on the floor after the song had finished talking, laughing and kissing and everyone else sat down. We all just waited in silence for them to finish.—reluctant to break the moment".

There have been times at our group where the person with dementia has chosen to dance with someone other than their partner, a new acquaintance who presumably seems more interesting or maybe just a better dancer!

Songs to Dance to or Use Different Tempos with

The songs below are ones used in a sheltered housing complex where the people are generally more mobile but people who were less mobile were able to walk rhythmically with support.

Anniversary Waltz—Waltz

Are You Lonesome tonight—Waltz

Bill Bailey—Quickstep

Chanson D'Amour—Shuffle

Cokey Cokey—Party Time

Deep In The Heart Of Texas—2/4 March

Happy Days Are Here again—Quickstep

Macnamara's Band—6/8 March

Memories are made Of this—Quickstep

On A Slow Boat To china—Foxtrot

Rock Around The Clock—Rock

Underneath The Arches—Foxtrot

We'll Meet Again—Foxtrot

You Are My Sunshine—Quickstep.

(Kerr, 2015).

As a Personal Playlist Individually Tailored to Each Person, Enhancing Their Sense of Self and Individual Choice

What is a personal Playlist?

It is a way of capturing all that has previously been written in this chapter about the power of music but distilling it and tailoring it to each person's memories and history.

Of course all the interventions described tap into people's memories and histories and they are all valuable tools to reach people. Playlists are a way of doing this but using more recent technology. Playlists use iPods onto which a person's significant and emotionally meaningful music can be downloaded.

At this point, it is worth acknowledging that many if not most carers that I know would not know how to use an iPod and would be put off pursuing this intervention for that reason.

There is a website http://www.playlistforlife.org.uk/wordpress/use-thetechnology/ which could be accessed to help carers develop the skill.

Children or grandchildren can be consulted to show the way. Of course, many carers are perfectly competent to use an iPod, and presumably in the future, this will not be an issue.

Much of the publicity and uptake of this technology is the result of the work of Sally Magnusson (2014) who wrote a book "Where memories go" about her journey with her mother, who had dementia. Her experience of the power of individualized music stored and played using an iPod led her to found a charity "playlist for life".

It is not straightforward to develop a person's playlist. Family and friends may not have been around at the time of the most significant music in someone's life. Children will probably not know anything about their parent's musical choice before they were themselves in their teens and probably even then did not register much. Life partners have often only been partners since their twenties. So, it requires patience, not making assumptions, listening to the person with dementia, playing music and seeing how they respond.

The following case study is long but it captures well the many positives that can be gained from using playlists; it also highlights a few problematic areas.

Katherine

My mother, Kathrine, at the age of ninety-two, experienced a decline in her cognitive performance that was alarmingly fast. Initially, she stayed at home, supported by family and carers but as her confusion progressed, it was necessary to move her into a care home for twenty four hour care.

Kathrine was a very private and proud person and she found the transition to "communal" living very difficult and bewildering. She did not feel comfortable taking part in group activities and preferred the privacy of her own room. Interestingly, the exception to this was when the staff organized a church service; obviously, Kathrine did not see group singing of hymns as an alien concept. However, her concentration span was short and so watching television and her hobbies of painting, sewing and knitting, sadly, no longer held her interest. Consequently, Kathrine was in need of comfort and stimulation in her room. Music had always been an important part of our family life and it was through music that I found a breakthrough in Kathrine's care and well-being. The website "Playlist for Life" gave me the inspiration I needed. I bought an iPod and recorded a selection of music onto it. My mother liked classical music and I looked through her CD collection to help find her favourites. I then explored music from her youth and also war time songs. If I could remember even one line of songs that I had heard my parents sing, it was easy to use Google to find and download the appropriate piece. Then, there was the music from my youth. For example, The Beatles songs which initially she tolerated and then came to love! It took Kathrine a while to get used to wearing headphones but it was worth persevering as they provided an all-round sound and eliminate other noise distractions. The effect of the music was both astounding and magical. If Kathrine was agitated, some soothing classical music calmed her down, lowered her blood pressure and helped to release pent up emotions. On many occasions, she conducted the orchestra with a smile on her face. On other days, when she was quite clearly feeling sad, I played songs from the past. She became animated and she sang along, amazingly, word perfect. This sometimes led on to lively conversation about past, happy memories. I also used a splitter cable myself so that I could listen or sing along with her.

Conversation with a person living with dementia can be challenging but watching Kathrine enjoying the music or sharing it with her was helpful during visits.

It was also useful to put on some relaxing music for her to enjoy when I was leaving. I cannot overestimate the beneficial effects of music for Kathrine. It relaxed her and seemed to release her from the anxiety and confusion of dementia. There was, however, a problem as to who would set up the playlist in my absence. Kathrine was unable to do this for herself. The staff in the care home saw for themselves, the pleasing effect it had on her but were reluctant to be involved, claiming it would be extra work for them. Clearly, staff have to be convinced of the benefits and trained to use the equipment in order to appreciate that it could be of equal benefit to the resident and to them. This is not a "one size fits all" programme. There would be no advantage in "plugging in a group of residents together. The usage has to be at the right time with the right genre of music on an individual basis.

Kathrine has now died but I cherish the memories of when her personal playlist" wiped fear and sadness from her face and clearly gave her both pleasure and peace.

(My thanks to Sue Denholm for this poignant and powerful endorsement of the use of playlist).

As highlighted earlier in relation to the amount of time background calming or stimulating music should be played, it is important not to simply plug people into their playlist and leave them for an unspecified time. Periods of 15 min structured throughout the day have been shown to increase the beneficial effect (Playlist for life, 2014).Longer sessions can also be beneficial but this needs to be judged and take into account the person with dementia, the carers and the environment. Just as with the background music discussed above, if the music is left on all the time, it just becomes noise clutter and adds to distress.

As a Way to Give People an Opportunity to Sing Alongside Others

Listening to music is undeniably an emotional and physical experience which enhances and enriches our lives. Creating our own music, either through singing or playing, adds another layer to the enjoyment and personal involvement. Playing and singing with other people, however, adds another whole new dimension to the emotional and physical experience.

The level of enjoyment can be quite exhilarating. Singing releases feel good hormones, endorphins and the anxiety busting hormone, oxytocin. It is a stress reducer so it is not surprising that research has found that singing together not only improves people's sense of well-being but that it is actually good for your health. Beck et al (2000) found that being a member of a choir improved the immune system of the singers. There is evidence that singing in a choir affects our heart rates; choirs have been shown to synchronize their heart and breathing rates, increasing and decreasing them in response to the music (Vickhoff 2013). Singing together also appears to help reduce high blood pressure (Valentine & Evans 2001).

In addition, during singing and making music, a series of connections are made in the brain, and if these are accessed regularly, this helps our creativity and our problem-solving skills (Noice & Kramer, 2013, Welch, 2010).

Aldridge (2000) writes "For many (people living with dementia) singing provides the longest duration of time where they are focused and clearly engaged in purposeful, meaningful activity". I would add to this that they are equally engaged in a fun, familiar, joyful activity.

People living with dementia and their carers can easily become isolated and feel abandoned by friends and neighbours. This feeling of social isolation coupled with a realization of the loss of ability to carry out many of the tasks of daily living can make the world a frightening, lonely and emotionally difficult place. Browning wrote "who hears music feels his solitude peopled at once" how much better if that music involves real people.

Singing with others in a group means that people are not isolated, they share experiences with others and they engage in an activity which is accessible and achievable, unlike many other activities. When we sing with others, we enter and share the same emotional space.

As stated earlier, 'one of the important aspects of supporting people with dementia is to minimize the impact of their losses and play to their strengths' (Kerr, 2015).If people can sing, then we should be encouraging this, helping them to maintain the skill and sense of achievement and joy that goes with that (Kerr, 2015).

Choirs

Most people's experience of singing with others will be in a choir. This can be an exhilarating, healthy, enjoyable and demanding experience. Choirs that have been going for a long time may well have people in them who develop dementia but who are able for a while to maintain a presence. It is always very sad when the time comes and people can no longer cope and have to leave. It is clear, however, that for some time the person with dementia has continued to derive pleasure from the choir.

This raises a question about the place of choirs in the lives of people with dementia.

If the inevitable outcome is that someone will reach a stage where they cannot any longer cope, or indeed the other members of the choir cannot cope, then the question needs to be asked about the place of "dementia choirs" that are either ordinary choirs that include people with dementia or choirs that are set up solely for people with dementia.

Clearly people who develop dementia while in a choir will have to face the horrible realization that they can no longer cope. This is going to be the case in choirs that people with early stage dementia join. This raises an important and difficult dilemma. Of course we want to encourage people to have meaningful joyful activities, but it is also the case that being asked to leave a choir presents the person with another experience of loss. Some choirs do have an "exit interview" to help people leave, but this must be so painful and the balance between the pleasure of the choir and the increasing difficulties and eventual rejection needs to be weighed.

For some people, the benefits are sufficient to outweigh the negative aspects. If the conductor and other members of the choir are sensitive to the needs of the person with dementia, if the conductor does not give long explanations and does not expect people to be able to cope with lengthy instructions, then things can work but this takes awareness and tolerance all round.

The added pressure of giving a performance needs to be weighted in the balance as well. It is a thrill to perform with others and performing in front of others adds another dimension and sense of achievement; however, this can also be a stressful experience for anyone but more so for people coping with dementia.

There is a possibility that the raison d'etre of choirs including people with dementia, that is that they are inclusive, actually becomes undermined as people are excluded when the choir becomes too demanding.

Inevitably, of course, choirs will meet the needs of some people with dementia particularly in the early stages and so are to be encouraged but if the maximum number of people living with dementia are going to experience the joy of singing with others, then a more inclusive, relaxed and undemanding setting and focus is required.

Singing Groups

Setting up either a community singing group or a singing group in a residential setting is a way to enable people with dementia to engage in an unthreatening and joyful social activity.

It is important to put in the caveat that there are some people for whom being in a group would be anathema. There are also people for whom the type of songs sang in social groups would be boring, unfamiliar and even an insult. People who have sung all their life in a Bach choir, singing a classical repertoire may well find the type of songs incongruous and even annoying. So, singing groups are not for everyone and there should not be an assumption that they are; however, they do hit the spot for most people living with dementia.

For most people living with dementia, the opportunity to come together with others and sing familiar songs provides an experience that is sadly lacking in much of the rest of their lives. They are enabled to engage in an activity that they can still contribute to, an activity that brings them into a social setting in which they can experience the joy of company and enter the same emotional space as others.

We know that singing and listening to music can make us happy. Even if the person with dementia forgets that they were singing soon after the event, this does not negate its worth. They will still feel good even if they cannot remember why. People with dementia live much in the moment so we should be trying to make as many moments as possible good ones. Music and particularly singing together is undoubtedly a way to achieve this.

The following two stories are wonderful illustrations of this and of many of the other benefits of music and singing discussed earlier in this chapter.

Malcolm and Isobel

Malcolm heard of a singing group being set up in his local community for people with dementia and their carers. His wife Isobel, who had dementia, had not spoken for the last five months. She no longer showed any facial expression and seemed to be in a world of her own.

With great trepidation, Malcolm brought Isobel to the group.

To begin with Isobel appeared agitated and just sat and stared as the singing got underway. Slowly her hand started to tap her thigh, then she moved her body, then she began to make some humming noises. By the end of the session, Isobel was singing and smiling.

The next morning she woke up and said to her husband 'I think we should bake a cake today' and they did!

Kate

As a young woman, Kate sang in a Band and continued to sing throughout her life. Her dementia was at a stage that conversation was limited and she did not sing at all, even with the encouragement of her family.

Her husband and daughter brought Kate to the Singing Group in the hope that she might sing. The Singing Leader that day had brought hats which resembled the hats worn by the Andrews Sisters which would have been the era when Kate was singing in the Band. She of-fered the hats out to anyone who might want to wear them whilst we sang "Don't sit under the Apple tree".

With encouragement, Kate took a hat and along with another member of the group and the singing leader started to sing and slowly parade around the hall whilst everyone sang. It seemed that the setup had triggered memories for Kate and it became obvious that she had moved into "performance mode". Not only did she sing but she took a bow at the end of her "performance".

It would be foolish and wrong to pretend that music always leads to such dramatic events. The above stories are, however, two of many that could be told. Some not as dramatic but for the person involved still a very worthwhile and life enhancing experience (Kerr, 2015).

Both of these case studies involve a carer. The involvement of carers in singing groups can have a significant impact on the relationship between the carer and the person with dementia and consequently on the well-being of both. The benefits of singing in a group are experienced by carers as well as the person they are caring for. Carers are enabled to relax, enjoy others company and also connect with the person they are caring for in a way that does not require talking.

This connection is particularly relevant to carers of people in residential settings.

Relatives often find it hard to maintain a conversation with their loved one with dementia and visits can become painful and hard to maintain.

The following case study is an illustration of how singing can heal and facilitate a meaningful relationship, even if only for the duration of the singing group.

Mary

Mary had been visiting her mother for five months since her admission to a residential care home. Mary found the visits stressful and was increasingly reluctant to visit. This was because when she was with her mother it was very hard to sustain any sort of conversation, in fact at times her mother did not even recognize Mary as her daughter.

When the activities coordinator started a singing group, she asked Mary and other relatives if they wanted to come along.

This was a turning point in Mary's involvement with her mother and indeed with the home generally.

Mary was able to sing along with her mother who sang songs with joy and recognition. Some of the songs were ones she had sung to Mary as a child and this was particularly poignant.

Mary had a focus for her time with her mother but also it was a focus that was often joyful and had meaning for them both.

Mary got to know other residents in the group and the other staff who were also involved.

Important Aspects of a Singing Group for People with Dementia and Their Carers

The first thing to acknowledge is that many people reading this chapter will see the benefits of singing groups for people with dementia but will think themselves insufficiently musical or talented.

I have been involved in setting up and supporting many singing groups for over ten years. A high proportion of the hundreds of people involved in these groups had serious doubts about their musical abilities. Such doubts disappeared once they realized that the point of the exercise is for the people with dementia to sing and for them to support this. There is an undeniable need for an accomplished pianist or player of another suitable instrument but once you have that person the rest is down to being able to lead a group, being able to assess the abilities and mood of the people singing and to be able to laugh at your mistakes because for sure the people in the group will laugh at them.

Choosing suitable songs will depend on the people in the group their age, their community and ethnicity being three important considerations.

When choosing songs, it is important to remember that as we age our voices get lower so many songs are written too high for older people. The D in the octave above middle C is probably as high as you should expect people to sing. Either choose songs that are low enough or transpose songs down a bit. This may sound daunting to those people not knowledgeable about music but if you have a pianist they will know what to do. The best thing is to have an electric piano or keyboard which has a transpose button!!

It is, however, important not to have all your songs in the same key and pitch as this is not good for the voice or mood. A good variety is what you are looking for.

If no pianist or other instrumentalist is available then the use of backing tracks can be used to accompany the singing. The disadvantage of using recorded music is that it is unforgiving. If people sing more slowly or change the tempo, the track will not accommodate this as would be the case with a live accompanist.

It is important that the choice of songs reflects the preferences of the participants. Leaders may well have favourites but these may not be shared by the participants.

The list of requested songs will, of course, get longer as time goes on.

The nature of the group means that the membership changes so there is a need to be alert to each new member's response to songs and also add new songs that new people may request or respond to well.

Try to keep a variety within the session with calmer melodic songs such as "Somewhere over the rainbow" contrasted with something more upbeat such as "Alexander's rag time band". It is a good idea to build up to more stimulating songs throughout the session and then towards the end return to calming songs (Kerr, 2015).

The length of the session will determine how many songs are used. The number of songs will also be determined by how often a song is sung through (repeat a song that goes well), how many verses are used in some songs, how much talk there is between songs, how able people are to concentrate and how comfortable the chairs are.

Introductions to songs need to be clear so that everyone knows when to start singing. The keyboard player will often play the last line of the song. If the introduction is too long, people will often think they have to start singing and then it is confusing. The leader needs to indicate clearly the moment that the group should join in and start singing.

It is a good idea to break up the session with various formats for singing the songs. Different formats will have different purpose and effect. By getting to know the group, it will be possible to determine when and what to use. For rounds and partner songs, it is helpful to have two people leading. For actions and walking /dancing songs, the capabilities of participants needs to be taken into account and also, for the walking songs, there needs to be sufficient help from others to support people who maybe unsteady on their feet. People using wheel chairs should still be encouraged to join in the movement round the room. Sitting on the side lines watching others maybe a bit lonely.

Below is a summary of the various benefits of using the different song types and formats. This should help with developing a balanced, enjoyable, purposeful, and varied session. The following are ways to do this:-

1. Rounds and canons
2. Action songs
3. Partner songs
4. Call and response
5. Divide men and women
6. Songs to walk and dance to.

Summary of the Different Formats and Their Benefits

Format/type	Involves	Examples	Benefit
Opening Songs	Same song's each session	"Here we are again" "It's a good time to get acquainted" "Hello Dolly"(saying the name of person sitting next to you'	Familiarizes participants with the format and expectations of the session Promotes group feeling
Themed Songs	Using a theme such as a season, a musical, a holiday or cities (linked to a story of travel)	**Seasonal songs** "June is busting out all over" "When it's Spring again" "Easter Bonnet" "Summer holiday" "Winter Wonderland" **Cities** "Wonderful Copenhagen" "I love Paris" "Tulips form Amsterdam"	Aids recognition and continuity. Enables story to be woven through the songs
Rounds	Singing a song in parts, from simple to complex	"Bella Mama" "Frere Jacques" "Something inside me says "time for my tea"	Creates sense of achievement when it works Encourages laughter when it does not Can create some lovely harmonies. These should be commented on Enables greater involvement of carers
Partner songs	A different song for each side of the circle sung at the same time	"Swing Low sweet Chariot" and "When the Saints" (Note the side singing "Swing Low "come in on "Saints") "Michael Finnegan" and "Bobby Shafto"	Promotes sense of achievement, competition and laughter

(continued)

(continued)

Format/type	Involves	Examples	Benefit
Call and response	Songs requiring question and answer	"Gilly Gilly Ossenfeffer" "There's a hole in my bucket" "Down by the riverside"	For some participants, this is easier than straight lyrics as they are prompted for the answer. People may look at other people in the group rather than the leader. This helps with group feelings of belonging and involvement
Dividing men and women	Making sure enough men!	"Que Sera Sera" "Daisy Daisy/Henry, Henry"	Encourages fun of competition Can enable more direct contact, e.g. between a male participant and a female carer
Walking/dancing songs	Providing opportunity to exercise. Make sure people are properly supported	"I could have danced all nigh" "Beautiful Sunday" "Tip toe through the tulips"	Movement stimulates the body, brain and emotions Involves contact with others Often after the walking/dancing songs the group will sing even more lustily
Final section	Rousing songs full of energy	"Alexander's ragtime band" "Roll out the barrel" "I could have danced all night"	Aids breathing and muscle use Creates another level of enjoyment
Finale songs	Quieter songs	"Somewhere over the rainbow" "Edelweiss"	Helps people wind down a bit
Final song	Same song	"Wish me luck as you wave me goodbye"	Acknowledgement that the session is over

Adapted from Kerr (2015)

The above is a guide and in no way prescriptive. Each group will respond to the members and setting and develop its own favourite songs and culture.

Entertainment From Musicians

Perhaps, the most common way in which people with dementia get access to music is through entertainment either in care and nursing homes or in public arenas.

An increasing number of dementia friendly concerts are a welcome development. People living with dementia can exercise choice about whether they attend and the performers know about the needs of people living with dementia.

What is not acceptable is the appearance of individuals or groups in care/nursing homes playing to a captive audience stuck in a room where they are exposed to music that is too loud, and not of their choosing. There are, of course, excellent musicians who play in homes who are aware of the needs of the audience and these need to be encouraged. I have had the privilege of working with students at the local conservatoire who provide music of a professional standard and are sensitive to the needs of people with dementia. Such students need to be encouraged to venture into the world of people living with dementia.

Conclusion

Music can awaken our connection to the world and our history and this is true for people living with dementia.

This chapter has explored some of the many ways in which music can be used to improve the lives of people living with dementia. No one intervention is right all the time, how the person with dementia responds will depend on their own personality, their history, stage of the condition, the time of day, the relationship with the provider of the music, the environment and the choice of music.

Not to be using music as something to listen to, dance to, or sing along with is to deprive people living with dementia of a wonderful, core human activity that will enrich their lives.

References

Aldridge, D. (2000). *Music therapy in dementia care* (p. 95). Jessica Kingsley Publishers.

Ashida, S. (2000). The effect of reminiscence music therapy sessions on changes in depressive symptoms in elderly persons with dementia. *Journal of Music Therapy, 37*, 170–182.

Beck, R. J., Cesario, T. C., Yousefi, A., & Enamoto, H. (2000). Choral Singing performance perception and immune system changes in Salivary immunoglobulin A and cortisol. *Musical Perception, 18*(1), p87-106.

Brontons, M., & Koger, S. (2000). The impact of music therapy on language functioning in dementia. *Journal of Music Therapy, 37*, 183–195.

Casby, J., & Holm, M. (1994). The effect of music on repetitive disruptive vocalisations of persons with dementia. *American Journal of Occupational Therapy, 48*(10), 883–889.

Clair, A. A. (1996). *Therapeutic uses of music with older adults*. Health professions Press Baltimore.

Crowe, B. (2004). *Music and soulmaking: Toward a new theory of music therapy*. Oxford Scarecrow press.
Clark, M. E., Lipe, A. W., & Bilbrey, M. (1998). Use of music to decrease aggressive behaviours in people with dementia. *Journal of Gerontological Nursing*, 24(7), 10–17.
Denny, A. (1997). Quiet music: An intervention for mealtime agitation. *Journal of Gerontological Nursing, 23*(7), 16–23.
Groene, R. (1993). Effectiveness of music therapy intervention with individuals having senile dementia of the Alzheimer's type. *Journal of Music Therapy, 30*(3), 138–157.
Gregory, D. (2002). Music listening for maintaining attention of older adults with cognitive impairment. *Journal of Music Therapy XXX1X, 244–264*
Kerr, D. (2015). *Singing groups for people with dementia. A guide to setting up groups in community and residential settings*. The Choir press.
Lamont, A. M. (2001). Infants preferences for familiar and unfamiliar music: A socio-cultural study. *Paper read at Society for Music Perception and Cognition*, August 9th 2001 Kingston Ontario.
Levitin, D. (2008). This is your brain on music. *Atlantic Books* (Introduction pp. 5–6, 232, 169).
Lord, T., & Garner, J. (1993). Effects of music on Alzheimer patients. *Perceptual and Motor Skills, 76*, 451–455.
Magnusson, S. (2014). Where memories go' Hodder and Stoughton Ltd.
Nietzsche, F. (1888/1974). 'The Gay Science', translated by Walter Kaufmann, *New York Vintage Books*.
Noice, T. H., & Kramer, A. F. (2013). Participatory arts for older people: A review of benefits and challenges. *The Gerontologist*
Nystrom, K. (2002). Dans pa gransen. Demensdrabbades kommunikation I dansterapi' (Swe)' Dance on the border. Communication in dance therapy with persons afflicted with dementia'. Thesis. Stockholm Universitet, Edsbruk.
Playlist For Life. (2014). www.playlistforlife.org.uh
Rader, J. (2002). Organisational issues impacting practice related to bathing. In A. Barrick (Ed.), *Bathing without a battle: personal care of individuals with dementia*. Springer Publishing.
Ragneskog, H., Brane, G., Karlsson, I., & Norberg, A. (1996). Influence of dinner music on food intake and symptoms common to dementia. *Scandinavian Journal of caring Science, 10*, 11–17.
Rio, R. (2009). *Connecting through music with people with Dementia* (p. 95). Jessica Kingsley Publishers.
Sacks, O. (2007). 'Musicophilia' Picador (Preface p X1), p. 347.
Valentine, E., & Evans, C. (2001). The effects of solo singing, choral singing and swimming on mood and psychological indices. *British Journal of Medical Psychology, 74*, 115–120.
Vickhoff, B. (2013). Music structure determines heart rate variability of singers. *Frontiers in Psychology, 4*, 334.
Welch, G. (2010). *The impact of singing; an independent research-based evaluation—the story so far*. Institute of Education, University of London.

Diana Kerr has over 30 years experience as practitioner, researcher and educator in the field of dementia. She was Research Fellow at the University of Edinburgh. Prior to that she was the Course Director for the M.Sc. in Dementia Studies at the University of Stirling .Diana has a lifelong interest and involvement with music making. She is now involved with the provision of singing groups for people with dementia and their carers. She is the author of a number of publications on dementia and on learning disability and dementia, on the night time care of older people and on music and dementia. Her work has been translated into Dutch, Chinese, Norwegian, German, Polish and Spanish.

Chapter 4
The McDonaldization of Long-Term Care

Christopher J. Johnson and David Dietrich

Abstract The organization of long-term care homes has changed significantly over the past decades. Local mom and pop nursing homes (NH) of the past have long since given way to large bureaucratic NH chains, creating an industry whose business is the production of "care". This paper explores how Ritzer's theory of "McDonaldization" can be used to strategically analyse the impact of these structural changes in American NH and assisted living homes in the care of elders. This analysis applies Ritzer's four central concepts of McDonaldization to NH's in America: (1) efficiency, the best method to achieve the two key goals of passing inspections and filling beds that transcend the goal of optimum care; (2) calculability, the assumption that quantity equals quality, such as staff to resident ratios; (3) predictability, extremely repetitive, highly structured activities of daily living (ADLs); and (4) control, taking skills away from people and incorporating them into technology, such as electronic medical charts, medication monitoring software, and "call lights". As with other McDonaldized industries, this creates an "irrationality of rationality", resulting in dehumanizing conditions for care partners. These strategies would rationally seem to reduce financial costs by decreasing the number of workers. However, these strategies create greater human costs and often more financial issues in the long run from staff burnout, high turnover, poor quality of care and decline in the residents over all well-being.

Introduction

> I work as a nurse aide in a nursing home. Let me walk you through our little world that thrives within our sealed people-container. The economy of the building, with its flat metal roof and concrete block walls spread out low and long, is void of ambition or pretence. Like an industrial complex or self-storage unit, it is clean, efficient and functional. Our inventory here is medical specimens. Our product is time. We give old people a little more time. And if we lighten the burden of some family members, all the better. As long as we keep these faint

C. J. Johnson (✉) · D. Dietrich
Department of Sociology, Texas State University, 601 University Drive, San Marcos, TX 78666, USA

M. K. Shankardass (ed.), *Dementia Care*,
https://doi.org/10.1007/978-981-16-3864-0_4

> hearts pumping, the Medicaid dollars and the life savings keep rolling in. With 130 beds, my boss grosses $3 million a year. He drives elegant cars and lives in a house big enough to have its own name. As a relatively well-paid aide, my cut is $6.20 an hour. (Gass, 2004, p. 1)

Although spoken close to two decades ago, this quote still reflects the economic discrepancy in pay of long-term care (LTC) staff and the financial success of owners of for-profit homes. LTC is an industry whose goal is to earn profits through the production of health care to the chronically ill and most vulnerable (Heinemann, 2015). Following Ritzer (2015) we argue that this healthcare production process most often operates according to similar principles as fast-food (FF) enterprises. Like McDonald's, LTC [e.g. continuing care retirement communities (CCRC), nursing home (NH) and assisted living (AL)] homes often can be considered as health-care industries who most often achieve enormous profits for owners. In pursuit of these profits, the LTC industry usually employs minimally trained, cheap labour while meeting minimum or less staffing requirements. Furthermore, as chains have expanded over the past decade, many LTC industries have increasingly employed repetitive structures and technologies to lower costs, which again reflects similar patterns within the FF industry.

This trend is especially concerning given the rise in diseases of dementia among both the young and old (Alzheimer's Disease International, 2019). At the same time, sociological and demographic developments over the past century have created the greying of American—a culture which worships youth—suffering from ageism (Gaugler, 2016; Thomas, 2006). Additionally, the global increase in dual-career couples along with mobility and industrialization has negatively impacted younger generation's desires and abilities to offer aid to their parents. LTC for-profit homes were spawned within this ageist society and are rapidly expanding. And just as the first McDonald's restaurant was transformed from a mom and pop burger joint into a corporate mega-chain; in the last forty years, most local mom and pop NH's have been purchased and incorporated into larger bureaucratic LTC chains, making each homeless personal while mass-producing health care (Duffy, 2011; Dunlap, 1979). Both FF and LTC chains reduced the costs of production through a process of hyper-rationalization—a process coined by Ritzer (2015) as "McDonaldization". In this paper, we strategically analyse and compare LTC and FF industries in America using Ritzer's four principles of McDonaldization (efficiency, calculability, predictability and control) to demonstrate how LTC has become McDonaldized and how this process has resulted in detriments for both residents and workers within the LTC industry.

Rationalization Processes: Bureaucratization and McDonaldization

Ritzer (2015) employs Max Weber's theory (1947) of modernity and the process of rationalization—a complex process where antiquated philosophies are changed by the goals and process of efficiency and formalized social control. For Weber (1947), the archetypical manifestation of this process was the *bureaucracy*, a large, formal organization characterized by hierarchical structured authority, well-established division of labour, written policies and procedures, impersonality and a concern for technical competence. This is the logos of many modern capitalist industries, including FF and LTC chains. Bureaucratization of LTC "communities" has become a potent organizational trend extending the process of the rationalization to an expanding and vulnerable elder population on fixed incomes.

Bureaucracies offer advantages over the more traditional methods of organization, most notably in their ability to manage larger tasks far more efficiently. But bureaucracies can bring with them a number of detriments, including actually creating inefficiencies due to a tangle of complex rules, decreasing concern for the quality of work, and depersonalization of employees due to replacing human judgement with the rules, regulations and dictates of the bureaucratic structure (Weber, 1947). When bureaucracies become depersonalized emotions such as love, hatred, any other irrational and incalculable feelings are excluded from the execution of official tasks. While this has the advantage of ensuring that emotional bias is eliminated from the decision-making process, Ritzer (2015) fears depersonalization can also lead to the management and removal of all emotion from the bureaucratic process, including feelings of compassion or sympathy. We argue that bureaucracy in LTC and FF chains can impose this depersonalized structure by commodifying labour, families and residents.

But Ritzer (2015) goes further than Weber in arguing that the world has become not only rationalized, but in many cases hyper-rationalized through the process of McDonaldization. McDonaldized systems are successful because they follow four rationalization principles: efficiency, calculability, predictability and control. And while this process began in the FF industry and was perfected by McDonalds, as the name implies, Ritzer argues that McDonaldization has spread beyond the FF industry to affect many aspects of modern life. This is illustrated by research that demonstrates the influence of McDonaldization while applying it to: medicine (Dorsey & Ritzer, 2016), hospitality management (Taylor & Lyon, 1995), tourism (Zegre et al., 2012) and digital consumption (Belk, 2013) as well as mixed "bricks-and-clicks consumption sites" (Ritzer & Miles, 2019, p. 4). The LTC industry has likewise adopted such principles in a historic approach to medicalizing and commodifying care. Thus, it is not surprising that the central concepts utilized in FF industry have infiltrated LTC and MC industries as well. Comparative studies of developments in for-profit LTC chains and in five countries have shown these trends, such chains: are more and more being owned by private equity investors; have a revolving door of different owners over time; and have complex organizational structures that suggest increasing demands making higher and higher profits (Harrington et al., 2017). Large chains buy smaller LTC homes–increasingly controlling the market (Harrington et al., 2011). Their policies include the separation of property from operations, diversification, expanding locations, lowering costs through technology, cutting staff and/or supplies to minimum levels, while finding tax havens that the lack of adequate public information about the ownership, costs and quality of services provided by NH chains—which is a global trend (Harrington et al., 2017).

LTC and FF owner's rational choice of expanding business to make more profit—a central principle of capitalism—often has negative consequences. McDonaldization manifests the logic of capitalism, driven in some cases by misplaced priorities of elevating profits over residents and staff's well-being. This manifests in what Ritzer (2015) called the "irrationality of rationality", where the very same measures designed to create a hyper-efficient, hyper-rational organization causes irrationality within the system. "Irrationality means that rational systems are unreasonable systems that deny the basic humanity, the human reason, of the people who work within them or are served by them" (Ritzer, 2015, p. 132). While LTC mega-chains may have large revenues with high profits, they suffer from quality issues with high levels of staff turnover, inadequate staffing and lack of supplies (Johnson, 2014; Spilsbury et al., 2011). Additionally, the rational need to have minimum staffing and supplies (Johnson, 2014) to increase profit margins often sabotages certified nursing assistants (CNAs) abilities to establish person-centred care (PCC) with residents due to time constraints and can lead to abuse (Ho, 2007).

Efficiency

Efficiency ("the optimum method for getting from one point to another") is applied to both customers and employees (Ritzer, 2015, p. 14). Customers have their needs met in expedient ways facilitated by employees that follow prescribed steps in the production of care, food and services. Efficiency in McDonaldization means that every aspect of the organization is geared toward the minimization of time to get this done and to get customers serviced quickly and easily (Ritzer, 2015). Owners of NH businesses, like owners of FF businesses, know that to make a profit they must find an inexpensive workforce and make them productive and efficient at little cost. Despite owner and management "cream-skimming" of substantial profits, NH's 2.1% profit margins are the least desirable of the large array of lucrative LTC homes (Lord et al., 2018). NHs suffer from low Medicaid reimbursements, state regulatory stringency, staffing issues, falling occupancy rates and high liability insurance costs (Lord et al., 2018; Mukamel et al., 2012). In LTC, NH residents' higher amounts of dementia, frailty and comorbidities (Highfill & Johnson, 2016) raises the costs of doing business. Moreover, MC and AL for some years have been giving NH indirect competition for beds (Lord et al., 2018). Thus, residents often become commodities packaged in containers that deliver speed of care by CNAs paid at minimum wage with few benefits and high turnover rates (National Citizens' Coalition for Nursing Home Reform,

2001; Rosenzweig, 2012). As for training, Kim's (2019) study confirms our assertion that direct care assistants in AL (compared to NH aides) had a serious deficiency in dementia and gerontological training. Yet aides in both NHs and AL care for persons with dementia (PWD) through the entire disease process until death.

Incentives for delivering efficient care pull LTC in the opposite directions toward limited staffing with reduced time to get to know the residents (Stuckey, 1997; Lopez, 2014). As one would expect with heavy workloads, they can often objectify residents (Johnson, 2014). This practice adversely impacts both the CNAs and the residents (Ho, 2007; Johnson, 2014). Thus, the LTC industry may succeed in conveying what Goffman (1959) terms "impression management" to casual visitors, policymakers and the public that they are attentive to delivering "person-centred care" (PCC) to their residents. The reality is that CNAs are seriously constrained in performing PCC in nursing, even of the most intimate kind with advanced dementia care. Consequently, the LTC industry has been slow to adopt and implement PCC and often ascribe unique definitions for operationalizing PCC (Behrens et al., 2019). The variation in the operational definitions of PCC and its core domains are barriers to adopt and create an accurate (inclusive) definition of it that can be used to assess and improve individualized care partnering (Johnson et al., 2017).

Management

The profile of the average McDonald's manager or LTC administrator is that of a bureaucrat. Most managers are not hired to be creative nor proactive but rather efficient and predictable. The McDonalds managers' job with regard to efficiency is simple: (1) hire workers at minimum wage; (2) make sure workers do not contaminate the food; (3) pass inspections and keep customers pacified; and (4) sell a lot of products. Similarly, LTC management tends to look for the most efficient route for two key goals: passing inspections and filling beds (Gass, 2004; Vladeck, 1980; Harrington et al., 2011). This results in similar efficiency expectations of LTC administrators: (1) hire CNAs and other workers at very low wages (usually with little or no benefits); (2) make sure residents and their rooms are clean and food is not contaminated; (3) pass inspections and keep families happy customers; and (4) fill the beds, while keeping residents alive as long as possible.

The promise of making progressively greater profits (Highfill & Johnson, 2016) motivates many owners to hire cheap and minimally educated labour to reduce costs. But having low staffing and inadequate supplies increases worker discontent (Johnson, 2014). Hence, the lives of both the McDonald's workers and CNAs in LTC have similarities. Owners of many LTC chains and FF franchises often make it clear to workers they are expendable and replaceable. Both industries management styles are driven by desire for profits being prioritized over workers which can result in alienated labour and turnover (Castle et al., 2006; Parsons et al., 2003).

LTC is growing at very high rates due to global ageing, more people reaching old age and the "tsunami" of dementia (Zeilig, 2014). While research indicates the gerontology education and credentials for LTC administrators matters in the quality of care (Castle & Ferguson, 2010; Castle et al., 2015), this is not reflected in the expectations of most LTC managers and owners. Wanting to fill beds and effectively handle "challenging behaviour's" of persons with dementia (PWD) with staff and management who have minimal gerontology and dementia education is a daunting task. LTC personnel labour to earn wages to enable them to survive while satisfying the owners' needs to fill beds, make families happy, pass inspections and make money. This labour most often becomes a means to an end of most owners. Turnover rates in NHs can run anywhere from 100 to as high as 400% (Johnson, 2014), which is even higher than a peak staff turnover rate of 150% in the FF industry (Rosenbaum, 2019). When employees do leave someone else is hired on the spot, often in the same day. "Any warm body will do" is the hiring philosophy of many owners in LTC (Johnson, 2014) and FF franchises (Pyrillis, 2014).

One area where the FF and NH's differ is working conditions. In the case of McDonald's, the risks associated with the working environment are fewer than the dangers CNAs face in LTC. A qualitative study of CNAs at "Heartland" nursing home by Lopez (2014, p. 63) reveals on the connection between efficiency and quality of care (e.g. delivering PCC):

> The positive features of the managerial approach at Heartland, however, did not address the most basic and fundamental problem of the nursing home: the fact that workers could not complete their work tasks in the time allotted if they followed official care rules and procedures. This was particularly the case on the dayshift during the crucial period of "a.m. care," between the beginning of the shift at 6:30 a.m. and breakfast at 8:00 a.m., when residents were awakened, washed, dressed, and readied for the day. Especially during this intense period (but also at other points throughout the day), aides at Heartland were forced to devise their own work methods, violating official rules in order to match the workload with the time available. This resulted in an ironic situation: all of the workers justly felt that there was too much work and not enough time, even though experienced workers could complete their work in the given time by using shortcuts.

CNAs are provided with gloves but still are exposed to a wide range of life-threatening workplace hazards (e.g. diseases, viruses and bacteria) (Walton & Rogers, 2017). In addition, Ye et al. (2015) reveal that current laws of patient (resident) confidentiality result in situations where CNAs are not informed of residents having been diagnosed with communicable diseases. Consequently, health risks in NH bed and bodywork are as high as they have ever been with the 2020 Coronavirus pandemic and staff and mask shortages. A Washington Post article raised concerns about NHs (Jacobs et al., 2020) indicating they had run short on staffing and protective supplies. Senator Elizabeth Warren (D-Mass.) and two lawmakers addressed the "systemic failures" of NHs in keeping both staff and residents safe (Jacobs et al., 2020). Caspar et al. (2020) outlines a litany of issues, many of which are directly related to the McDonalization of LTC:

> Specifically, low job satisfaction and high turnover among resident care aides (RCAs, also referred to as Certified Nurses Assistants—CNAs) has been linked to poor quality of care (Chou et al., 2002), poor quality of life (Pekkarinen at al., 2004), and poor end-of-life quality indicators and resident behavior deficiencies (Castle & Anderson, 2011; Lerner et al., 2014). Research also indicates that care staff members who are dissatisfied with their work-life often show signs of burnout in the form of an unreliable work ethic (e.g., taking unscheduled

days off) (Castle et al., 2006). Finally, evidence suggests that some dissatisfied care staff show greater aggression towards residents (Parsons et al., 2003) and other workers (Spector, 1997).

Workers

Both NH and FF workers are trapped in what Ritzer terms "McJobs", which involve minimal training, low pay, and with few exceptions, thin opportunity for upward mobility. Residents are packaged in LTC homes that deliver caregiving by CNAs paid minimum wages with few benefits (Glasser, 2019). Inadequate staffing and high turnover cause LTC workers to feel overworked and trapped, frequently seeing no other alternatives but to sacrifice their unmet needs and sometimes even their lives for the production of health care (Johnson, 2014; Lopez, 2006). RN's become efficient bureaucrats tucked away in their offices almost all day filling out paperwork in order to pass inspections (Vogelsmeier et al., 2011). Few have the time or energy to adequately walk the halls conducting medical checks of residents (Lopez, 2006, 2014). LPN's jobs are usually restricted to efficiently administering medications to the residents. Turnover rates and low staffing make it virtually impossible for CNAs to have the time and energy to deliver PCC to residents and efficiently carry out their duties (Harrington et al., 2017).

LTC Activity Directors are required to have anywhere from two to four years of college education. Idealistically, it would empower them in their jobs if they had needed recreation therapy or gerontology undergraduate degree, and/or some dementia courses but they rarely do. McDonaldization mandates that they operate efficiently, with historically limited budgets. This makes offering PCC activities a daunting task. Thus, most LTC has historically engaged in "warehousing activities" (one activity fits all residents) for residents (Zarit et al., 1998; Johnson et al., 2017). Activity calendars are designed to be efficient, usually displaying similar daily and weekly activities (e.g. bingo, singing or watching a movie, etc.). The first author has observed planned activities can amount to as little as offering residents food (e.g. "popcorn" or "hot dog" parties, etc.). This efficient, easy approach to activities involves little or no participation other than having residents eat rather than help prepare food as in "therapeutic kitchens" in some progressive MC homes. Without skilfully crafted PC activity plans, residents can suffer from loneliness, depression and boredom–leading to challenging behaviours (Power, 2017; Thomas, 1996).

Design

Most MC neighbourhoods need more attention being given to evidence-based designs and details than what is seen (Eijkelenboom et al., 2017). Design issues are described by some authors (e.g. Buchanan, 1992; Cleeve et al., 2020; Coyne, 2005; Krippendorff, 2006; Rittel & Webber, 1973; Treadaway et al., 2018) as "wicked" problems cutting across several disciplines and beset with challenges. Just as LTC managers usually lack gerontology and dementia education, most interior decorators and architects of LTC rarely have formal gerontology and/or dementia education or credentials as well (Grabinski et al., 2014). Consequently, too often there is a lack of attention to details of evidence-based design for dementia in MC. For example, AD persons in Stage 6 do not see in 3-D (Reisberg, 1986), yet MC milieus might have furniture the same colour as the tile or carpet instead of contrasting colours (Phelps, 2017). Similarly, AD persons go through trips back in time causing them to revisit earlier ages and people of their distant past (Johnson & Johnson, 2000). MC neighbourhoods frequently have problems with the content of "Memory Boxes" (Gulwadi, 2013) used to cue residents to their rooms, overlooking what Johnson and Johnson (2000) refer to as an AD person's "time travel" with appropriate cueing pictures (Johnson et al., 2017). Hence, the McDonalization of LTC attempts to achieve greater efficiency in the production of MC but such efforts often result in irrational interior designs that do not function for PWD (Gulwadi, 2013; Johnson et al., 2017).

Calculability

Calculability enables efficiency when management quantifies all aspects of staffing and production—whether it be health care or food. In FF, the sizes of portions, the cost of meals and the time necessary to complete the transaction are but a few of the elements assigned numbers and compared to ensure maximum profits. Customers, too, have become trained to pay attention to such quantitative aspects, such that more is better, bigger is better and faster is better, regardless of the actual quality of the products being consumed.

Whether it be LTC or FF chains, rational calculation is manifest at every stage. LTC has historically been dominated by the "medical models" of care (Schwarz, 1996). Minimum-staffing standards and euphemistically labelled PCC initiatives claim to meet individual needs of residents while making a profit. To achieve this, management's objective is to figure out and limit the costs of labour and overhead. They do this by excluding relational work from CNAs pay, making it difficult if not impossible for them to continue providing this necessary attention to residents (Fossey et al., 2014; Glasser, 2019; Heinemann, 2015). The performance of each FF or LTC employees' workloads are mathematically measured. Hence, each employee is essentially reduced to a number, becoming replaceable (Mills et al., 2018).

One key metric for evaluating worker quality in many industries is hours of training. However, even by such purely quantitative standards, the minimum requirements for many LTC homes worker training are inadequate. For example, it takes roughly 1500 hours to become a beautician, 400 hours to be a dog groomer, and only 75 hours to be a CNA (Johnson, 2014). The issue is even more concerning about LTC administrators, who with some exceptions have only two to four years of college and little or no gerontology or dementia education, manage LTC homes (New York State Department of Health, 2017; Texas Health & Human Services, 2011).

As NH employees are quantitatively evaluated, so are the residents. One of the requirements for NHs to receive medicare funding is to implement specific assessments by a "Resident Assessment Instrument" of residents at intervals (Stone, 2005). This assessment examines several aspects of resident health and functioning, including cognitive abilities, communication and visual acuity, mood, and behaviour. But of relevance here is that this information is calculated to a few numbers then recorded on a form that comes to describe a resident. The danger lies in condensing down numerous qualitative observations over time into one of only a few numeric categories that describe the resident as no longer as a person but as a calculated set of measurements. One reason for using such forms is to serve an additional dimension of McDonaldization, predictability, as all facilities are expected to conform to the same set of measures. LTC residents need for relationships with CNAs are calculated and ultimately hampered by lack of staffing and time (Johnson, 2014; Johnson et al., 2017; Fossey et al., 2014). Stone's study addresses the role of owners in minimalizing the number of CNAs and cost-cutting, while increasing profits (2005, p. 282):

> In the private sector, once patients (or children or clients or anybody else) become the raw material out of which profits are made the entrepreneur is dedicated to keeping costs down. In an economic sense the cost of care is almost all in the labor, so keeping costs down means suppressing care. When care is calculated as a market good--its essence--the human component and the relationships with residents has to be minimized because efficiency requires getting the most output for the least input.

Predictability

Predictability enabled McDonalds to expand from one simple burger joint in California to a global franchise by ensuring that as many aspects of the product and experience are identical, no matter where or when you visit. Customers rest secure in the knowledge that the hamburger they get today will be the same as the one tomorrow and the day after, even if they travel to another town or even to another country. To ensure this predictability and efficiency McDonalds maintains strict control over virtually every aspect of the process, from the production of the ingredients (which are painstakingly weighed and measured), the preparation of individual items, to the dress and behaviour of employees.

In FF or LTC bureaucratic settings, it is important that the business operations proceed like "clockwork". LTC chains offer daily routines which often become rigid and inflexible, becoming "total institutions" with predictable and efficiently running parts (Goffman, 1957, 1961, p. 4). In for-profit NH chains, all actions of staff and residents in delivering activities of daily living (ADL's) are under complete control and thus made predictable (Harrington et al., 2011; Shadish & Bootzin, 1984). "Total institutions" usually require all residents to cooperate and eat at designated meal and snack times (Shadish & Bootzin, 1984). Such restrictions can result in homes becoming semi-prisons for residents in distress, triggering "wandering" and other "problem" behaviours (Dewing, 2011). Rather than address these structural issues based upon unmet needs—which would create a less predictable environment—NH often employs electronic "wander guards" to prevent "problems" such as "wandering". Thus, residents get daily predictable "health care", eschewing uncertainty.

In LTC, predictability is critical to passing inspections for administrators whose jobs depend on it. But, in a potentially positive application of the tenet of predictability, it is also critical to PWD who need daily predictable routines and milieu (Johnson & Johnson, 2000). Unfortunately, this aspect of predictability is often overlooked by many LTC homes, often in the service of other aspects of McDonaldization, such as efficiency. Staff turnovers cause distress among PWD and create increasing unpredictable behaviours (Williams-Burgess et al., 1996). Residents who exhibit challenging behaviours are frequently labelled "problems" and are often just avoided. Loneliness and lack of attention to unmet needs can cause PWD to be candidates for polypharmacy or geropsych evaluations for "medication adjustments". Such challenges in overly medicalized homes can cause PWD to be stigmatized and even discharged in lieu of their challenging behaviours (Rosenzweig, 2012), thus ensuring a predictable environment via an "out of sight, out of mind" approach.

Control

Administrators manage control in McDonaldization by employing the ways and means to gain more efficiency and predictability from their staffs and customers/clients. This includes, but is not limited to, the use of robots and technology to reduce or replace some workers. The salience of control can be seen in production, labour, and in the consumer experience, and it can be witnessed in virtually all areas of both the workplace and all characteristics of social life.

McDonalds has used mechanization to improve food quality, efficiency, service, while using less labour. Predictability, efficiency and control are achieved by technology using automated equipment in processing customers' orders. Automated cooking helps to remove inconsistencies. McDonalds adoption of handwritten typeface is an attempt to create an economic advantage by transmitting a sense of humanism and love in the restaurant's menus (Liu et al., 2019). This impression management creates favourable consumer responses including positive attitudes toward the menu, enhanced perceived healthiness of the product and higher social media engagement (Liu et al., 2019). In terms of total customer traffic, McDonalds has driven "a monumental shift" toward digital ordering with incentives given to ensure this shift speeds up (Purtill, 2019). Average ticket size from a digital order is higher than a traditional order. The app or kiosk "suggestively sells" based on order data (Purtill, 2019). McDonalds controls customers in ways they often never realize, manipulating perceptions and behaviours to encourage more efficient consumer behaviour to pay, eat and leave satisfied (Belk, 2013).

Similarly, LTC is employing more high technological "help" to speed up and control "health care" (Niemeijer et al., 2014). This process is manifested by high tech thermometers, "call lights", electronic medical charts, medication monitoring software, robots and electronic "memory boxes" by their rooms for cueing and wayfinding, and the use of "wander guards" to "protect" residents from "wandering". But when the necessary roles and responsibilities are not present, the technology becomes a burden rather than a support (Spilker & Norby, 2019). For instance, during a study on prompting technology, researchers found incomplete tasks led to frustration and decreased usefulness for caregivers and PWD (Braley et al., 2019). Similarly, the success of GPS tracking as a service to increase predictable tracking of LTC residents depends upon establishing a techno-organizational network with established practices, one which is flexible enough to handle new individuals and possesses competencies required to handle human and non-human elements (Spilker & Norby, 2019). While not as prevalent as in the FF industry (Rosenbaum, 2019; Purtill, 2019), robots are being used to a small extent in LTC. Robotic technology in LTC includes robots performing nurse aide work, social robots who serve as "pets" while offering biofeedback for residents, and biconically engineered nursing (LeGrand, 2017).

Control is also a goal in design in MC "Neighbourhoods", and McDonalization has created what can be called, "McMemory Care". Like McDonalds employees handing out hamburgers in styrofoam containers, the MC can become "people-sealed containers" –a term linked to NHs frequently commodifying care (Gass, 2004). Just as McDonald's controls its restaurant "curb appeal" through colours and cartoon characters which attract attention and stimulate appetite, LTC homes also engage in similar practices for the purpose of controlling perceptions. LTC buildings will design fancy front facades to make them seem more pleasing and inviting, like the "Golden Arches" of McDonalds (Alford, 1993; Gass, 2004). Similarly, industry brochures offer flowery descriptions of their homes, including the use of terms designed to portray "Memory Care" (MC) as "homelike", such as "MC Communities" or "Neighbourhoods". (Jacobs et al., 2019).

The Impacts of McDonaldization

The consequences of McDonaldization in the LTC industry are emblematic of Ritzer's "irrationality of rationality". Over a decade ago, these consequences on LTC staff were addressed by Ho (2007) as "structural violence". Structural violence is a term attributed to Johan Galtung who described a social institution that causes harm to workers and those they serve by preventing both from meeting basic needs. Low staffing and poor pay set staffs up for failure to effectively carry out their duties and to raise themselves up out of poverty. Low levels of positive worker regard by some managers are problematic for some homes. Direct physical and verbal abuse of residents by over-stressed CNAs can occur and is symptomatic of an industry that "undervalues both its staff and residents" –harming them both (Brophy et al., 2019).

The effect of structural violence on individual agency results in a gap between potential and actual fulfilment of human rights. CNAs are America's working poor who suffer from the unequal share of power to decide over the distribution of resources (salaries, supplies, time spent with residents) as the pivotal causal factor of structural inequalities (Bruni et al., 2019; Johnson, 2014). Stone (2005) suggests that love may conquer self-interest for many CNAs but ultimately this may not matter when corporate structures restrict their staffing and supplies which sets them up for failure to carry out their jobs and establish PCC (Johnson, 2014, 2018). Even non-profit homes can fall prey to McDonaldization. Some non-profit homes are run by for-profit "management groups", resulting in McDonaldized homes creating lost human rights for care partners (Stevenson et al., 2013).

Even trends such as "person-centred care" (PCC) (Kitwood, 1997) and "memory care" (MC) communities (Johnson, 1989) often fall prey to these same issues, despite appearing to counter McDonaldized practices. MC is a phenomenon that quickly became McDonaldized as the solution to controlling what then was referred to as "catastrophic behaviour's" of persons with dementia (PWD) living in NHs. In the 1980s, it was common to have PWD trying to navigate from various areas of NHs that had similar looking and confusing hallways in various shapes. Johnson (1989) described redesigning one hallway into a self-contained "dementia wing". These wings or "units" evolved to be labelled dementia "neighbourhoods", "communities" or pods. However, over three decades ago, such "dementia wings" were described by Alford (1993) as varying extensively in quality. Alford's paradigm identified four types as: "Ideal", "Uncultivated", "Heart of Gold" and "Rotten to the Core".

Similarly, when promoting PCC, the LTC industry markets its "homelike" milieus where residents are said to be treated "like family". In contrast to such claims for PCC, gerontological research (Diamond, 2009; Gubrium, 2012; Harrington & Carrillo, 1999; Kayser-Jones, 1990; Mendelson, 1974; Moss & Halamandaris, 1977; Shield, 2014, 2018; Stannard, 1973; Vladeck, 1980; Yates, 2006) conducted from 1973 up to the present find a litany of serious issues in such LTC homes. R. H. Johnson's (2014) three-country (America, Canada and Scotland) comparative study of certified nursing assistants (CNAs) is significant, it found that they were considered the "working poor" in America. Her examination of for-profit homes revealed extremely high staff turnover rates in America but in all three countries for-profits suffered from low staffing and supplies—evidence of a flawed industry. In researching CNAs in NH's, Roberts et al. (2019) concluded that the ability to maintain the same care partner/resident ratios over time considerably varied. Although some homes were consistent, there was a wide range between minimum and maximum ratios, suggesting residents might experience wide swings in staffing–in some homes doubling from one week to another. These staffing fluctuations data raise questions about the most salient measures of the quality of PCC and assignments from a resident perspective.

Such practices are substantially similar to what Ritzer (2015) calls "Starbuckization", in which FF chains, most notably Starbucks, take explicit steps to hide the hyper-rationalized, McDonaldized nature of an institution by engaging in theatrical displays. For Starbucks, this involves providing higher quality products, creating a more inviting atmosphere through store design, comfortable seating and creating an impression of good treatment of employees—with minimal commitments to causes (Ritzer, 2015). LTC homes are designed with enticing colours and interiors, coupled with a warm brick façade to exteriors to greet families. Upscale homes offer food choices to residents. Through a combination of curb appeal and fancy websites, owners employ Goffman's (1959) "impression management". Such elegant labels and facades deflect from the commodification of care—a "McDonaldized" home. Impressions are also controlled by alluring and elegant homelike labels (e.g. "Hillside Manor", "The Oaks", "Avalon Place", etc.) to NHs. Beautiful brochures are used to portray happy elders engaged in PCC (ArentFox.com, 2008; Harrington et al., 2017). Yet many LTC homes today remain "total institutions" which control most every aspect of a resident's life. Such homes are designed to be functional for observation and control of residents, with some level of elegance (Shadish & Bootzin, 1984). Nevertheless, too much control can stifle residents from self-actualizing and receiving PCC (Johnson, 2018).

Despite the many negative consequences of McDonaldization (cf. National Employment Law Project, 2013), as described above, there are some positive outcomes created through this process in the LTC industry. For instance, studies have shown that health information technology (HIT) can improve the quality of care and resident safety. Purtill (2019) discusses how automation and information technologies are being used to address current LTC healthcare challenge, where greater efficiency, predictability and control results in positive outcomes. For example, new electronic health records can reduce human errors and provide efficient recordkeeping while allowing more staff time for relationship building (Felix et al., 2020). Management is also using new data mining technologies, sensors for health monitoring and advanced recognition of anomalous activities and diseases, new hardware/software technologies, artificial intelligence and machine learning. Technology is also being used to empower PWD with independence while being efficient as well (Braley et al., 2019). Technology customizes the system for care partners, offering predictability and control for interventions (Mitchell et al., 2018). Cueing technology offers predictable and controlled outcomes of wayfinding for PWD. It has been shown to be most successful as evidenced by predictable completion of tasks (Braley et al., 2019; Pang et al., 2018). Consistent success in outcomes is thought to reduce disability, while adding to the QOL of PWD (Scholl & Sabat, 2008).

Furthermore, there is an increasing recognition of the negative effects of McDonaldization, resulting in some positive global trends in both FF and LTC (Duan et al., 2020; Power, 2017). Globally, gerontologists point to an expanded version of Rogers notion of "personhood" by Kitwood to PWD (Kitwood, 1997). Physicians insights into NH resident's lack of social engagement, empowerment and PCC activities have also triggered what has been called "culture change" in LTC (Port et al., 2011). Ownership greed, coupled with antiquated LTC management styles, is being questioned by this "culture change Movement" (Corazzini et al., 2015). It is based on PCC which empowers elders, PWD, and LTC staffs—at least more so than in the past (Caspar et al., 2020). Duan et al. (2020) argue that such "culture change" has the potential to transform LTC from impersonal institutions into PCC homes with more quality of life (QoL) focus, displaying love and care for residents. For example, LTC American geriatricians have developed new progressive models of social engagement in NHs like the "Eden Alternative" movement (Thomas, 1996) and experiential learning strategies (Power, 2017) for staffs in LTC. The industry manages the struggle between PCC being difficult in nursing arenas and the need for efficiency and calculability by allowing more PCC in the activity department (Lopez, 2006, 2014). Research measures of QoL measures vary across studies, ranging from overall perception of life to specific measures associated with well-being. Yet, Duan et al. (2020) research concluded that "culture change" had positive effects on resident satisfaction and personal freedom.

Evidence-based designs for cognitive disabilities are sociologically significant. They have demonstrated that changing architecture to more open-styles, including community "therapeutic" kitchens and public couches, has been associated with upsurges in social connectivity of residents (Morgan-Brown et al., 2013). In MC, it is becoming more common to see wall-mounted "memory boxes" with proper height and content relating to an AD person's "time travel". Memory boxes preserve identity, cue the PWD to their rooms, allow for reminiscence, while helping staff to engage with residents (Gulwadi, 2013). Additionally, Zarem (2010) found MC communities moving to creative dementia pod designs with varying levels of care (e.g. ambulatory pods, non-ambulatory or wheelchair pods, skilled nursing areas, etc.). These changes suggest a transition from medical to social models of care in America and globally (Brune, 2011; Duan et al., 2020).

There are also newer management styles based on decentralized authority, where CNAs have more of a voice in the LTC community they serve. In saying this, their low pay and turnover remain obstacles to this movement (Banerjee & Armstrong, 2019; Brune, 2011) but developing paradigms shifts are occurring. There is also a developing paradigm shift toward profit-sharing in progressive and worker-oriented LTC firms (Corazzini et al., 2015; Filmore et al., 2014). Competition has helped develop LTC communities that have eliminated middle management, offer insurance and childcare to direct care workers, some companies pay for continued schooling and try to use low-cost PCC methods (Filmore et al., 2014; Roberts et al., 2019). In addition, there are more progressive homes that hire M.A.s in Gerontology or related degree as trained administrators (Castle et al., 2015), although that should be a requirement. Being progressive can attract healthcare consumers. Some of the homes mentioned have waiting lists for beds and their staffs make reasonable earnings as well.

Conclusion

Nations are facing the LTC challenges of global ageing and the need for gerontologically educated staffs who earn living wages. The McDonaldization of LTC, based on greed and ageism, has stifled the provision of PCC and full dementia citizenship. The low pay and lack of benefits, inadequate training and career development opportunities, and organizational and societal cultures all operate in tandem to create a plethora of unmet needs in residents and care partners challenges (Alexander & Madsen, 2017). McDonaldization disempowers elders with a combination of social pathologies: ageism, structural violence, rationalization of medicine (polypharmacy), sabotaging PC activities and care, creating high staff turnover and lack of supplies, impoverishment of workers and dehumanizing death (Berger, 2017; Castle et al., 2015).

Historically, incentives of McDonaldization have pulled LTC toward limited staffing and time to establish relationships with residents (Lopez, 2014). The LTC struggles between providing time-consuming PCC and its need for efficiency and calculability in nursing and activities (Lopez, 2014). In terms of calculated costs, providing PC Activities is visible (impression management) but less costly (calculability) vis-a-vis providing PC nursing (Lopez, 2014). The demands for efficient nursing care can adversely impact both CNAs and residents (Ho, 2007; Johnson, 2014). Adequate staffing with low turnover, who are given enough time to establish relationships with residents is needed. If all this happens then CNAs can deliver PCC and be "best friends" with residents and PWD (Stuckey, 1997).

Gerontologists recognize the detriments of McDonaldization of LTC. Negative public opinions of NHs have over the past decade driven them to seek a better image through humanizing their care (Duffy, 2011; Lopez, 2014). Gerontologists have been catalysts for change, elderhood, staff retention and PCC for persons with cognitive disabilities in LTC (Berridge et al., 2018). The "Culture Change" movement in the LTC industry took hold of the PCC idea, striving to place residents' empowerment and autonomy at the centre stage (Corazzini et al., 2015; Koren, 2010). The Baby Boomers and the media are not settling for maintaining the status quo (Brune, 2011). Growing numbers of LTC homes strive to create true homelike, evidence-based environments, while moving toward deinstitutionalization of elders quicker and providing high-quality care (Doty et al., 2008; Behrens et al., 2019).

Sapkota and Subedi (2019) and the World Alzheimer Report (Alzheimer's Disease International, 2019) suggest that LTC and our communities should advocate for human rights for the cognitively disabled. We as a global LTC advocacy community should attempt to decrease the stigma of dementia by valuing the abilities of a PWD which are so often overlooked (Harding & Palfrey, 1997; Mukadam & Livingston, 2012; Power, 2017). There is evidence of social change. A growing number of LTC homes have higher expectations for their delivery of health care. Social models of care are driven by the high expectations of Baby Boomers and growing competition in LTC. For example, increasing numbers of owners are investing in more homelike and PCC homes in the USA, matching global trends. Engaging household designs to socially connect with PWD has been supported by studies in many countries, including in Ireland (Morgan-Brown et al., 2013).

Sociological models of care like the Green House Project and the Eden Alternative expands our awareness to fulfil unmet needs for full human rights, equality, citizenship, social inclusion and relationships with both LTC staff and residents. New innovative LTC homes offer socially engaging, therapeutic and PC activities in such areas as music, art, gardening, pet therapy, intergenerational connections, Namaste advanced dementia care and many other programs (Johnson et al., 2017). Technologies are developing for LTC at home with redesigning homes for ageing in place and cognitive disabilities (Yassine et al., 2017).

The need for LTC increases with the greying of the world but the McDonaldization of health care must be addressed. While we identify irrationalities that result from the adoption of the McDonaldization of LTC, future studies are called for to ascertain the full extent of both its present and potential consequences for sabotaging PCC. While the innovative approaches described above offer potential avenues to avoid or ameliorate the negative effects of McDonaldization, further studies are vital to figure out mine whether these approaches will be successful in providing PCC and dementia citizenship to the LTC community and still be reasonably profitable for owners.

References

Alexander, G. L., & Madsen, R. W. (2017). A report of information technology and health deficiencies in U.S. Nursing Homes. *Journal of Patient Safety*. Advance online publication. https://doi.org/10.1097/PTS.0000000000000390

Alford, D. M. (1993). Dementia units in long-term care. *Journal of Gerontological Nursing, 19*(3), 47–47.

Alzheimer's Disease International. (2019). *World Alzheimer Report 2019: Attitudes to dementia*. https://www.alz.co.uk/research/WorldAlzheimerReport2019.pdf

ArentFox.com. (2008). *Dangers of nursing home and health care facility advertising*. https://www.arentfox.com/newsroom/alerts/dangers-nursing-home-and-health-care-facility-advertising

Banerjee, A., & Armstrong, H. (2019). Models for long-term residential care: A summary of the consultants' report to long-term care homes and services, City of Toronto.

Behrens, L. L., McGhan, G., Abbott, K. M., Fick, D. M., Kolanowski, A. M., Liu, Y., Buck, H. G., Roes, M., Heid, A. R., Spector, A., & Van Haitsma, K. (2019). Mapping core concepts of person-centered care in long-term services and supports. *Journal of Gerontological Nursing, 45*(2), 7–13.

Belk, R. (2013). Cultures, consumers, and corporations. In L. Visconti, L. Peñaloza, & N. Toulouse (Eds.), *Marketing management* (pp. 53–67). Routledge.

Berger, R. (2017). Aging in America: Ageism and general attitudes toward growing old and the elderly. *Open Journal of Social Sciences, 5*, 183–198.

Berridge, C., Tyler, D. A., & Miller, S. C. (2018). Staff empowerment practices and CNA retention: Findings from a nationally representative Nursing Home Culture Change Survey. *Journal of Applied Gerontology, 37*(4), 419–434.

Braley, R., Fritz, R., Van Son, C. R., & Schmitter-Edgecombe, M. (2019). Prompting technology and persons with dementia: The significance of context and communication. *The Gerontologist, 59*(1), 101–111.

Brophy, J., Keith, M., & Hurley, M. (2019). Breaking point: Violence against long-term care staff. *New Solutions: A Journal of Environmental and Occupational Health Policy, 29*(1), 10–35.

Brune, K. (2011). Culture change in long term care services: Eden-greenhouse-Aging in the community. *Educational Gerontology, 37*(6), 506–525.

Bruni, M. L., Mammi, I., & Verzulli, R. (2019). Quality in nursing homes. In *Oxford research encyclopedia of economics and finance*.

Buchanan, R. (1992). Wicked problems in design thinking. *Design Issues, 8*(2), 5–21.

Caspar, S., Phinney, A., Spenceley, S., & Ratner, P. (2020). Creating cultures of care: Exploring the social organization of care delivery in long-term care homes. *Journal of Long-Term Care, 2020*, 13–29.

Castle, N. G., & Anderson, R. A. (2011). Caregiver staffing in nursing homes and their influence on quality of care: Using dynamic panel estimation methods. *Medical Care, 49*(6), 545–552. https://doi.org/10.1097/MLR.0b013e31820fbca9

Castle, N. G., Degenholtz, H., & Rosen, J. (2006). Determinants of staff job satisfaction of caregivers in two nursing homes in Pennsylvania. *BMC Health Services Research, 6*, 60.

Castle, N. G., Furnier, J., Ferguson-Rome, J. C., Olson, D., & Johs-Artisensi, J. (2015). Quality of care and LTC administrators' education: Does it make a difference? *Health Care Management Review, 40*(1), 35–45.

Castle, N. G., & Ferguson, J. C. (2010). What is nursing home quality and how is it measured? *The Gerontologist, 50*(4), 426–442.

Chou, S. C., Boldy, D. P., & Lee, Y. H. (2002). Measuring job satisfaction in residential aged care. *International Journal for Quality in Health Care, 14*(1), 49–54.

Cleeve, H., Borell, L., & Rosenberg, L. (2020). (In) visible materialities in the context of dementia care. *Sociology of Health & Illness, 42*(1), 126–142.

Corazzini, K., Twersky, J., White, H. K., Buhr, G. T., McConnell, E. S., Weiner, M., & Colón-Emeric, C. S. (2015). Implementing culture change in nursing homes: An adaptive leadership framework. *The Gerontologist, 55*(4), 616–627.

Coyne, R. (2005). Wicked problems revisited. *Design Studies, 26*(1), 5–17.

Dewing, J. (2011). Dementia care: Assess wander walking and apply strategies. *Nursing & Residential Care, 13*(10), 494–496.

Diamond, T. (2009). *Making gray gold: Narratives of nursing home care*. University of Chicago Press.

Dorsey, E. R., & Ritzer, G. (2016). The McDonaldization of medicine . *JAMA Neurology, 73*(1), 15–16.

Doty, M. M., Koren, M., & Sturla, E. L. (2008, May 1). *Culture change in nursing homes: How far have we come? Findings from the Commonwealth Fund 2007 National Survey of Nursing Homes*. https://www.commonwealthfund.org/publications/fund-reports/2008/may/culture-change-nursing-homes-how-far-have-we-come-findings

Duan, Y., Mueller, C. A., Yu, F., & Talley, K. M. (2020). The effects of nursing home culture change on resident quality of life in U.S. nursing homes: An integrative review. *Research in Gerontological Nursing*, 1–15. Advance online publication. https://doi.org/10.3928/19404921-20200115-02

Dunlap, W. R. (1979). How do parents of handicapped children view their needs. *Journal of the Division for Early Childhood, 1*(1), 1–10.

Duffy, M. (2011). *Making care count: A century of gender, race, and paid care work*. Rutgers University Press.

Eijkelenboom, A., Verbeek, H., Felix, E., & van Hoof, J. (2017). Architectural factors influencing the sense of home in NH: An operationalization for practice. *Frontiers of Architectural Research, 6*(2), 111–122.

Felix, H., Dayama, N., Morris, M. E., Pradhan, R., & Bradway, C. (2020). Organizational characteristics and the adoption of electronic health records among nursing homes in one southern state. *Journal of Applied Gerontology: The Official Journal of the Southern Gerontological Society*. Advance online publication. https://doi.org/10.1177/0733464820906685

Filmore, H., DuBard, C. A., Ritter, G. A., & Jackson, C. T. (2014). Health care savings with the patient-centered medical home: Community care of north Carolina's experience. *Population Health Management, 17*(3), 141–148.

Fossey, J., Masson, S., Stafford, J., Lawrence, V., Corbett, A., & Ballard, C. (2014). The disconnect between evidence and practice: A systematic review of person-centred interventions and training manuals for care home staff working with people with dementia. *International Journal of Geriatric Psychiatry, 29*(8), 797–807.

Gass, T. (2004). *Nobody's home: Candid reflections of a nursing home aide*. Cornell University Press.
Gaugler, J. E. (2016). Innovations in long-term care. In L. K. George, & K. F. Ferraro (Eds.), *Handbook of aging and the social sciences* (pp. 419–439). Academic Press.
Glasser, A. L. (2019). Rationalized aging. *Medicine Anthropology Theory, 6*(2), 79–92.
Goffman, E. (1957). Characteristics of total institutions. In *Symposium on Preventive and Social Psychiatry* (pp. 43–84). United States, Army Medical Services.
Goffman, E. (1959). *The presentation of self in everyday life*. Doubleday.
Goffman, E. (1961). *Asylums*. Doubleday.
Grabinski, C. J., ABD, F., & Niles-Yokum, K. (2014). *101+ careers in gerontology*. Springer Publishing Company.
Gubrium, J. F. (2012). *Living and dying at Murray Manor*. University of Virginia Press.
Gulwadi, G. B. (2013). Establishing continuity of self-memory boxes in dementia facilities for older adults: Their use and usefulness. *Journal of Housing for the Elderly, 27*(1–2), 105–119.
Harding, N., & Palfrey, C. (1997). *The social construction of dementia: Confused professionals?* Macmillan.
Harrington, C., & Carrillo, H. (1999). The regulation and enforcement of federal nursing home standards, 1991–1997. *Medical Care Research and Review, 56*(4), 471–494.
Harrington, C., Hauser, C., Olney, B., & Rosenau, P. V. (2011). Ownership, financing, and management strategies of the ten largest for-profit nursing home chains in the United States. *International Journal of Health Services, 41*(4), 725–746.
Harrington, C., Jacobsen, F. F., Panos, J., Pollock, A., Sutaria, S., & Szebehely, M. (2017). Marketization in long-term care: A cross-country comparison of large for-profit nursing home chains. *Health Services Insights, 10*, 1178632917710533.
Heinemann, L. (2015). Accommodating care: Transplant caregiving and the melding of health care with home life in the United States. *Medicine Anthropology Theory, 2*(1), 32–56.
Highfill, T., & Johnson, D. (2016). *Measuring nursing home price growth between 2000–2009*. U.S. Bureau of Economic Analysis. https://www.bea.gov/system/files/papers/WP2015-8.pdf
Ho, K. (2007). Structural violence as a human rights violation. *Essex Human Rights Review, 4*(2), 1–17.
Jacobs, J., Mulcahy, S., King, S., & Cenziper, D. (2020, April 29). Number of NH with publicly reported cases of the coronavirus soars. *Washington Post*. https://www.washingtonpost.com/investigations/number-of-nursing-homes-with-publicly-reported-cases-of-the-coronavirus-soars/2020/04/29/7375941c-898c-11ea-ac8a-fe9b8088e101_story.html
Jacobs, M. L., Snow, A. L., Allen, R. S., Hartmann, C. W., Dautovich, N., & Parmelee, P. A. (2019). Supporting autonomy in long-term care: Lessons from nursing assistants. *Geriatric Nursing, 40*(2), 129–137.
Johnson, C. J. (1989). Sociological intervention through developing low stimulus Alzheimer's wings in NH. *American Journal of Alzheimer's Care and Related Disorders & Research, 4*(2), 33–41.
Johnson, C. J., & Johnson, R. H. (2000). Alzheimer's disease as a "trip back in time." *American Journal of Alzheimer's Disease, 15*(2), 87–93.
Johnson, C. J. (2018). Dementia citizenship the need for dementia and gerontology education in hospice and palliative care. *Hospice & Palliative Medicine International Journal, 2*(2), 344–345.
Johnson, C., Kelch, J., & Johnson, R. (2017). Dementia at the end of life and family partners: A symbolic interactionist perspective on communication. *Behavioral Sciences, 7*(3), 42.
Johnson, R. H. (2014). *Experiences of dementia care workers in nursing homes: An exploratory study comparing Canada, Scotland, and the United States*. [Unpublished doctoral dissertation]. University of Stirling.
Kayser-Jones, J. S. (1990). *Old, alone, and neglected: Care of the aged in Scotland and the United States* (Vol. 4). University of California Press.
Kim, J. S. (2019). *Implementation of function focused care to change practice behaviors of staff*. [Unpublished doctoral dissertation]. University of Maryland Baltimore.
Kitwood, T. (1997). *Dementia reconsidered: The person comes first*. Open University Press.

Krippendorff, K. (2006). *The semantic turn: A new foundation for design.* CRC/Taylor & Francis.

Koren, M. J. (2010). Person-centered care for nursing home residents: The culture-change movement. *Health Affairs, 29*(2), 312–317.

LeGrand, M. (2017, October 30). *The use of robotics in long-term care facilities: A step into the future?* The upside to aging: Caregiving with dignity. https://theupsidetoaging.com/2017/07/11/the-use-of-robotics-in-long-term-care-facilities-a-step-into-the-future/

Lerner, N. B., Johantgen, M., Trinkoff, A. M., Storr, C. L., & Han, K. (2014). Are nursing home survey deficiencies higher in facilities with greater staff turnover. *Journal of the American Medical Directors Association, 15*(2), 102–107.

Liu, S. Q., Choi, S., & Mattila, A. S. (2019). Love is in the menu: Leveraging healthy restaurant brands with handwritten typeface. *Journal of Business Research, 98*, 289–298.

Lopez, S. H. (2006). Culture-change management in long-term care: A shop-floor view. *Politics and Society, 34*(1), 55–80.

Lopez, S. H. (2014). Culture change and shit work: Empowering and overpowering the frail elderly in long-term care. *American Behavioral Scientist, 58*(3), 435–452.

Lord, J., Davlyatov, G., Thomas, K. S., Hyer, K., & Weech-Maldonado, R. (2018). The role of assisted living capacity on nursing home financial performance. *INQUIRY: The Journal of Health Care Organization, Provision, and Financing, 55*, 0046958018793285.

Mendelson, M. A. (1974). *Tender loving greed: How the incredibly lucrative nursing home" industry" is exploiting America's old people and defrauding us all.* Knopf.

Mills, W. L., Pimentel, C. B., Palmer, J. A., Snow, A. L., Wewiorski, N. J., Allen, R. S., & Hartmann, C. W. (2018). Applying a theory-driven framework to guide quality improvement efforts in nursing homes: The LOCK model. *The Gerontologist, 58*(3), 598–605.

Mitchell, S. L., Shaffer, M. L., Cohen, S., Hanson, L. C., Habtemariam, D., & Volandes, A. E. (2018). An advance care planning video decision support tool for nursing home residents with advanced dementia: A cluster randomized clinical trial. *JAMA Internal Medicine, 178*(7), 961–969.

Moss, F. E., & Halamandaris, V. J. (1977). *Too old, too sick, too bad: Nursing homes in America.* Aspen Systems Corporation.

Morgan-Brown, M., Newton, R., & Ormerod, M. (2013). Engaging life in two Irish nursing home units for people with dementia: Quantitative comparisons before and after implementing household environments. *Aging and Mental Health, 17*(1), 57–65.

Mukadam, N., & Livingston, G. (2012, August). Reducing the stigma associated with dementia: approaches and goals. *Aging Health, 8*(4), 377–386.

Mukamel, D. B., Weimer, D. L., Harrington, C., Spector, W. D., Ladd, H., & Li, Y. (2012). The effect of state regulatory stringency on nursing home quality. *Health Services Research, 47*(5), 1791–1813.

National Employment Law Project. (2013, October). *Super-sizing public costs: How low wages at top fast-food chains leave taxpayers footing the bill.* https://www.nelp.org/wp-content/uploads/2015/03/NELP-Super-Sizing-Public-Costs-Fast-Food-Report.pdf

National Citizens' Coalition for Nursing Home Reform. (2001). *The nurse staffing crisis in nursing homes.* Retrieved October 29, 2017, from http://theconsumervoice.org/uploads/files/issues/Consensus_Statement_Staffing.pdf

New York State Department of Health. (2017). *Nursing home administrator licensure program: Qualifications.* Retrieved October 29, 2017, from https://www.health.ny.gov/professionals/nursing_home_administrator/licensure_program/qualifications.htm

Niemeijer, A. A., Depla, M., Hertogh, C., Frederiks, B., & Francke, A. L. (2014). The use of surveillance technology in residential facilities for people with dementia or intellectual disabilities: A study among nurses and support staff. *AJN American Journal of Nursing, 114*(12), 28–38.

Pang, Z., Yang, G., Khedri, R., & Zhang, Y. T. (2018). Introduction to the special section: Convergence of automation technology, biomedical engineering, and health informatics toward the healthcare 4.0. *IEEE Reviews in Biomedical Engineering, 11*, 249–259. https://doi.org/10.1109/RBME.2018.2848518

Parsons, S. K., Simmons, W. P., Penn, K., & Furlough, M. (2003). Determinants of satisfaction and turnover among nursing assistants: The results of a statewide survey. *Journal of Gerontological Nursing, 29*(3), 51–58.

Pekkarinen, L., Sinervo, T., Perälä, M. L., & Elovainio, M. (2004). Work stressors and the quality of life in LTC units. *The Gerontologist, 44*(5), 633–643.

Phelps, R. (n.d.) *Visual changes in dementia patients*. Retrieved November 11, 2017, from https://www.agingcare.com/articles/visual-changes-in-dementia-patients-205658.htm

Port, A., Barrett, V. W., Gurland, B. J., & Perez, M. (2011). Engaging nursing home residents in meaningful activities. *Annals of Long Term Care: Clinical Care and Aging, 19*(12), 20–26.

Power, A. (2017). *Dementia beyond drugs*. Health Professions Press.

Purtill, C. (2019). Stop me if you've heard this one: A robot and a team of irish scientists walk into a senior living home. *Time Magazine*. Retrieved October 10, 2019 from https://time.com/longform/senior-care-robot/

Pyrillis, R. (2014). Long-tem-care providers earn a little respect–and minimum wages. *Workforce, 93*(2), 16.

Reisberg, B. (1986). Dementia: A systematic approach to identifying reversible causes. *Geriatrics, 41*, 30–46.

Rittel, H., & Webber, M. (1973). Dilemmas in general theory of planning. *Policy Sciences, 4*, 155–169.

Ritzer, G. (2015). *The mcDonaldization of society* (8th ed.). Sage.

Ritzer, G., & Miles, S. (2019). The changing nature of consumption and the intensification of McDonaldization in the digital age. *Journal of Consumer Culture, 19*(1), 3–20.

Roberts, T. J., Nolet, K., & Bowers, B. (2019). Exploring variation in certified nursing assistant assignments from the perspective of nursing home residents: A comparison of adopters and nonadopters of consistent assignment. *Journal of Applied Gerontology, 38*(11), 1583–1594.

Rosenbaum, E. (2019, August 29). Panera is losing nearly 100% of its workers every year as fast-food turnover crisis worsens. *CNBC*. https://www.cnbc.com/2019/08/29/fast-food-restaurants-in-america-are-losing-100percent-of-workers-every-year.html

Rosenzweig, A. (2012). *The harsh realities of Alzheimer's care: An insider's view of how people with dementia are treated in institutions*. Praeger.

Sapkota, N., & Subedi, S. (2019). Dementia as a public health priority. *Journal of Psychiatrists' Association of Nepal, 8*(2), 1–3.

Shadish, W. R., & Bootzin, R. (1984). Nursing homes: The new total institution in nursing home policy. *International Journal of Partial Hospitalization, 2*(4), 251–262.

Scholl, J. M., & Sabat, S. R. (2008, January). Stereotypes, stereotype threat and ageing: implications for the understanding and treatment of people with Alzheimer's disease. *Ageing and Society, 28*(1), 103–130.

Schwarz, B. (1996). *Nursing home design: Consequences for employing the medical model*. Garland Science.

Shield, R. R. (2014). Jurkowski, Elaine T. Implementing culture change in long-term care: Benchmarks and strategies for management and practice. *Anthropology & Aging, 35*(1), 73–76.

Shield, R. R. (2018). *Uneasy endings: Daily life in an American nursing home*. Cornell University Press.

Spector, P. E. (1997). *Job satisfaction: Application, assessment, causes, and consequences* (Vol. 3). Sage.

Spilker, H. S., & Norby, M. K. (2019). Understanding the role of technology in care: The implementation of GPS-technology in dementia treatment. *Ageing International, 44*(3), 283. https://doi.org/10.1007/s12126-018-9340-z

Spilsbury, K., Hewitt, C., Stirk, L., & Bowman, C. (2011). The relationship between nurse staffing and quality of care in NH: A systematic review. *International Journal of Nursing Studies, 48*(6), 732–750.

Stannard, C. I. (1973). Old folks and dirty work: The social conditions for patient abuse in a nursing home. *Social Problems, 20*(3), 329–342.

Stevenson, D. G., Bramson, J. S., & Grabowski, D. C. (2013). Nursing home ownership trends and their impacts on quality of care: A study using detailed ownership data from Texas. *Journal of Aging & Social Policy, 25*(1), 30–47.

Stone, D. (2005). For love nor money: The commodification of care. In M. Ertman, & J. Williams (Eds.), *Rethinking commodification: Cases and readings in law and culture* (Vol. 52 , pp. 271–90). NYU Press.

Stuckey, J. C. (1997). The best friends approach to Alzheimer's care. *Journal of Applied Gerontology, 16*(3), 367–369.

Taylor, S., & Lyon, P. (1995). Paradigm lost: The rise and fall of McDonaldization. *International Journal of Contemporary Hospitality Management, 7*(2/3), 64–68.

Texas Health and Human Services. 2011. *NFA, Subchapter B, Requirements for Licensure.*

Thomas, W. H. (1996). *Life worth living: How someone you love can still enjoy life in a nursing home.* VanderWyk & Burnham.

Thomas, W. H. (2006). *In the arms of elders: A parable of wise leadership and community building.* VanderWyk & Burnham.

Treadaway, C., Fennell, J., Prytherch, D., Kenning, G., Prior, A., Walters, A., & Taylor, A. (2018). Compassionate design: How to design for advanced dementia–a toolkit for designers.

Vladeck, B. C. (1980). *Unloving care: The nursing home tragedy.* Basic Books.

Vogelsmeier, A. A., Scott-Cawiezell, J. R., & Pepper, G. A. (2011). Medication reconciliation in nursing homes: Thematic differences between RN and LPN staff. *Journal of Gerontological Nursing, 37*(12), 56–63.

Walton, A. L., & Rogers, B. (2017). Workplace hazards faced by nursing assistants in the United States: A focused literature review. *International Journal of Environmental Research and Public Health, 14*(5), 544.

Weber, M. (1947). *The theory of social and economic organization* (Henderson, A.M. & Parsons, T., Trans.). The Free Press.

Williams-Burgess, C., Ugarriza, D., & Gabbai, M. (1996). Agitation in older persons with dementia: A research synthesis. *Worldviews on Evidence-Based Nursing Presents the Archives of Online Journal of Knowledge Synthesis for Nursing, 3*, 97–107.

Yassine, A., Singh, S., & Alamri, A. (2017). Mining human activity patterns from smart home big data for health care applications. *IEEE Access, 5*, 13131–13141. https://doi.org/10.1109/ACCESS.2017.2719921

Yates, M. D. (2006). Capitalism is rotten to the core. *Monthly Review, 58*, 1–51.

Ye, Z., Mukamel, D. B., Huang, S. S., Li, Y., & Temkin-Greener, H. (2015). Healthcare-associated pathogens and nursing home policies and practices: results from a national survey. *Infection Control & Hospital Epidemiology, 36*(7), 759–766.

Zarem, J. (2010). *Today's continuing care retirement community (CCRC)* [White paper]. American Seniors Housing Association. https://www.seniorshousing.org/filephotos/research/CCRC_whitepaper.pdf

Zarit, S. H., Dolan, M. M., & Leitsch, S. A. (1998). Interventions in nursing homes and other alternative living settings. In Nordhus, I. H. (Ed.), *Clinical geropsychology* (pp. 329–343). American Psychological Association.

Zegre, S. J., Needham, M. D., Kruger, L. E., & Rosenberger, R. S. (2012). McDonaldization and commercial outdoor recreation and tourism in Alaska. *Managing Leisure, 17*(4), 333–348.

Zeilig, H. (2014). Dementia as a cultural metaphor. *The Gerontologist, 54*(2), 258–267.

Chapter 5
Delivering Multidisciplinary and Cross-Cultural Education in Dementia Studies

Louise McCabe and Heather Comstock

Abstract This book chapter considers the role that education plays in the development of a competent workforce that can drive forward a research-led, innovative approach to treatment, care and support of families with dementia living around the world. Discussion focuses on postgraduate level education that aims to develop leaders in dementia care and draws from over twenty years of experience in delivering education to professionals from a wide range of disciplines and from around the world. As the prevalence of dementia rises and counties around the world face the challenge of supporting families to live well with dementia, the need for specialist knowledge grows. In order to promote the best quality of life for families facing dementia, it is important that professionals and practitioners are educated about dementia and, further, that they understand the wider disciplinary and cultural setting for families. Dementia is a multifaceted condition, and families require support from a range of professionals and practitioners from across disciplinary fields including (but not limited to) medicine, nursing, social work, psychology and care. Education that draws together these different disciplines to learn together provides a setting for knowledge exchange and learning across disciplinary boundaries and between countries and cultures. This cross-disciplinary, cross-national approach to education and learning offers a richer experience and supports the development of a multifaceted understanding of dementia and dementia care. Further, as knowledge about dementia and dementia care grows there is a need to share learning globally to maximize impact for research and educational programmes, such as the one that is the focus of this chapter, provide an effective route for knowledge exchange. A focus on reflective learning helps to support the transfer of new knowledge into practice. This chapter

---

L. McCabe (✉)
Dementia Studies, University of Stirling, Stirling, UK
e-mail: louise.mccabe@stir.ac.uk

H. Comstock
Dementia Care Educator, Dementia Friendly Wyoming, Sheridan, USA
e-mail: heather@heathercomstock.com

M. K. Shankardass (ed.), *Dementia Care*,
https://doi.org/10.1007/978-981-16-3864-0_5

reflects on experiences of delivering dementia education, provides feedback from students learning within a multidisciplinary and cross-cultural student group and concludes on the crucial role to be played by education and knowledge exchange in supporting the increasing numbers of people with dementia and their families around the world.

Introduction

This book chapter considers the role that education plays in the development of a competent workforce that can drive forward a research-led, innovative approach to treatment, care and support of people and families living with dementia around the world. Discussion here focuses on postgraduate level education that aims to develop leaders in dementia studies and draws from over twenty years of experience in delivering education to professionals from a wide range of disciplines and from around the world. A case study is presented of a dementia-specific programme where professionals working across health and social care, and in other related professions such as architecture, education and business, as well as family carers, learn together.

As the prevalence of dementia rises and countries around the world face the challenge of supporting people to live well with dementia, the need for specialist knowledge grows. In order to promote the best quality of life for families facing dementia, it is important that professionals and practitioners are educated about dementia and, further, that they understand the wider disciplinary and cultural settings the individuals and families live within. Dementia is a multifaceted condition, and families require support from a range of professionals and practitioners from across disciplinary fields including (but not limited to) medicine, nursing, social work, psychology and care. Education that draws people together provides a setting for knowledge exchange and learning across disciplinary boundaries and between countries and cultures. Further, as knowledge about dementia and dementia care grows there is a pressing need to share learning globally to maximize impact from research and practice; and educational programmes, such as that discussed here, provide fertile ground for knowledge exchange.

This chapter was written during the lockdown put in place in response to the COVID-19 pandemic during 2020. Due to this, the chapter reflects on some of the specific experiences of students during that time and the positive impact of belonging to an online community of practitioners and professionals from around the world.

The chapter is co-authored by an academic educator, working in a city in the UK, and a postgraduate student, working in a rural part of the USA, who is herself an educator and community worker. As such, the chapter offers examples and reflections on dementia education from these two complementary perspectives. In each section, the boxed text provides reflections from the specific perspective of the student.

The Need for Dementia Education

People living with dementia receive treatment, care and support from paid and unpaid professionals and carers across different disciplines and it is crucial that they are provided with specialist training and education to guide and support their work. The need for a competent dementia workforce is growing rapidly yet at the same time demographic and societal changes are impacting negatively on its quality and availability.

The ageing of the world's population is well-documented and is leading to continued and significant increases in the number of people living with dementia particularly in low- and middle-income countries (LMIC). This increase in numbers alongside changes in the balance of paid and unpaid care is increasing demand for a trained dementia care workforce. The balance between unpaid and paid care for people living with dementia differs across regions of the world with more paid care provided in Europe, North America and Australia while in the Global South the family is still the main provider of care. Formal homecare and care home sectors are less developed in Africa, Asia and Latin America than in high-income countries with only around 6% of people living with dementia in LMICs cared for in care homes (Alzheimer's Disease International, 2013). However, these different patterns of care are changing and traditional systems of unpaid care by family, friends and community are increasingly coming under threat because of processes of globalization leading to changes in family structures, improved education for women and urbanization. These different factors mean that education of workers within the paid care sector is crucial.

Dependency ratios are used as a way of illustrating the ratio of working-age people available to the workforce compared to the number of older people who may need access to paid care. These ratios paint a stark picture in many countries where population ageing has led to a significant imbalance. While dependency ratios are still higher in regions like Europe these ratios are also gradually changing over time around the world with the global ratio rising from 8.59 in 1960 to 13.95 in 2019 (World Bank, 2020). Dependency ratios suggest a shrinking population of people of working age, impacting the availability of people to take on paid dementia care roles; however, caution needs to be applied in interpreting these figures. More complex models have been developed (e.g. Wachs et al., 2020) that suggest that a more nuanced approach is needed to understand the links between ageing and productivity. Many older people contribute significantly through their role as unpaid carers, often of other older people, and many older people continue to be a part of the workforce.

The provision of good quality dementia care is also impacted by stigma and ageism (Carney & Gray, 2015). Ageism is present globally and is often casually accepted in everyday life, just take a look at a selection of birthday cards for the over 50s in your local shop. Ageism was brought into stark focus in the recent COVID-19 crisis where we saw many more older people dying of COVID-19 and a significant lack of investment in care homes saw deaths there contributing to around half of all deaths in several European countries (Bell et al., 2020; Comas-Herrera et al., 2020). Further, there were disturbing trends in social media with hashtags describing COVID-19 such as *#boomerremover* that reflected everyday ageism (Lichtenstein, 2020).

Ageism and the findings from the care home studies above reflect the lack of value placed on working with older people across the spectrum from geriatrics to frontline health and social care (Clark et al., 2009). The dementia workforce in Australia, UK and USA is made up mainly of women who are often low paid, lacking qualifications and working part-time (Elliott et al., 2012). There are systemic issues in countries such as the UK where part-time, agency workers are over-represented among those who care for people living with dementia. In many LMICs, the field of geriatrics is still relatively new and under-resourced raising important questions about how this growing new workforce will be educated and a lack of dementia specialists is noted across different regions. The USA (Cottingham et al., 2014), Australia, (Chenoweth et al., 2014), UK (Poulter, 2014), the Netherlands (Vernooij-Dassen et al., 2009), and many LMICs (Bruckner et al., 2011) report significant shortages in the workforce for mental health, dementia and care for older people with the situation predicted to worsen in coming years.

This leaves a significant need for specialist education in dementia care that can equip the existing and growing workforces that will provide care for families and individuals around the world. This chapter argues that offering online, cross-national, multidisciplinary education enables and supports knowledge exchange to make the best use of current knowledge and expertise to challenge and address these issues. Further, the provision of a postgraduate qualification can add status and value to practitioners and elevate the work that they do, challenging the current lack of status and value afforded to this work.

A Case Study

The question addressed in this chapter is about the role that postgraduate education can play in ensuring the development of a competent workforce that delivers high-quality care and support to families and individuals when they need it.

The postgraduate programme that provides the case study underpinning the reflections offered in this chapter is an online, part-time programme for professionals working in dementia care. Professionals join from countries in all parts of the world and from a wide range of disciplinary backgrounds. Students use their practice experience as a basis for reflection during their studies and new knowledge from the programme to enhance their practice. The programme emphasizes the expertise and knowledge of the student body, encouraging knowledge exchange and peer-to-peer learning that takes place between countries and cultures and across disciplines. This chapter argues that drawing upon the students' experiences and emphasizing the need for cross-cultural, interdisciplinary learning supports the development of leaders in dementia care who can cascade learning; informing, up-skilling and adding value to the dementia workforce.

Knowledge Exchange and Peer to Peer Learning

The inclusion of a diverse student population who are all working in a related field means that students hold their own expertise and knowledge of dementia treatment and care that is based within their cultural and professional context. Students, therefore, bring a rich resource of skills and knowledge that can be shared. The learning approach taken on the programme provides guidance and encouragement for students to share knowledge and support others. The programme provides an online community of practice for students where their learning is facilitated by the lecturers but where knowledge is drawn from a range of resources, key among these being the other students on the programme. To facilitate this, an adult learning approach is taken that supports and encourages experiential and peer-to-peer learning.

Experiential learning recognizes the expertise and knowledge that each student brings and provides a framework within which they can reflect on and share their experience and then develop new knowledge and skills that build from the experience of the student community of practice (Merriam & Bierema, 2014). Each of the assignments within the programme asks students to critically reflect on their practice experience in the light of their new learning, thus prioritizing experiential learning as a mechanism to embed new learning.

Peer-to-peer learning has been an important part of practice education for a number of years, particularly in nursing education (Stone et al., 2013) and has demonstrated positive outcomes for students including improved communication, critical thinking, reflective practice and self-confidence (Gray et al., 2019). Peer-to-peer learning can take place in different forms and within this programme it has involved structured peer-review activities and less formal interaction on online discussion forums. Structured peer-review approaches involve students providing feedback to one other student on a particular piece of work. During the online discussion forums that form the main mechanism for online student interaction students are encouraged to comment and respond to postings made by other students.

The text presented in boxes within each section provides specific examples and reflections from the chapter co-author, a current student on the programme.

In a conversation with my tutor, I mentioned that I desired to speak with fellow students also actively engaged in implementing dementia-specific community services to have a better understanding of challenges I was encountering. It was then that I was referred to a fellow student in Scotland who was able to share how she was translating the education from the programme to design and implement dementia-specific training in her community. This student pairing proved to be one of the most influential peer-to-peer experiences early in my postgraduate education for two reasons. First, this immediate student referral gave me confidence that, if I had a need and requested assistance from the teaching staff, my needs would be addressed. Second, I now had a fellow student—who offered to mentor me as I continued my education. Although she has graduated, we still discuss theoretical ideas and exchange suggestions on community work through emails and Zoom, I value her professional perspective and continued friendship.

The online discussion boards ask open questions designed to lift key issues out of the modules and enhance our understanding of the course material. On more than one occasion, peer communication added far more depth in complexity while lending immediate relevance to issues with examples that allowed learning to take lively shape in a real-world context. One discussion board had an unexpected personal and professional bearing. A fellow student reflected, as a professional trainer how important it is to have "authentic relationships" with persons living with dementia. The discussion further explored the vulnerable position and inner fears of someone who earlier in life identified as transgender and is now also living with dementia. One of the complexities raised from the perspective of persons living with dementia centred on the anxiety of being mistreated or discriminated against by care staff. The layered complexities of this specific conversation made perfect sense to me, as my own stepparent came out late in life as transgender in a small rural town. The discussion posts allowed me the space to recognise how valuable the professional carer support was in the retention of this person's gender identity as she experienced cognitive decline. I saw with new eyes the "authentic friendship" that allowed the highest levels of empathy to inform these professional actions.

The impact of this particular discussion thread gave me spacious ground to reflect, producing fresh clarity on my own lived experience and how I could apply it for the benefit of training and coaching opportunities. In this example, I am compelled to encourage more curiosity about those we serve, instead of judgement. When we honour *all* life choices, not just the choices we agree with, we also improve our ability to be "present" as professional and personal carers to what persons living with dementia really need.

Cross-Cultural Learning

The experience of learning about dementia care within a group of students from different countries offers an opportunity to learn both from one another but also together, developing new ideas and knowledge that draw upon perspectives from different places. The cultural contexts of students on the programme vary dramatically, from students working in remote rural island communities in the UK to students working in large urban areas such as Hyderabad in India. These contexts offer a huge variety in examples to learn from, and this can be both a benefit and a challenge. Learning how high-quality dementia care is delivered in a setting very different to our own can push us out of our comfort zone and challenge perceived ideas of how things can be done. However, there are difficulties in transferring ideas between countries and regions if the difference between the contexts is too significant. Challenges to cross-national learning can arise due to cultural or societal differences between countries but learning together within cross-cultural groups of students provides a rich experience that offers opportunities for new ideas to emerge. The experiences of our students mirror those described by Commander et al. (2016) who provide an analysis of the cross-national learning taking place between American and Chinese students working together online. Students studying cross-nationally often find that they face similar challenges as students in different countries but that these challenges are addressed in different ways. This study confirms experiences in the dementia studies programme of students sharing similar challenges, such as how to support people living with dementia to continue to live in a place of their choosing. The differences in students' specific experiences and cultural perspectives enable them to learn and develop new perspectives within that cross-cultural learning setting.

A significant challenge as a student and as a professional in implementing excellent dementia-specific community work has been eliciting and including the voices and perspectives of persons living with dementia. Unique, cross-national insight into practical ways to address this inclusion gap is presented by the developing work, published insights and lessons learned by DEEP (https://www.dementiavoices.org.uk/) and the Scottish Dementia Working Group (https://www.alzscot.org/our-work/campaigning-for-change/have-your-say/scottish-dementia-working-group). Historically in my culture in North America, it is not uncommon for public opinion or carers to diminish or completely dismiss the ideas and notions of a person following their dementia diagnosis. Grassroots efforts, such as those of DEEP and the Scottish Dementia Working Group offer a realistic road map for communities like ours, who desire first to initiate and second, to implement cocreative conversations with persons living with dementia. These efforts are now allowing a higher quality of programming to develop and continue through deep listening and revisiting the perspectives of those receiving these dementia-specific services. "Nothing about us without us", is now a familiar refrain in our meetings and we endeavour to sustain this core value as the most excellent programming standard.

During the peak of the COVID-19 pandemic in the second and third quarters of 2020 many of the programme's students were engaged with providing care and support to individuals and families. Some students faced extremely stressful situations, such as caring for older people at the end of their lives in care homes during exceptional circumstances where usual practices and rituals were constrained. During this time, the programme offered a safe space for students to share experiences and offer peer support. Further, students were able to share learning from their experiences and offer resources and information developed by their own organizations or national bodies. Often students from different countries were experiencing changes at different times and could benefit from experiences shared by students whose countries were at different points on the COVID-19 curve, as reflected below.

One of the most unexpected impacts of cross-national learning emerged during the COVID-19 pandemic. Early in March of 2020, as the pandemic became known the world over, it was apparent to me that my fellow students in the UK were experiencing COVID-19 in a much more direct and impactful way than my rural community. The peer support online discussion boards developed just days after the publicly acknowledged viral outbreak detailed in real-time how other students were presented and dealt with everyday COVID-19 challenges. Posts exposed truths such as issues with securing enough and appropriate PPE, rapidly increasing rates of COVID-19 infection in staff and care home residents and the heart-wrenching realities of escalating social isolation and death occurring in mounting numbers as the days and weeks unfolded. Posts also mentioned the positive power that communities showed through notes and small gifts, how even the smallest acts of kindness translated to a perception of hope and emotionally supported people to do their best, despite all the uncertainties. The largest COVID-19 wave did not reach my community until July of 2020 and even then, we did not experience anything like that which my professional peers did in those early months regarding numbers of infection or deaths due to virus exposure. While I am grateful that our rural community has been buffered from the extreme initial and immediate effects that so many of my fellow students faced, these cross-national conversations gave and continue to give me insight into the drastically different lived experiences of my fellow students and those they endeavour to support during this continuing pandemic. I took my peers' reflections on those small acts of kindness to heart. In discussing the power of simple phone check-ins and handwritten notes, we ordered 2500 note cards, precisely so that our staff would have plenty of branded stationery to remind our community that their local senior centre is here and remains available to support them, despite all our uncertainties as well.

Inter-Disciplinary Learning

Dementia is a complex condition which has biomedical, psychological, sociological and environmental aspects; the experiences of an individual with dementia are affected by the neurological changes in their brain, their own life history and personality, their social networks and connections, their spirituality, and the physical environments that they live within. Thus any, educational programme aiming to provide students with a full understanding of dementia and the meanings assigned to it must engage across disciplines. Current and past students on the programme represent a range of professions including: social worker, occupational therapist, speech and language therapist, nurse (both general and mental health and working in the community, care homes and in hospitals), doctor, psychologist, architect, lawyer, educationalist, trainer, manager, physiotherapist. Each of these professions brings their own theories, practice and ways of understanding dementia and offer new learning and understanding for everyone. Thompson Klein (2012) uses the concept of critical interdisciplinarity to explain how we further knowledge through bringing together different disciplines to debate and discuss their different perspectives and thus enable new, more complex understandings to emerge. Within the programme, debate and discussion are encouraged to support students to deconstruct what they think they know about dementia and for them to build new, more nuanced and complex understandings of the condition and the experiences of individuals with dementia.

Hearing from fellow students across a wide range of professions allowed the introduction of what initially seemed like "radical new ideas" to take root and eventually shape as self-evident truths. Prior to this course, I did not have many creative ideas to suggest, even when asked for alternative paths to dementia care beyond the overwhelmingly negative biomedical narratives. I often felt frustrated by fear-based commercials and documentaries that were repeatedly referenced, further calcifying stigma and depriving our community of more recent psycho-social perspectives and approaches. As the course work continued and peers' ideas were shared, I began to see creative ways our dementia culture could evolve by continually eliciting the reflections of medical practitioners, persons living with dementia and carers in innovative ways in our community. These collaborative approaches meant higher levels of community inclusion, which offered more opportunities to challenge previously ensconced negative stereotypes about our community-wide ability to offer more proactive and supportive dementia care.

As an example, in the student discussion boards, professionals would mention the need for referral of occupational therapist, physical therapist or speech and language therapist services for persons living with dementia. Initially, as I am not a professional "therapist" of any kind, I did not fully understand the magnitude nor the scope of the need for initial and reoccurring referral of these services, especially early in the process of a lived dementia experience. During research for an essay for class, I happened to be interviewing a person living with dementia. Let's call her Sue. She had become nearly incapacitated by a sudden reduction in cognitive processing and a shockingly shrinking speech ability. It turned out that she was one of the lucky individuals who was given the absolutely needed immediate referral to a skilled cognitive-speech therapist. Sue explained how grateful she was to have access to this critically needed rehabilitative therapy, which allowed her to regain a significant portion of her cognitive processing and speech ability. Sue was quite convinced that without that skilled cognitive-speech therapist's efforts, she would have lost her ability to live independently and would have been immediately transferred to a full-skill nursing home. At first, the very idea of reoccurring referral rehab services was not top-of-mind in our community for persons living with dementia. The need and practicality of expanding early intervention strategies make financial sense to delay the exorbitant costs of institutional care. Additionally, current COVID-19 conditions continue to limit many non-essential medical visits, which now allows these preventative measures towards sustaining abilities to be understood as a straightforward opportunity to save money while offering a chance to increase the quality of life for persons living with dementia.

Cascading Learning Through Training and Evaluation

One key area of skills and knowledge the programme aims to confer on students is the ability to deliver education and training to others to share and cascade their new learning. Understanding of adult learning theory as well as knowledge on what works for whom in terms of training and education equips students with the ability to design appropriate training for different groups. This supports wider impact from the programme ensuring that learning from the programme is cascaded well beyond the student group, helping to address wider workforce issues.

While research findings on the impact of training in dementia care are mixed (Surr et al., 2017), there are many examples of innovative training approaches that increase knowledge and skills and lead to better outcomes for people living with dementia and family carers that emphasize the need for educators to be skilled and imaginative in how training is delivered. Research shows positive outcomes for carers and for people living with dementia from different training approaches such as cognitive behaviour therapy programmes to reduce depression and anxiety for carers (Vernooij-Dassen et al., 2011) and training carers to promote better sleep for the person with dementia they care for (Tewary et al., 2018). Thus, an important aspect of the programme is providing students with an understanding of adult learning theory and how they might cascade new knowledge within their workplaces and to the individuals and families they work with and support.

One of my primary professional responsibilities is to work with, educate and prepare personal and professional carers with dementia-specific skills. The second year of the programme added multiple layers of information, which increased my perspective and understanding of ongoing challenges and successes in my design and implementation of dementia-specific skills training with personal and professional carers. A significant challenge is educating and tooling nursing and frontline healthcare staff with dementia-specific awareness and applied skills. The broad research and practical advice on ways to increase training and dissemination have been instrumental in boosting my confidence and professional educational ability. Knowledge about experiential learning has allowed me to see the areas I needed to immediately change to increase employer and staff satisfaction within my training or educational experiences. For example, in nursing staff training opportunities I needed to research what was working, what was not, and what they had already tried to remedy any challenging situations. The course work emphasized meet your students, where they are! This should have been obvious before and, after reflection, indeed became hard to miss. Developing a training programme as part of my course work helped me identify I needed less instructional theory and more hands-on application during certain types of training. It became clear that a well-designed work scenario with professionals wearing a virtual dementia suit can be quite helpful in allowing staff to experience another's reality. I have also observed the success of facilitating groups through reflective practices after a hands-on scenario, how powerful simple questions designed to guide staff through what they just learned, can be. It is gratifying to witness previously held assumptions and hardened-perceptions of "challenging-behaviours" become abruptly understood through a group discussion. Hearing professionals commit to more productive, positive and proactive intentions refuelled by their sudden understanding and renewed empathy for the lived experience of those living with dementia makes all my training preparations worth every minute!

Discussion

The start of this chapter provided a brief overview emphasizing the current and evolving need for a well-trained workforce to care for people living with dementia within a wider global context where the numbers of people living with dementia are increasing while demographic changes impact on the availability of this workforce. Specialist postgraduate education offers the opportunity for people to become experts and leaders in dementia care and support and provides them with the skills and knowledge to cascade their learning in their local and national contexts. This process promotes cross-national knowledge exchange and encourages learning from a place of experience that has application in the real world. The online, flexible nature of the education offered enables a very diverse student group, crossing cultural and disciplinary boundaries, to learn together.

The focus on knowledge exchange and peer-to-peer learning within the programme offer opportunities for people to learn from other countries and other disciplines. There is a huge amount of excellent work going on in different parts of the world and by different disciplines that can be usefully adapted and adopted by others. As noted, students often face similar challenges but in different contexts. The ability to learn from others who have already found solutions for shared challenges allows change to happen faster and reduces the need for resources to develop unique solutions in every place.

This chapter has evidenced the crucial role that can be played by particular forms of education and knowledge exchange in supporting good quality care and support for the increasing numbers of people living with dementia and their families around the world. The key mechanisms used in this programme: flexible online learning, cross-national knowledge exchange, peer-to-peer learning and providing skills to cascade knowledge could be adapted to support intercultural and multiprofessional groups both in formal education but also as wider mechanism for knowledge exchange and learning that improves the lives of people with dementia living around the world.

References

Alzheimer's Disease International. (2013). *World Alzheimer Report 2013: An analysis of long-term care for dementia.* Alzheimer's Disease International.

Bell, D., Comas-Herrera, A., Henderson, D. et al. (2020). *COVID-19 mortality and long-term care: a UK comparison*, International Long Term Care Policy Network. Available at: https://ltccovid.org/2020/08/28/covid-19-mortality-and-long-term-care-a-uk-comparison/. Accessed September 2020.

Bruckner, T., Scheffler, R., Shen, G., Yoon, J., Chisholm, D., Morris, J., Fulton, B., Dal Poz, M., & Saxena, S. (2011). The mental health workforce gap in low- and middle-income countries: A needs-based approach. *Bulletin of the World Health Organisation, 89*, 183–194.

Carney, G. M., & Gray, M. (2015). Unmasking the elderly mystique': Why it is time to make the personal political in ageing research. *Journal of Aging Studies, 35*, 123–134.

Chenoweth, L., Merlyn, T., Jeon, Y.-H., Tait, F., & Duffield, C. (2014). Attracting and retaining qualified nurses in aged and dementia care: Outcomes from an Australian study. *Journal of Nursing Management, 22*, 234–247.

Clark, A., Hayes, R., Jones, K., & Crosby, G. (2009) *Ageism and age discrimination in social care in the United Kingdom: A review from the literature.* Centre for Policy on Ageing for the Department of Health. Available: http://www.cpa.org.uk/information/reviews/CPA-%20ageism_and_age_discrimination_in_social_care-report.pdf. Accessed September 2020.

Comas-Herrera, A., Zalakaín, J., Litwin, C. et al. (2020). *Mortality associated with COVID-19 outbreaks in care homes: Early international evidence,* International Long Term Care Policy Network. Available at: https://ltccovid.org/2020/04/12/mortality-associated-with-covid-19-outbreaks-in-care-homes-early-international-evidence/?subscribe=success#blog_subscription-3. Accessed September 2020.

Commander, N., Zhao, Y., Gallagher, P., & You, Y. (2016). Cross-national online discussions: International learning experiences with American and Chinese students in higher education. *Innovations in Education and Teaching International, 53*(4), 365–374.

Cottingham, A., Alder, C., Austrom, M., Johnson, C., Boustani, M., & Litzelman, D. (2014). New workforce development in dementia care: Screening for "caring": Preliminary data. *Journal of the American Geriatrics Society, 62*(7), 1364–1368.

Elliott, K., Scott, J., Stirling, C., Martin, A., & Robinson, A. (2012). Building capacity and resilience in the dementia care workforce: A systematic review of interventions targeting worker and organizational outcomes. *International Psychogeriatrics, 24*(6), 882–894.

Gray, S., Wheat, M., Christensen, M., & Craft, J. (2019). Snaps+: Peer-to-peer and academic support in developing clinical skills excellence in under-graduate nursing students: An exploratory study. *Nurse Education Today, 73*, 7–12.

Lichtenstein, B. (2020). From "Coffin Dodger" to "Boomer Remover": Outbreaks of ageism in three countries with divergent approaches to coronavirus control. *Journals of Gerontology b, Online First View.* https://doi.org/10.1093/geronb/gbaa102

Merriam, S., & Bierema, L. (2014). *Adult learning: Linking theory and practice.* Jossy-Bass.

Poulter, D. (2014). Investment is crucial to strengthen the nursing workforce. *Nursing Standard, 28*(45), 8.

Stone, R., Cooper, S., & Cant, R. (2013). The value of peer learning in undergraduate nursing education: A systematic review. *International Scholarly Research Notices.* https://doi.org/10.1155/2013/930901

Surr, S., Gates, C., Irving, D., et al. (2017). Effective dementia education and training for the health and social care workforce: A systematic review of the literature. *Review of Educational Research, 87*(5), 966–1002.

Tewary, S., Cook, N., Pandya, N., & McCurry, S. (2018). Pilot test of a six-week group delivery caregiver training program to reduce sleep disturbances among older adults with dementia (Innovative practice). *Dementia, 17*(2), 234–243.

Thompson Klein, J. (2012). *The Oxford handbook of interdisciplinarity.* Oxford University Press.

Vernooij-Dassen, M., Draskovic, I., McCleery, J., & Downs, M. (2011). Cognitive reframing for carers of people with dementia. *Cochrane Database of Systematic Reviews, 11* (CD005318).

Vernooij-Dasssen, M., Faber, M., Olde Rikkert, M., Koopmans, R., Achterberg, T., Braat, D., Raas, G., & Wollersheim, H. (2009) Dementia care and labour market: The role of job satisfaction. *Aging & Mental Health, 13*(3), 383–390.

Wachs, D., Roman-Urrestarazu, A., Brayne, C. et al. (2020). Dependency ratios in healthy ageing. *BMJ Global Health, 5,* e002117.

World Bank. (2020). *Age dependency ratio, old (% of working-age population).* https://data.worldbank.org/indicator/SP.POP.DPND.OL. Accessed September 2020.

Louise McCabe I have over twenty years of experience in research, education and practice on ageing and dementia, starting my journey as a frontline care worker in residential care for people with dementia in 1996. I have been at the University of Stirling since 2000 where I have built a significant portfolio of ageing and dementia research with a focus on the potential of services to improve the lives of families with dementia. My research includes user-focused evaluations and cross-national studies of ageing and dementia care services in the UK, Europe and India. My research mainly utilises participatory, qualitative methods to engage with the experiences and views of older people, their family carers and practitioners. Over this time, I have also been involved in the development and delivery of a range of programmes and short courses in dementia studies including an internationally renowned MSc programme that is delivered online to a multidisciplinary student body around the world.

Chapter 6
Language Impairments in Dementia: *From Word-Finding Difficulties to Everyday Conversation in a Dementia-Friendly Community*

Daria Smirnova, Tatiana Smirnova, and Paul Cumming

Abstract Dementia syndromes can include language impairments (LIs) of severity extending from lexical access difficulties within anomic aphasia to non-fluent effortful speech and semantic aphasia, depending on the stage and etiology of the underlying disease. Relevant etiologies include neurodegenerative Alzheimer's disease (AD) and non-AD dementias, such as frontotemporal lobar degeneration (FTD), Parkinson's (PD) and Lewy body diseases, vascular and toxic alcohol-related dementia, depressive pseudodementia and mixed type dementia. Irrespective of the underlying disease, LIs interfere with social contacts and personal relationships, thus substantially reducing the quality of life and daily functioning of patients, while increasing their need for supervision and care. Socio-linguistic discourse describes such patients as experiencing "loss of self", "no meaningful present", "active presence of the past in the body itself", and as the "long goodbye" (Snyder in Dementia: Mind, meaning and the person, Oxford University Press, p. 268, 2006), highlighting the stigmatization and low quality of life of dementia sufferers. In this chapter we summarize the similarities and differences in clinical and linguistic presentations of LIs in AD and the most commonly occurring types of non-AD dementias, emphasizing the decisive diagnostic and prognostic roles of LIs, as well as their implications for choice of treatment. We present an account of the neuropsychological and psycholinguistic approaches to assess LIs occurring in dementia through evaluation of language functions/domains, such as sound-based domain and lexis

D. Smirnova (✉)
Department of Psychiatry, Narcology, Psychotherapy and Clinical Psychology, Samara State Medical University, Samara, Russia
e-mail: d.a.smirnova@samsmu.ru

D. Smirnova · P. Cumming
International Centre for Education and Research in Neuropsychiatry, Samara State Medical University, Samara, Russia

T. Smirnova
Department of Neurology, Family Outpatient Clinic Kosma, Samara, Russia

P. Cumming
Department of Nuclear Medicine, Inselspital, Bern University, Bern, Switzerland

School of Psychology and Counselling, Queensland University of Technology, Brisbane, Australia

M. K. Shankardass (ed.), *Dementia Care*,
https://doi.org/10.1007/978-981-16-3864-0_6

(naming, reading, writing), syntax (repeating, composing sentences), and semantics, pragmatics, and discourse (comprehension—auditory, semantic knowledge, understanding commands). We discuss research findings on the protective properties of cognitive reserve, second language acquisition (L2), and multilingualism, all of which can delay the onset of dementia symptoms. We make note of the available interventions in the management of LIs, which include pharmacotherapy (acetylcholinesterase inhibitors such as donezepil, galantamine, and rivastigmine), cognitive interventions (lexical-semantic therapy, action-language therapy, language socialization), and other options of person-centered care (e.g., narrative care). We also review the benefits of destigmatization activities that can be obtained through building a dementia-friendly community environment.

Abbreviations

AD	Alzheimer's disease
FTD	Frontotemporal lobar degeneration
LIs	Language impairments
PPA	Primary progressive aphasia
PD	Parkinson's disease

Introduction

The term dementia (from Latin, *de*—loss, *mentos*—mind) refers to various syndromes entailing a loss of intelligence relative to the premorbid condition, which is characterized by significant polyfunctional cognitive disturbances in domains of attention, gnosis, memory, praxis, and language/speech. Generally occurring due to brain disease or in the aftermath of brain trauma, dementia also manifests in behavioural and personality changes and affective disturbances (McKhann, 2011; Yakhno et al., 2013). The ICD-10 (1992) criteria characterize the dementias as a range of chronic and progressive disorders that lead to psychosocial or occupational impairment due to the compromising of higher cortical functioning within one or more domains such as memory, thinking, judgment, orientation, comprehension, calculation, learning capacity and language. The DSM-5 (2013) defines dementia as a neurocognitive disorder bringing a decline from the previous level of performance in one or more cognitive domains, which may include complex attention, executive function, learning and memory, language (both expressive and receptive), and perceptual-motor or social cognition. Together, these impairments substantially affect the ability to cope with daily activities, such that the afflicted individual requires ever more assistance from caregivers.

Patients with dementia are often readmitted to hospitals as they struggle to manage their self-care due to apraxia, and due to language and communication problems within a-/dysphasias. Communication disturbances, which are interrelated with the language impairments (LIs), are among the earliest symptoms of incipient dementia (Stanyon et al., 2016). Since 2020, patients with dementia are also at elevated risk for COVID-19 infection, due to their advanced age, multi-morbidity, and incapacity to process accurate information about the virus pandemic. General cognitive decline entails an inability to follow health warnings, remember safety procedures, take sufficient self-quarantine measures, and maintain physical distancing. Moreover, approximately 30% of dementia patients with COVID-19 undergo an exacerbation of cognitive decline, and as a group have a generally higher risk of death associated with the pandemic (Canevelli et al., 2020; Livingston et al., 2020; Wang et al., 2020). Social communication is recognized as a protective factor against dementia, which enhances and maintains cognitive reserve, while conversely, lacking social contact, isolation, or loneliness are all associated with the worsening of language and communication disturbances (Evans et al., 2019).

Indeed, social isolation is an independent risk factor for developing dementia, such that vulnerable patients have been apt to experience a weakening integrity of their cognitive state during the COVID-19 outbreak (Canevelli et al., 2020; Wang et al., 2020). According to the Lancet Commission and Alzheimer's Disease (AD) International reports, some 50 million people around the world are now suffering from dementia, but these numbers are expected to increase three-fold by the middle of this century. Nonetheless, only 25–50% of dementia cases are officially diagnosed and documented in primary care, such that best estimates may fail to capture the real and burgeoning need for professional care (Alzheimer's Disease International, 2019; Livingston et al., 2020; Patterson, 2018).

The clinical syndrome of dementia presents with cognitive decline, with diagnosis based on repeated medical follow-up assessments. These can include clinical interviews with patients, provision of information from caregivers, and detailed cognitive assessment, which should properly entail a precise focus on LIs. Among dementia patients, LIs occur at different levels of language organization, extending from lexical access (word-finding) difficulties within anomic aphasia (anomia), to a lack of cohesion in discourse presenting within non-fluent effortful speech. Other language disturbances prominently include comprehension deficits within semantic aphasia, which are related to the stage and etiology of the underlying organic cause of the dementia. Chief among the etiological reasons of dementia are neurodegenerative AD and fronto-temporal lobar degeneration (FTD), Parkinson's (PD) and Lewy body diseases, vascular and prion diseases, HIV-infection, traumatic brain injury and toxic alcohol-related dementias, depressive pseudodementia, and mixed type dementia (Kempler & Goral, 2008; Klimova & Kuka, 2016; Reilly et al., 2011; Yakhno et al., 2013). However, compared to aphasias arising from focal brain damage, LIs in dementia manifest within the broader variety of multiple cognitive impairments (Kempler & Goral, 2008). Nevertheless, the emergence of specific patterns of language decline in late life can serve as a sensitive marker of the risk for developing overt dementia, thus predicting its onset and clinical course or progression (Kemper

& Altmann, 2009). Failure to note the presence of LIs during a patient interview can lead to misdiagnosis (Le Rhun et al., 2006; Tang-Wai & Graham, 2008). Typically, semantic disturbances are prominent in dementias compared to the relatively intact phonology and syntax (Banovic et al., 2018; Murdoch et al., 1987; Reilly et al., 2011). Linguistic examinations using machine learning and natural language processing approaches have demonstrated that LIs presenting in written and spoken speech of patients differed among dementia variants, and could significantly differentiate important forms of dementia, notably AD and subtypes of non-AD frontotemporal dementias (Field et al., 2017; Hardy et al., 2016; Marsani, 2018). However, the majority of dementia patients have difficulties in performing normal communication. For example, individuals with vascular dementia can suffer from language abnormalities that include dysarthria extending over a wide range of severity. In contrast, individuals with AD can have deficits in fluent language production and comprehension within verbal communication, while having good preservation of nonverbal communication in their use of gestures and understanding mimics and pantomimics (Banovic et al., 2018; Rouseeaux et al., 2010). Clarifying the language profiles of dementias is a useful component of evidence-based diagnostics, but can also define objective markers of therapeutic efficacy and help to decide upon further treatment strategies. These strategies can include target-specific language remediation approaches, and psychosocial rehabilitation interventions tailored to the specific language and communication abnormalities by staff with appropriate training in LI remediation.

Clinical and Neuropsychological Assessments of Language Impairments in Dementia: Biomedical Diagnostic Approaches

Many studies have shown that speech and language pathology, or LIs, are among the most reliable objective markers of behavioral disturbances supporting the differential diagnostics of dementia types (Cycyk & Wright, 2008; Grossman et al., 1996; Reilly et al., 2010). In an ongoing scientific discussion, the clinical phenomenon of stroke-related aphasia represents a disorder of impaired access to relatively intact conceptual knowledge. On the other hand, the LIs of dementia constitute a distinct category of language disorder, which is due to degraded memory storage as distinct from impaired retrieval (Adlam et al., 2006; Rapp & Caramazza, 1993; Reilly et al., 2010; Rogers et al., 2006; Warrington & Shallice, 1979). The studies that gave rise to this theoretical dichotomous interpretation of LIs do not unequivocally explain the entire phenomenology of these clinical states, since patients with AD often demonstrate impaired language access with preservation of modality-specific functions, for example when naming pictures better than words (Ober & Shenaut, 1999). In contrast, patients with post-stroke aphasias such as Broca's aphasia or transcortical sensory aphasia fail to manifest non-verbal conceptual knowledge, which clearly relates to their degraded memory storage (Chertkow et al., 1997; Jefferies & Lambon, 2006). Another important aspect of LIs in dementia is the phenomenon of primary progressive aphasia (PPA), first defined by Mesulam (1982, 2003) as a slowly progressive LI manifesting in anomia, word finding difficulties, impaired word comprehension and reduced syntactic complexity of speech. These symptoms can present for at least two years before the onset of frank dementia, and can occur seemingly without a precipitating cause. Mesulam (2001) described such LIs as "slowly progressive aphasia without generalized dementia" or "language-based dementia". A further aspect of PPA is its frequent comorbid acalculia and ideomotor apraxia, which occur in the absence of amnestic disorders or personality changes that affect daily functioning (Mesulam, 2001). Semantic dementia, as described by Warrington (1981), is another form of language decline occurring within the primary semantic impairment of dementia. As described in a review by Reilly et al. (2010), semantic dementia is the next stage of PPA, which is termed fluent PPA (Adlam et al., 2006; Snowden & Neary, 1993). Nevertheless, Mesulam et al. (2009) differentiated between semantic dementia and the semantic variant of PPA, informed by the knowledge that semantic dementia is the primary aspect of FTD, but is extremely rare in AD. In contrast, the semantic variant of PPA entails behavioural manifestations common to various neurodegenerative diseases including AD, Parkinson's disease, and Creutzfeld–Jacob disease. Gorno-Tempini et al. (2011) classified PPA into three variants according to a triad of clinical syndromes: non-fluent/agrammatic dementia associated with taupathies, semantic PPA represented in FTD, and logopenic PPA, as occurs in AD. To this triad, Mesulam (2013) added the categories of the anomic variant of PPA and the mixed variant.

Apart from a clinical interview focusing on basic LIs associated with dementia, there are specific neuropsychological tests and instruments to clarify the diagnosis of dementia through estimating the extent and variety of language disorders. Nevertheless, strict reliance on neuropsychological examination is insufficient for diagnosis, because it is administered through verbal instructions, commands, responses, and reasoning. Therefore, such tests can support invalid conclusions related to the verbal behaviour assessments, or can overestimate the severity of the LIs. This risk of exaggeration is particularly high when using the Mini Mental State Examination in individuals with sparing of performance in certain cognitive domains (Folstein et al., 1975; Osher et al., 2007). For example, PPA patients have difficulties in remembering word lists and in following tasks involving reasoning, while being quite able to recall daily events, demonstrate intact social skills, and participate in conversations (Mesulam, 2013). The standard approach to measure language domains includes attentive listening to the patients' spontaneous speech in response to their being asked open questions or requested to describe complicated pictures in the Boston Diagnostic Aphasia Examination (Goodglass et al., 2001). Interviewers should take account of diverse metalinguistic and pragmatic aspects of language, including articulation, prosody, fluency, grammatical accuracy, syntactic complexity, lexical variety and word use. Put another way, there is a need to distinguish fluent, grammatically correct speech from dysfluent and effortful speech, with difficulties in finding the correct words or other sorts of errors. Therefore, it is important for interviewers to check their client's naming ability by asking them to name commonplace objects around the room or to identify line drawings in the Boston Naming Test, while noting any phonemic or semantic paraphasias/errors and potential anomia (Kaplan et al., 2001). Another formal approach to examination relates to the repetition of words and phrases (e.g., unfamiliar sentences), aiming to thus differentiate a transcortical aphasic disorder from other LIs. Comprehension, which is an index of the semantic language domain, is tested by asking patients to perform one-, two-, three-step commands of increasing grammatical complexity. An example of such sequentially complex commands is: ''Touch your nose", "Touch your nose and then point to the door", "Take this piece of paper in your left hand, fold it in half, and put it on the floor", or "Before touching your nose, point to the floor". Performance of these instructions calls upon the patient to interpret the definition of words or brief stories, with subsequent testing by asking questions about their content (Tang-Wai & Graham, 2008). Reading and reading comprehension are assessed by asking the client to read sample paragraphs, as well as reading instructions to perform certain tasks and the verbal requests to follow them. This kind of task allows the evaluation of sound-based language domains such as normal prosody or effort in speaking and pronunciation difficulties. Performance of these tasks requires preserved functioning at the semantic level and the retained ability to decode semantic knowledge or an ability to understand meaning, commands, or figurative language within humor or proverbs. Writing of sentences or delivery of answers to questions may unveil errors in grammar, spelling, punctuation, or repetition of syllables.

There is an abundance of standardized clinical and neuropsychological batteries designed to test LIs in dementia (Mesulam, 2013). In the present context, we emphasize the tests most commonly used to assess LIs in dementia research and in clinical practice. The Western Aphasia Battery (WAB) (Kertesz, 2007) measures the severity of aphasia based on the summary Aphasia Quotient, which is calculated based on the assessment of a spontaneous speech sample, comprehension of words and sentences as syntactic constructs, as well as naming, reading, and writing tasks. The WAB repetition subtest serves to estimate the performance of repetition tasks (Mesulam et al., 2012). Among the most famous LI tests are the Aachen Aphasia Test (AAT), which also rates spontaneous speech, the Token test, and other language subtests (Huber et al., 1984). The Boston Naming Test (BNT) is widely used to measure the capacity for object naming (Kaplan et al., 2001). Assessing the more complex aspects of language performance, i.e., the grammatical accuracy of sentence production, calls for exclusion of the interferences arising from deficits in working memory. Procedures for avoiding this interference include using the Northwestern Anagram Test (NAT), as well as with the Sentence Production and Priming Test (SPPT) of the Northwestern Assessment of Verbs and Sentences (NAVS) (Thompson, 2011; Weintraub et al., 2009). The Peabody Picture Vocabulary test, specifically its subset of 36 items of moderate level of difficulty, measures word association and comprehension, as well as picture naming, insofar as each item requires the patient to match an auditory word representing an object, action, or one out of four pictures (Dunn & Dunn, 2006). The Sydney Language Battery (SYDBAT) is a computer-based test of language developed to assess the patterns of deficits in expressive and receptive language abilities at the single word level through scoring of subtests of naming, repetition, comprehension and semantic association (Savage et al., 2013). The Functional Linguistic Communication Inventory (Bayles & Tomoeda, 1994) along with its further revisions is a standardized combination of tests, which evaluates language function in patients with moderate or severe dementia, based on measuring their communication skills in relation to the composite of all language domains. This testing system includes greeting and naming, answering questions, writing, comprehension of signs/object-to-picture matching, word reading and comprehension, reminiscing, following commands, pantomime, gesture, and conversation. The Psycholinguistic Assessment of Language Processing in Aphasia (PALPA) is an instrument that measures the level of dyslexia and dysgraphia (Kay, 1992). There are also tests to measure nonliteral language or the plan of pragmatics in dementia (Hur & Caixeta, 2013); these include the Screening Test for Alzheimer's Disease with Proverbs (STADP, Santos et al., 2009), the metaphor test of language competence (Wiig & Secord, 1989), and the Irony test (Eviatar & Just, 2006). Thus, information received from clinical interview and neuropsychological assessments about LIs can together contribute to the diagnostic process in dementia patients.

Variety of Language Impairments Across Clinical States of Dementia: Psycholinguistic Approaches

We now review the similarities and difference of LIs at different levels of language organization, extending from the sound-based domain, to lexis, syntax, semantics, pragmatics, and discourse in AD and the most commonly occurring types of non-AD dementias, such as frontotemporal lobar degeneration (FTD), Lewy body and Parkinson's disease (PD) and vascular dementia. We summarize the relevant data in Tables 6.1 and 6.2.

Alzheimer's disease (AD) is the most common cause underlying dementia and one of the most notorious diseases in society. In the majority of cases AD symptoms are revealed after the age of 65 years, with inexorable progression leading to death within 8 to 10 years of diagnosis (Alzheimer's Disease International, 2019). LIs are a core marker that progress over the course of the disorder (Weiner et al., 2008). At early AD disease stages, language seems to be relatively preserved at the phonetic level (Lambon Ralph et al., 1995), but proceed to manifest in frequent phonological confusions related to the recognition of spoken words (Sommers, 1998). Other LIs typical of AD include poor accommodation of acoustic variability in the patient's speech perception and deficits in the so-called speaker normalization abilities, which entail a poor capacity to decode pronunciation of different speakers, phonetic errors such as pseudoword production, word initialization errors and phonemic substitutions (Biassou et al., 1995; Glosser et al., 1998). A key feature for clinical diagnosis of AD is its characteristic disruption in the naming pathway at the lexical level, commonly known as anomia or anomic aphasia. These LIs arise because of

Table 6.1 Summary of language impairments based on the literature about Alzheimer's dementia

Category	Language impairments	Sources
Alzheimer's dementia (moderate-severe stages)		
Sound-based	Echolalia, phonetic paraphasias, verbal apraxia, loud speech	Biassou et al. (1995), Glosser et al. (1998), Klimova and Kuca (2016), Sommers (1998)
Lexis	*Action (verb) fluency deficit, anomic aphasia/anomia (since early stages), impaired lexical access and retrieval/word finding difficulties, lexical repetition/verbal stereotypies/verbigerations, neologisms, simplified language/diminished vocabulary*	*Cappa et al. (1998), Cummings et al. (1985), Feldman and Woodward (2005), Garrard et al. (2005), Grossman et al. (2003), Hodges et al. (1996), Reilly et al. (2011)*
Syntax	Agrammatisms, reduced syntactic complexity	Grober and Bang (1995), Kempler et al. (1998)
Semantics	*Selective semantic fluency deficit, sentence comprehension deficit, semantic paraphasias, unclear references*	*Aronoff et al. (2006), Ehrlich et al. (1997), Gonnerman et al.,(1997), Whatmough et al. (2003)*
Pragmatics	Failure to understand figurative language/to decode nonliteral meaning (metaphors, irony)	Hur and Caixeta (2013)
Discourse	*Empty, incoherent discourse, topic shifting, weakening of goal*	*Chapman et al. (1998), Ehrlich et al. (1997), Garrard et al. (2005), Tomoeda and Bayles (1993), Tomoeda et al. (1996)*

difficulties in object recognition and problems in lexical retrieval. Indeed, some 76% of AD patients demonstrate significant naming impairments (Hodges et al., 1996). AD patients also manifest a selective semantic category fluency deficit, for example in naming biological natural kinds (e.g., animals, fruits) worse than manufactured artefacts (Gonnerman et al., 1997), a phenomenon that is present across all AD stages (Whatmough et al., 2003). AD patients also have greater difficulty in understanding and processing of verbs rather than dealing with the nouns (Cappa et al., 1998; Grossman et al., 2003). These phenomena are hypothetically a consequence of differing neurobiological underpinnings of these semantic categories. AD patients demonstrate syntactic processing deficits, as well as difficulties in sentence comprehension irrespective of their grammatical complexity (Grober & Bang, 1995; Kempler et al., 1998). However, AD patients with classic agrammatism are characterized by more prominent deficits across the comprehension of long sentences, as compared to other types of semantic deficits (Aronoff et al., 2006). Garrard et al. (2005) found lexical impoverishment, repetitions, and a significantly reduced index of lexical diversity, designated as the type-token-ratio, which is calculated as the number of distinct words divided by the total number of words pronounced by the test subject. Hur and Caixeta (2013) reported that AD patients fail to understand

Table 6.2 Summary of language impairments based on the literature about non-Alzheimer's dementias

Category	Language Impairments	Sources
Non-Alzheimer's dementias		
Frontotemporal dementia (i) progressive nonfluent aphasia/agrammatical variant (ii) semantic variant (iii) frontal/behavioral variant		
Sound-based	*Late mutism (i), phonetic paraphasias (i), stuttering (i), surface dyslexia/ "reading without semantics" (ii), verbal apraxia (i)*	*Neary et al. (1998), Clark et al. (2005), Patterson and Lambon Ralph (1999), Shallice et al. (1983), Weintraub et al. (1990)*
Lexis	Agraphia (i), alexia (i), anomia (i, ii), impaired lexical retrieval and word processing deficits (ii), verbal fluency deficit (i)	Clark et al. (2005), Pulvermüller et al. (2010), Weintraub et al. (1990)
Syntax	*Agrammatisms (i)*	*Grossman and Moore (2005), Neary et al. (1998), Peelle et al. (2007)*
Semantics	Semantic/conceptual deficit /word comprehension deficit (ii), selective semantic fluency deficit (concrete concepts) (ii)	Grossman and Ash (2004), Kempler and Goral (2008), Vesely et al. (2007)
Discource	*Empty discourse/omission of information (i), incoherence (iii), disorganized discourse (iii)*	*Ash et al. (2006), Reilly et al. (2010), Rascovsky et al. (2007)*
Lewy body dementia		
Lexis	*Impaired lexical access and retrieval/word finding difficulties, verbal fluency deficit*	*Doubleday et al. (2002), Klimova and Kuca (2016), Reilly et al. (2010)*
Discourse	Incoherence, perseverations	Doubleday et al. (2002), Galasko et al. (1996), Gibb et al. (1989)
Parkinson's disease dementia		
Sound-based	Dysarthria	Klimova and Kuca (2016)
Lexis	*Action (verb) fluency deficit, anomic aphasia (anomia), verbal fluency deficit*	*Piatt et al. (1999)*
Syntax	Agrammatism	Klimova and Kuca (2016), Reilly et al. (2010)
Semantics	*Concept formation deficit*	*Emre et al. (2007)*

(continued)

Table 6.2 (continued)

Category	Language Impairments	Sources
Pragmatics	Failure to understand figurative language (metaphors)	Monetta and Pell, (2007)
Discource	*Incoherence, perseverations*	*Emre et al. (2007)*
Vascular dementia		
Sound-based	*Flattened intonation/dysprosody, phonetic fluency deficit*	*Jones et al. (2006), Klimova and Kuca (2016)*
Lexis	Anomic aphasia (anomia), verbal fluency deficit	Lukatela et al. (1998), Tierney et al. (2001)
Syntax	*Reduced syntactic complexity*	*Looi and Sachdev (1999)*

figurative language and to comprehend or extract nonliteral meanings during their performance of such listening tasks. Moreover, there is an impairment of discourse across a number of domains in patients with AD. Some patients are unable to connect with global meaning to build up a cohesive train of statements, while others are deficient at the level of semantic propositional knowledge (Ehrlich et al., 1997). These deficits could include unclear references to people or places (e.g., he, there) or over-repetition of the language content and topic changes, typically along with poor organization and an abundance of circumlocutions. In this context, the patients describe things rather than naming them—the so-called "fluent, but empty" discourse—with a general lack of coherence and information context (Chapman et al., 1998; Tomoeda & Bayles, 1993; Tomoeda et al., 1996) (see Table 6.1).

Frontotemporal dementia (FTD) is the second most common dementia following AD, with mean age of onset at about 60 years, thus half a decade earlier than AD, and with a characteristically more rapid progression, usually leading to death within two to eight years after onset (Forman et al., 2006; Hodges et al., 2003). Neary et al. (1998) described three FTD linguistic variants: (i) progressive nonfluent aphasia, (ii) semantic dementia, and (iii) frontal variant FTD. *Progressive nonfluent aphasia variant* is associated with the traits and prominent markers of dysfluency, such as agrammatisms, phonemic paraphasias, and anomia. Speech production in this LI variant characteristically entails stuttering, oral/verbal apraxia, impaired repetition, alexia, agraphia, with early preservation of word meaning, but progressing to late mutism (Neary et al., 1998). These patients' speech is often slow, effortful, and full of phonemic paraphasias, thus being reminiscent of Broca's aphasia (Reilly et al., 2010). Anomia represents the primary deficit in conversational speech in FTD. Deficits are evident in naming, repetition, reading tasks, and lexicon, as well as phonological versus semantic paraphasias, which mostly relate to impaired access to phonological representations rather than semantic deficit per se (Clark et al., 2005; Weintraub et al., 1990). Sentence comprehension impairments in FTD appear during different tasks, e.g., sentence-picture matching with grammatically complex sentences, with presentation of syntactic impairment through an insensitivity to grammatical violations and the use of short phrases (Grossman & Moore, 2005; Peelle et al., 2007).

Their discourse entails limited language production and a lack of significant information (Ash et al. 2006). *Semantic variant* of FTD is characterized by lexical (anomia) and semantic (word comprehension) deficits, first appearing at an early disease stage and progressing over the course of the disease (Grossman & Ash, 2004; Kempler & Goral, 2008). Patients demonstrate difficulties in performing tasks related to naming objects, defining words, sorting objects, and correctly performing the proper use of objects (Hodges & Patterson, 1996; Hodges et al., 2000; Warrington, 1975). The semantic variant of FTD is also characterized by a pattern of surface dyslexia or "reading without semantics", where patients read easily and correctly pronounce orthographically regular words (e.g., cat), while experiencing difficulties with the pronunciation of irregular words (e.g., yacht) (Patterson & Lambon Ralph, 1999; Shallice et al., 1983). Semantic/conceptual deficits in this variant of FTD appear much earlier and across more tasks as compared to AD (Kempler & Goral, 2008), although there is evidence that concrete objects (e.g., fruits) are managed worse than abstract concepts (e.g., numbers) (Vesely et al., 2007). However, there is sparing of lexical processing, grammatical and syntactic structures, as well as discourse in this disorder (Kave et al. 2007). Compared to grammatical and semantic variants, *frontal/behavioural variant of FTD* does not entail a prominent language decline. Due to the associated changes in personality, emotions, and behavior, speech content may shift towards violating moral rules and telling inappropriate jokes (Reilly et al., 2010). However, due to their poor inhibitory control across the language domain, FTD patients can demonstrate a sequencing deficit, incoherence, or thematic disorganization within the discourse (Ash et al. 2006; Rascovsky et al., 2007) (See Table 6.2 for LIs in non-AD dementias).

Dementia caused by Lewy body disease, the third most common form of dementia, has a mean onset at 68 years, leading to death in 2 to 10 years (mean 6 years). Initial disturbances progress to include levodopa refractory parkinsonism, severe cognitive decline, and disability with total dependence (Geser et al., 2005; Reilly et al., 2010). Cognitive decline typically fluctuates over the course of the disease, which is marked by primary language complaints in only 5% of cases, but entailing a combination of memory problems, visuospatial symptoms, and LIs (Doubleday et al., 2002). The content of language is disrupted, thus including confabulations, and demonstrates incoherence and perseveration in attempts at communication, object naming difficulties, and a verbal fluency deficit (Doubleday et al., 2002; Galasko et al., 1996; Gibb et al., 1989).

Parkinson's disease (PD) is a primary motor disturbance with mean age of onset of 60 years, but including 10% of cases with onset aged 45 years or younger (Korchounov et al., 2004; Rao et al., 2006). There is an emerging appreciation that cognitive and communication disturbances are an early feature of PD (Riedel et al., 2008). Indeed, one of the core diagnostic criteria of LIs in PD is impaired "verbal free recall" (Reilly et al., 2010). These patients demonstrate LIs within verbal fluency, comprehension of sentences, and spontaneous speech processing (Grossman et al., 2002; Henry & Crawford, 2004; Hochstadt et al., 2006). In particular, fluency of action naming is disproportionately distorted as compared to semantic or phonemic fluency, showing that patients encounter more difficulties when naming actions rather than objects (Cotelli et al., 2007). The specific deficits in action or functionally related language are seemingly responsive to levodopa treatment, implying a primary association between dopaminergic/noradrenaline degeneration of processing of action items in language (Johari et al., 2019). A decline in action (verb) fluency predicts progression of dementia in these patients (Piatt et al., 1999), perhaps implying a crossover between primary motor symptoms and processing of action words. Patients with PD also speak slowly, struggle with articulation, and have difficulty in understanding metaphors, and figurative language, as well as impaired perception of language at the phonological level and emotional prosody (Monetta & Pell, 2007) along with perseveration and deficits in concept formation (Emre et al., 2007).

Vascular dementia, a frequent dementia type, with an incidence nearly as high as FTD, is associated with cerebrovascular disease (Fratiglioni et al., 2000), but almost half of such patients have mixed vascular and AD components of their dementia (Rockwood et al., 1999). Vascular dementia patients demonstrate decreased phonemic fluency, reduced grammatical complexity in sentence production, and flattening of prosody in conversational speech (i.e., dysprosody) (Jones et al., 2006; Looi & Sachdev, 1999; Lukatela et al., 1998; Tierney et al., 2001). Though some recent studies show that LIs in vascular dementia occur at all levels of language organization, these findings are not specific and are reminiscent of those occurring in uncomplicated AD (Freitas et al., 2018; Klimova & Kuca, 2016).

Thus, the corpus of research on LIs in different dementias reveals differences and overlapping similarities, doubtless reflecting the mixture of organic etiologies. The preponderance of research has focused on LIs in AD and FTD, thus calling for a more detailed investigation of the discriminative aspects of language disturbances in other non-AD dementias.

Socio-linguistic Approaches to Understand the Discourse, Language and Message of Dementia

Dementia is not simply a neuropsychiatric condition that affects linguistic abilities, but is a disorder with pervasively detrimental effects on the quality of life of patients and their caregivers and families. This draws attention to the need to disseminate contextual knowledge about dementia throughout society and introduce a so-called linguistic discourse related to dementia as a social phenomenon. As an object of social knowledge, dementia falls within the field of linguistic narratives with their particular meanings, and occupies a space in the entire socio-cultural matrix. By definition, dementias are a class of disorder, but occurring so commonly among the elderly as to qualify as an aspect of human experience. The problematic behavior of dementia patients, and their need for extensive care, however, call for a broader discourse about the wide-ranging effects on dementia on society (Villar et al., 2019). The socio-linguistic implications of dementia arise in everyday conversations with patients and with their family-members and caregivers, as well as in the numerous press reports and narratives appearing in social media (Van Dijk, 2016). This dissemination of knowledge through many channels involves an internalization and conceptual transformation by society. Narratives about dementia arising from biomedical terminology (e.g., "incapacity for self-care", "loss of autonomy"), research communities (e.g., "social, health and economic burden", "attempts to attack or combat the disease"), and caregivers (e.g., emotional burden related to diagnosis and disease progression, stress related to giving care) are mostly pessimistic and of negative valence; furthermore, descriptions of dementia cases in the public media are often overly dramatic and stereotyped (Hinton et al., 2006; Muñoz et al., 2009; Zeiling, 2013; Zimmermann, 2017). Stigmatization and discrimination of dementia patients arise mostly form these negative narratives. This also occurs within the socio-constructionists' paradigm, when society elaborates the criteria to categorize people as "us versus they", "different" or "sick", thus referring dementia to undesirable states and lower social status (Bosco et al., 2019). These narratives can centralize the core socio-linguistic meaning of the "dissolution of self" and the semantics of "zombification"; sadly, this encourages the perception of people with dementia as the "living dead"/"socially dead" occupying "selfless bodies" or "empty shells", or as "strangers who lost their status of being a person", all of which invokes the imagery of "the long good-bye" (Aquilina & Hughes., 2006; Behuniak, 2011; Goffman, 2006; Snyder, 2006 p. 268; Sweeting & Gilhooly, 1997). All too often, dementia is depicted as a relentlessly threatening, aggressive, and devastating disease, which affects not only patients but also their families and caregivers. In recent years, we are finally starting to see the emergence of a more respectful societal narrative about this clinical state, and a greater tolerance and appreciation of the value and needs of patients living with dementia (Behuniak, 2011; Bosco et al., 2019).

To retransform negative narratives about dementia, as well as to destigmatize the dementia state in the community, it is important to revise the meaning of dementia by refocusing from the labelling from a biomedical perspective (patient with dementia) to a person-centered discourse (person with dementia) (Villar et al., 2019). This approach has been elaborated in the so-called the "personhood movement", and has succeeded in raising general interest in how best to help people living with dementia, aiming to maximize their social engagement and self-esteem (Hamilton, 2008; Leibing, 2006). While medicine remains nearly helpless in treating the underlying brain disorders leading to dementia (with a few exceptions such as dementia due to normal pressure hydrocephalus), there is much that can be done "to promote health and well-being", to convey coping strategies, and to create friendly and supportive environments for individuals with dementia (Downs et al., 2006; Hamilton, 2008). The medical community is making increasing attempts to reconfigure from dementia to the person with dementia—"the person within—the reflexive, immaterial, communicable essence of a person that is located deep within the body, but that is sometimes veiled by symptoms" (Leibing, 2006, p. 243). This involves an increasing appreciation of the "capacities of the feeling person and not only on his or her losses" (Leibing, 2006, p. 255).

Individuals with dementia often lose their significant social relationships and withdraw from their former activities in the community. For example, for residents in nursing homes or long-term care institutions, 75% of social interactions occur with visitors, 25% with staff, and only 10% with other people (Brannelly, 2011; Ward et al., 2008). People with dementia encounter the problem of social neglect, when the staff minimize their contact or when cohabiting family members do not communicate sufficiently, such that patients often seem depersonalized from the perspective of others (Doyle & Rubinstein, 2013). Persons with dementia experience "narrative dispossession", when they stop telling their stories and forgetting (in part through lack of practice) how to understand the stories of others (Baldwin, 2006). This involves a vicious cycle, which tends to worsen language processing and comprehension skills and promote further cognitive decline. With this in mind, we must focus on bolstering the communication with others, especially in terms of understanding that "us" and "them" can reunite as "we" in the context of a friendly conversation. Caregivers, friends, and relatives must never forget that the person who is now suffering from dementia is a grandmother or father, or a beloved friend. Thus, we should start moving away from the concept of disability towards seeing new opportunities. What some may perceive as a hopeless decline in narrative can be reconfigured as an opportunity for empathetic conversations and storytelling, supported by an attentive and proactive audience (Villar et al., 2019). Narrative care provided by trained staff and properly informed family members can help to preserve and restore meaning in discourse with the patient, and this meet the emotional needs of patient and those in contact with the patient.

We hope that this brief plea for person-centered and humanizing care may diffuse across the community, via the media, and in aspects of the daily life of each patient, ultimately to enable formulation of a more dementia-friendly community and empathetic environment (Villar et al., 2019). Practical approaches include attracting patients to different forms of activities, including participation in different organizations (e.g., Dementia Alliance International, moto "Nothing about us, without us") (Bartlett, 2014), and as much as possible, integration back into society.

Another important socio-linguistic approach we would like to emphasize is the relationship between dementia and second language acquisition, especially in terms of its protective role to delay the symptoms of dementia. Second language has a relation with cognitive reserve, such that people with high educational and language attainment are apt to retain better performance of complex mental activities in the face of dementing illness (Kemper et al., 2001; Stern, 2009, 2013). Bialystok et al. (2007, 2014) reported that premorbid proficiency in and daily use of a second language are protective factors that delay the onset of the first symptoms of dementia by a mean of 4.7 years for mild cognitive impairment and 7.3 years for AD. Alladi et al. (2011) demonstrated that the strongest protective effect of bilingualism was observed in FTD (6 years), followed by vascular dementia (3.7 years), and AD (3.2 years). In a 2017 comparison among FTD variants, Alladi et al. showed that the age of onset for the behavioral variant was delayed by a mean of 6.1 years in bilinguals compared to monolinguals, while there was no difference in onset age for the aphasic variant. This suggests that the effect of bilingualism relates to specific behavioral symptoms, in line with the data showing that bilingualism plays a significant role in the preservation of executive functions (Bialystock et al., 2012). Furthermore, Perquin et al. (2013)

showed that polyglot individuals (with three or more languages) showed a reduced incidence of cognitive impairment or dementia, correcting for confounding factors such as educational attainment and immigration history (Bak et al., 2014).

Understanding this socio-linguistic context calls for a greater appreciation of the importance of positive and tolerant attitudes towards dementia and the need for dementia-friendly environments. In the present context of LIs, we see a need for maintaining relationships and communication with persons with dementia. Involving such individuals in different forms of activities including second language exposure could help to create an enriched environment amenable to the maintenance of cognitive reserve.

Conclusions: Treatment Approaches to Restore Language Functions in Dementia

While there is a general lack of effective treatments for dementias, we would be remiss in not mentioning the experience with LIs management through pharmacotherapy (e.g., the acetylcholinesterase inhibitors donezepil, galantamine, and rivastigmine), cognitive interventions (e.g., lexical-semantic therapy, action-language therapy, language socialization, and second language use, as noted above), and other treatment options of person-centered care (e.g., narrative care).

In an early conception of AD pathology, much emphasis was placed on the importance of degeneration of the cholinergic neurons of the basal forebrain, and the concept of enhancing cholinergic transmission with acetylcholinesterase inhibitors. The US Food and Drug Administration has approved three acetylcholinesterase inhibitors for the symptomatic treatment of AD in clinical practice. Donepezil treatment improved scores in the Language subscale of the Severe Impairment Battery when administered at a dose of 23 mg/day, with lesser benefits at a dose of only 10 mg/day. Despite their identical mechanism of action, neither rivastigmine nor galantamine showed significant benefits in the language domain, compared to placebo (Ferris & Farlow, 2013). Thus, LIs in AD are not very responsive to pharmacotherapy with acetylcholine esterase inhibitors, despite having a certain efficacy in other cognitive domains at early disease stages. Morello et al. (2017) reviewed the literature on language and communication (non-pharmacological) interventions for dementia: (i) Intervention involving training in several cognitive skills, including language; (ii) Language activities integrated with physical activities; (iii) Lexical-semantic therapy, in which the individual with dementia is asked to define semantic relationships between items, name figures, interpret the meaning of isolated words, as well as words as parts of the sentences constructs and within storylines; (iv) Face-name association intervention; (v) Instrumental communicative activities in daily living; (vi) Interventions based exclusively on conversational interactions; and (vii) Use of memory cards during conversation. These various approaches all demonstrated a certain efficacy in at least one aspect of language functions, among which the most efficacious interventions were lexical-semantic therapy, which engages several cognitive skills at the same time. Intensive Language Action therapy was proven effective in controlled trial studies of post-stroke aphasia patients, reliably bringing an improvement in a variety of language functions (naming, auditory comprehension, reading, and verbal repetition) already in two weeks (3–4 h daily) of intensive treatment (Berthier et al., 2011). There has as yet been no test of this approach for the treatment of LIs in dementia.

Hamilton (2008) suggested using the approach of language socialization, i.e., "Talk-In-Interaction", thus aiming to focus patients on the interactional contexts. This line of research could yield new treatment strategies by attempting to bridge the gap between neurobiological, psycholinguistic and socio-linguistic approaches to provide person-centered patient care. Several studies have focused on patients' activities and "unpacking the social engagement black box", aiming to show that voluntary participation in different social activities have a more positive impact on cognitive functions than do the daily interactions with family or staff; these activities include playing games, doing volunteer work, and activity in business or political societies, any of which can reduce the patients' feelings of "no meaningful present", and help in"maintaining positions grounded in reality" (Gallacher et al., 2005; Glei et al., 2005; Snyder, 2006). While support groups and different kinds of interactional activities can be helpful for those with mild dementia, they may simply cause worsened confusion in individuals with severe dementia. Here, interactive methods that use visual stimuli/images, ask open-end questions, and invite patients to express their opinion can all be helpful to restore their sense of autonomy (Basting, 2006).

Apart from focusing on activities within the approach "Talk-In-Interaction", verbal communication with use of lexical items, such as pronouns, syntactic structures, and nonverbal communication with of gestures and laughter in the conversation, can help to engage patients with LIs. Hamilton (2005) used the metaphors of *prism*, *soliloquy*, *couch*, and *dance* to describe four main approaches to the analysis and remediation of language and communication impairments. The *prism* places its focus on selected linguistic or discursive phenomena, without relation to the speaker or the context of speech. Maclagan and Mason (2005) noted that patients had "good days and bad ones" in terms of indicators of lexical richness/empty language. This observation guides the remediation of declining discourse within the day-to-day fluctuations in language performance. The *soliloquy* provides and analysis of "how individuals with dementia integrate their linguistic and communicative phenomena identified by the *prism* approach into a whole discourse" Hamilton (2008) noted that individuals with dementia used an excessive number of pronouns and primitive lexical items such as *people*, *do*, and *make*, as well as predominantly expressing narrative traces (73%) versus conversational clauses (27%). These markers could serve as the targets for linguistic treatment interventions. In the Hamilton model, the *couch* refers to the use of language for creating and sustaining the making of meaning and identity of the person. The final term *dance* refers to "the interaction between language users and the ways in which interlocutors influence each other—both socially and linguistically" (Ryan et al., 2005). This entails introducing health conversational strategies for patients with dementia, aiming to provide proper nouns instead of pronouns and allowing long pauses to await responses from patients, which might be termed as "warm reciprocity" and "conversation as care".

By focusing on the particular language patterns used by dementia patients, it is possible to identify targets for further linguistic interventions and approaches tailored for needs of different categories of persons with dementia. Our main objective in this chapter is to draw appropriate professional attention to the various implications of LIs occurring in association with dementia. Our review of the literature gives rise to three summary statements: (i) LIs in different types of dementia present with distinct clinical, psycholinguistic and sociolinguistic aspects; (ii) Language skills provide an important component of cognitive reserve and are a therapeutic resource to restore healthy language in dementia; and (iii) It is critically important to listen to the message of dementia and to enhance the conversation between patients, family, caregivers, and society in general, all for the greater good of people living with dementia.

References

Adlam, A., Patterson, K., Rogers, T. T., Nestor, P. J., Salmond, C. H., Acosta-Cabronero, J., et al. (2006). Semantic dementia and fluent primary progressive aphasia: Two sides of the same coin? *Brain, 129*, 3066–3080.

Alladi, S., et al. (2011). Subtypes of dementia: A study from a memory clinic in India. *Dementia and Geriatric Cognitive Disorders, 32*(1), 32–8. https://doi.org/10.1159/000329862.

Alladi, S. et al. (2017) Bilingualism delays the onset of behavioral but not aphasic forms of frontotemporal dementia. *Neuropsychologia, 99*, 207–212. https://doi.org/10.1016/j.neuropsychologia.2017.03.021.

Alzheimer's Disease International. (2019). *World Alzheimer Report 2019: Attitudes to dementia.* Alzheimer's Disease International.

American Psychiatric Association. (2013). *American psychiatric association: Desk reference to the diagnostic criteria from DSM-5.*

Aronoff, J. M., Gonnerman, L. M., Almor, A., Arunachalam, S., Kempler, D., & Andersen, E. S. (2006). Information content versus relational knowledge: Semantic deficits in patients with Alzheimer's disease. *Neuropsychologia, 44*, 21–35.

Aquilina, C., & Hughes, J. C. (2006). The return of the living dead: Agency lost and found? In J. Hughes, S. Louw, & S. R. Sabat (Eds.), *Dementia: Mind, meaning and the person* (pp. 143–161). Oxford University Press.

Ash, S., Moore, P., Antani, S., McCawley, G., Work, M., & Grossman., M. (2006). Trying to tell a tale: discourse impairments in progressive aphasia and frontotemporal dementia. *Neurology, 66(9)*, 1405–1413. https://doi.org/10.1212/01.wnl.0000210435.72614.38.

Bak, T. H., Nissan, J. J., Allerhand, M. M., Deary, I. J. (2014). Does bilingualism influence cognitive aging? *Annals of Neurology, 75*(6), 959–63. https://doi.org/10.1002/ana.24158. Epub 2014 Jun 2. PMID: 24890334; PMCID: PMC4320748.

Baldwin, C. (2006). The narrative dispossession of people living with dementia: Thinking about the theory and method of narrative. In J. Milnes, C. Horrocks, N. Kelly, B. Roberts, & D. Robinson (Eds.), *Narrative, memory and knowledge: Representations, aesthetics and contexts* (pp. 101–109). University of Huddersfield.

Banovic, S., et al. (2018). Communication difficulties as a result of dementia. *Materia Socio-medica, 30*(3), 221–224. https://doi.org/10.5455/msm.2018.30.221-224.

Bartlett, R. (2014). Citizenship in action: The lived experiences of citizens with dementia who campaign for social change. *Disability Society, 29*, 1291–1304.
Basting, A. D. (2006). Creative storytelling and self-expression among people with dementia. In A. Leibing & L. Cohen (Eds.), *Thinking about dementia: Culture, loss, and the anthropology of senility* (pp. 180–194). Rutgers University Press.
Behuniak, S. M. (2011). The living dead? The construction of people with Alzheimer's Disease as Zombies. *Ageing & Society, 31*, 70–92.
Berthier, M. L., García-Casares, N., Walsh, S. F., Nabrozidis, A., Ruíz de Mier, R. J., Green, C., Dávila, G., Gutiérrez, A., & Pulvermüller, F. (2011). *Discovery Medicine, 12*(65), 275–289.
Bialystok, E., et al. (2007). Bilingualism as a protection against the onset of symptoms of dementia. *Neuropsychologia, 45*(2), 459–64. https://doi.org/10.1016/j.neuropsychologia.2006.10.009.
Bialystok, E., et al. (2012). Bilingualism: consequences for mind and brain. *Trends in Cognitive Sciences, 16*(4), 240–50. https://doi.org/10.1016/j.tics.2012.03.001.
Bialystok, E., et al. (2014). Effects of bilingualism and aging on executive function and working memory. *Psychology and Aging, 29*(3), 696–705.https://doi.org/10.1037/a0037254.
Biassou, N., Grossman, M., Onishi, K., Mickanin, J., et al. (1995). Phonologic processing deficits in Alzheimer's disease. *Neurology, 45*(12), 2165–2169.
Bosco, A., Schneider, J., Coleston-Shields, D. M., Higgs, P., & Orell, M. (2019). The social construction of dementia: Systematic review and metacognitive model of enculturation. *Maturitas, 120*, 12–22.
Brannelly, T. (2011). Sustaining citizenship: People with dementia and the phenomenon of social death. *Nursing Ethics, 18*, 662–671.
Bayles, K. A., & Tomoeda, C. K. (1994). Functional Linguistic Communication Inventory. Tucson, Arizona: Canyonlands Publishing Inc.
Caballol, N., Marti, M. J., & Tolosa, E. (2007). Cognitive dysfunction and dementia in Parkinson disease. *Movement Disorders, 22*(S17), S358–S366. [PubMed: 18175397].
Canevelli, M., et al. (2020). Facing dementia during the COVID-19 outbreak. *Journal of the American Geriatrics Society, 68*(8), 1673–1676. https://doi.org/10.1111/jgs.16644.
Cappa, S. F., Binetti, G., Pezzini, A., Padovani, A., Rozzini, L., & Trabucchi, M. (1998). Object and action naming in Alzheimer's disease and frontotemporal dementia. *Neurology, 50*(2), 351–355.
Chapman, S. B., Highley, A. P., & Thompson, J. L. (1998). Discourse in fluent aphasia and Alzheimer's disease: Linguistic and pragmatic considerations. *Journal of Neurolinguistics: Special Issue, 11*, 55–78.
Clark, D. G., Charuvastra, A., Miller, B. L., Shapira, J. S., & Mendez, M. F. (2005). Fluent versus nonfluent primary progressive aphasia: A comparison of clinical and functional neuroimaging features. *Brain and Language, 94*, 54–60. [PubMed: 15896383].
Chertkow, H., Bub, D., Deaudon, C., & Whitehead, V. (1997). On the status of object concepts in aphasia. *Brain and Language, 58*(2), 203–232.
Cotelli, M., Borroni, B., Manenti, R., Zanetti, M., Aravalo, A., Cappa, S. F., et al. (2007). Action and object naming in Parkinson's disease without dementia. *European Journal of Neurology, 14*(6), 632–637. [PubMed: 17539940].
Cummings, J. L., Benson, D. F., Hill, M. A., et al. (1985). Aphasia in dementia of the Alzheimer type. *Neurology, 35*, 394–397.
Cycyk, L. M., & Wright, H. H. (2008). Frontotemporal dementia: Its definition, differential diagnosis, and management. *Aphasiology, 22*(4), 422–444.
Doubleday, E. K., Snowden, J. S., Varma, A. R., & Neary, D. (2002). Qualitative performance characteristics differentiate dementia with Lewy bodies and Alzheimer's disease. *Journal of Neurology, Neurosurgery, and Psychiatry, 72*(5), 602–607.
Downs, M., Clare, L., & Mackenzie, J. (2006). Understandings of dementia: Exploratory models and their implications for the person with dementia and therapeutic effort. In J. C. Hughes, S. J. Louw, & S. R. Sabat (Eds.), *Dementia: Mind, meaning and the person* (pp. 235–258). Oxford University Press.

Doyle, P. J., & Rubinstein, R. L. (2013). Person-centered dementia care and the cultural matrix of othering. *The Gerontologist, 54*, 952–964.

Dunn, L. A., & Dunn, L. M. (2006). Peabody picture vocabulary test-4: Pearson.

Ehrlich, J. S., Obler, L. K., & Clark, L. (1997). Ideational and semantic contributions to narrative production in adults with dementia of the Alzheimer's type. *Journal of Communication Disorders, 30*(2), 79–99.

Emre, M., Aarsland, D., Brown, R., Burn, D. J., Duyckaerts, C., Mizuno, Y., et al. (2007). Clinical diagnostic criteria for dementia associated with Parkinson's disease. *Movement Disorders, 22*(12), 1689–1707. [PubMed: 17542011].

Evans, I. E. M., et al. (2019). Social isolation and cognitive function in later life: A systematic review and meta-analysis. *Journal of Alzheimer's Disease: JAD. 70*(s1), S119–S144.https://doi.org/10.3233/JAD-180501.

Eviatar, Z., & Just, M. A. (2006). Brain correlates of discourse processing: An fMRI investigation of irony and conventional metaphor comprehension. *Neu- Ropsychologia, 44*, 2348–2359.

Feldman, H. H., & Woodward, M. (2005). The staging and assessment of moderate to severe Alzheimer disease. *Neurology, 65*, S10–S17.

Ferris, S. H., & Farlow, M. (2013). Language impairment in Alzheimer's disease and benefits of acetylcholinesterase inhibitors. *Clinical Interventions in Aging, 8* 1007–1014. https://doi.org/10.2147/CIA.S39959.

Field, T. S., Masrani, V., Murray, G., & Carenini, G. (2017). Improving diagnostic accuracy of Alzheimer's disease from speech analysis using markers of hemispatial neglect. *Alzheimer's & Dementia: THe Journal of the Alzheimer's Association, 13*(7), 157–158.

Folstein, M., Folstein, S., & McHugh, P. R. (1975). Mini-mental state: A practical method for grading the cognitive state of patients for the clinician. *Journal of Psychiatric Research, 12*, 189–198.

Forman, M. S., Farmer, J., Johnson, J. K., Clark, C. M., Arnold, S. E., Coslett, H. B., et al. (2006). Frontotemporal dementia: Clinicopathological correlations. *Annals of Neurology, 59*(6), 952–962. [PubMed: 16718704].

Fratiglioni, L., Launer, L. J., Andersen, K., Breteler, M. M., Copeland, J. R., & Dartigues, J. F. et al. (2000). Incidence of dementia and major subtypes in Europe: A collaborative study of population-based cohorts. *Neurologic Diseases in the Elderly Research Group. Neurology, 54*(11 Suppl 5), S10–15. [PubMed: 10854355].

Freitas, M. I. D., et al. (2018). Linguistic abilities in major vascular cognitive impairment: a comparative study with Alzheimer's disease. *Acta Neurologica Belgica, 118*(3), 465–473. https://doi.org/10.1007/s13760-018-0977-x.

Galasko, D., Katzman, R., Salmon, D. P., & Hansen, L. (1996). Clinical and neuropathological findings in Lewy body dementias. *Brain and Cognition, 31*(2), 166–175. [PubMed: 8811993].

Gallacher, J., Bayer, A., & Ben-Shlomo, Y. (2005). Commentary: Activity each day keeps dementia away: Does social interaction really preserve cognitive function? *International Journal of Epidemiology* [advance access May 23, 2005], 1–2.

Garrard, P., Maloney, L. M., Hodges, J. R., & Patterson, K. (2005). The effects of very early Alzheimer's disease on the characteristics of writing by a renowned author. *Brain, 128*(2), 250–260.

Geser, F., Wenning, G. K., Poewe, W., & McKeith, I. (2005). How to diagnose dementia with Lewy bodies: State of the art. *Movement Disorders, 20*(Suppl 12), 11–20. [PubMed: 15455447].

Gibb, W. R., Luthert, P. J., Janota, I., & Lantos, P. L. (1989). Cortical Lewy body dementia: Clinical features and classification. *Journal of Neurology, Neurosurgery, and Psychiatry, 52*(2), 185–192.

Glei, D. A., Landa, D. A., Goldman, N., Chuang, Y.-L., Rodriguez, G., & Weinstein, M. (2005). Participating in social activities helps preserve cognitive function: An analysis of a longitudinal, population-based study of the elderly. *International Journal of Epidemiology* [advance access March 11, 2005], 1–8.

Glosser, G., Freidman, R., Kohn, S. E., Sands, L., & Grugan, P. (1998). Cognitive mechanisms for processing nonwords: Evidence from Alzheimer's disease. *Brain and Language, 63*(1), 32–49.

Goffman, E. (2006). *Estigma. La Identidad Deteriorada.*
Gonnerman, L. M., Andersen, E. S., Devlin, J. T., Kempler, D., & Seidenberg, M. S. (1997). Double dissociation of semantic categories in Alzheimer's disease. *Brain and Language, 57*(2), 254–279.
Goodglass, H., Kaplan, E., & Barresi, B. (2001). *Boston diagnostic aphasia examination* (3rd ed.). Lippincott Williams & Wilkins.
Gorno-Tempini, M., Hillis, A. E., Weintraub, S., Kertesz, A., Mendez, M., Cappa, S. F., et al. (2011). Classification of primary progressive aphasia and its variants. *Neurology, 76*, 1006–1014. https://doi.org/10.1212/WNL.0b013e31821103e6.
Grober, E., & Bang, S. (1995). Sentence comprehension in Alzheimer's disease. *Developmental Neuropsychology, 11*(1), 95–107.
Grossman, M., & Ash, S. (2004). Primary progressive aphasia: A review. *Neurocase, 10*, 3–18. [PubMed: 15849155].
Grossman, M., D'Esposito, M., Hughes, E., Onishi, K., Biassou, N., White-Devine, T., et al. (1996). Language comprehension profiles in Alzheimer's disease, multi-infarct dementia, and frontotemporal degeneration. *Neurology, 47*(1), 183–189.
Grossman, M., Kalmanson, J., Bernhardt, N., Morris, J., Stern, M. B., & Hurtig, H. (2000). Cognitive resource limitations during sentence comprehension in Parkinson's disease. *Brain and Language, 73*(1), 1–16. [PubMed: 10872635].
Grossman, M., Koenig, P., DeVita, C., Glosser, G., Moore, P., Gee, J., et al. (2003). Neural basis for verb processing in Alzheimer's disease: An fMRI Study. *Neuropsychology, 17*(4), 658–674.
Grossman, M., Lee, C., Morris, J., Stern, M. B., & Hurtig, H. I. (2002). Assessing resource demands during sentence processing in Parkinson's disease. *Brain and Language, 80*(3), 603–616. [PubMed: 11896660].
Grossman, M., & Moore, P. (2005). A longitudinal study of sentence comprehension difficulty in primary progressive aphasia. *Journal of Neurology, Neurosurgery and Psychiatry, 76*, 644–649.
Hamilton, H. E. (2005). Epilogue: The prism, the soliloquy, the couch, and the dance: The evolving study of language and Alzheimer's disease. In B. H. Davis (Ed.), *Alzheimer talk, text and context: Enhancing communication* (pp. 224–246). Palgrave Macmillan.
Hamilton, H. E. (2008). Language and dementia: Sociolinguistic aspects. *Annual Review of Applied Linguistics, 28*, 91–110. https://doi.org/10.1017/S0267190508080069.
Hardy, C. J. D. et al. (2016). The language profile of behavioral variant frontotemporal dementia. *Journal of Alzheimer's Disease: JAD, 50*(2), 359–371. https://doi.org/10.3233/JAD-150806.
Henry, J. D., & Crawford, J. R. (2004). Verbal fluency deficits in Parkinson's disease: A meta-analysis. *Journal of the International Neuropsychological Society, 10*(4), 608–622. [PubMed: 15327739].
Hinton, L., Flores, Y., Fraz, C., Hernández, I., & Mittenes, L. (2006). The borderlands of primary care. Physician and family perspectives on "Troublesome" Behaviors of people with dementia. In A. Liebing & L. Cohen (Eds.), *Thinking about dementia: Culture, loss, and the anthropology of senility* (pp. 43–63). Rutgers University Press.
Hochstadt, J., Nakano, H., Lieberman, P., & Friedman, J. (2006). The roles of sequencing and verbal working memory in sentence comprehension deficits in Parkinson's disease. *Brain and Language, 97*(3), 243–257. [PubMed: 16332387].
Hodges, J. R., Bozeat, S., Lambon, R. M. A., Patterson, K., & Spatt, J. (2000). The role of conceptual knowledge in object use evidence from semantic dementia. *Brain: A Journal of Neurology, 123*, 1913–1925. [PubMed: 10960055].
Hodges, J. R., Davies, R., Xuereb, J., Kril, J., & Halliday, G. (2003). Survival in frontotemporal dementia. *Neurology, 61*(3), 349–354. [PubMed: 12913196].
Hodges, J. R., & Patterson, K. (1996). Nonfluent progressive aphasia and semantic dementia: A comparative neuropsychological study. *Journal of International Neurological Society, 2*, 511–524.
Hodges, J. R., Patterson, K., Graham, N. L., & Dawson, K. (1996). Naming and knowing in dementia of Alzheimer's type. *Brain and Language, 54*(2), 302–325.

Huber, W., Poeck, K., & Willmes, K. (1984). The aachen aphasie test. In F. C. Rose (Ed.), *Advances in neurology* (vol. 42). *Progress in Aphasiology*. Raven Press.

Hur, M. R. & Caixeta, L. (2013). Non-literal language and semantic dementia. *Dementia & Neuropsychologia, 7*(4), 424–427. https://doi.org/10.1590/S1980-57642013DN74000011.

Jefferies, E., & Lambon, M. A. R. (2006). Semantic impairment in stroke aphasia versus semantic dementia: A case-series comparison. *Brain, 129*(8), 2132–2147.

Johari, K., et al. (2019). A dissociation between syntactic and lexical processing in Parkinson's disease. *Journal of Neurolinguistics, 51,* 221–235.https://doi.org/10.1016/j.jneuroling.2019.03.004.

Jones, S., Laukka, E. J., & Backman, L. (2006). Differential verbal fluency deficits in the preclinical stages of Alzheimer's disease and vascular dementia. *Cortex, 42*(3), 347–355. [PubMed: 16771040].

Kaplan, E., Goodglass, H., & Weintraub, S. (2001). *Boston Naming Test* (2nd ed.). Lippincott Williams & Wilkins.

Kavé, G., Leonard, C., Cupit, J., & Rochon, E. (2007). Structurally well-formed narrative production in the face of severe conceptual deterioration: A longitudinal case study of a woman with semantic dementia. *Journal of Neurolinguistics, 20*, 161–177.

Kay, J. L. R. (1992). *Psycholinguistic Assesment of Language Processing in Aphasia.* Lawrence Erlbaum Associates.

Kemper, S., Greiner, L. H., Marquiz, J. G., Prenovost, K., & Mitzner, T. L. (2001). Language decline across the life span: Findings from the nun study. *Psychology and Aging, 16*, 227–239.

Kemper, S. & Altmann, L. J. P. (2009). Dementia and language. In R. Larry (Ed.), *Squire, encyclopedia of neuroscience* (pp. 409–414). Academic Press. https://doi.org/10.1016/B978-008045046-9.01873-8.

Kempler, D., Almor, A., Tyler, L. K., Andersen, E. S., & MacDonald, M. C. (1998). Sentence comprehension deficits in Alzheimer's disease: A comparison of off-line vs. on-line sentence processing. *Brain and Language, 64*(3), 297–316.

Kempler, D., & Goral, M. (2008). Language and Dementia: Neuropsychological Aspects. *Annual Review of Applied Linguistics, 28*, 73–90. https://doi.org/10.1017/S0267190508080045.

Kertesz, A. (2007). *The Western aphasia battery-revised*. The Psychological Corporation.

Klimova, B., & Kuca, K. (2016). Speech and language impairments in dementia. *Journal of Applied Biomedicine, 14*(2) (Apr 2016), 97–103.

Korchounov, A., Schipper, H. I., Preobrazhenskaya, I. S., Kessler, K. R., & Yakhno, N. N. (2004). Differences in age at onset and familial aggregation between clinical types of idiopathic Parkinson's disease. *Movement Disorders, 19*(9), 1059–1064. [PubMed: 15372596].

Lambon Ralph, M. A., Ellis, A. W., & Franklin, S. (1995). Semantic loss without surface dyslexia. *Neurocase, 1*(4), 363–369.

Leibing, A. (2006). Divided gazes: Alzheimer's disease, the person within, and death in life. In A. Leibing & L. Cohen (Eds.), *Thinking about dementia: Culture, loss, and the anthropology of senility* (pp. 240–268). Rutgers University Press.

Le Rhun, E., Richard, F., & Pasquier, F. (2006). Different patterns of Mini-Mental Status Examination responses in primary progressive aphasia and Alzheimer's disease. *European Journal of Neurology, 13*, 1124–1127.

Livingston, G. et al. (2020). Dementia prevention, intervention, and care: 2020 report of the Lancet Commission. *Lancet (London, England), 396*, 10248; 413–446. https://doi.org/10.1016/S0140-6736(20)30367-6.

Looi, J. C., & Sachdev, P. S. (1999). Differentiation of vascular dementia from AD on neuropsychological tests. *Neurology, 53*(4), 670–678. [PubMed: 10489025].

Lukatela, K., Malloy, P., Jenkins, M., & Cohen, R. (1998). The naming deficit in early Alzheimer's and vascular dementia. *Neuropsychology, 12*(4), 565–572. [PubMed: 9805326].

Maclagan, M., & Mason, P. (2005). Bad times and good times: Lexical variation over time in Robbie Walters' speech. In B. H. Davis (Ed.), *Alzheimer talk, text and context: Enhancing communication* (pp. 146–166). Palgrave Macmillan.

Masrani, V. (2018). Detecting dementia from written and spoken language. Electronic Theses and Dissertations (ETDs) 2008+. T, University of British Columbia.https://doi.org/10.14288/1.0362923.
Mesulam, M. M. (1982). Slowly progressive aphasia without generalized dementia. *Annals of Neurology, 11*, 592–598.
Mesulam, M. M. (2001). Primary progressive aphasia. *Annals of Neurology, 49*(4), 425–432.
Mesulam, M. M. (2003). Primary progressive aphasia - A language-based dementia. *New England Journal of Medicine, 349*(16), 1535–1542.
Mesulam, M. (2013). Primary progressive aphasia. A dementia of the language network. *Dement Neuropsychol, 7*(1), 2–9.
Mesulam, M. M., Rogalski, E., Wieneke, C., Cobia, D., Rademaker, A., Thompson, C., et al. (2009). Neurology of anomia in the semantic variant of primary progressive aphasia. *Brain, 132*(9), 2553–2565.
Mesulam, M.-M., Wieneke, C., Thompson, C., Rogalski, E., & Weintraub, S. (2012). Quantitative classification of primary progressive aphasia at early and mild impairment stages. *Brain, 135*, 1537–1553.
McKhann, G. M., et al. (2011). The diagnosis of dementia due to Alzheimer's disease: Recommendations from the National Institute on Aging-Alzheimer's Association workgroups on diagnostic guidelines for Alzheimer's disease. *Alzheimer's & Dementia, 7*(3), 263–269.
Monetta, L., & Pell, M. D. (2007). Effects of verbal working memory deficits on metaphor comprehension in patients with Parkinson's disease. *Brain and Language, 101*(1), 80–89. [PubMed: 16875726].
Morello, A. N. da C. et al. (2017). Language and communication non-pharmacological interventions in patients with Alzheimer's disease: a systematic review. Communication intervention in Alzheimer. *Dementia & Neuropsychologia, 11*(3) 227–241. https://doi.org/10.1590/1980-57642016dn11-030004.
Muñoz, M., Pérez-Santos, E., Crespo, M., & Guillén, A. I. (2009). *El estigma de la enfermedad mental*; Ed. Complutense.
Murdoch, E. B., Chenery, J. H., Wilks, V., & Boyle, S. R. (1987). Language disorders in dementia of the Alzheimer type. *Brain and Language, 31*(1), 122–137.
Neary, D., Snowden, J. S., Gustafson, L., Passant, U., Stuss, D., Black, S., et al. (1998). Frontotemporal lobar degeneration: A consensus on clinical diagnostic criteria. *Neurology, 51*(6), 1546–1554. [PubMed: 9855500].
Ober, B. A., & Shenaut, G. K. (1999). Well-organized conceptual domains in Alzheimer's disease. *Journal of the International Neuropsychological Society, 5*(7), 676–684.
Osher, J., Wicklund, A., Rademaker, A., Johnson, N., & Weintraub, S. (2007). The Mini- Mental State Examination in behavioral variant frontotemporal dementia and primary progressive aphasia. *American Journal of Alzheimer's Disease and Other Dementias, 22*, 468–473.
Patterson, C. (2018). *World Alzheimer Report 2018 - The state of the art of dementia research: New frontiers*. Alzheimer's Disease International.
Patterson, K., & Lambon Ralph, M. A. (1999). Selective disorders of reading? *Current Opinion in Neurobiology, 9*(2), 235–239.
Peelle, J. E., Cooke, A., Moore, P., Vesely, L., & Grossman, M. (2007). Syntactic and thematic components of sentence processing in progressive nonfluent aphasia and nonaphasic frontotemporal dementia. *Journal of Neurolinguistics, 20*(6), 482–494. [PubMed: 18978927].
Perquin, M. et al. (2013). Lifelong exposure to multilingualism: New evidence to support cognitive reserve hypothesis. *PloS one 8*(4), e62030. https://doi.org/10.1371/journal.pone.0062030.
Piatt, A. L., Fields, J. A., Paolo, A. M., Koller, W. C., & Troster, A. I. (1999). Lexical, semantic, and action verbal fluency in Parkinson's disease with and without dementia. *Journal of Clinical Experimental Neuropsychology, 21*(4), 435–443.
Pulvermüller, F., et al. (2010). The word processing deficit in semantic dementia: all categories are equal, but some categories are more equal than others. *Journal of Cognitive Neuroscience, 22*(9), 2027–2041. https://doi.org/10.1162/jocn.2009.21339.

Rao, S. S., Hofmann, L. A., & Shakil, A. (2006). Parkinson's disease: Diagnosis and treatment. *American Family Physician, 74*(12), 2046–2054. [PubMed: 17186710].

Rapp, B., & Caramazza, A. (1993). On the distinction between deficits of access and deficits of storage: A question of theory. *Cognitive Neuropsychology, 10*(2), 113–141.

Rascovsky, K., Hodges, J. R., Kipps, C. M., Johnson, J. K., Seeley, W. W., Mendez, M. F., et al. (2007). Diagnostic criteria for the behavioral variant of frontotemporal dementia (bvFTD): Current limitations and future directions. *Alzheimer's Disease and Associated Disorders, 21*(4), S14-s18.

Ryan, E. B., Spykerman, H., & Anas, A. P. (2005). Writers with dementia: The interplay among reading, writing, and personhood. In B. H. Davis (Ed.), *Alzheimer talk, text and context: Enhancing communication* (pp. 190–198). Palgrave Macmillan.

Reilly, J., et al. (2010). Cognition, language, and clinical pathological features of non-Alzheimer's dementias: an overview. *Journal of communication disorders, 43*(5), 438–452. https://doi.org/10.1016/j.jcomdis.2010.04.011.

Reilly, J., et al. (2011). Anomia as a marker of distinct semantic memory impairments in Alzheimer's disease and semantic dementia. *Neuropsychology, 25*(4), 413–26. https://doi.org/10.1037/a0022738.

Riedel, O., Klotsche, J., Spottke, A., Deuschl, G., Forstl, H., Henn, F., et al. (2008). Cognitive impairment in 873 patients with idiopathic Parkinson's disease: Results from the German Study on Epidemiology of Parkinson's Disease with Dementia (GEPAD). *Journal of Neurology, 255*, 255–264. [PubMed: 18204803].

Rockwood, K., Bowler, J., Erkinjuntti, T., Hachinski, V., & Wallin, A. (1999). Subtypes of vascular dementia. *Alzheimer's Disease and Associated Disorders, 13*(Suppl 3), S59-65.

Rogers, T. T., Ivanoiu, A., Patterson, K., & Hodges, J. R. (2006). Semantic memory in Alzheimer's disease and the frontotemporal dementias: A longitudinal study of 236 patients. *Neuropsychology, 20*(3), 319–335.

Rouseeaux, M., Seve, A., Vallet, M., Pasquier, F., & Mackowiak- Cordoliani, M. A. (2010). An analysis of communication in conversa- tion in patients with dementia. *Neuropsychologia, 48*(13), 3884–3890.

Santos, M. T. F., Sougey, E. B., & Alchieri, J. C. (2009). Validity and reliability of the screening test for Alzheimer's disease with proverbs (STADP) for the elderly. *Arquivos De Neuro-Psiquiatria, 67*, 836–842.

Savage, S., et al. (2013). Distinguishing subtypes in primary progressive aphasia: Application of the Sydney language battery. *Dementia and Geriatric Cognitive Disorders, 35*(3–4), 208–218. https://doi.org/10.1159/000346389.

Shallice, T., Warrington, E. K., & McCarthy, R. A. (1983). Reading without semantics. *Quarterly Journal of Experimental Psychology a: Human Experimental Psychology, 35A*(1), 111–138.

Snowden, J. S., & Neary, D. (1993). Progressive language dysfunction and lobar atrophy. *Dementia, 4*(3–4), 226–231.

Snyder, L. (2006). Personhood and interpersonal communication in dementia. In J. C. Hughes, S. J. Louw, & S. R. Sabat (Eds.), *Dementia: Mind, meaning and the person* (pp. 259–276). Oxford University Press.

Sommers, M. S. (1998). Spoken word recognition in individuals with dementia of the Alzheimer's type: Changes in talker normalization and lexical discrimination. *Psychology and Aging, 13*(4), 631–646.

Stanyon, R. M., Griffiths, A., Thomas, A. S., & Gordon, A. L. (2016). The facilitators of communication with people with dementia in a care setting: An interview study with healthcare workers. *Age and Ageing, 45*(1), 164–170.

Stern, Y. (2009). Cognitive reserve. *Neuropsychologia, 47*(10), 2015–2028. https://doi.org/10.1016/j.neuropsychologia.2009.03.004.

Stern, Y. (2013). Cognitive reserve: Implications for assessment and intervention. *Folia Phoniatrica Et Logopaedica : Official Organ of the International Association of Logopedics and Phoniatrics (IALP), 65*(2), 49–54. https://doi.org/10.1159/000353443.

Sweeting, H., & Gilhooly, M. (1997). Dementia and the phenomenon of social death. *Sociology of Health & Illness, 19*, 93–117.
Tang-Wai, D. F., & Graham, N. L. (2008). Assessment of language function in dementia. *Geriatr Aging, 11*, 103–110.
Tierney, M. C., Black, S. E., Szalai, J. P., Snow, W. G., Fisher, R. H., Nadon, G., et al. (2001). Recognition memory and verbal fluency differentiate probable Alzheimer disease from subcortical ischemic vascular dementia. *Archives of Neurology, 58*(10), 1654–1659. [PubMed: 11594925].
Thompson, C. K. (2011). Northwestern assessment of verbs and sentences (NAVS). In I. L. Evanston (Ed.), Northwestern University.
Tomoeda, C. K., & Bayles, K. A. (1993). Longitudinal effects of Alzheimer disease on discourse production. *Alzheimer Disease & Associated Disorders, 7*(4), 223–236.
Tomoeda, C. K., Bayles, K. A., Trosset, M. W., Azuma, T., et al. (1996). Cross-sectional analysis of Alzheimer disease effects on oral discourse in a picture description task. *Alzheimer Disease & Associated Disorders, 10*(4), 204–215.
Van Dijk, T. (2016). *Discurso y Conocimiento. Una Aproximación Sociocognitiva.*
Vesely, L., Bonner, M. F., Reilly, J., & Grossman, M. (2007). Free association in semantic dementia: The importance of being abstract. *Brain and Language, 103*, 154–155.
Villar, F., et al. (2019). Giving them a voice: Challenges to narrative agency in people with dementia. *Geriatrics (Basel, Switzerland), 4*(1), 20. February 12, 2019. https://doi.org/10.3390/geriatrics4010020.
Wang, H. et al. (2020). Dementia care during COVID-19. *Lancet (London, England), 395*(10231), 1190–1191. https://doi.org/10.1016/S0140-6736(20)30755-8.
Warrington, E. K. (1975). The selective impairment of semantic memory. *Quarterly Journal of Experimental Psychology, 27*, 635–657. [PubMed: 1197619].
Ward, R., Vass, A. A., Aggarwal, N., Garfield, C., & Cybyk, B. (2008). A different story: Exploring patterns of communication in residential dementia care. *Ageing & Society, 28*, 629–651.
Warrington, E. K., & Shallice, T. (1979). Semantic access dyslexia. *Brain, 102*, 42–63.
Warrington, E. K. (1981). Concrete word dyslexia. *British Journal of Psychology, 72*(2), 175–196.
Weiner, M. F., Neubecker, K. E., Bret, M. E., & Hynan, L. S. (2008). Language in Alzheimer's disease. *Journal of Clinical Psychiatry, 69*(8), 1223–1227.
Weintraub, S., Rubin, N. P., & Mesulam, M. M. (1990). Primary progressive aphasia. Longitudinal course, neuropsychological profile, and language features. *Archives of Neurology, 47*, 1329–1335. [PubMed: 2252450].
Weintraub, S., Mesulam, M.-M., Wieneke, C., Rademaker, A., Rogalski, E. J., & Thompson, C. K. (2009). The Northwestern Anagram Test: Measuring sentence production in primary progressive aphasia. *American Journal of Alzheimer's Disease and Other Dementias, 24*, 408–416.
Whatmough, C., Chertkow, H., Murtha, S., Templeman, D., Babins, L., & Kelner, N. (2003). The semantic category effect increases with worsening anomia in Alzheimer's type dementia. *Brain and Language. Special Issue: Meaning in language, 84*(1), 134–147.
Wiig, E. H., & Secord, W. (1989). *Test of language competence—Expanded Edition.* The Psychological Corporation.
World Health Organization. (1992). *International classification of diseases and related health problems, 10th revision.* World Health Organization.
Yakhno, N. N., Zaharov, V. V., Lokshina, A. B., Koberskaya, N. N., & Mhitaryan, E. A. (2013). *Dementias: Guidelines for doctors* (p. 264). 4th Eds, MEDpress-infrom.
Zeiling, H. (2013). Dementia as a cultural metaphor. *The Gerontologist, 54*, 258–267.
Zimmermann, M. (2017). Alzheimer's disease metaphors as mirror and lens to the stigma of dementia. *Literature and Medicine, 35*, 71–97.

Chapter 7
Psychosocial Rehabilitation of Persons with Dementia in Indian Context

S. Shaji

Abstract Dementia is a disease characterized by progressive deterioration. A therapeutic nihilism exists regarding interventions even among the healthcare professionals. So the focus of psychosocial rehabilitation is to help the persons with dementia to lead a better life and to reduce the distress of the caregivers. The main aim of management is to treat the symptoms which are amenable to treatment and to stabilize the condition of the patients by controlling the factors contributing to the deterioration. Psychosocial interventions include cognitive rehabilitation, management of behavioural problems, stimulation and emotion-oriented interventions to improve functioning and to reduce the distressing symptoms. Empowering the family members to look after their relatives is important in rehabilitation activities. Educating family members regarding the nature, course and prognosis of the disease, developing skills and attitudes necessary for caring people with dementia and developing strategies to manage the distressing symptoms of dementia are the major components. Distress among the caregivers can be reduced by individual counselling and participation in support groups. Training of professionals, paraprofessionals, health workers and community volunteers is an essential prerequisite for the implementation of community-based programmes. A multidisciplinary team approach is essential. Services for dementia are non-existent or far from adequate in Indian context. Services like memory disorders clinic, day care centres, respite care and institutions for long-term care are needed. Culturally appropriate, cost-effective models of care suited to Indian context is important.

Dementia is emerging as an important health problem in India along with the growing population of elderly people. Prevalence studies conducted in different parts of India confirmed its presence (Chandra et al., 1998; Mathuranath et al., 2010; Rajkumar et al., 1997; Shaji et al., 1996, 2005; Vas et al., 2001). India is a populous country so that the number of people affected with dementia is significant rather than its percentage. Clinicians, researchers and other experts working in this area have a general consensus regarding following points.

S. Shaji (✉)
Bethsada Hospital, Perumbavoor, Kerala, India

M. K. Shankardass (ed.), *Dementia Care*,
https://doi.org/10.1007/978-981-16-3864-0_7

1. Dementia is not identified as a disease condition which requires intervention. There is a commonly held belief in the community that it is an inevitable state associated with aging so that people allow it to progress in its own way (Dementia-a Public Health Priority, 2012; India & Report, 2010; Jacob, 2007; Shaji, 2002).
2. Lack of awareness about the disease leads to stigmatization, misinterpretation of symptoms and mismanagement. The abnormal and aberrant behaviours exhibited by people with dementia are seen as psychiatric or mad, so family members attempt to hide it. Socio-economic and cultural variables contribute to shaping knowledge, attitudes and beliefs about dementia (Garvey, 2011; Low & Anstey, 2009; Shaji, 2010).
3. The healthcare system in India is not sensitive enough to address the problems associated with dementia. There exists a therapeutic nihilism even among the healthcare personnel with a general feeling that nothing can be done (Benbow & Reynolds, 2000; Clare et al, 2006; Quantitative & Qualitative Research in to Stigma & Dementia, 2008). The primary care physicians and health workers in the community are neither equipped nor empowered to meet the challenges associated with this illness (Dementia. Out of the Shadows, 2008; Jolly & Benbow, 2000).
4. Specific services for people with dementia are rare or non-existent in India. Developing services attains significance in the context of increasing number of elderly people in our country along with the rapid social changes that is going on.

Planning and implementation of any rehabilitation activities in relation to dementia must consider the above-mentioned ground realities. The concept of rehabilitation is usually applicable to incurable illness and is aimed to maximization of recovery opportunities and minimization of disabilities. The aim of rehabilitation in the context of dementia is to enable them to participate in everyday activities in a way that is meaningful to them (Clare, 2017).

The major components of rehabilitation of people with dementia include:

1. Early diagnosis
2. Monitoring the general health and well-being
3. Maintaining the existing abilities and preventing decline in functioning
4. Controlling the distressing behavioural problems and psychiatric symptoms associated with dementia
5. Caregiver interventions aimed to reduce distress and long-term support to cope up with the disease process.

The Importance of Making Early Diagnosis

The early diagnosis of the disease contributes significantly to effective management. There are a lot of barriers which make early diagnosis difficult which include low awareness about dementia, misconceptions, stigmatization and lack of institutions for making the diagnosis. Better understanding of the disease, its course, prognosis and its various implications will promote better adjustment, adaptation and coping strategies from the family members.

Monitoring the General Health and Well-Being

Models of intervention developed in the west or other cultures may not be suitable in Indian context. It should be adapted or modified to suit the sociocultural context of our country (Dias et al., 2008). The intervention should be flexible, tailor suited or individualized considering the characteristics of the person and the environment. Deterioration of the general health condition can contribute to decline of functioning. Periodic medical checkup, controlling the blood pressure, blood sugar level, blood cholesterol levels, correcting thyroid dysfunction, nutritional deficiencies and electrolyte imbalance can be beneficial for maintaining the health status. Promoting a healthy lifestyle with adequate exercise and activities, avoidance of alcohol and smoking can contribute to the well-being of the persons.

Once the diagnosis is made, it is appropriate to provide the person with a structured and predictable environment. This may provide more comfort, for example a room with essential and minimum number of furniture with adequate lighting and safety precautions is preferable than a dim lit room with crowded furniture. Establishing and maintaining a routine may avoid confusion. Though a person is diagnosed as having dementia, encourage and support him or her to live an independent life as long as possible. Cognitive stimulation, activities and social stimulation may help to maintain the self-esteem and contribute to the sense of well-being. The persons may be helped to utilize their existing abilities to the maximum extent. The caregivers are supposed to provide the required assistance needed by the persons for an independent living. The family members and caregivers should understand that the behavioural problems and psychological symptoms exhibited by the persons with dementia are products of the disease process so that they can avoid a confrontational approach.

The persons with dementia may not be able to carry out many of the complex activities that had been done in the past. Simplifying the tasks to fit to their existing abilities can be comfortable to the affected person. Persons with memory problems are vulnerable to accidents, falls and injuries so that utmost care should be given to prevent such untoward happenings. Correction of the sensory problems, modification of the environment in a dementia friendly way and provision of support aimed at the overall safety of the person is essential. General physical health and fitness should be maintained, and prompt treatment of infections and early intervention of disease conditions will be of help to preserve the quality of life.

There is a common error in judgement from the part of the caregivers to fall in to the trap of concentrating on the illness and ignoring the person who was there before the illness. Many assessments concentrate on things person cannot do rather than the things which they do or perform. It is necessary to know the activities a person can no longer perform without assistance to create a proper care plan, but the emphasis should be on the things a person can still do and concentrate on them so that the person may feel better himself and will help to preserve self-esteem. A good care provider is supposed to know as much about the person as possible (Tibbs, 1998).

The communication skills and language abilities are progressively affected in many cases of dementia, and caregivers should take into consideration the deficits and impairments in communication while interacting with the person (Waal et al., 2013). Many factors like sensory impairments, neurological language and speech problems, depression and confusion can make it difficult. Communication is vitally important to the well-being of the person with dementia, and it depends largely on the caregiver. Listening to them, giving time to communicate and care and attention given to the person are important.

Specific Interventions

Behavioural disturbances and psychiatric symptoms are seen in a significant proportion of persons with dementia which make the caregiving difficult. Behavioural disturbances and psychiatric symptoms can take various forms which include psychiatric symptoms like delusions, hallucinations and activity disturbances like wandering, purposeless activities, aggressiveness, verbal outbursts, violence, agitation, sleep disturbances, affective disturbances, anxieties and phobias. Managing the behavioural disturbances is most difficult and distressing part of managing patients for majority of the caregivers. Specific interventions can be of help in this context (Deutsch & Rowner, 1991; Rabins et al., 1982; Shaji et al., 2009a; Steele et al., 1990).

The interventions can be broadly categorized as different approaches.

a. Cognition-oriented approaches
b. Emotion-oriented approaches
c. Behaviour-oriented approaches
d. Stimulation-oriented approaches.

Cognition-Oriented Approaches

Cognitive deterioration is inevitable in most of cases of dementia. Enhancing the existing abilities and maintaining it to prevent early deterioration is the principle behind this approach. Various types of memory aids are useful in this context. Traditional memory aids include calendars, diary, wall charts and notice boards in a visible area. Electronic memory aids (assistive technology) like alarm clocks, kitchen timer, computer, tablets, various apps, mobile phones, locators and GPS device can be used in this context. This method is applicable in mild cases and is aimed to help the person to lead an independent life as long as possible. Reality orientation can be tried in institutions.

Emotion-Oriented Approaches

Persons with dementia tend to live in the past as their remote memory is comparatively preserved. Life review process is a common psychological phenomenon that is associated with ageing in a lifespan perspective. Robert Butler, a geriatrician, incorporated life review process in psychotherapy of elderly in 1960s. Validation theory states that seemingly irrelevant and abnormal behaviour exhibited by people with dementia has hidden meanings that are significant to them. Naomi Feil introduced validation therapy. The therapist tries to enter the phenomenological field of person through empathy, establishes communication, utilizes the techniques of reassurance, distraction, emotional support and validates the emotional ties with the past.

Behaviour-Oriented Approaches

ABC model of behavioural management is one in which A is the antecedent, B is behaviour itself and C is the consequence. It involves observation and analysis of a particular behaviour exhibited by persons with dementia. Antecedents are environmental events, triggers or situations immediately preceding the target behaviour. Identification of the causes of a particular maladaptive behaviour will help to control the behaviour and its consequences. Comorbid stressors are also to be taken into consideration.

Stimulation-Oriented Approaches

Cognitive, physical and social stimulations are important in the prevention, promotion and preservation of cognitive abilities and overall well-being of the person. This involves sensory stimulation like massage, aroma therapy, pet therapy, doll therapy, etc. Therapeutic activities involve recreational activities, exercise, music, games, gardening and spiritual activities like prayer and meditation. Various art forms like painting, creative writing, singing and handicrafts can be utilized for therapeutic purposes.

There exists a wide range of therapeutic techniques that can be effectively utilized in the psychosocial rehabilitation of person with dementia. But the approach selected should be tailor suited to fit to the rehabilitation needs of the persons and caregivers considering the context or situation in which it is implemented.

Psychosocial Rehabilitation—An Indian Experience

A psychosocial rehabilitation project for dementia was initiated by Alzheimer's and related disorders Society of India (ARDSI) in 1992. Dr. K. Jacob Roy, a pediatrician, based on his personal experience as a caregiver initiated the activities of the Society.

The rural project started in 1992 in Thiruvaniyoor, Ernakulam District. The main activities were an epidemiological survey to identify people suffering from dementia, home care for the identified cases, a geriatric clinic for the elderly people and training programme for geriatric health workers.

The Urban Community Dementia Care Services started functioning in 1996 in South Kalamassery, in a suburb of Kochi. The third project Comprehensive Dementia Care services, Kochi, started its functioning in 2000. All these projects provided various services and conducted research on Epidemiology and psychosocial interventions.

The essential prerequisite for the implementation of any community-based rehabilitation programme is:

1. A multidisciplinary team
2. Training of community volunteers or health workers
3. Participation of the community.

Comprehensive management of persons with dementia is a complex task with multiple components. Coordinated work of a team consisting of specialists like neurologist or psychiatrist, primary care physicians, nurses, geriatric health workers or community volunteers is necessary for the successful implementation of the program.

India is a country in which the number of elderly people with various morbidities is really high. It is difficult to depend on specialists, medical doctors or trained nurses to provide care for the people in the community. With the aim of empowering people in the community to take care of person with dementia, a training programme was initiated by ARDSI in 1992 with following objectives:

1. To equip them with skills and attitudes needed to take care of elderly people with memory problems
2. To prepare the family members to take up the challenging task of care giving and to provide them with necessary help as and when it is needed.

The training programme was certificate course of 12 months duration. Minimum qualification for joining the course was pass in tenth standard. The course received recognition from Christian Medical Association of India (CMAI). This programme continued in the urban centre also. Many persons providing care from homes for the elderly and other caregiving institutions received training, as a part of Comprehensive Dementia Care project volunteers selected from the community, were given short-term training and entrusted them with the responsibility of identification and primary interventions. The training programmes with the sponsorship of National Institute of Social Defiance (NISD) were also conducted.

Identification of dementia

Early identification of dementia is essential to effective management of dementia in any setting.

WHO survey identified five major themes in relation to raising awareness (Dementia-a Public Health Priority, 2012), these are;

a. Raising public awareness and understanding of dementia
b. Reducing the stigma of dementia and challenging discriminatory behaviour
c. Recognizing the early signs of dementia to aid early diagnosis
d. Living well with dementia
e. The importance of healthy lifestyle and reducing the risk.

ARDSI started its campaigns to raise the awareness since its beginning in 1992. This included various training programmes and workshops for primary care physicians, nurses, paraprofessionals, health workers and general public. Community health workers were instrumental in disseminating information in the community. They kept with them "A handbook for caregivers for dementia and other IEC materials with them while visiting the households. A global focus for awareness raising is World Alzheimer's Day (21 September) and World Alzheimer's month (September). ARDSI took up the initiative of observing the day with various programmes since its beginning like memory walk, walkathon, poster presentations, exhibitions, seminars and public meetings. Different media channels such as film, television, Internet and social media provide opportunities for disseminating positive images and messages about dementia". The Malayalam movie "Thanmathra" directed by Blessy made a great impact to create awareness about dementia in Kerala society. Similarly, "Asthu" a Marathi movie and "Black" in Hindi also created similar effects.

Domiciliary Care

Home Care is Aimed to Provide Better Care in a Familiar Setting

The geriatric health workers visited the households periodically and empowered the family members to look after the persons with dementia in a better way. An individualized care plan was formulated with the help of medical staff working in association with the project. The frequency of visits depended on the severity of illness and needs associated with caregiving for a particular person. The main responsibilities of the geriatric health workers are to educate the family members about the disease process, helping the affected persons to carry out their day today activities in a better way, providing practical suggestions to contain the behavioural problems and monitoring the physical and psychological status of the persons and interventions to reduce the caregiver's distress. Epidemiological survey conducted in the rural project identified sixty-seven cases of dementia. These cases were included for intervention. This pilot project revealed that community-based interventions are needed in our country. The main feedback was that such intervention was possible and feasible in Indian community.

Day Care

The day care services for persons with dementia have been going on since 1996 and catering it for persons in need of it. People preferred day care services in urban areas where most of the family members are working outside home. The day care provided a setting of supervised care in which physical and psychological status of the people was monitored closely. The activities include prayer, newspaper reading, reality orientation sessions, reminiscence, relaxation, refreshment, light exercises, group singing, etc.

Memory Clinic

Memory problems or forgetfulness is a common complaint among the elderly people. Many factors like depression, anxiety, physical diseases, endocrine, metabolic and nutritional problems can contribute memory problems. Memory clinics provide an easy access for the people to clarify their doubts and misunderstandings about dementia and differentiate whether it is due to a genuine disease process. It facilitates early identification of cases. Primary care physicians, health workers and community volunteers are the usual source of referral. The initial screening was done by a trained social worker. The identified cases were subjected detailed assessment, investigations and provided with comprehensive management which included family intervention.

Support Groups

Organizing caregiver's meeting is an important component of community-based rehabilitation of persons with dementia. Caregiver's meetings are organized in a venue that is convenient for the caregivers. Commonly held misconceptions about the disease, lack of understanding of the peculiarities of the disease condition, lack of strategies to deal with disturbing behaviours make caregiving a difficult task. Keeping unrealistic expectations about improvement may lead to frustration. Many caregivers may suffer psychological distress and physical exhaustion due to long-term care. Support group meeting provides an opportunity for the caregivers to come together. The caregiver understands that they are not the only one who is suffering and there are many other people who are struggling. This feeling of universality that occurs in group sessions may bring comfort. It is an opportunity to get scientific information about the disease process and to make better plans and strategies for care. Caregivers can share their experiences and feelings which will have a cathartic effect. Better problem solving strategies can be formulated from the inputs received from the group setting. A caregiver who is in need of individual counselling will be referred for further intervention. The whole sessions will be monitored by experts from the centre.

Help Line

A 24-h help line was introduced on the World Alzheimer's Day 2001. This provided easy access to the family members to receive needed help in time and facilitate the utilization of services.

There has been a remarkable change in the awareness and attitudes toward dementia across India during the past two decades. The activities initiated by ARDSI became the pioneering attempts in this field. Alzheimer's Disease International conference held in Kochi made significant impact. The first meeting of 10/66 Dementia research group was held in association with this conference was instrumental in starting many psychosocial research projects in India. The community-based identification of dementia cases utilizing multipurpose health workers in Thrissur and psychosocial intervention projects held in Goa are worth mentioning (Shaji et al., 2002). ARDSI has chapters all across the country especially in major cities and carries out activities for the welfare of persons with dementia which include memory clinics, day care centres, respite care and institutions for long-term care. The major limitation is that most of the dementia-related programmes are urban centred.

The possible solution in Indian setting is to integrate the geriatric care with the primary healthcare system. A reorientation of priorities with importance of geriatric or dementia care is required. It necessitates a change in policies and programmes. A public health approach based on the principles of inclusion, integration, equity and evidence is needed in this context (Ortiz et al., 2012).

References

Benbow, S. M., & Reynolds, D. (2000). Challenging the stigma of Alzheimer's Disease. *Hospital Medicine, 61* (3), 174–177.

Chandra, V., Ganguli, M., Pandav, R., et al. (1998). Prevalence of Alzheimer's Disease and other dementias in Rural India. *The Indo-US Study, Neurology, 51*(4), 100–1008.

Clare, L. (2017). *Public Library of Science Medicine, 14*(3), e1002245.

Clare, L., Goater, T., & Woods, B. (2006). Illness representation in early stage Dementia: A preliminary investigation. *International Journal of Geriatric Psychiatry, 21*, 761–767.

Dementia- a Public Health Priority. (2012). World Health Organization.

Dementia India Report. (2010). Alzheimer's and related disorders society of India.

Dementia. Out of the shadows. (2008). *Alzheimer's society.*

Deutsch, L. H., & Rowner, B. W. (1991). Agitation and other non-cognitive abnormalities in Alzheimer's disease. *Psychiatric Clinics of North America, 14*, 341–351.

De Waal, H., Lyketsos, C., Ames, D., & O'Brien, J. (2013). *Designing and delivering dementia services.* Wiley Online Library.

Dias, A, Dewey, M. E., & D-Souza, J., et al. (2008). Effectiveness of a homecare programme for supporting caregivers of persons with dementia in developing Countries: A randomized controlled trial from Goa. *Public Library Science One, 3*(6), e238.

Garvey, G., et al. (2011). Making sense of dementia: Understanding among indigenous Australians. *International Journal of Geriatric Psychiatry, 26*(5), 649–656.

Jacob, K. S. (2007). Can health workers diagnose Dementia in the Community? *Acta Psychiatrica Scandinavica, 8.*

Jolly, D. J., & Benbow, S. M. (2000). Stigma on Alzheimer's Disease, causes, consequences and a constructive approach. *International Journal of Clinical Practice, 64*(2), 117–119.

Low, L. F., & Anstey, K. J. (2009). Dementia literacy: Recognition and beliefs on dementia of the Australian Public. *Alzheimer's and Dementia, 5*(1), 43–49.

Mathuranath, P. S., Cherian, P. J., & Mathew, R. et al. (2010). Dementia in Kerala, South India: Prevalence and influence of age, education and gender. *International Journal of Geriatric Psychiatry, 25*, 290–297.

Ortiz, S., Luisa, A., Acosta-Castillo, I., & Prince, M. J. (2012). Epidemiology of dementias and Alzheimer's disease. *Archives of Medical Research.*

Quantitative and Qualitative Research in to Stigma and Dementia. (2008). *The gate.*

Rabins, P. V., Mace, N. L., & Lucas, M. J. (1982). The impact of dementia on the family. *Journal of American Clinical Association, 248*, 333–335.

Rajkumar, S., Kumar, S., & Thara, R. (1997). Prevalence of dementia in a rural setting a report from India. *International Journal of Geriatric Psychiatry, 12*(7), 702–707.

Shaji, S. (2002). Developing dementia care services in the community. In K. Baghi (Ed.), *Alzheimer's disease in India*, Society for Gerontological Research.

Shaji, S. (2010). Developing Dementia care in developing countries. In J. Hughes (Ed.), *Supportive care series on Dementia.* Oxford University Press.

Shaji, S., Pramodu, K., Abraham, T., Roy, J. K., & Varghese, A. (1996). An epidemiological study of Dementia in a rural community in Kerala, India. *British Journal of Psychiatry, 168*(6), 745–749.

Shaji, K. S., Arun Kishore, N. R., Lal, K. P., & Prince, M. (2002). Revealing a hidden problem. An evaluation of a community case finding programme from the Indian 10/66 dementia research network. *International Journal of Geriatric Psychiatry, 17*(3), 222–225.

Shaji, K. S., Smitha, K., Lal, K. P., & Prince, J. (2003). Caregivers of people with Alzheimer's Disease: A qualitative study from India 10/66 Dementia Research Network. *International Journal of Geriatric Psychiatry, 18*(1), 1–6.
Shaji, S., Bose, S., & Varghese, A. (2005). Prevalence of dementia in an Urban population in Kerala, India. *British Journal of Psychiatry, 186*, 136–140.
Shaji, S., Bose, S., & Kuriakose, S. (2009a). Behavioral and psychological symptoms of dementia: A study of symptomatology. *Indian Journal of Psychiatry, 58*, 38–41.
Shaji, K. S., George, R. K., Prince, M. J., & Jacob, K. S. (2009b). Behavioral symptoms and caregiver burden in Dementia. *Indian Journal of Psychiatry, 51*(i), 45.
Steele, C., Rovner, B., Chase, G. A., & Folstein, M. (1990). Psychiatric symptoms and nursing home placement of patients with Alzheimer's Disease. *American Journal of Psychiatry, 147*, 1045–1051.
Tibbs, M. A. (1998). Discovering the person, People with dementia Benson, S (ed). *Journal of Dementia Care*, 3rd Edition.
Vas, C. J., Pinto, C., & Paniker, D., et al. (2001). Prevalence of dementia in an urban Indian population. *International Psychogeriatrics, 13*(4), 439–450.

S. Shaji is working as Chief Psychiatrist in Bethsada Hospital, Perumbavoor. His main areas of research interests are Geriatric Psychiatry and Community Psychiatry. He worked as principal investigator in various research projects on Dementia. He was the Editor of Indian Psychiatric Society, Kerala Chapter during 2000–2004. He secured The Eapen-Samuel State award of IMA-Academy of medical specialties in 2005. At present he is the President of ARDSI-Cochin Chapter.

Chapter 8
Family Caregiving in Dementia in India: Challenges and Emerging Issues

Mathew Varghese, Upasana Baruah, and Santosh Loganathan

Abstract This chapter would provide an overview of the caregiving scenario in India with a focus on families as the mainstay of support and care for people with dementia. The various aspects of caregiving in dementia would be discussed in the light of the Indian sociocultural context. The impact of caregiving and challenges faced by the family and informal caregivers would be described in the light of the changing demographics and urbanization in India. The need for different kinds of caregiver education and training programmes tailored to the domiciliary and socioeconomic status of the family would be discussed. The resources available for family caregivers, like existing programmes for psychoeducation, family self-care, online and other resources for support would be described. We would discuss the challenges faced in developing culturally appropriate interventions for India that can be delivered within existing resources, such as supporting families in their role as caregivers and providing training and support for them. The chapter would discuss the emerging issues in the models of care for low- and middle-income countries like India, where the care is primarily home-based.

Keywords Dementia · Family caregiving · Family-based interventions · Emerging issues · Challenges in caregiving · India

Introduction

According to World Population Prospects 2019, 1 in 6 people in the world will be over the age of 65 by 2050, up from 1 in 11 in 2019 (United Nations, 2019). As the demographic transition unfolds in India, an increase from 6.4% in 2019 to 8.4% by 2030 in the proportion of older adults aged 65 and older is projected (United Nations, 2019). This transformation will have a significant impact on age-related conditions

M. Varghese (✉) · U. Baruah · S. Loganathan
Geriatric Clinic and Services, Department of Psychiatry, National Institute of Mental Health and Neurosciences (NIMHANS), Bangalore 560029, India

U. Baruah
Department of Psychology, Christ (Deemed to be University), Bangalore 560029, India

M. K. Shankardass (ed.), *Dementia Care*,
https://doi.org/10.1007/978-981-16-3864-0_8

such as dementia. Dementia mainly affects older people, with most cases starting past the age of 65 years. As the prevalence doubles approximately every 5 years after 65 years (Jorm & Jolley, 1998; Ritchie et al., 1992), the rapid demographic ageing will result in more people living with dementia. A recent report of the Alzheimer's and Related Disorders Society of India (ARDSI) says that there are over 4.4 million people with dementia in 2015 in India, a number which is expected to double by 2030 (ARDSI, 2018). India will, therefore, host one of the largest populations of people with dementia. This rise in cases would affect the formal and informal health, social and financial systems in general, with specific increase in the social and economic demands on individuals and families taking care of people with dementia.

Family Caregiving in Dementia in the Indian Sociocultural Context

In India, as a society, families have always assisted and provided support to older family members when they lose the ability to function independently. The role of family is paramount in dementia care. Family caregiving is the foundation of informal care as a large proportion of people with dementia are cared for by their immediate families and extended families (Ganesh et al., 2016). In India, caregiving is rooted in the traditional extended family system and constitutes a part of their responsibilities as members of the family. Family members taking on the role and responsibilities of caregiving can be attributed to the cultural beliefs like kinship obligation, filial piety (respect for elders and parents) and the preeminent role that family takes in the in the decision-making process. Family caregiving in dementia is more exhaustive and demanding than caregiving for older adults without dementia, and most of the time the family members are not sufficiently equipped to take on the caregiving role. Caregiving roles are either assigned or accepted, with most caregivers undertaking the task due to family obligation, expectations from the community, or just because no one else would take on the responsibility (Narayan et al., 2015).

Families have always placed great values on filial responsibility in Asian societies. It is expected of adult children to care for their ageing parents and to provide material, healthcare and emotional support (Chan, 2010). The effect of dementia, in such societies, is experienced by the whole family. Family caregivers in low- and middle-income countries, like India, hardly receive any support in taking care of persons with dementia. Lack of mental health infrastructure and scarcity of other resources like short-term or long-term care facilities also contribute to the fact that families are most of the time the only line of support. This exacts tremendous costs from the caregivers, physically, emotionally and financially (Prince et al., 2015; World Federation of Mental Health, 2010). Family caregivers must take care of various physical, emotional and behavioural challenges in the persons with dementia in addition dealing with their own personal and professional demands and commitments. Thus, caregiving is can gradually exert tremendous stress and is associated with burden and negative physical and mental health (Papastavrou et al., 2007; World Federation of Mental Health, 2010; World Health Organization [WHO], 2017). However, it is seen that despite these difficulties which are associated with caregiving, many caregivers willingly continue to care for the person with dementia (Guerra et al., 2016).

Culture has been attributed with providing a framework for understanding actions and experiences and has played an important role in how caregiving is perceived in India. Culture has also been influential in how the family assigns meaning to dementia (Gilbert et al., 2007; La Fontaine et al., 2007). One of the common cultural norms and beliefs that are prevalent around caregiving and dementia in India include viewing dementia as part of normal ageing. This results in decreasing the stigma associated with dementia but delays the time to seeking treatment. Cultural expectations of ageing as "second childhood" normalizes the symptoms of early dementia. Owing to the family's low expectations from the older adults, they have a high tolerance to the disability caused due to cognitive impairment in their affected relatives. Ultimately this results in lower recognition of dementia (Jacob et al., 2007). Symptoms of dementia are popularly attributed to multiple causes, which range from the nature and personality of the individual to neglect by the family. A prevalent belief in Asian countries is that problems should be solved within the family and anything that affects the "good name" of the family in a negative way is generally kept a secret (Tseng et al., 1995). In Indian culture this belief along with the fear that the family member with dementia will be labelled mentally ill keeps families away from seeking help.

Cultural gender norms play an important role in family caregiving in India. A vast majority of the caregivers in India are women who live in extended households (Narayan et al., 2015; Prince, 2004). These caregivers, who are mostly the wives and daughters-in-law of the person with dementia, are expected to adhere to the cultural norms of care and obligations that are ascribed to women in the society. Their time and effort in the caregiving more often go unacknowledged and unnoticed and they themselves turn a blind eye to their problems or underrate their contributions due to culturally grounded norms of gender roles (Ashwin et al., 2013; Lai, 2009).

Overall, dementia care in India is ingrained in the concept of "family centeredness", where individuals place family needs over their own (Mehta, 1997). Another important feature is the multiple layers of care (*seva*), which is the centre of intergenerational relations (Brijnath, 2014). The care is often compared to that of offering divine worship to the elderly. It includes taking physical care of the frail elder and to reveal power: that the ageing body is in the process of weakening. *Seva* also disciplines younger family members about their roles and responsibilities (Brijnath, 2014). The interpretations of the cultural conceptualizations of care are instrumental in helping caregivers deal with the negative impact of caregiving and help them derive beneficial outcomes from the experiences. Beneficial outcomes have been documented to encompass a range of positive feelings like a finding a purpose in life, understanding meaning in life, experiencing pride and satisfaction, a feeling of achievement and mastery; gaining renewed perspectives on life and enhanced relationships with family members (Ducharme et al., 2011; Lane et al., 2012; Sanders, 2005). Thus, it appears that, despite the stress, burden and hardships associated with caregiving, those who strive to find meaning and growth from the role, are able to describe the caregiving process as a positive experience (Ducharme et al., 2011).

Considering the progressive nature of the illness and need for long-term care, understanding the beliefs, norms and customs that influence the caregivers will help in identifying caregiving needs and difficulties, so that appropriate strategies can be planned (Alzheimer's Disease International [ADI], 2016; Chan, 2010). Considering the importance of family as the primary long-term caregiver, family-oriented care should be a priority with family members as care partners.

Emerging Issues and Challenges for Family Caregiving in India

Though family caregiving is recognized as the mainstay of care and support for persons with dementia in India, it has been affected by several social and demographic trends which has led to unique challenges. Key issues include demographic changes like the ageing of the world's population, increasing mobility of people, rapid urbanization, changing family patterns including increase in nuclear family and changes in gender balance of the caregivers. These factors are likely to have an impact on dementia care and the family caregiving system (Bloom et al., 2010).

Change in traditional family structure and altering family patterns: Sweeping changes are altering life and family arrangements across Asia which is becoming evident in India too. In previous generations, large, extended families could be relied on for help with health and functioning needs. Traditional multigenerational households have till recently ensured care for ageing relatives including persons with dementia. However, a shift has been observed in family-based care, which is less prevalent than before in India (Arokiasamy et al., 2012). With the family networks coming under stress and being subjected to the changing norms, India is undergoing a disintegration of the traditional "the great big Indian family" (Bloom et al., 2010; Pal, 2007). Fertility rates in India have also seen a huge fall, from nearly 6 children per woman in 1950 to 2.6 children per woman in 2010. Declining fertility rates, delayed childbearing, increase in childlessness, smaller numbers of siblings, changes in the traditional structure of the family, higher rates of divorce and an increasing number of people who are not married have led to smaller family sizes, resulting in a decreasing number of potential caregivers (Redfoot et al., 2013; United Nations Economic Commission for Europe [UNECE], 2019). The increase in nuclear families and high rates of migration of the younger generation is leading to changes in living arrangements for older adults. Evolving complex family structure and non-traditional households like families with children who are divorced, families with stepchildren and families with unmarried children are more common in the current scenario. These changes have significant implications as research has indicated that having adult step children as caregivers have negative impacts on the quality of care for older step parents (Silverstein & Giarrusso, 2010; van der Pas et al., 2013). These changes in family structures along with demographic ageing would increase the probability for most individuals becoming informal caregivers at some point in their life and also alter the existing family caregiving characteristics (Pickard & King, 2012). Caregiving for older adults and specifically for persons with dementia will soon, in all probability, depend on whether the family members are willing, available or have the capacity to undertake the role. There would be fewer family members that older adults would be able to depend on, with adult children more likely to be geographically distant from their parents, who in all probability would be living alone. Brijnath (2014) in her book, reveals how dementia interrupts the power dynamic when the elderly person feels that their physical and cognitive ability is worsening, and they are becoming dependent on their children. There is also a sense of "dependency anxiety" that develops that they would be a burden on the resources of the family. However, to understand in greater detail how these emerging trends are going to impact people with dementia, especially in the Indian context, further research is warranted.

Changes in the traditional gendered role in caregiving: In India, the primary caregiving role has always been fulfilled by women in the family, as a spouse, daughter, daughter-in-law or sibling. Traditionally in India, it is seen that once a daughter is married, her obligations are shifted to her family of procreation, though in the event that there is no male sibling available, or when a parent is widowed, daughters have been seen to take up the responsibility of her family of origin too (Gupta et al., 2009; Lamb, 2005). However, in recent times, there has been a change in the gender balance of informal caregivers. With rising mobility, there has been a significant increase in the number of women joining the workforce outside home, which is leading to them gradually becoming less available for caregiving roles. Men, who were less engaged in caregiving roles at the working age, are becoming more involved in caregiving, at an older age, particularly for their spouses. This could signify the initiation of a change in family dynamics including change in assigned role in the traditional family system (World Federation of Mental Health, 2010). There is a need to focus on male caregiving role in Asian countries as there is a paucity of research in this area.

Unpreparedness for the role of caregiver: With the recent trends and changes that are evolving in the caregiving scenario, lack of proper understanding and inadequate information often results in caregivers feeling overwhelmed or not being sufficiently prepared, especially when caring for a person with dementia, which in itself is a challenging task. Not being prepared for the role could be due to inadequate knowledge and skill and not being familiar with the process of caregiving. Many caregivers do not have a real choice when they become caregivers and might assume the role due to various reasons like carrying out legal duties, fulfilling family obligations, unavailability of other care facilities or lack of information about such alternatives. They may not know when they need help and how to avail and make use of whatever resources are available in the community (Given et al., 1994). When the care needs of the person with dementia exceeds the caregiver's capacity for caring it can sometimes lead to a negative evaluation of the relationship and become a burden (Reid et al., 2005).

Hidden costs of family caregiving in dementia: Though the cost of dementia has always been a significant issue, the hidden costs of informal caregiving for family caregivers have become more apparent (Rodrigues et al., 2013). These hidden costs encompass various aspects like loss of job and career, the loss of associated income and most importantly the negative impact on health of the caregivers which results from caregiving burden and stress. Studies from India have indicated that 23% of caregivers quit their job or decrease working hours and one-fourth of them must deal with financial difficulties because of change in job situation (ARDSI, 2010). Costs were sometimes direct in nature, such as with specific requirements like drugs, diapers, catheters. Sometimes, they were indirect hidden costs with subjective feelings such as sadness, fatigue, guilt and *seva*. Admission into psychiatric facilities and hiring attendants to manage caregiving were additional costs (Brijnath, 2014). The 10/66 group found that families from the low socioeconomic status in the Indian society also utilized the services of private medical care resulting in increased risk of poverty (Prince, 2009). There is also an ever-looming risk on families of suffering from ill-health and being socially excluded without adequate support. High levels of caregiving activities may result in the caregiver not getting adequate rest or care for themselves which could happen when dementia progresses and the person with dementia is totally dependent. Low incomes and limited financial resources are predictors of increased stress. With inadequate amount of money, they are unable to utilize the resources normally available to provide relief to caregivers, like opting for respite care or other longer care alternatives.

The balancing act—incompatibility of care, employment and personal life: With advancement in education and greater mobility in India and other developments, most caregivers are part of the workforce. In this scenario, managing caregiving roles and responsibilities with employment obligations becomes extremely stressful. In today's competitive job market, it is often exceedingly hard to find the flexibility in a workplace that would allow reconciling work with caregiving duties. It becomes increasingly difficult for caregivers to attempt to balance caregiving and work-related commitments with other activities, such as family and leisure, which often lead to negative consequences including caregiving burden (Stephens et al., 2001). This leads many primary caregivers to leave their jobs due to the stress and inability to juggle work and care in the absence of appropriate support (Eurofound, 2015; WHO, 2012). Loss of employment adds to the hidden costs in dementia caregiving. On the other hand, being employed serves as a buffer at times to provide respite from caregiving activities, and it also serves as a source of financial support (Pinquart & Sörensen, 2003). This adds on to the dilemma of a family caregiver.

Need for recognition and risk of being socially excluded: Caregiving in dementia requires a lot of patience and skills to deal with the challenging behavioural problems associated with the illness. In the process of caregiving, dementia caregivers, develop and learn many skills which elevate them to the level of experts in dealing with situations. However, the recognition of the family caregivers' work is low, both by society and family circles, and they are neither supported nor properly acknowledged (UNECE, 2019; WHO, 2012). It is also plausible that the changes that take place in the social life of a primary caregiver of a person with dementia in terms of engagement with social networks may increase risk of loneliness and social isolation (Victor et al., 2020). Caregivers felt avoided and discounted along with the person with dementia and often a sense of martyrdom loomed around their sacrifices, though they felt that the sacrifices were not their choices. Being treated with pity and unsolicited advice from friends, family and strangers were common experiences for caregivers. A common advice that caregivers received was to place the person with dementia in a care home (ADI, 2012). This is another important aspect of family caregiving in dementia that needs to be explored in India.

Need for information and training and access to services: As many family caregivers enter their new role unprepared, there is an utmost need for training which would equip them with knowledge to deal with the challenges in dementia caregiving, provide information about available resources, welfare benefits and other services available (UNECE, 2019). Though information is available, the process of obtaining it can be time-consuming. The scarcity of support services like day-care facilities, at-home nursing services and short-term residential care, which could provide respite to informal caregivers, constitute another challenge for family caregivers in India. The services that are available may not be accessible or affordable. Quality support services in the community, which are affordable for all, is the need of the hour. Caregivers face additional barriers when seeking services for themselves and very often they do not give due attention to their own mental and physical well-being (WHO, 2012). In many parts of India as in many other low- and middle-income countries, support services for family caregivers are not routinely available.

Supporting Family Caregivers

It has become increasingly important to acknowledge caregiving in dementia as an imperative public health issue. Caregiving should not be construed as the sole responsibility of the family, but it should be part of community welfare programmes and government's public health agenda (WHO, 2012). The first step towards that can be public acknowledgment of caregiving by families. Family caregivers, especially in India, are the pillar of support for people with dementia and they require support from all avenues in the process of providing care and in maintaining their own health and well-being. To achieve this objective, there is a need for more inclusive public policies and supportive community practices catering to the needs of people with dementia and their caregivers.

Research has elucidated that caregiver interventions which include education and skills training, caregiver skill building and environmental modifications, have beneficial implications for the person with dementia as well as the caregivers. (National Academies of Sciences and Medicine, 2016). Caregiver interventions which were multicomponent, required active involvement, was skill or application based; and continued for a longer period or provided periodic support throughout the process of caregiving have been found to be more effective.

Acknowledgement, information, support and skill building: The need for caregiver interventions focussing on information and skill building has been exhaustively studied and talked about, but the development and implementation of such programmes in India have been slow and not adequate. There is a need to focus on the family and skills training for caregivers as essential programmes in dementia care in research and in clinical practice as well (Murthy, 2016). Though the immediate family members or next of kin are most often the primary caregivers of people with dementia, there are friends, neighbours and other non-formal paid caregivers in middle class households in India. When people provide informal care, whether by choice or by default, they perform the important function of informal care and contribute to the caregiving process (UNECE, 2019). It is important to give them their due recognition, acknowledge their role as carers and extend support to them as in the case of family caregivers.

Facilitating access to information and support services: Family caregivers must be encouraged to be proactive in their interaction with the healthcare system to avail services. Having knowledge about services available could boost their confidence and improve their mental well-being (Toseland et al., 2004). Additionally, primary caregivers should develop skills to enlist the help of family and friends and mobilize resources. These interventions can be part of caregiver-support programmes that include psychological assessment and treatment of mental health issues for the caregiver, along with respite, education and training; and can be delivered individually or in groups by community health workers or by experienced caregivers (Patel et al., 2009). It is important for mental health professionals to address "self-care" of caregivers in all their interactions with them (Murthy, 2016). Technology-based interventions or self-care programmes delivered through online modules are also important and gradually gaining momentum in India.

Culturally competent interventions and programmes: Research has time and again underscored the need for pragmatic family educational and skills training interventions which are user friendly in different modalities (Isaac, 2016). However, these intervention modules or manuals need to be culturally competent. Direct application of interventions which have been developed in high-income countries or in countries with a different sociocultural background might not be effective or acceptable in India. India being a cultural melting pot of beliefs, values, customs, traditions, lifestyles and languages, the "one-size-fits-all" concept would not be very effective. Interventions need to be adapted to each cultural community, ethnic and other minority groups to address the cultural norms that may be impacting the behaviours or decision-making processes of caregivers (Barrera et al., 2013).

Enabling family caregivers to reconcile caregiving responsibilities with employment: It is important to create an environment in which it is possible for working caregivers to remain in employment along with their care responsibilities. This would be possible through a supportive work environment with flexible schedules and mechanism to access care leaves (UNECE, 2019). A balanced life for family caregivers is again intricately associated with the support available in terms of community resources as discussed earlier. There is also an urgent need to redefine, relax and minimize bureaucratic and organizational procedures associated with availing services for persons with dementia including provision for accurate documentation and prompt referral.

Community and intergenerational solidarity: Raising awareness about dementia in the community including caregivers and the general population has been a priority all over the world (Graham & Brodaty, 1997). Awareness can help reduce stigma, decrease social exclusion and foster community solidarity, which in turn can bring forth supportive policies and programmes for dementia care and hold policy makers accountable through advocacy (Prince et al., 2008). Though India is experiencing a change in the traditional multigenerational family structure, cohesion and cooperation between the generations is still a fundamental aspect of dementia care in India. The receding numbers of potential caregivers and increase in nuclear families is a concern, but the younger generation are still the future flag bearers of dementia caregiving, especially in countries like India. Awareness campaigns and workshops in the community, schools and colleges, through which children and young adults are easily accessed, can be instrumental in promoting intergenerational solidarity (Prince et al., 2008).

Family Interventions and Support for Caregivers of People with Dementia

Needs of families of persons with Dementia: Before caregiver interventions are planned for the family, it is important to assess what needs are met or unmet and what is expressed by the family. Needs of the family members will help in determining the kind of interventions that are required. Assessment of the needs will also enable the treating team to give personalized interventions to the patients and their families. The needs of persons with dementia and their families are varied and extensive and not very well studied. Much of the work done from India shows that patients and families have many unmet needs in the course of caregiving. Most of the family members want information about the illness and its treatment, practical advice about management of the behavioural symptoms and help with daily activities. Some caregivers, especially those from a lower socioeconomic status, want to know how to obtain a senior citizen's certificate from the government. This is linked to the benefits and financial support like disability and elderly pensions that are given by each state government. Caregivers who are spouses also have a need to know what will happen to the patient in case the caregiver dies before the patient. Hence, they ask for provision of short-term or long-term care homes for elders and other provisions for respite of caregiving after their time. Research studies that have looked at needs have also studied the links with knowledge about the illness and coping mechanisms. These studies have shown that caregivers with adequate knowledge about the illness had better acceptance of dementia and exhibited better coping patterns. Caregivers who coped well used problem-focused and emotional-focused coping approaches, and they had better physical and psychological well-being.

Family caregiver intervention and support programmes: Caregiver intervention programmes for dementia in the West and in India have been in existence for over 30 years. Most of the caregiver interventions designed used the psychoeducational model of intervention which uses a cognitive behavioural framework (Cheng et al., 2019). Most intervention programmes use an information giving and educative format where all aspects of the illness like the clinical features, staging of dementia, course and outcome, possible aetiology of the illness, treatment, community services and care options are discussed. The other major component of the intervention focuses on how to manage difficult behavioural and psychological symptoms in the patient usually using behavioural techniques. These interventions enable the caregiver to learn better skills and use problem-solving skills to gain mastery over the different challenges they face. More elaborate behavioural interventions have additional components of communication skills training. This focuses on better communication between the family and the patient using simple, direct, non-critical, verbal communication. Self-care of the caregiver is also an important focus of interventions to avoid or limit caregiver burnout. Since 30–50% of caregivers who are in this role for a long time are prone to depression and anxiety disorders, intervention programmes also offer relaxation training, mindfulness meditation techniques and individual or group therapy in case of clinical depression or anxiety. More elaborate and complex interventions are available to intervene with the whole caregiving system using a multi-component intervention. In addition to education and skills training, these interventions also offer problem-solving skills, counselling, support groups and home modification. The different components of these psychosocial intervention programmes given above have been recently reviewed by Cheng et al. (2019).

Most of these intervention programmes have been manualzsed for research. There are numerous randomized control trials conducted in different regions of the world to look at the efficacy and effectiveness of these family caregiver interventions. There is good evidence that these interventions work well from the early stages of the dementing illness and right through the process of caregiving to end stages of dementia. These family-based interventions reduce burden of care, improve mastery and coping and improve the quality of life of the caregivers. A review of many of these psychoeducational intervention programmes (Beinart et al., 2012) concluded that interventions that are person-centred and individually tailored to the caregiver are particularly useful in reducing caregiver burden. They found that women caregivers were more prone to suffer from depression or loneliness, while caregivers who were men had physical illnesses and cardiovascular disease because of caregiving. The caregivers who were children were in general less affected compared to a spouse caregiver. The burden of caregiving was more in patients who had problematic behavioural symptoms, more cognitive deficits, impaired activities of daily living and those who resisted medication and treatments. In general, the duration of dementia was correlated with higher burden.

Modality of delivery of the caregiver intervention programmes: The caregiver intervention programmes are delivered in different modalities depending on the setting, educational status and availability of caregivers. For caregivers with low literacy, the interventions are usually delivered in a face-to-face format either individually or in a group setting with many caregivers provided they can speak a common language. The 10/66 Caregiver Education and Training Intervention was developed in India, with inputs from other developing country experts (10/66 Dementia Research Group, 2002). This manualized intervention was meant for use by basic health workers after undergoing minimal training. It targeted the main caregiver and sometimes included members of the extended family. The aim was to provide basic education about dementia and specific training to manage problem behaviours. The intervention comprised of three modules, which were delivered over five, weekly, half hour sessions and has been studied in some randomized control trails (Gavrilova et al., 2009). For family and public awareness about mental illness and its management, street plays have been used effectively to impart knowledge about the illness, problems faced by the patient and their families and possible solutions to the problems (Loganathan & Varghese, 2019).

A much more sophisticated, accessible and cost-effective intervention may be possible for highly literate caregivers who may also be savvy with technology by using computer-based programmes or programmes delivered by telephone or an IVRS. In recent years, Web-based interactive programmes and portals are also available wherein caregivers can independently access the psychoeducational programme. A few of these programmes are accessible on the Internet namely the iCare programme (Kajiyama et al., 2013) and the Caregiver Skill Building Intervention (CSBI) programme (Farran et al., 2017) and the WHO iSupport for dementia caregivers (Pot et al., 2019). The WHO iSupport programme was developed by a consultative group, and the randomized controlled trial to test its usefulness was carried out in Bangalore, India (Mehta et al., 2018).

Acceptability of psychosocial interventions: An important factor in imparting caregiver intervention programmes for persons with dementia maybe the acceptability of these programmes by the recipients. The reason for the equivocal nature and moderate effect sizes of the intervention programmes is possibly because the family caregivers may not have a felt need. In a systematic review of this issue, it was found that acceptability of interventions may be determined by various factors like appropriateness of the content, organization of the intervention, knowledge and professional nature of the staff members and the needs of the family caregivers (Qiu et al., 2019). Poor acceptability of interventions was related to various caregiver factors like low literacy and education, poor health, higher caregiver burden and perceived lack of need for the treatments. Physical interventions like exercise or daily scheduling were more acceptable than cognitive or psychosocial interventions. This would possibly be due the fact that readiness for change is important for much of the interventions which are based on cognitive behaviour therapy models that require behavioural change in the caregivers. Hence, for interventions to be accepted, it would be necessary to evaluate the needs of the caregivers, assess readiness to change and take them through the process of behaviour change. It is also necessary to improve the applicability of the intervention so that there is caregiver satisfaction.

Future research to support family caregivers of persons with dementia: Future research in the area of dementia caregiving should invariably focus on a diverse population of caregivers. Research should be undertaken keeping in consideration the changing trends and emerging issues in Indian context. Results should not be generalized, and studies should be designed specific to the population depending on the caregiver characteristics like gender, age, educational background, financial status, nature of caregiving responsibilities and ethnic and cultural background. More research is needed to elucidate the influences of family on caregiving practices and highlight the positive aspects of the caregiving process. The feasibility and effectiveness of technology-based interventions in developing countries need to be explored along with other innovative designs and interventions to prepare and support the family in the caregiving process (National Academies of Sciences and Medicine, 2016; Reinhard et al., 2008).

Conclusions

Family caregiving is a cornerstone in dementia care in India, and it would continue to remain so even though it is being challenged by various emerging issues and demographic trends. Recognition of family caregivers as partners in dementia care, generating awareness about dementia and the challenges it presents and promoting effective adaptation of family caregiving is crucial for dealing with this public health challenge.

In our experience of working with family caregivers, education of the families about various aspects of dementia is imperative. Management of challenging behaviours and dealing with everyday activities requires special skills. Intervention programmes should focus on skill building and should be tailored to the needs and sociocultural background of the caregiver, in addition to addressing caregiver stress. Future direction in supporting and strengthening family caregivers of persons with dementia in India would be to find the best possible response within the context of limitations, possibilities and emerging demographic and cultural changes. Knowledge and utilization of local resources should be enhanced to ensure sustainability.

References

Alzheimer's and Related Disorders Society of India (ARDSI). (2018). *Dementia India strategy report*. Accessed July 10, 2020. https://ardsi.org/pdf/Dementia%20India%20Strategy%20Report%202018.pdf.

Alzheimer's Disease International (ADI). (2012). *The world alzheimer report 2012: Overcoming the stigma of dementia*. Alzheimer's Disease International (ADI). https://www.alz.co.uk/research/WorldAlzheimerReport2012.pdf.

Alzheimer's Disease International (ADI). (2016). *World Alzheimer Report 2016: Improving health care for people living with dementia; Coverage, quality and costs now and in the future*. Alzheimer's Disease International (ADI). https://www.alz.co.uk/research/WorldAlzheimerReport2016.pdf.

Alzheimer's & Related Disorders Society of India (ARDSI). (2010). The Dementia India Report: prevalence, impact, costs and services for Dementia. In K. S. Shaji, A. T. Jotheeswaran, N. Girish, S. Bharath, A. Dias, M. Pattabiraman, & M. Varghese (Eds.), *ARDSI*. ISBN: 978-81-920341-0-2.

Arokiasamy, P., Bloom, D., Lee, J., Feeney K., & Ozolins, M. (2012). Longitudinal aging study in India: Vision, design, implementation, and preliminary findings. In J. P. Smith & M. Majmundar (Eds.), *National Research Council (US) panel on policy research and data needs to meet the challenge of aging in Asia*, Aging in Asia: Findings from New and Emerging Data Initiatives. National Academies Press (US). https://www.ncbi.nlm.nih.gov/books/NBK109220/.

Ashwin, S., Tartakovskaya, I., Ilyina, M., & Lytkina, T. (2013). Gendering reciprocity: Solving a puzzle of Nonreciprocation. *Gender & Society, 27*(3), 396–421.

Barrera, M., Jr., Castro, F. G., Strycker, L. A., & Toobert, D. J. (2013). Cultural adaptations of behavioral health interventions: A progress report. *Journal of Consulting and Clinical Psychology, 81*(2), 196–205.

Beinart, N., Weinman, J., Wade, D., & Brady, R. (2012). Caregiver burden and psychoeducational interventions in Alzheimer's disease: A review. *Dementia and Geriatric Cognitive Disorders Extra, 2*(1), 638–648.

Bloom, D. E., Mahal, A., Rosenberg, L., & Sevilla, J. (2010). Economic security arrangements in the context of population ageing in India. *International Social Security Review, 63*(3–4), 59–89.

Brijnath, B. (2014). *Unforgotten: Love and the culture of dementia care in India*. Berghahn Books

Chan, S. W. (2010). Family caregiving in dementia: The Asian perspective of a global problem. *Dementia and Geriatric Cognitive Disorders, 30*(6), 469–478.

Cheng, S.-T., Alma, Au., Losada, A., Thompson, L. W., & Gallagher-Thompson, D. (2019). Psychological interventions for dementia caregivers: What we have achieved, what we have learned. *Current Psychiatry Reports, 21*(7), 59–59.

10/66 Dementia Research Group. (2002). The 10/66 caregiver education and training intervention. Accessed on August 15, 2020. https://1066.alz.co.uk/caregiver_intervention_rct_intervent ion.php.

Ducharme, F. C., Lévesque, L. L., Lachance, L. M., Kergoat, M.-J., Legault, A. J., Beaudet, L. M., & Zarit, S. H. (2011). "Learning to Become a Family Caregiver" Efficacy of an intervention program for caregivers following diagnosis of dementia in a relative. *The Gerontologist, 51*(4), 484–494.

Eurofound. (2015). *Working and caring: Reconciliation measures in times of demographic change*. Publications Office of the European Union.

Farran, J. C., Zurawski, P., Inventor, B. R., Urbanic, J., & Paun, O. (2017). An evidence-based technological caregiver skill building intervention for dementia family caregivers: Pilot study. *Alzheimer's, Dementia & Cognitive Neurology*. https://doi.org/10.15761/ADCN.1000104

Ganesh, S., Dhanasekaran, S., Nirisha, L., Chatterjee, S. S., & Sivakumar, P. T. (2016). Care arrangements for persons with dementia: A review on formal care and its relevance to Indian context. *Journal of Geriatric Care and Research*, 3(2), 27–34.

Gavrilova, S. I., Ferri, C. P., Mikhaylova, N., Sokolova, O., Banerjee, S., & Prince, M. (2009). Helping carers to care–the 10/66 dementia research group's randomized control trial of a caregiver intervention in Russia. *International Journal of Geriatric Psychiatry, 24*(4), 347–354.

Gilbert, P., Bhundia, R., Mitra, R., McEwan, K., Irons, C., & Sanghera, J. (2007). Cultural differences in shame-focused attitudes towards mental health problems in Asian and Non-Asian student women. *Mental Health, Religion & Culture, 10*(2), 127–141.

Given, B. A., Given, C. W., Stommel, M., & Lin, C.-S. (1994). Predictors of use of secondary carers used by the elderly following hospital discharge. *Journal of Aging and Health, 6*(3), 353–376.

Graham, N., & Brodaty, H. (1997). Alzheimer's disease international. *International Journal of Geriatric Psychiatry, 12*(7), 691–692.

Guerra, S., Figueiredo, D., Patrão, M., & Sousa, L. (2016). Family integrity among older caregivers of relatives with dementia. *Paidéia (ribeirão Preto), 26*, 15–23.

Gupta, R., Rowe, N., & Pillai, V. K. (2009). Perceived caregiver burden in India: Implications for social services. *Affilia: Journal of Women & Social Work, 24*(1), 69–79.

Isaac, M. (2016). Cross-cultural differences in caregiving: The relevance to community care in India. *Indian Journal of Social Psychiatry, 32*(1), 25–27.

Jacob, K. S., Kumar, P. S., Gayathri, K., Abraham, S., & Prince, M. J. (2007). The diagnosis of dementia in the community. *International Psychogeriatrics, 19*(4), 669–678.

Jorm, A. F., & Jolley, D. (1998). The incidence of dementia: A meta-analysis. *Neurology, 51*(3), 728–733.

Kajiyama, B., Thompson, L. W., Eto-Iwase, T., Yamashita, M., Di Mario, J., Marian Tzuang, Y., & Gallagher-Thompson, D. (2013). Exploring the effectiveness of an internet-based program for reducing caregiver distress using the iCare Stress Management e-Training Program. *Aging & Mental Health, 17*(5), 544–554.

La Fontaine, J., Ahuja, J., Bradbury, N. M., Phillips, S., & Oyebode, J. R. (2007). Understanding dementia amongst people in minority ethnic and cultural groups. *Journal of Advanced Nursing, 60*(6), 605–614.

Lai, D. W. L. (2009). Filial Piety, Caregiving Appraisal, and Caregiving Burden. *Research on Aging, 32*(2), 200–223.

Lamb, S. (2005). Cultural and moral values surrounding care and (in)dependence in late life: Reflections from India in an era of global modernity. *Care Management Journals, 6*(2), 80–89.

Lane, G. W., Podgorski, C. A., Noronha, D. O., & King, D. (2012). "The Hidden Caregiver": Kinship caregivers and lessons learned from dementia family caregiving. *Clinical Gerontologist, 35*(2), 195–203.

Loganathan, S., & Varghese, M. (2019). Formative research on devising a street play to create awareness about mental illness: Cultural adaptation and targeted approach. *International Journal of Social Psychiatry, 65*(4), 279–288.

Mehta, K. (1997). Cultural scripts and the social integration of older people. *Ageing and Society, 17*(3), 253–275.

Mehta, K. M., Gallagher-Thompson, D., Varghese, M., Loganathan, S., Baruah, U., Seeher, K., Zandi, D., Dua, T., & Pot, A. M. (2018). iSupport, an online training and support program for caregivers of people with dementia: Study protocol for a randomized controlled trial in India. *Trials, 19*(1), 271.

Murthy, R. (2016). Caregiving and caregivers: Challenges and opportunities in India. *Indian Journal of Social Psychiatry, 32*(1), 10–18.

National Academies of Sciences, Engineering, and Medicine. (2016). *Families caring for an aging America.* In S. Richard & E. Jill. (Eds.), The National Academies Press.

Narayan, S. M., Varghese, M., Hepburn, K., Lewis, M., Paul, I., & Bhimani, R. (2015). Caregiving experiences of family members of persons with dementia in south India. *American Journal of Alzheimer's Disease and Other Dementias, 30*(5), 508–516.

Pal, S. (2007). *Effects of intergenerational transfers on elderly coresidence with adult children: Evidence from Rural India.* Institute of Labor Economics (IZA).

Papastavrou, E., Kalokerinou, A., Papacostas, S. S., Tsangari, H., & Sourtzi, P. (2007). Caring for a relative with dementia: Family caregiver burden. *Journal of Advanced Nursing, 58*(5), 446–457.

Patel, V., Simon, G., Chowdhary, N., Kaaya, S., & Araya, R. (2009). Packages of care for depression in low- and middle-income countries. *PLoS Med, 6*(10), e1000159.

Pickard, L., & King, D. E. N. (2012) Informal care supply and demand in Europe.

Pinquart, M., & Sörensen, S. (2003). Associations of stressors and uplifts of caregiving with caregiver burden and depressive mood: A meta-analysis. *Journals of Gerontology. Series b, Psychological Sciences and Social Sciences, 58*(2), P112–P128.

Pot, A. M., Gallagher-Thompson, D., Xiao, L. D., Willemse, B. M., Rosier, I., Mehta, K. M., Zandi, D., & Dua, T. (2019). iSupport: A WHO global online intervention for informal caregivers of people with dementia. *World Psychiatry, 18*(3), 365–366.

Prince, M. (2004). Care arrangements for people with dementia in developing countries. *International Journal of Geriatric Psychiatry, 19*(2), 170–177.

Prince, M. J. (2009). The 10/66 dementia research group—10 years on. *Indian Journal of Psychiatry, 51*(1) (Suppl1), S8–S15.

Prince, M., Acosta, D., Albanese, E., Arizaga, R., Ferri, C. P., Guerra, M., Huang, Y., Jacob, K. S., Jimenez-Velazquez, I. Z., Rodriguez, J. L., Salas, A., Sosa, A. L., Sousa, R., Uwakwe, R., van der Poel, R., Williams, J., & Wortmann, M. (2008). Ageing and dementia in low and middle income countries-Using research to engage with public and policy makers. *International Review of Psychiatry, 20*(4), 332–343.

Prince, M., Anders, M., Guerchet, Ali, G. C., & Wu, Y. T., Prina, M. (2015). World alzheimer report 2015: The global impact of dementia: an analysis of prevalence, incidence, cost and trends. Assessed July 12, 2020. http://www.alz.co.uk/research/world-report-2015.

Qiu, D., Mi, Hu., Yu, Yu., Tang, B., & Xiao, S. (2019). Acceptability of psychosocial interventions for dementia caregivers: A systematic review. *BMC Psychiatry, 19*(1), 23.

Redfoot, D., Feinberg, L., & Houser, A. (2013). The aging of the baby boom and the growing care gap: A look at future declines in the availability of family caregivers. Accessed July 20, 2020. https://www.aarp.org/content/dam/aarp/research/public_policy_institute/ltc/2013/baby-boom-and-the-growing-care-gap-in-brief-AARP-ppi-ltc.pdf.

Reid, C. E., Moss, S., & Hyman, G. (2005). Caregiver reciprocity: The effect of reciprocity, carer self-esteem and motivation on the experience of caregiver burden. *Australian Journal of Psychology, 57*(3), 186–196.

Reinhard, S. C., Given, B., Petlick, N. H., & Bemis, A. (2008). Supporting family caregivers in providing care.

Ritchie, K., Kildea, D., & Robine, J. M. (1992). The relationship between age and the prevalence of senile dementia: A meta-analysis of recent data. *International Journal of Epidemiology, 21*(4), 763–769.

Rodrigues, R., Schulmann, K., Schmidt, A., Kalavrezou, N., & Matsaganis, M. (2013). The indirect costs of long-term care. Brussels, DG employment, social affairs & inclusion. (Research note 8). Accessed on July 20, 2020. https://www.euro.centre.org/publications/detail/415.

Sanders, S. (2005). Is the glass half empty or full? Reflections on strain and gain in cargivers of individuals with Alzheimer's disease. *Social Work in Health Care, 40*(3), 57–73.

Silverstein, M., & Giarrusso, R. (2010). Aging and family life: A decade review. *Journal of Marriage and the Family, 72*(5), 1039–1058.

Stephens, M. A. P., Townsend, A. L., Martire, L. M., & Druley, J. A.. (2001) Balancing parent care with other roles: Interrole conflict of adult daughter caregivers. *The Journals of Gerontology: Series, B56*(1), P24–P34.

Toseland, R. W., McCallion, P., Smith, T., & Banks, S. (2004). Supporting caregivers of frail older adults in an HMO setting. *American Journal of Orthopsychiatry, 74*(3), 349–364.

Tseng, W. S., Lu, Q. Y, & Yin, P. Y. (1995). Psychotherapy for the Chinese: culture considerations. In T. Y. Lin, W. S. Tseng & E. K. Yeh (Eds.), *Chinese societies and mental health* Oxford University Press.

United Nations Economic Commission for Europe (UNECE). (2019). *The challenging roles of informal carers*. UNECE Policy Brief on Ageing No. 22. Accessed on July 20, 2020. https://www.unece.org/fileadmin/DAM/pau/age/Policy_briefs/ECE_WG1_31.pdf.

United Nations, Department of Economic and Social Affairs, Population Division. (2019). *World population ageing 2019: Highlights* (ST/ESA/SER.A/430).

van der Pas, S., van Tilburg, T. G., & Silverstein, M. (2013). Stepfamilies in later life. *Journal of Marriage and Family, 75*(5), 1065–1069.

Victor, C. R., Rippon, I., Quinn, C., Nelis, S. M., Martyr, A., Hart, N., Lamont, R., & Clare, L. (2020) The prevalence and predictors of loneliness in caregivers of people with dementia: Findings from the IDEAL programme. *Aging Ment Health*, 1–7.

World Federation of Mental Health. (2010). *Caring for the caregiver: Why your mental health matters when you are caring for others*. WFMH: Woodbridge. https://wfmh.global/wp-content/uploads/Caring-for-the-Caregiver-11_04_09-FINAL-2edit2018.pdf.

World Health Organization. (2017). Mental health of older adults. Assessed July 10, 2020. https://www.who.int/news-room/fact-sheets/detail/mental-health-of-older-adults.
World Health Organization (WHO). (2012). *Dementia: A public health priority*. World Health Organization.

Chapter 9
Local Strategies of Strengthening the Caregiving Potential of Families of People with Alzheimer's Disease—Good Practices Introduced in Poland

Beata Bugajska and Rafał Iwański

Abstract One of the consequences of population ageing is the growing number of people with dementias, including Alzheimer's disease where ageing is one of the main risk factors. Regardless of the level of their socio-economic development, the societies face the problem of providing care to the citizens with Alzheimer's—the condition which is progressive and for which there is no effective cure. The dynamic growth of the number of patients in the recent years requires new solutions in terms of organising and financing care for the patients. Institutional care provided by the social assistance and healthcare systems becomes more and more expensive and the demand for care, rehabilitation and medical services by far exceeds the supply. In many countries, including Poland, the main care force are family caregivers, usually women aged 55–64. The caregiving potential of families is often underestimated, the needs of the caregivers are not explored and there are no proper initiatives to support families in the caring process. For this reason, as the problem of ensuring adequate care for people with Alzheimer's disease, one of the priorities is to develop and implement strategies to strengthen the caregiving potential of families of people with Alzheimer's disease. This section presents the local programme of supporting caregivers of people with Alzheimer's disease implemented in Szczecin (Poland), designed based on the previous diagnosis of the needs and evaluation of the effectiveness of the solutions being implemented. The results of the analyses of the data collected during the survey ($n = 800$) among the family caregivers are presented, as well as the implemented forms of support such as financial support for the family caregivers and further development of the community-based care system. The solutions—designed and introduced locally—may inspire other communities which face the same problem.

B. Bugajska (✉) · R. Iwański
University of Szczecin, Szczecin, Poland
e-mail: beata.bugajska@usz.edu.pl

M. K. Shankardass (ed.), *Dementia Care*,
https://doi.org/10.1007/978-981-16-3864-0_9

Introduction

Population ageing, understood in the category of success in eliminating some risks which contributed to premature deaths, brings new challenges associated with the increased demand for long-term care over the dependent elderly. Unprecedented in the global demographic history, the dynamic growth of senior population translates into many spheres of functioning of societies both in social and economic perspective. Without a doubt, an ageing population will generate costs resulting, for example, from decreasing numbers of available workforce as well as expenditures connected with providing care services for the dependent older people. A significant part of this group are patients with dementias, including Alzheimer's disease. Duration of this disease and, consequently, care period is very long and last for years. Additionally, as the disease progresses, more advanced care is needed because Alzheimer's deprives patients of further abilities to take care of themselves. In its final stage, which may also last several years, the quality and comfort of life depends mainly on the resources, work and engagement of caregivers, mainly family members, and healthcare and social assistance professionals. Supporting carers of people with dementia is one of the key areas of activities identified by the World Health Organization (WHO).[1] Local strategies of supporting people with dementias should strengthen the caregiving potential of families. It should be done by identifying and meeting diverse needs of caregivers and their patients.

Epidemiological Risk of Alzheimer's Disease in the Perspective of Population Ageing

At the beginning of the 1960s, people aged 65 and more were 4.97% of the global population (150 million people), whereas in 2018, this proportion grew to 8.87% (637 million).[2] The differences in the percentage of senior citizens in certain countries are very big. This is because these countries are at different stages of their demographic development, what can be well illustrated by the theory of first and second demographic transition and epidemiological transition.[3,4] From 1960s to 2018, the percentage of older people in population in high-income countries increased from 8.4

---

[1] WHO, *Global action plan on the public health response to dementia 2017–2025*, Accessed May 1, 2020, 2020. https://www.who.int/mental_health/neurology/dementia/action_plan_2017_2025/en/.

[2] World Bank, *Health Nutrition and Population Statistics: Population estimates and projections*. Accessed February 20, 2020, from World Bank Open Data: https://databank.worldbank.org/source/population-estimates-and-projections#.

[3] R. Lee, The Demographic Transition: Three Centuries of Fundamental Change, *Journal of Economic Perspectives*, 17(4), 2003, pp. 167–190.

[4] A. Fihel, M. Okólski, Przemiany cywilizacyjne, ludnościowe i starzenie się społeczeństw, In M. Okólski, *Wyzwania starzejącego się społeczeństwa*, Warszawa: Uniwersytet Warszawski 2018, pp. 22–65.

to 17.1% for the OECD countries and from 9.5 to 20.2% for the EU member states. In 2018, the highest rate of citizens aged 65 and more was recorded in Japan (27.6%), Italy (22.8%) and Portugal (22.0%). According to demographic projections, in 2050, people aged 65 and more will constitute 15.9% of the global population. Populations of the high-income countries will become visibly older (27%), with estimated 26.1% in OECD and 30% in the EU.[5] And the percentage of people aged 80 and more will increase significantly. In the EU member states, they were 5.5% of the general population in 2016 and is it projected the share will reach 11.5% by the middle of the twenty-first century. It is worth noting that, according to demographic forecasts, in 2050, people aged 80 and more will constitute 39% of the senior population.[6]

In the case of Alzheimer's disease, age is one of the main risk factors that affect the likelihood of its occurrence. For people aged 65–69, the risk of dementia, out which two-thirds are Alzheimer's disease cases, is about 5%. The frequency of the disease occurrence increases with age. It almost doubles every five years and for individuals aged 85–90 and more, it may even be over 25%.[7,8,9]

Using the exponential methods and assuming that the risk of occurrence of dementias increases with age every 5.5 years, Brookmeyer experts have estimated that the number of seniors suffering from Alzheimer's disease will grow from 26.6 million (range 11.4–55.4 million) in 2006 to 106.8 million (range 47.2–221.2 million) in 2050.[10] Similar results were presented by Wimo et al., who estimated that in 2003, there were 27.7 million of people suffering from this disease.[11] According to the data and forecasts published in an OECD report, in 2017, the number of people with dementias per 1000 citizens was 23.3 in Japan, 22.5 in Italy, 9.9 in Poland. In the analysed group of countries, the least patients were found in Columbia 7.1 and Mexico 7.2.[12] It is projected that by 2037, the percentage of patients will grow in all countries and the main cause will be the ongoing population ageing. As for Japan,

[5] World Bank, *Health Nutrition and Population Statistics: Population estimates and projections*. Accessed: February 20, 2020, from World Bank Open Data: https://databank.worldbank.org/source/population-estimates-and-projections#.

[6] The European Commission, *The 2018 Ageing Report Economic & Budgetary Projections for the 28 EU Member States* (2016–2070) Institutional paper 079, Luxembourg: Publications Office of the European Union, 2018.

[7] T. Jonsson, et al., A mutation in APP protects against Alzheimer's disease and age-related cognitive decline, Nature, 488, 2012, 96–99. https://doi.org/10.1038/nature11283.

[8] C. Qiu, M. Kivipelto, E. von Strauss, Epidemiologia choroby Alzheimera: występowanie, determinanty i strategie interwencji. *DialoguesClin.Neurosci.* 11, 2009, 111–128.

[9] B. Plassman, et al., Prevalence of Dementia in the United States: The Aging, Demographics, and Memory Study, Neuroepidemiology 2007; 29:125–132. https://doi.org/10.1159/000109998.

[10] R. Brookmeyer, et al., (2007). Prognozowanie globalnego obciążenia chorobą Alzheimera.Alzheimer idemencja, 3 (3), 2007, 186–191, https://doi.org/10.1016/j.jalz.2007.04.381.

[11] A. Wimo, L. Jonsson, B. Winblad, An Estimate of the Worldwide Prevalence and Direct Costs of Dementia in 2003. *Dementia and Geriatric Cognitive Disorders,* 21(3), 2006, 175–181. https://doi.org/10.1159/000090733.

[12] OECD, *Life expectancy and healthy life expectancy at age 65 in Health at a Glance 2017.* Paris: OECD, 2017, Accessed: February 20, https://doi.org/10.1787/health_glance-2017-74-en.

as many as 38.4 per 1000 citizens will suffer from dementias, in Italy, it will be 33.7, in Poland 17, in Mexico 13.5 and in Columbia 15.6.[13]

Thus, population ageing will continue to affect the number of people with Alzheimer's disease. This in turn will translate into the growing demand for care services provided by family members as well as greater demand for local and central support programmes which will require financing.

Economic Aspects of Caring for People with Dementias

The growing number of people diagnosed with dementias results in the increasing costs of care, which include expenses connected directly with medical care and assistance provided by the social assistance sector. But the majority of these costs are family expenses, as family members are the main pillar of care systems. Depending on the models and solutions adopted in public policies, the economic and social costs will be born directly by families when they do not receive support or the costs will become public. In case of the latter, several solutions are possible, depending on the adopted model of financing: insurance-based, protection or care and on determining authorities at which level (central or local government) will assume the responsibility for organizing and possible cofinancing care over the dependent seniors.

According to the estimates from 2015, global costs related to dementia amounted up to 818 billion USD. During the last five years, the estimated costs grew by 35%. It is worth to point out that 86% of the global costs occur in high-income countries.[14] The costs of care and treatment of people with dementias may reach over 70 thousand USD per patient yearly; the average is about 30 thousand USD. The main cost elements are the expenditures connected with the informal, home-based care which, in case of this disease, is long-term. The share of medical expenses is smaller.[15] Depending on the level of development measured with GDP per capita, there are big differences between certain countries in terms of care costs. These differences result from different costs of administering certain medical procedures, drug refund systems and costs of professional and family caregiver labour costs which include income lost due to suspended or reduced professional activity. When analysing public and private healthcare expenses, even in OECD countries, there is a noticeable, large disparity in expenses per share of GDP. The countries which spent the most in 2018 were USA 16.9%, Switzerland 12.2% and Germany 11.2%. The least financing was in Turkey

---

[13] OECD, Care Needed: Improving the Lives of People with Dementia, *OECD Health Policy Studies*, OECD Publishing, Paris, 2018, Accessed: May 15 2020 https://doi.org/10.1787/9789264085107-en.

[14] A.Wimo, M. Guerchet, G. C. Ali, Y. T. Wu, A. M. Prina, B. Winblad, L. Jönsson, Z. Liu, M. Prince (2017). The worldwide costs of dementia 2015 and comparisons with 2010. *Alzheimer's & Dementia: The Journal of the Alzheimer's Association*, 13(1), 1–7. https://doi.org/10.1016/j.jalz.2016.07.150.

[15] S. Schaller, et al., The main cost drivers in dementia: a systematic review, *International Journal of Geriatric Psychiatry*, 30; 2014, pages 111–129, https://doi.org/10.1002/gps.4198.

4.2%, Mexico 5.4% and Poland 6.3%.[16] General healthcare expenses translate into forms, scope and availability of services for people who require long-term care, among them Alzheimer's disease patients.

Countries with advanced systems of care provided by social assistance and healthcare institutions can finance services from various sources. One is the insurance system based on the obligatory or voluntary (health or care) insurance premiums paid during activity period so that in case of health emergency, the payer is entitled to receive cash benefits for a defined period. Another model is based on a protection principle where individuals who stay within the country's territory for some time or hold the citizenship, are entitled to benefits financed from the public funds. The financial benefits are paid from the public budget. In the last model, benefits are financed by the country or local authorities and assistance and support are provided to everyone who needs them, based on the predefined criteria. It also allows for the possibility of cofinancing if a beneficiary's income exceeds the legally provided thresholds.

Care for People with Alzheimer's Disease in Poland

Polish model of care for the dependent elderly is based mainly on family supported by healthcare and social assistance institutions and private market of care services. In this system, there are no separate services for patients with dementias, including Alzheimer's disease. As for the services performed by social assistance institutions, it is local governments (municipalities and districts) that is responsible for organising and cofinancing benefits. The system is based on the principle of care, which means that everyone who is facing the risk or health crisis described in the legal provisions, is entitled to receive benefits. There are certain criteria regulating payments for the services. For example, in case of inpatient care, the beneficiary pays up to 70% of their pension or social benefit, then it is paid by the descendants or ascendants obliged to pay alimony (income criterion) and the missing amount is covered by the local government. According to research and analyses, local governments bear most of the costs of care.[17,18] In 2018, municipalities reimbursed 97% of the costs of patients staying in social assistance homes.[19]

[16] OECD, *Health expenditure and financing*, Retrieved from OECD. Stat, Accessed: May 14 2020 https://stats.oecd.org/Index.aspx?QueryId=30142.

[17] B. Bugajska, et al., The optimization of elderly care in Szczecin: research report. *Exlibris: biblioteka gerontologii społecznej*, 1(13), 2017, pp. 107–119.

[18] R. Iwański, *Opieka długoterminowa nad osobami starszymi*, Warszawa: Cedewu, 2016.

[19] Ministerstwo Rodziny, Pracy i Polityki Społecznej, 2020, Raport mrpips-03 za 2018 rok, Accessed: February 24 2020, https://www.gov.pl/web/rodzina/statystyka-za-rok-2018.

Social assistance sector provides care services in two forms: residential—social assistance homes (*DPS, domy pomocy społecznej*) and community—care services and day social assistance homes (*DDPS, dzienne domy pomocy społecznej*).[20]

When a family needs support in care, they turn to a social assistance centre which is the basic social assistance unit every Polish municipality is obliged to run. The social assistance centre sends their social worker to investigate the situation and collect the necessary documents, such as, certificate from the family practitioner confirming that care is necessary. Then, through administrative decision, care benefits are assigned, provided in the place of living of the patient, in a form and scope determined by the social assistance centre. The services are realized by a community assistant who is employed in the social assistance centre or works for an external institution (business or NGO) which implements tasks commissioned by the municipality. The caregiver provides nursing services (washing, changing hygiene measures) and helps in household activities (basic cleaning, shopping, administrative matters). In 2018, social assistance centres provided community care services for 112 thousands people. This group included patients with dementias.[21] In Poland, 19% of the municipalities or communes fail to organize and provide care services.[22] Some local governments decide to open and manage day social assistance homes. These are institutions where seniors, also those with Alzheimer's disease, can spend about 8 h daily on working days. They can take part in dedicated activities and benefit from care and rehabilitation services. In case of Alzheimer's disease patients, this form of assistance is available mainly for persons with first or second stage of the disease. In total, there are 362 such institutions in Poland, which can accommodate over 16 thousand people.[23] When a person has no family or their family is not able or willing to provide care over the dependent senior with Alzheimer's disease, the only solution is a social assistance home. There are seven main types of social assistance homes in the Polish social assistance system but in case of Alzheimer's disease, three are taken into account: for people with chronic somatic illnesses, for people with chronic mental conditions and for the elderly. In 2018, these three types of institutions together offered 53 thousand places.[24] The vast majority of the homes are located in buildings adapted to the needs of care institutions. In Poland, the standards in care institutions have been regularly improved but the number of places has been growing slowly. The present regulations allow to found facilities for maximum 100 residents; however, there are still many homes which function based on old provisions, accommodating more than 200 or

[20] Act of 12 March 2004 on Social Assistance, Journal of Laws 2004 nr 64 Item 593, Accessed: February 25 2020, http://prawo.sejm.gov.pl/isap.nsf/DocDetails.xsp?id=WDU20040640593.

[21] Ministerstwo Rodziny, Pracy i Polityki Społecznej, 2020, Raport mrpips-03 za 2018 rok, Accessed: March 14 2020, https://www.gov.pl/web/rodzina/statystyka-za-rok-2018.

[22] NIK, (2018). Usługi*opiekuńcze świadczone osobom starszym w miejscu zamieszkania.* Warszawa: Najwyższa Izba Kontroli, Accessed: February 20 2020, https://www.nik.gov.pl/kontrole/P/17/043/.

[23] Ministerstwo Rodziny, Pracy i Polityki Społecznej, 2020, Raport mrpips-03 za 2018 rok, Accessed: February 20 2020, https://www.gov.pl/web/rodzina/statystyka-za-rok-2018.

[24] Ministerstwo Rodziny, Pracy i Polityki Społecznej, 2020, Raport mrpips-03 za 2018 rok, Accessed: February 20 2020, https://www.gov.pl/web/rodzina/statystyka-za-rok-2018.

even 300 people. In the recent years, a rapid growth of costs of care provided in all types of assistance institutions has been noticed. It is associated mainly with the increased labour and facility maintenance costs. The average monthly cost is about 4000 PLN (1100 USD) while the average pension is 2179 PLN (550 USD).[25][26] It is worth emphasizing that the percentage of bedridden residents of social assistance homes has been growing with every year. In 2018, there were about 12% of them and the significant part of this group are people with Alzheimer's disease. In many facilities, actions are taken to adapt these institutions to care for the totally dependent. Even though the system does not identify separate social assistance homes for people with dementias, the share of patients with dementias has been regularly growing.

As for the healthcare sector, there are also two types of benefits available: inpatient and community-based. Financing is based on the insurance system (common and obligatory) and despite major deficits on the supply side, it does not allow for cofinancing. Community-based medical and care services are provided mainly by family practitioners and specialists, as outpatient treatment (GP centres), health visitors and long-term care nurses. Psychiatric treatment includes services provided by the community teams for environmental health treatment. In the residential form, the services are delivered in hospitals and therapeutic and nursing institutions. The last two types of institutions are for patients who finished their hospital treatment but require further care services provided by the healthcare sector representatives.

One of the major problems in supporting the elderly is reduced access to the diagnostics and treatment within the public healthcare system, limited by the level of central funding (National Health Fund). Another problem is the insufficient number of places in the therapeutic and nursing institutions as well as in geriatric departments. The demand for rehabilitation services has been constantly growing, including home rehabilitation and long-term nursing care. Separation of social and medical services impedes the coordination of activities. The legal and organizational regulations in Poland ensure access to diagnostics and treatment of Alzheimer's disease within both the healthcare and social assistance sectors but according to the reports of the Supreme Audit Office (*NIK, NajwyższaIzbaKontroli*), the existing solutions do not focus on early diagnosis. This, in turn, hinders the development of the community-based support system at the early stage of the disease. This way, development and implementation of standards of early diagnosis of the Alzheimer's disease symptoms and therapeutic and care procedures becomes one of the priorities.[27]

Most European countries have developed and introduced their national policies to address Alzheimer's disease.[28] In Poland, non-governmental organizations initiated the Polish Alzheimer's Plan which, however, has not been implemented as the official national solution. In terms of care, the Plan provides, among others, the development

[25] ZUS (2020).

[26] Narodowy Bank Polski 2020.

[27] NIK, *Opieka nad osobami chorymi na chorobę Alzheimera oraz wsparcie dla ich rodzin.* Warszawa, 2016, Accessed: May 20 2020, https://www.nik.gov.pl/plik/id,13694,vp,16130.pdf.

[28] Alzheimer Europe 2020. National Dementia Strategies. Accessed April 17, 2020 https://www.alzheimer-europe.org/Policy/National-Dementia-Strategies.

of standards of care over people with dementias, including palliative home-based care, obligatory early diagnostics, creating a network of specialist institutional care facilities dedicated to Alzheimer's disease patients and special rehabilitation holidays for patients and their caregivers. The project emphasizes the necessity to educate the whole society, family carers, medical and social assistance professionals. In terms of organization of the healthcare system, the Plan proposes founding model therapeutic centres for people with Alzheimer's disease, where day hospitalization would be provided.[29] Supported by the Commissioner for Human Rights, NGOs undertake activities to adopt the Polish Plan to Address Alzheimer's Disease as the nation-wide solution. At regional level, regional policies of supporting older people are implemented but problems of people with dementias are not much exposes in these strategies. Solutions introduced in municipalities result from the local strategies addressing social problems and are prepared based on the grassroots resources and concepts. They are the examples of good practices, and if a complex system of support for people with dementias is to be built, they should not only be continued but regularly expanded in terms of services and forms of assistance they offer. This, in turn, require securing a long-term financing by the local governments.

Local Programme of Financial Support for Caregivers of People with Alzheimer's Disease-Bon Alzheimer 75 (Care Voucher: Alzheimer 75)

In 2018 in Szczecin,[30] the first Polish programme of financial support for family caregivers of people with Alzheimer's disease aged 75 and more was introduced. It was financed entirely from the local government funds. The reason for developing this innovative benefit was demographic projections for the city of Szczecin, according to which by 2050, the number of citizens aged 85 and more is expected to triple. In 2030, there will be about 27 thousand of seniors aged 80 and more living in Szczecin, and by 2050, their population will grow to 36 thousand. According to the estimated for Poland, the main burden of caring for ill people is on their families that care for the dependent seniors for years and bear high health, social, economic or mental costs of such care. The programme team consisting of the representatives of the Department of Social Affairs of the Town Hall of Szczecin, universities and medical staff assumed that without systemic and adequate support mechanisms families will fail to provide care for their ill members or the carers will have to bear too much health, social, mental and economic costs. In consequence, caregivers, especially spouses of ill seniors, will need help themselves. In this context, a local programme of financial

[29] A. Szczudlik, M. Barcikowska-Kotowicz, A. Sadowska, Polski Plan Alzheimerowski - prezentacja założeń, In: Sytuacja osób chorych na chorobę Alzheimera. *Raport. Rzecznik Praw Obywatelskich*, Warszawa 2014.

[30] Szczecin—city with the district status, the capital city of Zachodniopomorskie region, located at Polish-German border; population of over 400 thousand.

support for caregivers of people with Alzheimer's disease, called *Bon opiekuńczy: Alzheimer 75 (Care voucher: Alzheimer 75)*, was prepared. It is addressed to family carers who look after an older family member aged 75 and more with Alzheimer's disease diagnosed by a psychiatrist, a neurologist or a geriatrician. It is a cash benefit. The programme provides payment of three instalments, 1000 PLN (252 USD) each (in 2018 there were two instalments). Due to growing interest in the programme, the budget was increased from 1.8 million (454,671 USD) in 2019 to 2.2 million (555,709 USD) in 2020. In total, the beneficiaries may receive up to 3000 PLN (757 USD) in 2020.[31]

There were 560 applications submitted in 2018 and 800 in 2019, out of which 731 were approved. Until 31 May 2020, there have been already 688 applications submitted, what proves that the demand for this type of support has been growing. The institution responsible for the implementation is a city unit—Szczecin Centre for Benefits. The benefit is based on the idea of social trust, which raised many discussions during the development phase. It means that there is no specific purpose the money received should be spent on. It was assumed that caregivers have diverse needs and it is their decision whether to buy a new bed or a mattress, medicine and hygienic products or, for example, cover some costs of adapting bathroom to the needs of person with limited mobility. There is also no income criterion, as the programme assumes that the benefit is beyond the Act of 12 March 2004 on Social Assistance, is addressed to everyone and is separate from the social assistance. From the technical point of view, verification and update information about income would increase the administration costs. The programme was created together with the specialist from different fields and consulted with the representatives of senior organizations and City Senior Council. Its assumptions have received a positive opinion of the members of the City Council, who voted in its favour regardless of the represented political option.

The basis of eligibility is a medical diagnosis indicating one of the diseases classified in ICD-10: F00.0, F00.1, F00.2, F00.9 or G30.0, G30.1, G30.8, G30.9.[32] Save for doctor's certificate (neurologist, psychiatrist or geriatrician), beneficiaries must provide the following documents: caregiver and senior personal data statement, a statement that senior is not a resident of 24-h care facility, a statement of being Szczecin resident, a declaration of family relationship and of providing care for the senior. In addition, at the application stage, the carers agree to take part in a survey which will be conducted to diagnose the situation of older people with Alzheimer's disease. The benefit maybe claimed by family members: spouses, children, grandchildren, siblings, daughters- and sons-in-law. The catalogue of beneficiaries has to be limited due to the lack of clear definition of family in the Act of 28 November 2003 on Family Benefits, which is the legal base of the benefit. In this way, despite

[31] Narodowy Bank Polski 2020.

[32] International Statistical Classification of Diseases and Related Health Problems F00.0—Dementia in Alzheimer disease with early onset (G30.0), F00.1—Dementia in Alzheimer disease with late onset (G30.1), F00.2—Dementia in Alzheimer disease, atypical or mixed type (G30.8), F00.9—Dementia in Alzheimer disease, unspecified (G30.9).

actually provided care, relatives who are not listed in the catalogue (e.g. nieces) had to receive negative decision. Further expanding the catalogue of carers requires regulations on the national level.

It was also necessary to set the age limit of people with the disease, entitled to receive the benefit. At the budgeting stage, it was difficult to estimate how many older citizens suffer from Alzheimer's disease. The main source of information was demographic estimates. Eventually, the age limit of 75 years was set due to the fact that the risk of suffering from multiple illnesses of both the sick individuals and their caring spouses increases with age and leads to greater expenses. People younger than 75 years can benefit from other forms of support.

One of the issues discussed at the development stage was the obligatory and quite complex evaluation survey which was anonymous and kept separate from the documents with personal data. However, it turned out that beneficiaries had no concerns regarding the survey. More, in the open-ended question section of the form, they shared their experiences and suggested solutions not mentioned in the closed-ended questions. The results of the survey allowed to explore the situation of the carers of people with Alzheimer's disease and have become the grounds for developing systemic solutions.

Results of Research Conducted Among the Beneficiaries of Alzheimer 75 Voucher Programme

Results presented in this section were collected during the second edition of the programme *Bon opiekuńczy: Alzheimer 75* in 2019. Upon submitting their applications, carers are obliged to fill in a questionnaire with questions about situation in their households, including: potential, problems, expected forms of support and other. In 2019, there were 800 applications submitted and 721 correctly filled questionnaires. That year, population of Szczecin was 402 thousand, including 33 thousand of citizens aged 75 and more (8.2%) (BDL 2020). People with Alzheimer's disease, whose carers applied for the benefit, were 2.4% of the population of Szczecin citizens aged 75 and more. Considering the fact that data on the number of people with illnesses in certain populations are based on estimates which assume large margin of error, these information maybe helpful when preparing estimates in populations with similar demographic characteristics.

The average age of the caregivers was 63.4 years (SD = 12.8), the oldest carer was 94 and women constituted 71.3% of the group. It is worth to emphasize that care period begins with at the end of the working age, which also entails the necessity to reduce or even suspend one's professional activity. Because Polish retirement system is based on the principle of insurance and the amount of pension depends on the premiums paid, premature professional deactivation results in lower pension in the future. It especially important in case of women because the retirement age in Poland was lowered in 2018 to 60 years for women and 65 for men. Given the average projected life expectance, it results in low pension benefits. The biggest group of carers were daughters (54.8%), then spouses (27.3%) and sons (16.9%). Daughters-in-law are 4.2% of the caregivers and sons-in-law 0.56%. Siblings are primary caregivers in 1.54% of the households. Engagement in care of grandsons and granddaughters is relatively low, 0.56% and 2.38%, respectively. The results of the survey confirm that women bear the main load of caring for the dependent elderly. Despite the fact that the number of men and women in population is similar, most often it is daughters who play the role of carers. This translates into many areas of family, social and professional functioning. Quite surprising is the low involvement of grandchildren who most often fulfil only supporting roles. In the original research conducted during previous editions of the programme, the respondents who were primary caregivers declared that they sometimes tried to protect their children or grandchildren against the hardships of care. They argued that they did want to add more responsibilities and that they can manage on their own. Another important issue is the group of carers in advanced age that is 85 years and more. In their case, sudden health-related emergency may lead to situations when two persons require community-based or residential assistance in the same time. It is particularly important during building support systems as it means that it is necessary to secure emergency beds, for example, in sheltered housing in social assistance homes, which are needed when there is a need to provide care over the dependent within hours. In addition, it is important that in the households where primary caregiver is in advanced age, social assistance workers supervise and provide support on demand or when they notice some deficits or negligence.

Analysis of the obtained empirical material indicates that a carer who applies for the benefit provides care from five years to seven months, 6.3% performs their care duties longer than ten years. 15.5% of the respondents considers placing their loved ones in a facility providing residential care. With Alzheimer's disease, the range and forms of care increase and the load becomes heavier. Extensive care period, which in some cases lasts over 10 years and maximum 20 years, may exhaust mental, economic and family resources. For this reason, it is extremely important that such local activities are undertaken that support carers not only economically but also in terms of counselling and therapy in case of lowered moods or depression. The subject matter literature describes carers of people with Alzheimer's disease as prisoners of care.[33] The name seems justified given the scope of responsibilities, sometimes 24-h care required and when there is no social support—the obligation to provide such care. Despite their load, only small percentage of caregivers considers placing their loved ones in residential care institutions. The main reason behind such attitude is cultural. Polish society is quite conservative and traditional when it comes to family life and roles. Placing older, dependent family member is not always accepted by the society. According to the prevailing cultural pattern, it is family that should ensure the care. There is a quite common belief that social assistance homes are for those

[33] M. Leszko, R. Iwański, B. Bugajska, Prisoners of Care: The Experience of Loneliness among Caregivers of Individuals with Alzheimer's Disease, *Paedagogia Christiana*, [S.l.], v. 45, n. 1, 2020, p. 171–184, May. 2020, https://doi.org/10.12775/PCh.2020.011.

seniors who have no family or were abandoned by their loved ones. In addition, when senior's pension is not enough to cover the costs of stay in a public facility, family members, including spouses, descendants and ascendants are obliged to pay the remaining fee calculated based on the income criterion.

The youngest sick beneficiaries were 75 years old and this age was the bottom eligibility level, the average age was 84.9 years (SD = 5.3) and the oldest person with the disease was 102 years old. Women constituted 71.6% of persons with Alzheimer's. This share results from the process of feminization of old age and, consequently, the growing risk of occurrence of the disease with age. Considering that demographic forecasts assume quite dynamic population ageing within the next years, the number of people suffering from Alzheimer's disease is expected to grow, especially among women aged 85 and more.

In the group of the respondents, 41.8% applied for the benefit during its previous edition in 2018. They were asked what had they bought for the money they had received. Most caregivers (46.5%) declared they had purchased medicines and hygiene products (42.7%). One in ten respondents (10.5%) had paid for doctor's appointments and private nurse (10.8%), and rehabilitation (10.5%). Some carers had also bought food products (14.7%) as well as clothes and bedding for the sick (14.4%). The benefit money enabled the purchase of special bed (7.8% of respondents), antibedsore mattress (8.3%) or care accessories (8.6%). For 7.2% of carers, the support received was used to adapt their apartments to the needs of people with limited mobility, whereas 4.3% financed holiday for themselves and sick seniors. The benefit was awarded as a lump sum and without any obligation to document the expenses. The programme was created based on the assumption that caregivers will now best what to spend the money on to meet the individual needs of their households. High amounts spent on medicines result from the fact that the latest pharmaceuticals are still relatively expensive, even if they are reimbursed by the National Health Fund. Additionally, some seniors suffer from conditions typical for their age and not related with Alzheimer's disease. Purchases of hygiene products for people who are bedridden result from the fact that reimbursement within the public health insurance is low and does not allow for purchasing the sufficient number of hygienic measures such as protective underwear, adult diapers and urology pads. More than 10% of the carers had spent their benefit money on medical services of doctors, rehabilitants or nurses. The healthcare service sector in Poland has been increasingly privatized. This is due to the growing demand for healthcare services and the conditions of supply which depend on the amount accumulated within the obligatory health insurance. Some caregivers bought beds and care accessories dedicated to people with Alzheimer's disease and bedridden patients. It is very important that caregivers are aware of how they can adapt their apartments to the needs of the dependent seniors and their situation so that they are appropriately cared for and safe. The ergonomics of work of carers must also be ensured. Thus, it is no surprise that some beneficiaries used the benefit to adapt their apartments or renovate bathrooms.

As *Bon opiekuńczy: Alzheimer 75* is only a part of the support system that is being created in the municipality, the carers were asked what other types of assistance they would be interested in. The most often selected form of support was 24-h relief care—41.7% of the respondents were interested in it. It is a possibility to place the sick person in a specially adapted apartment for a week or two, where professional care is provided 24 h a day. During this time, the caregiver can rest or has time, for example, to go to hospital or rehabilitation. Other popular form of support (38.3%) was the possibility to leave their loved ones in a specialist facility for several hours a week—the so called day relief care. This form of support allows carers to deal with other daily affairs (like shopping, doctor's appointment or rest) for several hours while the sick person remains under professional care. It is provided for people in the first and at the beginning of the second stage of Alzheimer's disease. Day social assistance homes were mentioned by 12.9% of the respondents. These institutions provide care five days a week, 8 h a day. Almost every fourth respondent (23.6%) is interested in trainings in care for a dependent person and 31.6% would like to have access to psychological-therapeutic consultations. The caregivers also mentioned transportation services as one of the desired forms of support (17.3%). It is particularly important in case of homebound people or people with limited mobility. Other meaningful forms of support involve meetings with other caregivers (14.1%), a helpline for the carers (9.15%) and practical help of volunteers (18.3%).

In order to build a system of supporting people with dementia and their caregivers, that is adequate to their diverse needs and independently of the evaluation performed in the previous editions of the programme, 30 qualitative interviews were conducted in 2018 with spouses who look after their husbands/wives. The respondents were between 72 and 83 years old (median = 78.3 years). The majority of the interviewees were women (73%). A semi-structured interview was used. All interviews were conducted in person, in the carers' homes and lasted from 90 to 120 min. The goal of the interview was to collect narratives about care-related experiences. The areas covered referred to the challenges connected with care, relationships between the caregiver and the sick individual, care needs, caregiver needs, response to diagnosis, expectations and fears connected with further care. In order to identify the common problems experienced by the respondents, a thematic (qualitative) analysis was performed. The quotations are marked to identify carers: W means a *woman*, M—a man and the number indicates the caregiver's age. C with a number means a caregiver and the serial number.

When asked what is the most difficult, the spouses-carers pointed out to the issues connected with physical, mental and social strain. The literature also mentions economic load.[34] The analysis of the answers to other questions indicates that financial situation of families is diverse and the costs of care are high enough so that they affect negatively the budget of patients and their families. However, when asked directly about the most difficult aspects of care, none of the spouses mentioned financial difficulties.

Physical load is the result of taking on too many responsibilities connected with managing the household and the necessity to help elders in their daily activities. As the disease symptoms aggravate with age, sick individuals are more and more dependent

[34] A. Sadowska, Organizacja opieki nad chorym na chorobę Alzheimera w Polsce. In: *Sytuacja osób chorych na chorobę Alzheimera*. Raport RPO. Warszawa 2014.

on their families and need help with the most intimate activities of personal hygiene. This maybe the source of stress and fatigue for the carers.[35] The respondents pointed out to this problem: "washing is the most difficult. I am happy when sometimes my daughter does it. It's a good day for me. Because the smell is sometimes unbearable" [C13-M84]; "Washing and dressing, I can't get enough sleep. I have to get up every two hours" [C15-W80]; "Right now it's the hygiene, that I have to watch over him" [C7-W71]. The necessity to help in daily hygiene, dressing and washing involves physical strain connected with lifting the sick person. One of the respondents who due to her advanced age is in the group of higher risk of health-related problems said: "Lifting my husband is the hardest" [C01-W68].

Lack of cooperation from the person with dementia, resistance against being nursed cause additional difficulty: "my husband's reactions to some of my requests or instructions are the most difficult thing; for example hygiene, that he must change or do something; he responds with hostility" [C05-W75]; "to force her to undress or dress, evenings are the worst, she is stubborn" [C17-M91]; "What is the most difficult… well, guarding… dressing… bathing and hygiene in general, nails, well, general hygiene. And I have to watch if he has dressed this or something else" [C18-W86].

Women who care for their husbands also mention problems with shaving: "Hygiene..bathing… cutting nails… shaving… and with men it is a tragedy… even more so he does not let me shave him" [C19-W82].

Only one respondent said feeding is the most difficult: "Well…now… I guess feeding is the most difficult…" [C20-W89]. This maybe connected with the stage of the disease the sick person experiences.

Another problematic area in care is the psychological stress connected with negative emotions experienced by the caregivers: "Staying calm is the most difficult. When he asks five times or when he cannot find the kitchen and goes to the bathroom instead" [W2]; "I can't leave him home alone, I mean I can and I do but when I leave I'm all stressed out because he may, for example, leave the tap running" [C07-W71].

Psychological load increases with time and may lead to the mental burnout of the carer.[36] According to the respondents, negative emotions result also from the lack of acceptance of the disease and its irreversible character: "The most difficult is the fact that the disease progresses, I mean is it getting worse. With almost every day. And this is so hard to accept [deep sigh] [C10-W83]; "That I have no contact with my husband and that he does not see, this awareness is horrible for me. That this is what has left of the man who worked with his intellect the whole life" [C16-W82].

Psychological load is also connected with communication problems, lack of possibility to interact and preserve the nature of relationship with person under care. Inability to communicate as the disease progresses often leads to negative emotions:

---

[35] E. Fauth, et al., Resistiveness to care during assistance with activities of daily living in non-institutionalized persons with dementia: associations with informal caregivers' stress and well-being, *Aging & mental health,* 20, no. 9, 2016. pp. 888–898.

[36] A. Sadowska, Organizacja opieki nad chorym na chorobę Alzheimera w Polsce. In: *Sytuacja osób chorych na chorobę Alzheimera.* Raport RPO. Warszawa 2014.

"The most difficult is that I don't have any logical contact with him, that he is cannot walk but the most it's the lack of logical interactions" [C08-W82]; "Simple communication… so that he understands what I'm saying… what he is saying. Because there are speech disorders… he is like unconscious, he lives in his own world and that's it… this the hardest that nothing can get through to him" [C11-W77].

Some caregivers do not cope with caring for their loved ones, everything is difficult, it is beyond their physical and mental abilities: "For me everything is hard" [C16-W82]; "How should I say this… well, it's all difficult… because I was not used to this happening. One illness, then another… and so the two happened together" [C14-W74]. This state of affairs suggests the need to provide assistance in different areas of daily life plus psychological or even therapeutic support.

Changes and difficulties maybe also associated with accepting the situation and be the evidence of the great caring potential of the caregivers. This is illustrated by the following statement: "The most difficult is constant repeating: throw waste to the bin, do not take it to the balcony… constant saying: Don't do this… Go, spit it to the sink, flush the toilet… saying this all the time. But… I became attached to him, I got used to it and… I keep going" [C15-W81].

Acceptance of the situation and the necessity to care for a spouse, as well as strategies used to cope with the situation of care is also visible in other statements: "I don't think about it at all. It just needs to be done and I do it. I don't think about it much" [C03-W78]; "Often when I'm tired, I feel this but at the same time I try to think rationally and not to give up to this, not to sit and cry, and I try not to burden my family because they are busy working" [C22-M85]; "There is nothing difficult" [C21-W84].

Another, rarely mentioned issue is social load connected with lack of possibility to participate in social and professional life, limited or non-existing social relationships and recreational, leisure activities: "I have no freedom. I can't go shopping alone, to buy something for myself. He's like a child, he wants to go with me everywhere" [C4-W72]; "Lack of possibility to go out and deal with my affairs (…) it's not that I would like to go out to relax but every time I leave the house, there must be some supervision—I can go out just for a little while or at all because I can't leave Mary. That is why I resigned from collaborating with the Pomeranian Medical University (…) [C09-M75]; "You know, the hardest thing is that I can't go for a walk with him, he doesn't go out and so do I" [C7-W71].

The analysis of the difficulties mentioned by the respondents shows that the caregivers experience physical, psychological and social strain. These loads are both objective and subjective. Objective difficulties are caused by the behaviours of people with Alzheimer's, changes in the daily rhythm and physical workload. Subjective load is connected with emotional reactions of carers to the situation of care. Both types of loads can result in psychosomatic symptoms and deteriorating health of the caregivers. People responsible for care over their loved ones, especially when they have no support from others, suffer from chronic fatigue, neglect their own needs and in consequence, experience physical, psychosomatic and psychosocial symptoms called the "caregiver syndrome".[37] Due to the fact that institutional forms of support of people with Alzheimer's disease are not able to provide care for the growing numbers of patients with dementias, local solutions are being sought that will strengthen the potential of families as the main source of care. The solutions are developed based on scientific research which are the integral element of the financial benefit introduced in Szczecin: Alzheimer 75.

Conclusion

In the Polish system of long-term care and assistance for the dependent persons, local governments are responsible for most of the tasks of organising and cofinancing the benefits. There is a noticeable, sizeable disproportion between the large cities which try to implement active senior policy in this regard and small municipalities which lack adequate financial and property resources to provide, for example, relief apartments. However, it is worth to emphasize that supporting families in caring for persons with Alzheimer's disease together with active development of community-based assistance (consistent with the idea of deinstitutionalization) is economically justified as it reduces the risk of placing seniors in the residential care facilities. Supporting caregivers translates also into their mental and physical wellbeing. Especially when after a long period of providing care, primary caregivers have to find themselves in their new role. Even though most people have general idea about Alzheimer's disease, it is important to support caregivers at certain stages of the disease to minimize the risk of physical and emotional exhaustion. Sometimes, it is enough to organize support groups, counselling points or trainings. Such initiatives may improve the quality of care as well as comfort and safety of the primary carers.

The local programme of financial support for caregivers of people with Alzheimer's disease called *Bon opiekuńczy: Alzheimer 75*, the integral part of which are the scientific research evaluation the programme, is a good practice in strengthening the care potential of families. This innovative benefit is part of regularly developed system of support of the elder citizens in Szczecin, who need assistance in their daily functioning and for family caregivers. The system consists of: social assistance homes, sheltered and supported apartments, day social assistance homes, system of care services provided at home and other. The analysis of the empirical data obtained from the caregivers of people with Alzheimer's aged 75 and more indicates that in most cases families feel responsible for providing care for their loved ones. However, they need systemic support from the public sources that would be adapted to their needs and caring potential. To meet the needs of the caregivers, the *Centre for support of caregivers of the dependent people with dementia "Forget-me-not"* was opened in 2020. It provides day relief care, consultations, counselling and trainings for the

[37] B. Grochmal—Bach, *Cierpienie osób z otępieniem typu Alzheimera Podejście terapeutyczne*, Wydawnictwo WAM. Kraków 2007.

caregivers, individual counselling, trainings in care, self-help and therapeutic groups. There is a rehabilitation and medical equipment rental and a mobile support team consisting of a psychologist, a carer and a social worker. The team provides assistance in home environment when the health condition of the caregiver prevents them from participating in the activities in the centre.

These intense activities to support families of people with Alzheimer's disease have met with positive social response. As there are no governmental programmes to support people with dementia, it is local authorities that have to face the challenge of developing local strategies of strengthening the care potential of families. This is particularly important in the light of projected growing demand for care over the dependent persons, including the ones with Alzheimer's disease, in the coming years. Local governments must initiate a discussion over the systemic and statutory solutions regarding long-term care insurance, system of geriatric care and the necessity to introduce proper standards, integrate social, healthcare and neighbourhood services. The above presented solutions which were designed and introduced locally may inspire other communities which face the same problem.

References

Act of 12 March 2004 on Social Assistance, Journal of Laws 2004 nr 64 Item 593, Accessed: February 25 2020. http://prawo.sejm.gov.pl/isap.nsf/DocDetails.xsp?id=WDU20040640593

Alzheimer Europe. (2020). *National dementia strategies*. Access April 17, 2020. https://www.alzheimer-europe.org/Policy/National-Dementia-Strategies

Bank Danych Lokalnych. (2020). https://bdl.stat.gov.pl/BDL/dane/teryt/tablica

Brookmeyer, R., Johnson, E., Ziegler-Graham, K., & Arrighi, H. M. (2007). Prognozowanie globalnego obciążenia chorobą Alzheimera. *Alzheimer i demencja, 3*(3), 186–191. https://doi.org/10.1016/j.jalz.2007.04.381

Bugajska, B., Iwański, R., Giezek, M., & Ciemniak, A. (2017). The optimization of elderly care in Szczecin: research report. *Exlibris: biblioteka gerontologii społecznej, 1*(13), 107–119. 2017.

European Commission. (2018). *The 2018 ageing report economic & budgetary projections for the 28 EU Member States (2016–2070) institutional paper 079*. Publications Office of the European Union.

Fauth, E. B., Femia, E. E., & Zarit, S. H. (2016). Resistiveness to care during assistance with activities of daily living in non-institutionalized persons with dementia: associations with informal caregivers' stress and well-being. *Aging & Mental Health 20*(9), 888–898.

Fihel, A., & Okólski, M. (2018). Przemiany cywilizacyjne, ludnościowe i starzenie się społeczństw. In M. Okólski, Wyzwania starzejącego się społeczeństwa (pp. 22–65). Uniwersytet Warszawski.

Grochmal – Bach B. (2007). Cierpienie osób z otępieniem typu Alzheimera Podejście terapeutyczne. Wydawnictwo WAM. Kraków.

Iwański, R. (2016). Opieka długoterminowa nad osobami starszymi. Cedewu.

Jonsson, T., Atwal, J., Steinberg, S., et al. (2012). A mutation in APP protects against Alzheimer's disease and age-related cognitive decline. *Nature, 488*, 96–99. https://doi.org/10.1038/nature11283

Lee, R. (2003). The demographic transition: Three centuries of fundamental change. *Journal of Economic Perspectives, 17*(4), 167–190.

Leszko, M., Iwański, R., & Bugajska, B. (2020, May). Prisoners of care: The experience of loneliness among caregivers of individuals with Alzheimer's disease. *Paedagogia Christiana,* [S.l.] *45*(1),

171–184. https://doi.org/10.12775/PCh.2020.011. ISSN 2451-1951. Available at: https://apcz.umk.pl/czasopisma/index.php/PCh/article/view/PCh.2020.011. Date Accessed June 19, 2020.

Ministerstwo Rodziny, Pracy i Polityki Społecznej. (2020). Raport mrpips-03 za 2018 rok, https://www.gov.pl/web/rodzina/statystyka-za-rok-2018

NIK. (2016). Opieka nad osobami chorymi na chorobę Alzheimera oraz wsparcie dla ich rodzin. Warszawa. Access May 20, 2020. https://www.nik.gov.pl/plik/id,13694,vp,16130.pdf

NIK. (2018). Usługi opiekuńcze świadczone osobom starszym w miejscu zamieszkania. Najwyższa Izba Kontroli. Retrieved from https://www.nik.gov.pl/kontrole/P/17/043/

Narodowy Bank Polski. (2020). Kursy średnie walut obcych. Kurs z dnia 19.06.2020 Accessed: June 19 2020. https://www.nbp.pl/home.aspx?f=/kursy/kursya.html

OECD. (2017). Life expectancy and healthy life expectancy at age 65 in Health at a Glance. Paris: OECD. https://doi.org/10.1787/health_glance-2017-74-en

OECD (2018). Care needed: Improving the lives of people with dementia, OECD Health Policy Studies, OECD Publishing, Paris. https://doi.org/10.1787/9789264085107-en

OECD. (2020). *Health expenditure and financing.* Retrieved from OECD Stat. https://stats.oecd.org/Index.aspx?QueryId=30142

Plassman, B. L., Langa, K. M., Fisher, G. G., Heeringa, S. G. Weir, D. R., Ofstedal, M. B., Burke, J. R., Hurd, M. D., Potter, G. G., Rodgers, W. L., Steffens, D. C., Willis, R. J., & Wallace, R. B. (2007). Prevalence of dementia in the United States: The aging, demographics, and memory study. *Neuroepidemiology, 29*, 125–132. https://doi.org/10.1159/000109998

Qiu, C., Kivipelto, M., & von Strauss, E. (2009). Epidemiologia choroby Alzheimera: Występowanie, determinanty i strategie interwencji. *DialoguesClin. Neurosci, 11*, 111–128.

Sadowska, A. (2014). Organizacja opieki nad chorym na chorobę Alzheimera w Polsce. In: Sytuacja osób chorych na chorobę Alzheimera. Raport RPO. Warszawa.

Schaller, S., Mauskopf, J., Kriza, C., Wahlster, P., & Kolominsky-Rabas, P. L. (2014). The main cost drivers in dementia: A systematic review. *International Journal of Geriatric Psychiatry, 30*, 111–129. https://doi.org/10.1002/gps.4198

Szczudlik, A., Barcikowska-Kotowicz, M., & Sadowska, A. (2014). Polski Plan Alzheimerowski - prezentacja założeń. In: Sytuacja osób chorych na chorobę Alzheimera. Raport. rzecznik Praw Obywatelskich, Warszawa.

WHO. (2017). *Global action plan on the public health response to dementia 2017–2025.* https://www.who.int/mental_health/neurology/dementia/action_plan_2017_2025/en/

Wimo, A., Jonsson, L., & Winblad, B. (2006). An Estimate of the Worldwide prevalence and direct costs of dementia in 2003. *Dementia and Geriatric Cognitive Disorders, 21*(3), 175–181. https://doi.org/10.1159/000090733

Wimo, A., Guerchet, M., Ali, G. C., Wu, Y. T., Prina, A. M., Winblad, B., Jönsson, L., Liu, Z., & Prince, M. (2017). The worldwide costs of dementia 2015 and comparisons with 2010. *Alzheimer's & Dementia : The Journal of the Alzheimer's Association, 13*(1), 1–7. https://doi.org/10.1016/j.jalz.2016.07.150

World Bank. (2020). *Health nutrition and population statistics: Population estimates and projections.* Retrieved Luty 20, 2020, from World Bank Open Data. https://databank.worldbank.org/source/population-estimates-and-projections

ZUS. (2020). Struktura wysokości emerytur i rent wypłacanych przez ZUS po waloryzacji w marcu 2018 roku, Accessed: Marc 29 2020, https://www.zus.pl/documents/10182/39637/Struktura+wysoko%C5%9Bci+emerytur+i+rent+wyp%C5%82acanych+przez+ZUS+po+waloryzacji+w+marcu+2018+r.pdf/2f377123-7bd5-4e3a-a511-013eed30671c

Chapter 10
Supporting Home-Based Dementia Care

Nimmy Chandran and K. S. Shaji

Abstract Majority of patients suffering from dementia live with their family members and get home-based care. This is more so in low- and middle-income countries, where institutional care is almost non-existent. Though home-based dementia care has many advantages, it is not easy to provide home-based dementia care for longer periods. Dementia care can adversely impact the physical and mental health of caregivers. Caregivers encounter many difficulties and challenges. Formal healthcare services are often unhelpful and do not reach out to support home-based care. The challenges involved in providing home-based care include difficulties in managing the behavioural and psychological symptoms of dementia, lack of financial support and difficulties in ensuring the safety of the person with dementia. Caregivers often do not have access to much needed information about dementia and its management. They often do not get the kind of support they very much need. A good home-based care model should be able to connect home-based care to formal healthcare delivery system by linking with community-based services. Outreach services can be provided by trained health workers who can make regular home visits and advise caregivers. These health workers can be supervised by primary healthcare teams. Clinicians and nurses who are part of this team can help by providing adequate knowledge and skills to manage people with dementia. They could also provide clinic-based services to support older people with dementia when needed. Home-based care can also take help from other available services like special healthcare services for older people or general healthcare services. We need a competent healthcare workforce who can take up the task of educating and supporting the caregivers to improve home-based care of older people with dementia. Technological assistance can be used to communicate with the caregivers, help in-home safety, and in monitoring the physical health aspects.

Keywords Dementia care · Home-based care · Caregiver health · Behavioural and psychological symptoms of dementia

---

N. Chandran (✉)
Department of Psychiatry, Government Medical College, Palakkad, Kerala, India

K. S. Shaji
Kerala University of Health Sciences, Thrissur, Kerala 680596, India

M. K. Shankardass (ed.), *Dementia Care*,
https://doi.org/10.1007/978-981-16-3864-0_10

Introduction

Most people with dementia receive home-based care. This is particularly so in low- and middle-income countries, where institutional care is almost non-existent. Around two-thirds of people with dementia live in their own homes in the USA.[1] The proportion of people with dementia receiving home-based care is about 94% in low- and middle-income countries.[2]

Dementia is a disabling chronic disease affecting older people. A prolonged period of care is often needed. Older people with dementia prefer to live with their family members at home, as long as possible.[3] Home-based care has many advantages. The home setting is capable of providing better social, environmental and emotional care. Studies have shown that community resident older people with dementia have better cognition, physical functioning and are more active socially when compared to those who receive institutional care.[4] Family members can share the responsibility of care among themselves. The cost of home-based care is less when compared to that of institutional care.[5] This maybe the only feasible option for most people with dementia in low and middle-income countries.

It is not easy to provide home-based dementia care over long periods. Dementia care can have adverse impact on the physical and mental health of caregivers. Caregivers encounter many difficulties and challenges.[6] Caregivers often do not get emotional support, adequate time for relaxation, proper education regarding the dementia and will not be able to care for their own medical and mental health needs. Their health, work and social life often suffer due to care-related responsibilities.

---

[1] Jennifer L. Wolff et al., "A National Profile of Family and Unpaid Caregivers Who Assist Older Adults With Health Care Activities", *JAMA Internal Medicine* 176, no. 3 (March 2016): 372–79, https://doi.org/10.1001/jamainternmed.2015.7664.

[2] "World Alzheimer Report 2015, The Global Impact of Dementia: An Analysis of Prevalence, Incidence, Cost and Trends", n.d., 87.

[3] Stein Erik Fæø et al., "'We Live as Good a Life as We Can, in the Situation We're in'—the Significance of the Home as Perceived by Persons with Dementia", *BMC Geriatrics* 19 (June 6, 2019), https://doi.org/10.1186/s12877-019-1171-6.

[4] Christine Olsen et al., "Differences in Quality of Life in Home-Dwelling Persons and Nursing Home Residents with Dementia—a Cross-Sectional Study", *BMC Geriatrics* 16 (11 2016): 137, https://doi.org/10.1186/s12877-016-0312-4; Yu-Chun Kuo et al., "Dementia Care Costs and the Patient's Quality of Life (QoL) in Taiwan: Home versus Institutional Care Services", *Archives of Gerontology and Geriatrics* 51, no. 2 (September 2010): 159–63, https://doi.org/10.1016/j.archger.2009.10.001.

[5] Ansgar Wübker et al., "Costs of Care for People with Dementia Just before and after Nursing Home Placement: Primary Data from Eight European Countries", *The European Journal of Health Economics: HEPAC: Health Economics in Prevention and Care* 16, no. 7 (September 2015): 689–707, https://doi.org/10.1007/s10198-014-0620-6.

[6] Alina de las Mercedes Campos-Puente et al., "Emotional Exhaustion in Housewives and Alzheimer Patients' Caregivers: Its Effects on Chronic Diseases, Somatic Symptoms and Social Dysfunction", *International Journal of Environmental Research and Public Health* 16, no. 18 (January 2019): 3250, https://doi.org/10.3390/ijerph16183250.

Other challenges include behavioural problems of patients with dementia, difficulties with administration of medication, lack of financial support and the difficulty in ensuring the safety of the person with dementia.[7] Often, there is a lack of coordination between healthcare services and community-based support systems. Formal healthcare services are often unhelpful and do not reach out to support home-based care. Services do not identify and make use of available community resources.

Supporting Home-Based Care

A good home-based care model should remain connected to the formal healthcare delivery system through existing community outreach services. Different services could be developed within the healthcare system to offer support for older people. A responsive community outreach service which provides support to those engaged in home-based care can be a game-changer. Many specialist teams can remain linked to home-based care. We could also consider primary care teams offering support for home-based dementia care.

Role of Specialist Care Providers

The specialist care teams can have psychiatrists, geriatricians, nurses and social workers. They can deliver services directly and also supervise the interventions delivered by non-specialist healthcare providers. Interventions shall aim to meet the needs of the person with dementia and their caregivers. This can lead to better management of physical/behavioural symptoms, reduction of polypharmacy, assistance with advance care planning, provision of psychosocial support and disease education for the caregivers. However, specialist care services are not feasible in low- and middle-income countries and the scenario is unlikely to change soon. Tertiary care centres may develop such services to supervise dementia care services provided by non-specialist healthcare providers.

One such treatment model is a consultative home-based dementia programme like Caring for Older Adults and Caregivers at Home (COACH) programme. This was developed and implemented at the Durham Veterans Affairs Medical Centre in Durham, North Carolina, in 2010.[8] It provides dementia care to the patient and support the caregivers of dementia to help them to live at home for as long as possible.

[7] Young Mi Lim et al., "Factors Affecting Burden of Family Caregivers of Community-Dwelling Ambulatory Elders with Dementia in Korea", *Archives of Psychiatric Nursing* 22, no. 4 (August 2008): 226–34, https://doi.org/10.1016/j.apnu.2007.12.005.

[8] Maria F. D'Souza et al., "Preliminary Data from the Caring for Older Adults and Caregivers at Home (COACH) Program: A Care Coordination Program for Home-Based Dementia Care and Caregiver Support in a Veterans Affairs Medical Center", *Journal of the American Geriatrics Society* 63, no. 6 (2015): 1203–8, https://doi.org/10.1111/jgs.13448.

A social worker and registered nurse with experience in geriatric care provide care through telephone and home visits. They also get support from an interdisciplinary team including the geriatrician, geriatric psychiatrist and geriatric pharmacist.

Role of Non-specialist Healthcare Providers

Services provided by non-specialist healthcare providers employed in primary and secondary care settings maybe much more feasible than specialist care services. A flexible stepped care model can be used. Outreach services can be provided by trained health workers who make regular home visits and give advice to caregivers. These health workers can be supervised by doctors and nurses from the primary healthcare teams. The goal of such services will be to improve the knowledge of the family caregivers, to give them emotional support, expand the caregiving resources and to improve their caregiving skills. Following are some of the important components of such services.

- Provision of information and education regarding the illness (nature of illness, long-term course, major symptoms and other associated features, etc.)
- Education about common behavioural symptoms associated with dementia and how to tackle them
- Emotional support to the caregiver
- Referral to specialists for severe behavioural problems; help to form support groups by connecting families caring for people with dementia
- Advice regarding existing government schemes for senior citizens.

An example of a similar intervention is a randomized controlled trial conducted in India to evaluate the effectiveness of a home-based intervention in 2008. The study was conducted in Goa, India. Subjects with dementia (excluding the patients with severe dementia) and their caregivers were randomized either to the intervention group or to a waiting list control group. The intervention was provided by the home care advisors (non-specialized healthcare workers in the community), who were trained specifically for the programme. They were managed by a counsellor and a psychiatrist. The intervention focused on supporting the caregiver by providing information on dementia, guiding behavioural management, psychiatric assessment and psychotropic medication if needed. That study demonstrated reduction in caregiver burden, mental stress and distress due to the behavioural and psychological symptoms of dementia.[9] A total of 59 patients and their caregivers completed the trial. They could find that the intervention led to a significant reduction of mental stress assessed through General Health Questionnaire and distress due to behavioural disturbances of

---

[9] Amit Dias et al., "The Effectiveness of a Home Care Program for Supporting Caregivers of Persons with Dementia in Developing Countries: A Randomised Controlled Trial from Goa, India", *PloS One* 3, no. 6 (June 4, 2008): e2333, https://doi.org/10.1371/journal.pone.0002333.

dementia assessed by Neuropsychiatric Inventory-D scores. They also reported non-significant reductions in the caregiver burden assessed with Zarit Burden Interview; behavioural symptoms assessed using Neuropsychiatric Inventory-Severity scores. They also found a non-significant decrease in the mortality among in the intervention group.

Another example of a care programme with non-specialist healthcare workers is Maximizing Independence (MIND) at Home dementia care coordination pilot trial, carried out in Baltimore, Maryland.[10] That trial recruited community resident older people with cognitive impairment and their caregivers. The intervention consisted of 18-month long care coordination provided by non-clinical community workers. They addressed the unmet care needs of patients by individually tailored care planning. Those who needed a referral to the dementia services were encouraged to take help as and when needed. Caregivers were provided with information about dementia and were assisted in skill building. Care progress monitoring was done. However, the investigators could not find any statistically significant difference between groups in caregiver burden measures, depression or quality of life. But they reported a reduction in the total time the caregivers spent with the person with dementia in the intervention group when compared with controls.

Other Models of Support

There is a growing realization among researchers and service providers about the need to strengthen home-based dementia care. We can learn many things from the ongoing efforts to support home-based care. Many interventions were tried, different outcome measures were used and that makes comparison difficult. However, we need to examine these interventions carefully to see if they can be used in home-based care.

One collaborative treatment model from Germany provided assistance to home-based care through geriatric nurses in collaboration with primary care physicians and specialists providing technological assistance. The effectiveness and safety of the dementia care management programme was then assessed by a general practitioner-based intervention.[11] The authors compared the intervention with the usual treatment at base line and after one year following the intervention. The intervention significantly decreased behavioural and psychological symptoms of patients with dementia and caregiver burden compared to the treatment as usual. It was also found to be

[10] Tanner Ja et al., "A Randomized Controlled Trial of a Community-Based Dementia Care Coordination Intervention: Effects of MIND at Home on Caregiver Outcomes". *The American Journal of Geriatric Psychiatry: Official Journal of the American Association for Geriatric Psychiatry* 23, no. 4 (August 13, 2014): 391–402, https://doi.org/10.1016/j.jagp.2014.08.002.

[11] Jochen René Thyrian et al., "Effectiveness and Safety of Dementia Care Management in Primary Care: A Randomized Clinical Trial", *JAMA Psychiatry* 74, no. 10 (01 2017): 996–1004, https://doi.org/10.1001/jamapsychiatry.2017.2124.

safe. The intervention arm patients had received more pharmacological treatment for dementia compared to others.

An example of another integrated model of care with a multidisciplinary team including case manager and specialists is Geriant in Netherlands. Geriant provides a combined set of dementia care services including diagnosis, management and continuous support. They have a multidisciplinary team, comprising of case managers, geriatricians, psychiatrists, clinical psychologist, dementia counsellor and one or more specialized home care nurses. The patients are referred to Geriant by their general practitioner, at the onset of the illness. The case management team will make a detailed assessment and treatment plan. They will then provide dementia-related acute and long-term care services throughout patients' life. Case manager coordinates the care services and provides additional support as and when needed. If specialized or intensive treatment needed, the case manager will inform and involve specialists. This model helps people with dementia to live at home for longer periods, by providing a supportive environment, and to improve the quality of life of both patients and their caregivers.[12] This is a well-integrated model of care for people with dementia, but may not be feasible in resource limited settings.

Access to tele consultations can be of great help to the patients and caregivers engaged in home-based care. A randomized trial (Care of Persons with Dementia in their Environments [COPE]) of patients with dementia and family caregivers (community-living dyads) was conducted in Pennsylvania. Up to 12 home or telephone contacts were provided over four months by health professionals.[13] They assessed the capabilities and deficits of people with dementia also did necessary laboratory investigations. They trained families to use in-home measures to ensure the safety of people with dementia, methods to simplify tasks, and reduce stress. Caregivers in the control group received only three telephone calls and educational materials. The investigators assessed the functional dependence, quality of life, frequency of agitated behaviours and engagement of people with dementia. The well-being, confidence and perceived benefits for the caregivers were also assessed. At the end of four months, people who received the COPE intervention had less dependence on their activities of daily living. The caregivers in the intervention group reported better engagement and improvement in their well-being.

A caregiver training programme in 2003 for supporting home-based dementia care from Taiwan showed that simple home-based interventions can improve the self-efficacy of caregivers. The caregivers received additional two sessions of individual training when compared to the control group who received only written materials for information and education about dementia. The experimental group showed significant improvement in overall self-efficacy for managing behavioural problems of

[12] Ludo Glimmerveen and HenkNies, "Integrated Community-Based Dementia Care: The Geriant Model", *International Journal of Integrated Care* 15 (September 23, 2015), https://www.ncbi.nlm.nih.gov/pmc/articles/PMC4628512/.

[13] Laura N. Gitlin et al., "A Biobehavioral Home-Based Intervention and the Well-Being of Patients With Dementia and Their Caregivers", *JAMA: The Journal of the American Medical Association* 304, no. 9 (September 1, 2010): 983–91, https://doi.org/10.1001/jama.2010.1253.

patients with dementia at three months after training.[14] Another randomized control trial in Taiwan used individualized training for managing the behavioural symptoms of dementia compared to written instructions and social telephonic follow up in controls.[15] The investigators found a significant reduction in physically aggressive behaviours of dementia patients and improvement in the caregiver's preparedness, self-efficacy and competence for managing the problem behaviours.

Newer integrated models of home-based care for dementia are being developed and getting evaluated at many parts of the world. Interdisciplinary Home-based Reablement Programme (I-HARP) is an integrated model of dementia care which integrates community-based care services and hospital-based community geriatric services.[16] It consists of 8–12 home visits by an occupational therapist, registered nurse and other health staff; technological assistance by providing minor home modifications/assistive devices and three individual caregiver education sessions.

Home-based dementia care is gaining more importance across the world. New and innovative models of home-based care approaches are emerging. It is difficult to compare the studies as most of the studies have different methodologies and have used different outcome measures. Feasibility and the cost-effectiveness of different models of home-based care need to be established. We have to make sure that home-based care remains both evidence-informed and user-friendly. Linking with existing services through outreach services is an option which needs to be exercised. Specialist inputs can complement existing general healthcare services.

Technological Assistance for Home Base Care

Many innovative assistive technologies are now available for helping people with dementia and their caregivers. The ENABLE project was an innovative project which made use of assistive technologies for the care of people with dementia. That project was conducted across the UK, Norway and Finland. They have looked into the

[14] Huei-Ling Huang et al., "A Pilot Study on a Home-Based Caregiver Training Program for Improving Caregiver Self-Efficacy and Decreasing the Behavioral Problems of Elders with Dementia in Taiwan", *International Journal of Geriatric Psychiatry* 18, no. 4 (April 2003): 337–45, https://doi.org/10.1002/gps.835.

[15] Huei-Ling Huang et al., "A Home-Based Training Program Improves Caregivers' Skills and Dementia Patients' Aggressive Behaviors: A Randomized Controlled Trial", *The American Journal of Geriatric Psychiatry* 21, no. 11 (November 2013): 1060–70, https://doi.org/10.1016/j.jagp.2012.09.009.

[16] Yun-Hee Jeon et al., "A Pragmatic Randomised Controlled Trial (RCT) and Realist Evaluation of the Interdisciplinary Home-BAsed Reablement Program (I-HARP) for Improving Functional Independence of Community Dwelling Older People with Dementia: An Effectiveness-Implementation Hybrid Design", *BMC Geriatrics* 19, no. 1 (29 2019): 199, https://doi.org/10.1186/s12877-019-1216-x.

difficulties involved in monitoring and supporting people with dementia.[17] They could then develop many useful devices for helping the people with dementia, such as gas cooker monitors, automatic night lights and smart flats.[18] They also played a major role in forming guidelines for the use of assistive technologies on the basis of field studies.[19]

A wide range of assistive technologies can support older people with dementia and their caregivers. This indeed is a welcome development for home-based dementia care.[20] Assistive technologies using smartphone-based applications like alarm techniques can remind people to take medications. Smartwatches and sensors can help with physical health monitoring and improve the safety of people with dementia. For example, the smartwatch can monitor the sleep patterns of the patient. Sensors can locate the person and prevent the person from wandering away. Technologies also help in facilitating social contact and communication. More than anything else, the use of technology would allow more quality rest time for the caregivers and this in turn can reduce the caregiver burden. Proper use of assistive technologies thus can improve the quality of lives of people with dementia and their caregivers.

More studies will have to evaluate the efficacy of assistive technologies. Supporting Family Caregivers with Technology for Dementia Home Care (FamTech-Care) is a randomized controlled trial to evaluate the effects of a tele health intervention on caregiver well-being and behavioural symptoms of people with dementia. It is being conducted at the University of Kansas and Iowa in the USA.[21] It provides individualized care strategies to caregivers based on video recordings of challenging care situations. Dementia care experts will watch the videos sent by caregivers weekly and provide instructions to improve the care. Control group will get advice over phone for improving care on a weekly basis and will receive feedback on their videos only at the end of the three-month study. The investigators will compare behavioural symptoms and caregiver burden after one and three months. More innovations are likely to follow and hopefully we are poised to have better solutions in the near future itself.

The major drawback with some of these assistive technologies is that they often are very costly, and thus most people cannot afford it. However, there are many mobile

---

[17] Timothy Adlam et al., "The Installation and Support of Internationally Distributed Equipment for People With Dementia", *IEEE Transactions on Information Technology in Biomedicine: A Publication of the IEEE Engineering in Medicine and Biology Society* 8 (October 1, 2004): 253–57, https://doi.org/10.1109/TITB.2004.834393; Sidsel Bjørneby et al., "Ethical Considerations in the ENABLE Project", *Dementia* 3 (October 1, 2004): 297–312, https://doi.org/10.1177/1471301204045162.

[18] Nina Evans et al., "Evaluation of an Enabling Smart Flat for People with Dementia" 15 (November 1, 2007): 33–36.

[19] Adlam et al., "The Installation and Support of Internationally Distributed Equipment for People With Dementia".

[20] Lefteris Koumakis et al., "Dementia Care Frameworks and Assistive Technologies for Their Implementation: A Review", *IEEE Reviews in Biomedical Engineering* 12 (2019): 4–18, https://doi.org/10.1109/RBME.2019.2892614.

[21] Kristine Williams et al., "A Randomized Trial Using Telehealth Technology to Link Caregivers with Dementia Care Experts for In-Home Caregiving Support: FamTechCare Protocol", *Research in Nursing & Health* 41, no. 3 (2018): 219–27, https://doi.org/10.1002/nur.21869.

applications which can be of help to the caregivers and persons with dementia. They are less costly, easily accessible and can be widely used. We need to educate the families engaged in home-based care about such mobile applications availability and potential usefulness.

Other Interventions for Support

Various complementary therapies like mindfulness-based cognitive therapy and music therapy have been tried for better dementia care at home.[22,23] Caregivers support groups are very helpful for the emotional support of caregivers. Advance care planning (ACP) is another useful intervention in dementia. ACP is defined "as a process of discussing and recording of wishes, values and preferences for the future care and treatment held between an individual, family members and their care provider".[24] A person with dementia may need to appoint a person for deciding on their future treatment of choice. ACP aims to ensure that people receive treatment and care according to their values and preferences during serious and chronic illness. ACP is estimated to occur with only 3%–39% of people with dementia internationally. As ACP is of particular significance to people with dementia, it should be considered and completed early in the course of illness before further cognitive decline. Cultural values, beliefs and practices of the home care system are important factors in decision-making for persons with dementia and caregivers. There are many challenges involved in planning the ACP of patient. Home-based care is an ideal setting for ACP.

Community-Based Services and Resources

There are many schemes and projects by governmental and non-governmental agencies which can assist home-based care. All of us should look for locally available resources and connect the families engaged in dementia care to such potentially useful resources. A project called "Vayomithram" is implemented by the Kerala Social Security Mission in the southern state of Kerala, India. This programme aims

---

[22] Patrick Pui Kin Kor, Justina YatWa Liu, and Wai Tong Chien, "Effects of a Modified Mindfulness-Based Cognitive Therapy for Family Caregivers of People with Dementia: A Pilot Randomized Controlled Trial", *International Journal of Nursing Studies* 98 (October 2019): 107–17, https://doi.org/10.1016/j.ijnurstu.2019.02.020.

[23] Hacer Gok Ugur et al., "Effects of Music Therapy on the Care Burden of In-Home Caregivers and Physiological Parameters of Their In-Home Dementia Patients: A Randomized Controlled Trial", *Complementary Medicine Research* 26, no. 1 (2019): 22–30, https://doi.org/10.1159/000490348.

[24] Rebecca L. Sudore et al., "Defining Advance Care Planning for Adults: A Consensus Definition from a Multidisciplinary Delphi Panel", *Journal of Pain and Symptom Management* 53, no. 5 (May 2017): 821–832.e1, https://doi.org/10.1016/j.jpainsymman.2016.12.331.

to provide services for elderly residing in the corporation/ municipal areas.[25] The services will soon be available in rural areas also. Vayomithram provides medicines free of cost by conducting multiple mobile clinics in various locations. It can be accessed by all regardless of their financial status. A Vayomithram unit comprises of a Medical Officer, Staff Nurse, Junior Public Health Nurse. Vayomithram coordinators are there in every district and they provide help desk and telephonic services to older people in need of help. Older people with dementia and their caregivers can get help from this project.

Palliative care services are providing home-based care at all panchayats, municipalities and corporations in the state of Kerala. It has a large network comprising of 1550 palliative care units, out of which 450 are run by various NGOs and community-based organizations. Efforts are being made to integrate dementia care into palliative care services. Many people with dementia, especially those who are bedridden are getting help and support from this community outreach service now. Palliative care services are getting scaled up in all parts of India and thus have the potential to provide wider support for home-based dementia care.

The Kerala government has a project called Aswasakiranam which support the caregivers of mentally and physically disabled bedridden people.[26] The scheme provides monthly financial assistance to informal caregivers who are unable to take up any employment due to their caregiving responsibilities. Caregivers of people with dementia can apply and get support. There is another project called Sayamprabha Home which offers help to older people and those in early stages of dementia.[27] Sayamprabha homes are basically day care centres which offer a stimulating environment for social interaction for senior citizens.

There are many underutilized resources in many states. There are many schemes which can potentially be of help to caregivers and families engaged in home-based care. Media and local self-governments can play an important role in raising public awareness about dementia and the need to support and strengthen home-based care. The Panchayati raj system of local governance can decide to help home-based care of all disabled people. We need to encourage people's participation in local governance and social welfare.

Conclusions

Strengthening home-based dementia care has become important, especially following the impact of the COVID 19 pandemic. Better home-based care models

[25] "VAYOMITHRAM", accessed June 27, 2020, http://www.socialsecuritymission.gov.in/scheme_info.php?id=7.

[26] "ASWASAKIRANAM", accessed June 27, 2020, http://www.socialsecuritymission.gov.in/scheme_info.php?id=5.

[27] "Social Justice, Kerala", accessed June 27, 2020, http://swd.kerala.gov.in/scheme-info.php?scheme_id=IDEyNw==.

are possible with advanced technology. Home-based should get connected with the existing healthcare services so that the unmet needs of people with dementia and their caregivers can be addressed easily. Effort should be taken in empowering people, families and communities so that they can make best use of the available community resources and support systems. Education regarding advanced care planning, assistance in planning legal and financial matters should get due attention. Technological advances offers hope for addressing challenging issues like ensuring the safety of the patient. The future of dementia care lies in the development and implementation of evidence-informed models for home-based care which meet the needs and aspirations of the families.

References

Adlam, T., Faulkner, R., Orpwood, R., Jones, K., Macijauskiene, J., & Budraitiene, A. (2004, October 1). The installation and support of internationally distributed equipment for people with dementia. *IEEE Transactions on Information Technology in Biomedicine : A Publication of the IEEE Engineering in Medicine and Biology Society, 8*, 253–57. https://doi.org/10.1109/TITB.2004.834393

ASWASAKIRANAM. Accessed June 27, 2020. http://www.socialsecuritymission.gov.in/scheme_info.php?id=5

Bjørneby, S., Topo, P., Cahill, S., Begley, E., Jones, K., Hagen, I., Macijauskiene, J., & Holthe, T. (2004, October 1). Ethical considerations in the ENABLE project. *Dementia, 3*, 297–312. https://doi.org/10.1177/1471301204045162

Campos-Puente, A. de las M., Avargues-Navarro, M. L., Borda-Mas, M., Sánchez-Martín, M., Aguilar-Parra, J. M., & Trigueros, R. (2019, January). Emotional exhaustion in housewives and alzheimer patients' caregivers: Its effects on chronic diseases, somatic symptoms and social dysfunction. *International Journal of Environmental Research and Public Health,* 16(18), 3250. https://doi.org/10.3390/ijerph16183250

Dias, A., Dewey, M. E., D'Souza, J., Dhume, R., Motghare, D. D., Shaji, K. S., Menon, R., Prince, M., & Patel, V. (2008, June 4). The effectiveness of a home care program for supporting caregivers of persons with dementia in developing countries: A randomised controlled trial from Goa, India. *PloS One, 3*(6), e2333. https://doi.org/10.1371/journal.pone.0002333

D'Souza, M. F., Davagnino, J., Nicole Hastings, S.., Sloane, R., Kamholz, B., & Twersky, J. (2015). Preliminary data from the caring for older adults and caregivers at home (COACH) program: A care coordination program for home-based dementia care and caregiver support in a veterans affairs medical center. *Journal of the American Geriatrics Society, 63*(6), 1203–8. https://doi.org/10.1111/jgs.13448

Evans, N., Orpwood, R., Adlam, T., Chadd, J., & Self, D. (2007, November 1). Evaluation of an enabling smart flat for people with dementia *15*, 33–36.

Fæø, S. E., Husebo, B. S., Bruvik, F. K., & Tranvåg, O. (2019, June 6). We live as good a life as we can, in the situation we're in'—The significance of the home as perceived by persons with dementia. *BMC Geriatrics, 19*. https://doi.org/10.1186/s12877-019-1171-6

Gitlin, L. N., Winter, L., Dennis, M. P., Hodgson, N., & Hauck, W. W. (2010, September 1). A biobehavioral home-based intervention and the well-being of patients with dementia and their caregivers. *JAMA: The Journal of the American Medical Association, 304*(9), 983–91. https://doi.org/10.1001/jama.2010.1253

Glimmerveen, L., & Nies, H. (2015, September 23). Integrated community-based dementia care: The geriant model. *International Journal of Integrated Care, 15*. https://www.ncbi.nlm.nih.gov/pmc/articles/PMC4628512/

Huang, H.-L., Shyu, Y.-I. L., Chen, M.-C., Chen, S.-T., & Lin, L.-C. (2003, April). A pilot study on a home-based caregiver training program for improving caregiver self-efficacy and decreasing the behavioral problems of elders with dementia in Taiwan. *International Journal of Geriatric Psychiatry, 18*(4), 337–45. https://doi.org/10.1002/gps.835

Huang, H.-L., Kuo, L.-M., Chen, Y.-S., Liang, J., Huang, H.-L., Chiu, Y.-C., Chen, S.-T., Sun, Y., Hsu, W.-C., & Shyu, Y.-I .L. (2013, November). A home-based training program improves caregivers' skills and dementia patients' aggressive behaviors: A randomized controlled trial. *The American Journal of Geriatric Psychiatry, 21*(11), 1060–70. https://doi.org/10.1016/j.jagp.2012.09.009

Jeon, Y.-H., Simpson, J. M., Low, L.-F., Woods, R., Norman, R., Mowszowski, L., Clemson, L. et al. (2019). A pragmatic randomised controlled trial (RCT) and realist evaluation of the interdisciplinary Home-BAsed Reablement Program (I-HARP) for improving functional independence of community dwelling older people with dementia: An effectiveness-implementation hybrid design. *BMC Geriatrics, 19*(1), 199. https://doi.org/10.1186/s12877-019-1216-x

Kor, P. P. K., Liu, J. Y. W., & Chien, W. T. (2019, October). Effects of a modified mindfulness-based cognitive therapy for family caregivers of people with dementia: A pilot randomized controlled trial. *International Journal of Nursing Studies, 98*, 107–17. https://doi.org/10.1016/j.ijnurstu.2019.02.020

Koumakis, L., Chatzaki, C., Kazantzaki, E., Maniadi, E., & Tsiknakis, M. (2019). Dementia care frameworks and assistive technologies for their implementation: A review. *IEEE Reviews in Biomedical Engineering, 12*, 4–18. https://doi.org/10.1109/RBME.2019.2892614

Kuo, Y.-C., Lan, C.-F., Chen, L.-K., & Lan, V. M. (2010, September). Dementia care costs and the patient's quality of life (QoL) in Taiwan: home versus institutional care services. *Archives of Gerontology and Geriatrics, 51*(2), 159–63. https://doi.org/10.1016/j.archger.2009.10.001

Lim, Y. M., Son, G.-R., Song, J.-A., & Beattie, E. (2008). Factors affecting burden of family caregivers of community-dwelling ambulatory elders with dementia in Korea. *Archives of Psychiatric Nursing, 22*(4), 226–34. https://doi.org/10.1016/j.apnu.2007.12.005

Olsen, C., Pedersen, I., Bergland, A., Enders-Slegers, M.-J., Jøranson, N., Calogiuri, G., & Ihlebæk, C. (2016). Differences in quality of life in home-dwelling persons and nursing home residents with dementia—A cross-sectional study. *BMC Geriatrics, 16*, 137. https://doi.org/10.1186/s12877-016-0312-4

Social Justice, Kerala. Accessed June 27, 2020. http://swd.kerala.gov.in/scheme-info.php?scheme_id=IDEyNw==

Sudore, R. L., Lum, H. D., You, J. J., Hanson, L. C., Meier, D. E., Pantilat, S. Z., Matlock, D. D. et al. (2017, May). Defining advance care planning for adults: A consensus definition from a multidisciplinary Delphi panel. *Journal of Pain and Symptom Management*, 53(5), 821–832.e1. https://doi.org/10.1016/j.jpainsymman.2016.12.331

Tanner, J. A., Black, B. S., Johnston, D., Hess, E., Leoutsakos, J. M., Gitlin, L. N., Rabins, P. V., Lyketsos, C. G., & Samus, Q. M. (2015). A randomized controlled trial of a community-based dementia care coordination intervention: Effects of MIND at home on caregiver outcomes. *The American Journal of Geriatric Psychiatry: Official Journal of the American Association for Geriatric Psychiatry, 23*(4), 391–402. https://doi.org/10.1016/j.jagp.2014.08.002

Thyrian, J. R., Hertel, J., Wucherer, D., Eichler, T., Michalowsky, B., Dreier-Wolfgramm, A., Zwingmann, I., Kilimann, I., Teipel, S., & Hoffmann, W. (2017). Effectiveness and safety of dementia care management in primary care: A randomized clinical trial. *JAMA Psychiatry, 74*(10), 996–1004. https://doi.org/10.1001/jamapsychiatry.2017.2124

Ugur, H. G., Orak, O. S., Aktas, Y. Y., Enginyurt, O., & Saglambilen, O. (2019). Effects of music therapy on the care burden of in-home caregivers and physiological parameters of their in-home dementia patients: A randomized controlled trial. *Complementary Medicine Research*, 26(1), 22–30. https://doi.org/10.1159/000490348

VAYOMITHRAM. Accessed June 27, 2020. http://www.socialsecuritymission.gov.in/scheme_info.php?id=7

Williams, K., Blyler, D., Vidoni, E. D., Shaw, C., Wurth, J., Seabold, D., Perkhounkova, Y., & Van Sciver, A. (2018). A randomized trial using Telehealth technology to link caregivers with dementia care experts for in-home caregiving support: FamTechCare protocol. *Research in Nursing & Health, 41*(3), 219–27. https://doi.org/10.1002/nur.21869

Wolff, J. L., Spillman, B. C., Freedman, V. A., & Kasper, J. D. (2016, March). A national profile of family and unpaid caregivers who assist older adults with health care activities. *JAMA Internal Medicine, 176*(3), 372–79. https://doi.org/10.1001/jamainternmed.2015.7664

World Alzheimer Report 2015, The Global Impact of Dementia: An Analysis of Prevalence, Incidence, Cost and Trends. n.d., 87

Wübker, A., Zwakhalen, S. M. G., Challis, D., Suhonen, R., Karlsson, S., Zabalegui, A., Soto, M., Saks, K., & Sauerland, D. (2015, September). Costs of care for people with dementia just before and after nursing home placement: Primary data from eight European countries. *The European Journal of Health Economics: HEPAC: Health Economics in Prevention and Care, 16*(7), 689–707. https://doi.org/10.1007/s10198-014-0620-6

Chapter 11
Rehabilitation in Dementia Care

Saadiya Hurzuk and Anand Ambali

Introduction

Dementia is a syndrome; it is a collective term used to describe various symptoms like cognitive decline interfering daily life and is progressive. It is an inclusive term for conditions characterised by memory disorders, impaired reasoning, personality changes and other skills that affect a person's ability to perform activities of daily living.

Globally, there are around 50 million people with dementia with an annual addition of 10 million new cases, out of which 60% live in lower-middle-income countries. Worldwide the total number of people with dementia is projected to reach 82 million by 2030 and by 2050 152 million. In India alone, it is estimated 5.3 million people living with dementia (Alzheimer's & Related Disorders Society of India, 2010), and only one person out of ten ever gets any diagnosis, treatment or care (Prince et al., 2015). Most of the increase in dementia cases is attributable to the rising number of people living in lower-middle-income countries (World Health Organization, 2020) Most of the dementia cost and care is out of pocket expenditure borne by families (Delavande et al., 2013).

Rehabilitation in dementia in this chapter are discussed predominantly into the below sections:

I. Pharmacological interventions
II. Non-pharmacological interventions for dementia

S. Hurzuk (✉)
Alzheimer's & Related Disorders Society of India (ARDSI), ARDSI Hyderabad Deccan Chapter, # 8-2-316/5/10, Ground Floor, Lane Behind Asha Hospital, Road No. 14, Resham Bagh, Telangana 500034 Banjara Hills, Hyderabad, India

A. Ambali
Geriatric Clinic, BLDE (Deemed to be University), Shri B. M. Patil Medical College Hospital and Research Centre, Vijayapura, Karnataka 586103, India

M. K. Shankardass (ed.), *Dementia Care*,
https://doi.org/10.1007/978-981-16-3864-0_11

- Rehabilitation in dementia
  - Cognitive rehabilitation
  - Functional rehabilitation

III. Importance of rehabilitation for BPSD
IV. Person-centred care
V. Caregivers support and training.

I. Pharmacological management of dementia

Introduction

The dementia is categorized into two groups. They are (1) primary and (2) secondary. The primary dementia is gradual in onset and progressive in nature without an apparent identifiable cause. In contrast, dementia which occurs abruptly the following stroke, head injury and sequelae to encephalitis is categorized as secondary dementia.

The aims of the pharmacological management will be to

(a) Maintain function
(b) Delay progression
(c) Facilitate independence and
(d) Ensure quality of life.

The treatment of established dementia is multidimensional. As on now, there are no treatments that have been shown to alter disease progression.

The medical management of a person with dementia can be provided in six contexts. The contexts can be concurrently applied to improve the quality of life and delay complications.

1. Effective pharmacological management of dementia
2. Management of comorbidities
3. Identifying the reversible cause of dementia and its management
4. Nutrition care and deficiency correction
5. Prevention of drug–drug interactions.

1. Effective pharmacological management of dementia

This constitutes treatment of dementia and its complications by the use of medicines. The goal will be for long-term amelioration of symptoms and to maintain quality of life.

When drugs are used, we need to monitor adverse drug reaction and act immediately when the side effect develops. Thorough scrutiny should be done for possible drug–drug interactions when older people are using drugs for comorbidities.

So far, the Food and Drug Administration (FDA) has approved only two classes of drugs for the treatment of dementia. The cholinesterase inhibitors (ChEIs) and N-methyl-D-aspartate receptor antagonists are the two classes of drugs that are approved only for Alzheimer's disease. These drugs have no clear benefit in Lewy body dementia (Rolinski et al., 2004) and front temporal dementia (Mendez et al., 2007); hence, these drugs are not recommended for treatment of dementia other than that of Alzheimer's disease.

The drugs available for the treatment of dementia are as follows

1. Cholinesterase inhibitors (ChEIs)

Donepezil-acetylcholinesterase inhibitor
Galantamine-nicotinic receptor agonist
Rivastigmine-acetylcholinesterase and butyryl-cholinesterase inhibitor.

2. N-methyl-D-aspartate (NMDA) receptor antagonist-memantine

The cholinesterase inhibitors have benefit in slowing progression of cognitive impairment in the early and moderate stage of dementia while NMDA antagonists have benefit in late stage. Another advantage of memantine is that it has beneficial effects on behaviour symptoms, and the person using these drugs shows less resistance to dressing and grooming during the late stage. The advantage of the use of rivastigmine is that it is also approved for the treatment of Parkinson's disease dementia (PDD).

It is to be understood that these drugs work to delay the progression of the disease and not to improve their cognition. The medications may not work for everyone. The benefits of drugs on function and decline in cognitive test scores are maintained up to three years. The limited efficiency both in disease progress and duration should be discussed with the caregiver before initiating the drugs. The caregivers will be having a high expectation of improvement of symptom once the drugs initiated.

All the three drugs used to delay the progression of symptoms both statistically and clinically for 6–12 months in mild to moderate stage, but not at three years (Raina et al., 2008; Petersen et al., 2005; Courtney et al., 2004).

An electrocardiogram is recommended to rule out bundle branch blocks before initiating cholinesterase inhibitors.

The details like dosage, route of administration and side effects of drugs used are as follows (Table 11.1).

The cholinesterase inhibitors should be initiated in person with mild-to-moderate dementia with MMSE score of 12 points and above. (NICE, 2001).

The dosage should be lowest, to begin with, and watch for the response and then increase the dose in increments as tolerated.

The advantage of using rivastigmine is that is shown to improve cognitive impairment and BPSD, including apathy, anxiety, delusions and hallucinations in dementia with Lewy bodies (McKeith et al., 2000).

How long can these medications be used?

There is no clear guidelines and consensus regarding the duration of use of these drugs in clinical practice. American Geriatric Society (AGS, 2014) recommends a trial of twelve weeks and observes for the benefits.

Table 11.1 Dosage, route of administration and side effects of drugs used in treatment of dementia

No.	Drug	Dosage	Route of administration	Side effects
1	Donepezil	5–10 mg/day	Oral	Nausea, diarrhoea, muscle cramps, altered sleep, vivid dreams, syncope (Gill et al., 2009) Hip fracture (Kim et al., 2011) bradycardia (Park-wyllie et al., 2009)
2	Rivastigmine	1.5–6 mg, in two divided doses	Oral	
		4.6–9.5 mg daily	Patch	
3	Galantamine	4–12 mg, in two divided doses	Oral	
4	Memantine	10 mg, in two divided doses	Oral	Headache, diarrhoea, confusion, cough dizziness

MMSE should be done after 2–4 months of establishing the maintenance dose, and if the score has improved or has remained the same as the baseline, then the medicines can be continued. The medicines can still be continued even if the behavioural or functional assessment shows improvement (NICE, 2001).

When to stop medications?

This is the most challenging decision to be taken by the clinician.

The concept of perceived benefits is applied in making a decision about continuing the drugs. Whether the medications should be continued or not, will depend upon the perceptions of the benefit by the patient and the caregivers.

According to NICE guidelines, the patient should be assessed every six months. Treatment should be continued if the Mini-Mental Status Examination (MMSE) score remains above 12 (NICE, 2001).

Can we restart medications?

Another essential aspect is can we restart the medications for the reappearance of presenting symptoms once it has been stopped? Yes, the medication is restarted if the symptoms of cognitive decline reappear after stopping the medication. The guidelines by AGS mention there is scope to restart the therapy again (AGS, 2013).

Does the combination of acetyl cholinesterase and memantine have any advantage?

There are schools of thought in using this combination. Howard et al. (2012) have not found adequate benefit and have not supported this combination, whereas Patel and Grossberg (2011) have shown proven clinical efficiency in using the combination of medicines. The combined drug therapy is effective when it is started in the initial stage (Wilkinson, 2012). However, there is a lack of guidelines regarding when to start and how long to use.

The beneficial effects of combined therapy on long-term use, preferably up to two years, are not apparent.

Precautions while using medicines

(1) Always start with a low dose, wait for a few days, observe the response and then increase the dose in increments.
(2) Stop the medications if it worsens existing cognitive state or new symptom develops.
(3) Avoid combination of allopathic medicines with alternative medicines.
(4) Change or add one medication at a time.

Safety issues regarding mediation in people with dementia to be followed by family members are

(a) Supervision of medication intake by family members.
(b) Medicines should be beyond reach from the person.
(c) Dispersible preparation or syrup-based formulas should be preferred for ease of administration.
(d) Adequate stock of regularly used medicines should be ensured.

(2) **Management of comorbidities**

The older people with dementia will have multiple comorbidities which lead to the use of multiple drugs. The symptoms of comorbidities which overlap make it challenging to identify whether it is behaviour disorder of dementia or drug-related adverse event.

The commonest comorbidities are hypertension, diabetes mellitus, ischemic heart disease, stroke, Parkinson's disease, malignancy, (Doraiswamy et al., 2002), frailty, renal failure, cardiac failure, depression and psychosis.

The exacerbations of comorbidities cause breathlessness, pain, polyuria, retention of urine, fall, electrolyte disturbances, hypoglycaemia, reduced sleep and constipation. Multiple symptoms can occur at the same time. This leads to behavioural symptoms in older people. When these symptoms occur in the person living with dementia, it adds to extremely complicated behaviour changes which are difficult to control.

Management of neurological and psychological comorbidities

The medicines are recommended as and when the symptoms are discovered (Table 11.2). Non-pharmacological measures should be used first, and concurrently one can use these drugs as add on. It is preferable to treat depression and seizures by drugs to improve the quality of life and safety of the person living with dementia.

The behaviour symptoms like disturbances in sleep, mood and perceptions can be well controlled by psychotropic drugs, but the long-term use of them is associated with increased mortality. Sedation of a person with behaviour disorder is not a substitute for good care by the caregiver (Steel & Lawrie, 2018).

The commonest comorbid condition present in the early stage of Alzheimer's disease is depression. Ballard C. G. et al. (2009) reported that depression was present

Table 11.2 Neurological and psychological comorbidities with medicine

No.	Comorbid	Drug used	Dosage
1	Depression	Escitalopram	5–10 mg/day
2	Seizures	Levetiracetam	500 mg/day. Maximum dose 1000 mg/day
3	Agitation	Olanzapine or quetiapine	5 mg/day. Maximum dose 10 mg/day
4	Delusions	Quetiapine	15 mg/day. Maximum dose 400 mg/day
		Olanzapine	2.5 mg/day. Maximum 10 mg/day
		Risperidone	0.5 mg/day. Maximum 2 mg/day
5	Anxiety insomnia	Clonazepam	0.25 mg/day. Maximum 2 mg/day
6	Stroke	Clopidogrel	75 mg/day life-long except for stroke due to haemorrhage

in apparently 50% of patients with Alzheimer's disease. Hence, screening for depression, especially in the early stage, should be carried out and treated accordingly. This will also prevent the occurrence of behaviour symptoms in the early stage.

When the person with dementia develops neurological or psychiatric symptoms, drugs like typical and atypical antipsychotics are used.

Both the typical and atypical antipsychotic group of drugs are associated with extrapyramidal symptoms, falls, sedation, confusion, stroke, increased cognitive decline, risk of cardiac arrhythmias and increased mortality (Ballard & Cream, 2004; Gill, 2007a, b).

The antipsychotic group of drugs is indicated only when the non-pharmacological measures have not been able to control the symptoms and when the person poses a serious threat to self or caregivers (AGS, 2013).

The treatment of metabolic and vascular comorbidities

The common metabolic and vascular conditions like diabetes, hyperlipidaemia, ischemic heart disease and hypertension need to be controlled with the use of regular oral hypoglycaemic, lipid-lowering, cardioprotective and antihypertensive drugs. This should ensure adequate and not rigid control of sugar and blood pressure levels so that the complications can be prevented. Monitoring of blood sugar, blood pressure and serum lipid levels should be carried out at regular intervals. The control of these chronic non-communicable diseases will not alter the course of dementia but will help in maintaining the quality of life.

Infections like pneumonia and urinary tract infections occur commonly. Judicious use of antibiotics is recommended for a short time. Macrolides like roxithromycin and azithromycin are avoided in the person receiving cholinesterase inhibitors.

(3) **Identifying the reversible cause of dementia and its management**

The reversible causes of dementia are hypothyroidism, Vitamin B12 deficiency, normal pressure hydrocephalus, human immunodeficiency virus infection and drugs

Table 11.3 Reversible causes of dementia

No.	Disease	Treatment
1	Hypothyroidism	Thyroid replacement
2	Vitamin B12 deficiency	Vitamin B12 Supplementation
3	Raised serum homocysteine	Vitamin B12, folic acid supplementation
4	Human immunodeficiency virus infection	Antiretroviral drugs
5	Normal pressure hydrocephalus	Ventricular shunt
6	Opportunistic infections of central nervous system	Antimicrobial
7	Cognition impairment drugs	Stop them

that may impair cognition. The investigations like thyroid profile, homocysteine levels, computed tomography of brain and blood for HIV infections should be carried out in all patients presenting with dementia. The specific disease when diagnosed is treated with specific drugs or surgical interventions, as mentioned in Table 11.3.

(4) **Nutritional care**

During the mild and moderate stage of dementia, intake of food will be good, whereas in advanced stage, the nutritional deficiencies and weight loss are commonly encountered.

The artificial nutritional support beyond oral supplementation has no effective role in improving the function.

The daily requirements of nutrients are met by providing them with oral feeds. The food fed should be what they desire and has been consuming for a long time. It should be locally available and palatable.

The feeding through a tube in person with advanced stage has not found to improve the overall quality and duration of life (Volkert, 2015).

The word of caution regarding oral feed-in advanced stage is that the risk of aspiration is very high, and in such a situation, tube feeding can be preferred (BAPEN).

The nutritional supplementation is to improve nutrition status and not cognitive decline.

Are nutritional supplementations beneficial?

There is no evidence on large-scale studies till date that supports the supplementations of vitamins, minerals, omega fatty acids and herbal preparations in delaying or preventing cognitive decline.

According to a study by (Petersen et al., 2005), the supplementation of vitamin E has shown no benefit in delaying the progress of dementia.

The study by Aisin, P. S. et al. did not show the effectiveness of vitamin B supplementation in slowing cognitive decline. It is also found that this supplementation leads to drug interactions and has potential adverse effects (Ernst, 2002). The use of

Ginkgo Biloba in the treatment has no enough evidence to be recommended (Ernst & Pittler, 1999).

(5) **Drug interaction**

People with dementia are at higher risk of drug interactions. A study by (Gustafsson et al., 2016) showed 41% of hospital admissions were attributable, or partially attributable, to drug-related problems.

The older people with dementia consume a higher number of drugs that people without dementia, hence the rate of drug–drug interactions increases (Lau et al., 2010).

The misinterpretation of adverse drug reactions leads to the addition of newly prescribed medications, and this further exaggerates the cascade (Mallet et al., 2007).

There are three categories wherein drug–drug interaction can happen.

The interaction between the

(a) Drugs used to treat dementia
(b) Drugs for dementia and comorbid conditions
(c) Drugs for dementia and non-allopathic medicine.

A detailed history is mandatory to know what are the medicines consumed by the person with dementia. The medicines are prescribed by doctors, herbal preparations and drugs for comorbid conditions.

The drug interaction should be suspected when there is sudden change in cognitive decline, falls, involuntary movements, agitation and behaviour changes. Most of the times identifying and withdrawing the culprit drug solve the problem.

(a) *Drugs used to treat dementia*

There are only two class of drugs approved for the treatment of dementia. The interaction between the two classes is very low. In fact, some studies have recommended the use of a combination of two classes of drugs in the initial stage of dementia.

(b) *Drugs for dementia and comorbid conditions*

The following drugs should be vigilantly avoided in the person taking medicine for dementia. The person receiving memantine should not receive the medicines like ketamine, barbiturates, amantadine, sodium bicarbonate and dextromethorphan. These drugs have potential and life-threatening interactions with memantine (Haas, 2012).

Phenothiazine and benzo diazepam should be avoided as it produces sedation, rigidity, dyskinesia and paradoxical disinhibition (Haas, 2012).

Diphenhydramine prescribed for insomnia and oxybutynin for incontinence should be avoided.

The person receiving galantamine should not receive medicines like ketoconazole and erythromycin. These drugs increase the bioavailability of galantamine (Bentué-Ferrer et al., 2003).

Citalopram used to treat depression should be avoided in patients receiving donepezil. This combination leads to a prolonged QT interval, which may lead to torsade's de points (Stockholm County Council, 2018).

(iii) *Drugs for dementia and non-allopathic medicine*

It is important to know what all the drugs are prescribed by various clinicians, the use of on the counter drugs, supplementations, herbal and Ayurveda preparations, tobacco addiction and alcohol abuse.

When alcohol and tobacco consumption is suddenly stopped, it may lead to behavioural symptoms, seizures and agitation. In such condition, lorazepam is the drug of choice and can be used safely.

The cognitive enhancers marketed in the form of herbal and Ayurveda preparations should not be used concurrently with allopathic medicines. This will lead to potential organ damaging interactions.

What new is expected in the next few years?

Agents to start newer molecular mechanisms are in the pipeline (Carlsson et al., 2009)

They are agents that

(a) Target beta amyloid metabolism
(b) Inhibit alpha beta fibrilization
(c) Selectively lower amyloid and
(d) Inhibit gamma secretage.

The other molecules being tried in treatment are Cerebrolysin and insulin.

All these agents are in various stages of development.

Conclusion

The drug treatment for dementia is effective during the initial stages and should be stopped during the severe stage.

The use of pharmacological agents should be initiated only after the non-pharmacological measures have not resolved the issue or when the symptoms are so severe that it may harm the person himself or caregivers. Medications must be initiated when depression, seizures and infection are diagnosed.

Caution should be applied when a new drug of categories likes vitamin supplementation, herbal preparations and antipsychotics are added. These groups of drugs are known to produce drug interactions or adverse drug reactions. This may lead to further aggravation of pre-existing cognitive decline and physical injuries like fall.

The pharmacological agents have a limited role in the management of dementia, its complications and behavioural symptoms associated with it. When there is a definite indication for the use of pharmacological agents, it should be for a short period and tapered and stopped once the symptoms are controlled.

So far, the guidelines have been very clear regarding the use of medicines for dementia, its complications and comorbidities, tube feeding. Research is being carried out on a large scale at a further level in identifying new drugs and vaccine.

As on now, there are limited drugs available in management and cure is not yet reported, the role is limited to delay the progress and prevent complications and be a part of holistic approach. This will add to provide a good quality of life until the end.

(II) Non-pharmacological interventions for dementia

The non-pharmacological intervention aims to improve or maintain cognitive abilities, empower an individual to continue to execute the daily activity, address behavioural symptoms that often accompany cognitive impairment, for example depression, sleep, agitation, wandering, restlessness, anxiety and aggression. Even though there is a requirement of extensive research in this field to support the efficiency of non-pharmacological interventions, there are findings that suggest that these interventions can decrease depressive symptoms while improving cognitive abilities, social integration and quality of life of the person (Berg-Weger & Stewart, 2017a, b) There are studies which support the belief cognitive and physical activities improve cognitive functioning, thus providing support for non-pharmacological interventions focusing on the person' abilities (Hertzog et al., 2008). It is an emerging treatment option for dementia as the non-pharmacologic psychosocial interventions are proving to be cost-effective than medical treatments since they do not comprise any prescribing of medication and do not have any side effects (Ballard et al., 2011). There is a need for further studies to understand the cost-effectiveness of non-pharmacological interventions in dementia care. One study of its kind reports constructive effectiveness of non-pharmacological interventions on behavioural management in individuals with dementia and reduced caregiver burden (Gitlin et al., 2010). There has been much encouragement from the various organization on the national and international level for the development of evidence-based non-pharmacologic interventions to address cognitive stimulation, BPSD and improving lives of persons with dementia and their caregivers. The WHO Mental Health Gap Action Programme (mhGAP) appeals for the development of interventions in which caregivers are educated to provide cognitive stimulation at home and in the community which can help with BPSD and depression. Non-pharmacological interventions come from various disciplines, all of which attempting to positively affect cognitive abilities, mood and challenging behavioural and psychological symptoms of dementia. Four categories that outline the interventions: brief psychotherapy, holistic methods, cognitive approaches and alternative methods. Authors have mainly focused on holistic approaches that have shown affirmative results and are well-acknowledged in the literature. The holistic interventions include

1. Cognitive stimulation therapy
2. Reality orientation therapy
3. Validation therapy
4. Reminiscence therapy.

Cognitive stimulation: The therapy is classically delivered in a social set-up mostly in small-size groups comprising cognitive-based tasks and activities, for instance, word games and puzzles. The recipient of cognitive stimulation exhibits improvement in cognition, quality of life and well-being compared to those receiving antipsychotic medications. Critical review of non-pharmacological interventions found cognitive stimulation therapy to be one of the most positively effective non-pharmacologic interventions for people it dementia (Cammisuli et al., 2016). Studies show cognitive stimulation therapy to be viable, cost-effective intervention for home-based care as well as institution care (Loraine et al., 2014).

Reality orientation therapy: Taulbee and Folsom first described it in 1966, and it is a method delivered with the help of different environmental cues in order to reduce disorientation, confusion, memory loss and challenging behaviour concerning time, person or place (Taulbee & Folsom, 1966). It helps the person with dementia engage within their environment by bringing their attention to the current reality of their location, surroundings, date and time. The caregiver plays an essential role in this by giving detail explanation to the person with dementia who needs much patience, proper communication and efforts from the carer's side.

Validation therapy: It is an approach therapy to communicate with people with dementia. It focuses on validating personhood and their feelings to provide comfort rather than concentrating on their deficit of abilities, mainly confusion. This therapy emphasizes on validating the person's feelings and emotions in the time of confusion rather than what is factually correct (Hitch, 1994).

Reminiscence therapy: It focuses on recalling the person's life events, reflecting on their own life experience with the aid of creative clues like old photograph albums, memory boxes, listening to old songs or watching movies, group talks. It is one of the most prevalent psychological interventions which is highly opted by professional carer's, families and persons with dementia. Studies show effective enhancement in mood and cognition as it is pleasurable to exchange memories within a structured session to find common ground to share a happy experience with others (Woods et al., 2018). It has been observed that reminiscence therapy has been beneficial to help self-esteem and allows dealing with painful memories in a supportive environment and build social well-being.

There is a slow advancement of pharmacological treatments to slow down the progression or to end dementia. Adequate evidence-based non-pharmacological interventions are available and can be incorporated to enhance the quality of life and well-being of the person and the caregiver. All the interventions mentioned above are cost-effective, with no side effects and can be delivered by trained, lay professional or family caregivers, or volunteers (Berg-Weger & Stewart, 2017a). There is a need for further study in this area primarily to discover strategies that can yield the best response as currently there are so many that can be confusing. Also, there is a need to simplify the concepts of delivering therapies that are easy for care providers to understand.

Rehabilitation model in dementia care could be familiarized in part through a reorganization of available capacities. This positive approach model can reduce the cost of care by averting difficulties, limiting that cost of management of challenging behaviours and delaying institutionalization. There is a need for implementing rehabilitative interventions on dementia care with proper focus on enhancing functioning and supporting relationships, self, meaningful engagement along with training of the staff or care provider. It is essential to ultimately involve the dementia person and carer dyad in the intervention to understand their perspectives (Clare, 2017).

Dementia affects a person with loss of memory, loss of planning and initiative, unable to comprehend, inability to do activities involving multiple steps, disorientation in time and place, inability to express what the feel or emotions, loss of procedural memory and function, poor coordination while doing tasks, fluctuation in mood and set of challenging behaviour. It impacts the functional decline of the person leading to low self-esteem/self-worth, helplessness, increased frustration, increased dependency on others, withdrawal, loss of role, increased problem behaviours, safety issues and concerns (Diagram 11.1).

1. **Cognitive rehabilitation**
2. **Functional rehabilitation**

1. **Cognitive rehabilitation**: It is a type of therapy to ease and help people with dementia generally in the early stages to manage their daily activities. Studies show that cognitive rehabilitation can help to sustain the individual's freedom and independence. The therapy is tailored to each individual in their homes. It can be defined as an individualized approach which concentrates on daily activities and real-life problems. In addition to cognition, it also helps in mood and

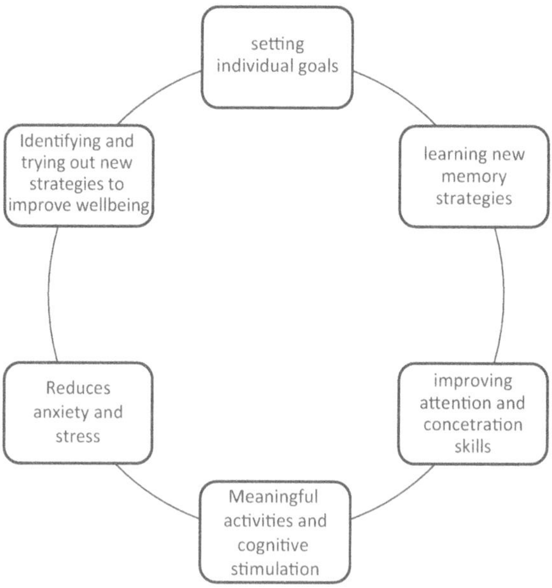

Diagram 11.1 Basics of rehabilitation in dementia care

behavioural difficulties; implementing it involves family and caregivers. Cognitive rehabilitation is generally goal-oriented involving evidence-based methods (Wilson et al., 2002). It is explicitly skill training, reality orientation or a mix of both and has been used widely by dementia therapist to manage cognitive impartment in persons with dementia. Carrion and colleagues in the systematic review of cognitive, psychosocial intervention in dementia highlight that cognitive stimulation can have improvement on people with dementia's daily functioning primarily through reality orientation (Carrion et al., 2013). There are various terms and techniques used for the same such as cognitive stimulation, cognitive training, cognitive rehabilitation which fall under the cognitive interventions. The primary purpose of this intervention is to use the various techniques to enhance the everyday functioning and well-being to lessen the excess disability for the person with dementia, decreasing the family and caregivers burden. The rationale of cognitive interventions is mainly to preserve the ability to learn and hold onto memory and abilities despite cognitive difficulties. There are some memories and cognitive abilities which are relatively preserved in the early stages of dementia. One of the goals of the cognitive intervention is to construct areas of relative strength and focus on the remaining abilities with the person. It aims to reimburse for deficiencies in the particular aspects of cognition that are significantly affected by the support of memory aids, environmental changes, reality orientation, minimizing additional cognitive demands.

Cognitive training

It is again a non-pharmacological form of dementia treatment that emphasizes on focused and guided set of standardizing tasks designed to reflect particular cognitive functions like memory, attention or problem-solving. It is a contrast to cognitive stimulation; it focuses on persons who have enough cognitive functions intact so that a therapist or a computer program can guide them to practices tasks designed in order to support them with impaired skills (Choi & Twamley, 2013). The training encourages adjustment of task difficulty in response to changes in the performance level of the person with cognitive disabilities. One assumption underlying cognitive training is that practice has the potential to improve or at least maintain functioning in the given domain. The cognitive training may have a negative impact on the mood of the person with dementia (Clare & Woods, 2004) There is a lack of literature with very few studies published so far on cognitive rehabilitation, and there is still no indication of any significant benefit derived from cognitive training (Bahar-Fuchs et al., 2013).

2. **Functional rehabilitation in dementia**

As we further proceed in this chapter to understand functional rehabilitation in dementia, it is essential to understand the concept of rehabilitation. Rehabilitation is a set of interventions provided to a person to overcome functional limitations in their daily activities which are caused due to any health condition, injuries and ageing. The purpose of rehabilitation is to enable a person to preserve or return to their activities of daily living, achieve significant roles and to attain the quality of life.

Until now, rehabilitation word has been perceived only as a physical intervention and not for cognitive intervention, rehabilitation and dementia until now has been considered as non-compatible. One possible reason why the two terms are alienated from each other is due to dementia generally presented as notions of helplessness, therapeutic nihilism, "the living death" or social death as described by Cahill and Dooley in their chapter historical context of rehabilitation and its applicant to dementia care. In contrast, rehabilitation is perceived as treatment, reclamation, empowerment and return to full functioning capacity (Cahill, 2005). One main principle of rehabilitation is a person-centred approach towards the individual. The intervention is generally delivered through trained experts in the respective domains. Rehabilitation has been integrated into many health programmes and services like primary health care, mental health, vision and hearing as an essential intervention. It should be taken care as it can lead to the misconception that Rehabilitation is delivered through specialized settings only, whereas in dementia care rehabilitation can be delivered by a well-trained carer in any facility or by a family caregiver can deliver it at home once they understand the concept. Functional rehabilitation in dementia can also be integrated into broader health programmes which can prove to be highly effective in providing quality care and cost-cutting. There is no pharmacological treatment available to cure cognitive impairment or any pharmacological interventions to slow down the progression of memory and functional deficits. The pharmacological interventions mentioned are mainly symptomatic treatments, or they tend to have limited benefits over the period of time (Berg-Weger & Stewart, 2017b) as there are no viable pharmacological treatments available the dementia professionals, persons with dementia and their carers have not much choice but to seek alternative interventions. At the same time, the perception and attitudes of medical professionals stating that there is no medicine or cure for dementia have put non-pharma interventions in the limelight for dementia care providers. The purpose of non-pharmacological rehabilitation is to help the person with dementia sustain or improve cognitive abilities, unable the individual to perform daily activities retaining their independence; also it helps in addressing behaviours and psychological symptoms due to dementia. The side effects of the medicines used dementia such as antipsychotics and benzodiazepines have made the caregivers opt for alternative treatments especially for challenging behaviours (Martini De Oliveira et al., 2015) Dementia leads to a family burden, decreases quality of life and well-being of the person with dementia and the family unit and can demand long-term institutional care resulting in a significant impact on health care and a financial burden (Seelye et al., 2012).

The impact of dementia further leads to functional decline in dementia with reverse ontogeny process, leading to semantic memory loss, loss of declarative memory and function, loss of procedural memory and function and loss of primitive reflex abilities. The decline in visual perceptual skills, motor apraxia, sensory-motor deficits, higher cognitive function decline such as problem-solving, judgement and sequencing are few challenges that are faced with the basic activity of daily living. Likewise, semantic memory loss leads to loss of independence with going outdoors, problems with judgement and problem-solving cause problems with cooking, loss of initiative cause problems with cleaning/laundry activities, memory loss adds to difficulty with financial management and the challenges in early stages with instrumental activities of daily living. The purpose of functional rehabilitation in dementia care is to regain maximum self-efficacy and normal function to the extent possible based on their functional abilities. The primary focus of functional rehabilitation in dementia care management is to attain the highest level of functional potential and maintain a lifestyle close to normal (Prizer & Zimmerman, 2018).

The areas of function that are concentrated are instrumental activities of daily living, basic activities of daily livings, the mobility and motor skills, communication and cognition. Interventions are graded into small and simple steps which can easily be followed by the persons with dementia. Interventions are mainly based on environmental modification, adjustments to reduce excess disabilities and maximize the independence of the person with dementia. Assistive devices are integrated in order to make it easier for the person to perform daily activities. Cueing to provide orientation clues are developed. To achieve the desirable outcomes, it requires setting goals and objectives such as understanding the person for whom the goal is being set, the objectives of why the goals are being set and by when it can be attained.

The standard principles of functional interventions identified are:

- Ability focused
- Within meaningful context
- Gradability of activities
- Long-term versus short-term benefits
- Ability to carry over the gains by the layperson.

It can immensely be a challenging procedure as the person may not follow what is expected. At the same time, the care provider may not have patience and time to support them learn the interventions repeatedly.

Providing rehabilitation to people with dementia to support for the loss of functional abilities that helps to person activities of daily living. Functional rehabilitation focuses on the goals of individual and their carers to reduce the care burden and retain independent functionality as long as possible. The term functional rehabilitation is not a popularly used in literature. In contrast, it is a cluster of multiple therapies like occupation therapy, environment and lifestyle modification, physiotherapy which promotes the functional independence for persons with dementia in their daily routine.

Need for functional rehabilitation in dementia

Dementia rips off a person of their self and independence due to loss of memory, loss of judgement making, unable to plan or initiate, unable to comprehend, inability to express needs or emotions, disorientation, poor motor skills and coordination, challenging neuropsychiatric behaviours, loss of role and unable to comprehend safety. Along with the general challenges of healthy ageing, it is a syndrome with a progressive degenerative nature, mainly effecting cognitive functions. Laterally with cognitive impairment, there is a deterioration in other significant aspects like motivation, emotion control and social well-being. The symptoms result in dependency on others and are one of the major causes of disability in old age. Providing care for loved ones with dementia can be very challenging as the course of the disease can last from 5 to 15 years. This disease which generally starts with cognitive inability eventually ends with a complete loss of functional independence resulting in complete dependency on care providers for necessary activities of daily living resulting in physical mistreatment of persons with dementia by their family caregivers (Vandeweerd et al., 2013). There seems to be significant interconnection between verbal aggression due to the caregiver's burden and dementia symptoms (Vandeweerd & Paveza, 2006).

In lower-income countries like India, only 10% of the total dementia population ever gets a diagnosis, treatment or any care for dementia (Prince et al., 2015) resulting in home-based care and complete responsibility on the family caregiver. At the same time, studies report that most family caregivers misconstrue symptoms of dementia and label these symptoms as deliberate misbehaviours towards them (Shaji et al., 2003). It is observed that the families seek professional help only in the situation where they are not able to understand and handle the challenging behaviours themselves. The elders commonly observe family neglect and abuse due to poor awareness of dementia, poor education and lack of skills about coping with symptoms arising at the different progressive stages. Understanding the nature of the dementia symptoms and training in skills about coping with dementia can help reduce the mistreatment by the care providers (Lee et al., 2018).

The various techniques of rehabilitation in dementia leads to a very positive ethos yet at the same time it should be taken in account that professional and family carers may try to incorporate it stubbornly without realizing the stage and state of the person with dementia and whether they are willing to participate. There is a high chance of misunderstanding in person with dementia that the efforts of rehabilitation are projected to aid recovery, and therapeutic "over-optimism" of the therapist or carer could lead to foster unrealistic anticipations about the progressive nature of the disease. In the long run, this can lead to adverse emotional and psychological repercussions. Although rehabilitation models do not focus on recovery, they are highly sought concepts in the field of dementia as they help people with dementia in maintaining abilities and independence. They also delay functional decline. Unfortunately, as recovery is not feasible, rehabilitation models are notably absent from the practical dementia care and in literature. Implementing rehabilitation in people with dementia depends heavily on the level of their impairment. The care plan and goals of rehabilitation have to be revisited and revised periodically as there are changes and decline in the abilities with a new set of symptoms emerging. Care providers should be willing to support and help people with dementia on every step to implement rehabilitation, which can be very time-consuming.

III. Importance of dementia rehabilitation for BPSD

Behavioural and psychological symptoms of dementia (BPSD) are defined as a cluster of neuropsychotic symptoms, disturbed thoughts, mood or behaviour that comprise depression, repetition of actions, psychosis, aggression, sleep problems and wandering. Studies show that non-pharmacological interventions provide positive outcomes in reducing BPSD, and these interventions have important and significant efficacy (Martini De Oliveira et al., 2015) evidence that shows non-pharmacological interventions result in less or no side effects making them safer options in dementia care (Gill et al., 2007a, b). Therapies including music therapy, sensory stimulation, cognitive therapy, behavioural therapy, reality orientation, individualized, tailored activities, physical exercise have shown favourable positive effects on the management of BPSD (Chen et al., 2014). The undesirable side effects of antipsychotics in dementia have led in the promotion of non-pharmacological treatment. Many experts advocate based on the evidence that these interventions should be considered as fist option interventions to treat BPSD. Also, findings suggest that non-pharmacological interventions are cost-effective as compared to the cost of medical management in dementia.

Case vignettes: non-pharma intervention for BPSD

Dr. Rajesh, 68 years old, was diagnosed with dementia three years back. He lives with his wife, who is also a medical doctor and is working. Currently, his dementia rating is mild to moderate with poor short-term memory. He visits a dementia day care on alternate days where cognitive stimulation, physical exercise and other therapies are conducted. He has eventually started enjoying the activities and the place. Dr Rajesh always relished singing old Hindi songs and has been a remarkable artist. He says if he was not pushed into medical school, he would have become an artist for sure. Generally, his wife drops him to the day-care centre and heads to her clinic. Dr Rajesh has developed separation anxiety and starts pacing up and down the day care, repeatedly asking when his wife would be back. The pacing and anxiety causes him much distress; and as an intervention for this situation, the staff plays his favourite music and arranges painting material on a nearby table in a way that catches his attention while pacing. After some time, he begins to notice the music and starts humming along. This elates his mood bringing him much needed comfort. The staff prompts him towards the painting material with pep talk and humour. All these helps divert his mind from the distressful thoughts, and he happily gets involved in painting and music. With a combination of psychological intervention and the knowledge of what Dr Rajesh prefers, the staff is successfully able to use music, reminiscence therapy and art therapy to manage his challenging behaviour.

IV. Person-centred care

The origin of the concept of person-centred care was started by Carl Rogers. It has been used as psychotherapeutic measure, and until recently it was unknown in dementia care. In 1988, Tom Kitwood, a social psychologist, introduced the person-centred approach in dementia care, which influenced in changing the perceptive towards dementia, and he challenged the idea that dementia leads to loss of a person. Kitwood proposed the term, person-centred care to separate the type of dementia care from medical and behavioural approaches. Furthermore, he explained that dementia could be best implicated between the relationship of cognitive impairment and psychological factors, namely individual psychology, the person's environment and their social well-being. He alleged that a person's environment has much influence on the brain as the brain has on one's abilities. One main fundamental principle to Kitwood's theory was the dismissal of the classical medical approach in dementia, with the belief that pharmacological treatment conveyed very negative and predictable consequences in dementia care. Individuals falling apart due to disabilities need to "remain in one piece" for which they need comfort and warmth in order to be connected with their past and self. The theory highlighted that the person with dementia should be included and involved in their care and life in order to attain fulfilment and self-satisfaction. As a final point, the theory emphasizes on people with dementia to have their own identity and caregivers to help them maintain their identity (Kitwood, 1997a, b). As Kitwood (1997a, b, p. 43) stated, "To have an identity is to know who one is, in cognition and in feeling. It means having a sense of continuity with the past; and hence a 'narrative', a story to present to others". Due to the cognitive loss of the person the carers need to "hold their story" in order to respond to their being and individuality. The theory focuses on the personhood of the individual with dementia rather than the care provided in accordance with the routine organized by the caregiver based on their convenience, efficacy or other standards. Providing a person, the love they need, affection, comfort and inclusion tends to offer them with self-worth reducing the challenging behaviour like agitation, apathy etc.

The term person-centred care is relatively new in the dementia scenario, but currently, it has become synonymous with good quality care within the dementia sphere (Brooker, 2004).

Defining person-centred care

Two different schools of thoughts can define person-centred care. Tom Kitwood's theory is generally defined in the context of dementia care mapping (DCM). Due to the untimely demise of Tom Kitwood, the definition of person-centred care is still left uncertain. However, the four major constituents involved in describing the person-centred care are through the equation (Røsvik et al., 2011)

$$\text{PCC(person} - \text{centred care)} = V + I + P + S$$

(1) "Valuing people with dementia and those who care for them (V)"
(2) "Treating people as individuals (I)"
(3) "Looking at the world from the perspective of the person with dementia (P)"
(4) "A positive social environment in which the person living with dementia can experience relative wellbeing (S)".

The VIPS acronyms to "very important persons" which is a useful aid to remember the fundamental principles of person-centred care. This approach can be practically complicated to implement in every individual care plan, as it can be challenging to comprehend by carers and can be time-consuming, yet it is persuasive in recognizing and enduring personhood in people with dementia (Kelly, 2010). However, with the progression of the disease, the person with dementia fails to express themselves, resulting in understanding selfhood, to address this selfhood approach can be described (Sabat, 2002). Sabat and Harre (1992) debated in contradiction of the assumption of loss of selfhood due to the progression of dementia, and they articulated their approach to include three Self (Self 1–3).

Self-1: defined as I, me, mine or our; as first-person pronounces it. The Self-1 remains intact in people with dementia.

Self-2: defined as a person characteristics of psychical mental or emotional self and is associated with the individuals' past life events such as accomplishments. Self-2 is intact with the cognitive decline but can negatively affect adverse interfaces of others.

Self-3: defined with relation to social circles like friends, colleagues or carer. It should be encouraged and fashioned by interaction with people surrounded. Self-3 has potential damage to Self-1 and Self-2 (Sabat, 2002).

The equation, PCC = $V + I + P + S$, does not support any specific form but gives in generalize the idea of the definition; however, the definition is not well received as it loses the focus and shape. It is implemented in the care plan with necessary information about the person with dementia; for instance, the care-assessment tool (P-CAT) considers five strategies to support person-centred care:

1. Collect personal information like the person's past likes, dislikes and incorporate
2. Facilitate reminiscence with appropriate stimuli
3. Consider the individual's interpretation of their subjective reality
4. To consider in the care plan that no task is placed higher than the subjective experience of the individual
5. Easy adaptable routines and activities according to the individual's likings.

With all the positive supporting studies and the outcome finding of the improvement in the quality of life this care plan has been widely accepted and tried to implemented throughout (Martínez et al., 2016).

Case vignette: developing a person-centred dementia care plan

Mrs. Singh was diagnosed with Alzheimer's disease at the age of 62. She retired as a school teacher and is living with her husband in their house of 20 years. After the diagnosis, they were referred to a dementia activity centre by her neurologist.

At the activity centre, after a preliminary observation of 4 weeks, the staff who is trained in dementia helped Mrs. and Mr. Singh in dementia care mapping

(DCM) and implementing person-centred approach in their care plan. The staff obtained Mrs. Singh's cognitive assessments from her neuropsychologist.

Further information was taken from her husband, who is also the primary caregiver. The staff, along with Mr. Singh, put in a lot of effort and time to collate as much information as possible in developing the care plan. It took them multiple sittings to gather detailed information about Mrs. Singh's life history including childhood, adolescence, education, family, marriage and job. Information was also gathered about her achievements, beliefs, religion and cultural background; including children and their life achievements, her retirement and the period postretirement leading up to the current diagnosis. Her personality, interests and the things she liked and disliked were taken into account.

Her assessments from memory clinics, dementia ratings, current abilities, behavioural insights, triggers and causes of challenging behaviours, socializing preferences were analysed. Based on all the information, suitable preferred activities were listed out to be incorporated in her daily routine. These included Mrs. Singh's choice of puzzles and board games, physical exercise, favourite music, outdoor games like throw ball and badminton, flower gardening, making *Rangoli* patterns and performing *Pooja*.

The staff from the centre made home visits to help them understand the benefits of environment modification and to implement functional rehabilitation for Mrs. Singh to help her manage activities of daily living independently.

Under the plan, Mr. Singh was advised to discuss long-term medication with the neurologist and the possible challenges of the condition they should be prepared for. Psychoeducation and counselling were delivered to Mr. Singh as he was the primary caregiver and also the spouse. He was recommended to join caregiver support groups and meetings. The DCM helped both Mrs. and Mr. Singh plan and decide long term dementia care and support. It was an intensive and time-consuming task for both professional and family caregivers with potential long term positive benefits. The care plan helped Mr. Singh handle the challenging phases of dementia including behaviour and psychological symptoms of dementia. As the plan was developed in the early stage of dementia, when her cognitive abilities and decision-making were intact, Mrs. Singh was able to make her own preferred decisions with regards to her care.

This helped reduce dependency on antipsychotics, cutting down caregiver's burden and care cost.

All these interventions helped slow down the progression of disease and Mrs. Singh continued to perform a lot of activities at the centre and even during the severe stages of dementia retained her personhood and enjoyed a decent quality of life.

Challenges encountered in person-centred care

It may sound like an ideal approach to dementia care and is considered by experts to be a synonym of quality care, yet practically it is not an ideal approach. Some underpinning issues that can make the person-centred care more practical and real:

1. **A dearth of lucidity in understanding the concept of person-centred care in dementia**

Brooker (2004) observed that the concept of person-centred care was found to be difficult to understand and implement by the healthcare professionals, researchers, person with dementia and their families, and it was observed that they all conceived it in different ways such as individualized care, value-based care, means of communication or phenomenological perspectives of care. Another challenge faced by many care providers is to implement this due to non-availability of the care plan in different languages. Person-centred care is defined as abstract and is yet to be implemented in guidelines to achieve quality care.

2. **Valuing the personhood**

Even though personhood is one of the main aspects in the person-centred care, studies suggest that most caregivers do not need an emphasis on the role of the person with dementia involving in developing their care plans (Kirkley et al., 2011).

3. **Person-centred care outcomes, lack of evidence and resources**

Extensive research is needed in this area which can provide evidence-based results of the implementation of person-centred care. There is need to train and sensate carers to understand the concept and how to implement; also scarcity there is of resources which acts as barriers for implementation of person-centred care approach (Kirkley et al., 2011).

To incorporate the person-centred plan for better functioning and outcomes, there is a need for greater availability of person-centred care. However, the lack of understanding of the philosophy, its implementation into day-to-day practice, inflexible dementia organization culture, scarcity of resources, lack of tools and their evidence block successes and constructive aspect of person-centred care approach. The dementia experts need to advocate and raise cognisance to comprehend and implement along with research evidence in order to make it more existent in practice than just keeping it superlative in literature.

XXII. **Role of caregivers in rehabilitation in dementia**:

Caregivers of people with dementia are generally called the invisible second patient. They are a susceptible group with an essential role in providing quality of life to the person with dementia, bearing high levels of caregiver's burden, social isolation, psychological and physical morbidity and financial hardship. People with dementia require high levels of care; most of the dementia care worldwide is provided by informal or family caregivers. Most of the time the negative aspects of caregiving receive more attention (Cohen et al., 2002). Caregivers face many challenges to balance caregiving with their life, and they are at increased risk of caregiver's burden, stress, depression and many health-related complications. The caregiver's strain can further manifest psychological morbidities like depression and anxiety.

Psychological interventions for caregivers have shown to reduce caregivers burden and depression. Comprehensive management of dementia includes a partnership between the dementia professional and the family caregiver. In the absence of family caregivers, there is a high possibility that people with dementia would have a poor quality of life and would need institutional care. This support from family caregivers comes at a high cost of caregiver's distress and reduced quality of life.

Cognitive rehabilitation not only involves the person with dementia, it also educates the family or the caregiver engaging them in the interventions. In case of progression of cognitive impairment where the person with dementia readily cannot comprehend or participate in the rehabilitation process, the rehabilitation approach may be used to help the family caregiver develop more operational strategies to support and empower the person with dementia (Kudlicka et al., 2019). Improving cognitive functioning person with dementia can delay hospitalization, reduce the cost of care and improve patients' and caregivers well-being (Carrion et al., 2013) Outcomes from a study conducted to understand the community occupation therapy intervention shows positive and cost-effectiveness for the person with dementia and caregiver especially in the informal caregiving setting (Graff et al., 2008).

It can be very tricky as a caregiver may not have so much time for learning and incorporating the rehabilitation techniques or some may not want to be involved. It is by default, expected from the families and caregivers to be engaged, involved with the best of their capabilities without giving much thought about their preferences. At times it can lead to stress and frustration for the person with and dementia caregiver creating the feeling of hopelessness and lows self-esteem through the training (Choi & Twamley, 2013). Psychological counselling and support shows sustained remunerations in reducing depressive symptoms in family caregivers and should be made extensively accessible to provide practical, evidence-based interventions for caregivers (Mittelman et al., 2004). Risk factors for dementia caregivers are generally associated with caregiver's burden and poor coping skills. Lack of dementia-friendly community leads in elder abuse and neglect which can be prevented with psychoeducation, caregivers support, training for community service professionals (Lee et al., 2018).

Training and support for dementia caregivers

Dementia training can be very crucial and beneficial intervention to provide information and better prepare caregivers by educating with valuable medical information, non-pharmacological strategies, managing BPSD, planning fiancé and decision-making for short-term and long-term care. The role of the caregiver continually evolves through the various stages of dementia. The progressive nature of the disease needs constant changes in approach and strategies of providing care. For a caregiver, it is essential to get educated in dementia and learn from fellow caregivers who can help in the daily routine of dementia living and care.

Caregiver's support to be extended in the form of instrumental support like helping with household tasks or daily living needs, communication and rapport building, emotional support and informational support. Informational support can be provided by professional and from other dementia caregivers who have experienced similar situations. Dementia caregivers support groups are very useful for reducing the caregiver's depression, social well-being, although not much effect on the burden of care (Chu et al., 2011). Dementia caregiving intervention is targeted explicitly at knowledge, skills and focused on providing dementia care; studies suggest benefits and positive effects on quality of life of people with dementia and their caregivers (Hepburn et al., 2001).

Conclusion: Rehabilitation is relatively a new term used in dementia care. It is a cluster of multidisciplinary interventions such as pharmacological and non-pharmacological interventions with a vast range of therapies. Rehabilitation in dementia can be generally delivered through cognitive interventions, functional rehabilitation, physiological and educational interventions for the person with dementia and caregivers. Person centre care is like an ideal approach to dementia care and is considered by experts to be a synonym of quality care. However, practically it is not an ideal approach as it has multiple challenges and can be challenging to incorporate in care. There is a need for understanding and standardized way of using of the person-centred care. All the interventions and therapies look good in the literature yet implementing them has been a challenge for practitioners, and there is a need for evidence-based, training, understanding and simplifying these approaches that can easily be incorporated in the daily life of dementia person to achieve quality of life.

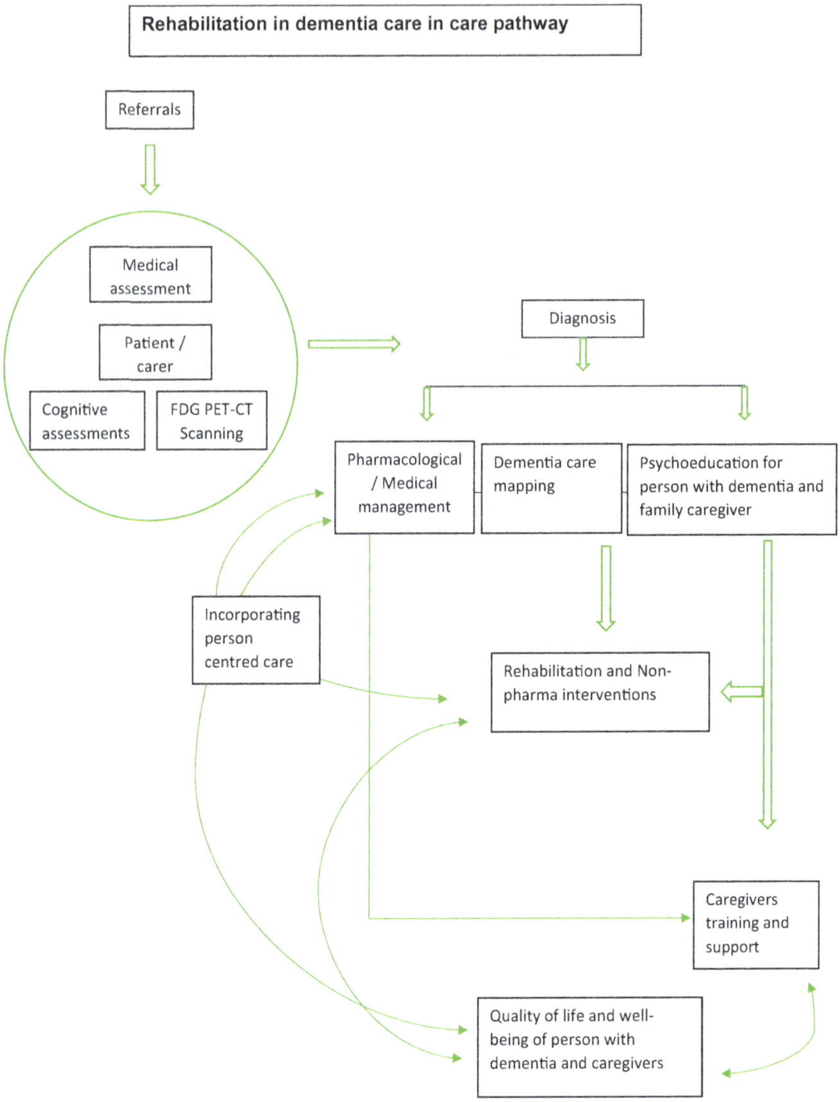
Rehabilitation in dementia care in care pathway
Referrals
Medical assessment
Patient / carer
Cognitive assessments
FDG PET-CT Scanning
Diagnosis
Pharmacological / Medical management
Dementia care mapping
Psychoeducation for person with dementia and family caregiver
Incorporating person centred care
Rehabilitation and Non-pharma interventions
Caregivers training and support
Quality of life and well-being of person with dementia and caregivers

References

AGS Choosing Wisely Workgroup. (2013). American Geriatric Society identifies five things that health care providers and patients should question. *Journal of the American Geriatrics Society, 61*(4), 622–631.

AGS Choosing Wisely Workgroup. (2014). American Geriatric society identifies five things that health care providers and patients should question. *Journal of the American Geriatrics Society, 62*(5), 950–961.

Alzheimer's & Related Disorders Society of India. (2010). The dementia india report 2010: Prevalence, impact, cost and services for dementia. New Dehli. http://ardsi.org/downloads/ExecutiveSummary.pdf

Ballard, C., & Cream, J. (2004). Drugs used to reduce behavioural symptoms in people with dementia or an unacceptable chemical cost? Argument. *International Psychogeriatrics, 17*, 4–12.

Ballard, C. G., Gauthier, S., Cummings, J. L., et al. (2009). Management of agitation and aggression associated with Alzheimer's disease. *Nature Reviews Neuralogy, 5*(5), 245–55.

Ballard, C., Khan, Z., Clack, H., Corbett, A. (2011). Nonpharmacological treatment of alzheimer disease. The Canadian Journal of Psychiatry, https://doi.org/10.1177/070674371105601004

Bahar-Fuchs, A., Clare, L., & Woods, B. (2013). *Cognitive training and cognitive rehabilitation for mild to moderate Alzheimer's disease and vascular dementia*. Wiley. https://doi.org/10.1002/14651858.CD003260.pub2

BAPEN. British Association for Parenteral and Enteral Nutrition guidelines. https://www.bapen.org.uk/resources-and-education/education-and-guidance/guidelines

Bentué-Ferrer, D., Tribut, O., Polard, E., & HervéAllain. (2003). Clinically significant drug interactions with cholinesterase inhibitors. A guide for neurologists. *CNS Drugs, 17*, 947–963.

Berg-Weger, M., & Stewart, D. B. (2017a). Non-pharmacologic interventions for persons with dementia. *Missouri Medicine, 114*(2), 116–19. http://www.ncbi.nlm.nih.gov/pubmed/30228557

Berg-Weger, M., & Stewart, D. B. (2017b). Non-pharmacologic interventions for persons with dementia. *Missouri Medicine, 114*(2), 116–19. http://www.ncbi.nlm.nih.gov/pubmed/30228557

Brooker, D. (2004). What is person-centred care in dementia? *Reviews in Clinical Gerontology*. Cambridge University Press. https://doi.org/10.1017/S095925980400108X

Cahil, S. (2005). The Historical context of rehabilitation and its application to dementia care [E-book]. In A. Dooley (Ed.), Perspectives on rehabilitation and dementia (pp. 30–39). Jessica Kingsley Publishers.

Cammisuli, D. M., Danti, S., Bosinelli, F., & Cipriani, G. (2016). Non-pharmacological interventions for people with Alzheimer's disease: A critical review of the scientific literature from the last ten years. *European Geriatric Medicine, 7*(1), 57–64. https://doi.org/10.1016/j.eurger.2016.01.002

Carlsson, C. M., Gleason, C. E., Puglielli, L., & Asthana, S. (2009). Dementia including Alzheimer's disease. In J. B. Halter, J. G. Ouslander, M. Tinetti, et al. (Eds.), *Hazzard's geriatric medicine and gerontology* (6th ed, pp. 797–812). McGraw-Hill Education.

Carrion, C., Aymerich, M., Baillés, E., & López-Bermejo, A. (2013). Cognitive psychosocial intervention in dementia: A systematic review. *Dementia and Geriatric Cognitive Disorders*. https://doi.org/10.1159/000354365

Cations, M., Laver, K., Crotty, M., & Cameron, I. (2017). Rehabilitation in dementia care. *Age and Ageing, 47*, 1–4. https://doi.org/10.1093/ageing/afx173

Chen, R.-C., Liu, C.-L., Lin, M.-H., Peng, L.-N., Chen, L.-Y., Liu, L.-K., & Chen, L.-K. (2014). Non-pharmacological treatment reducing not only behavioral symptoms, but also psychotic symptoms of older adults with dementia: A prospective cohort study in Taiwan. *Geriatrics & Gerontology International, 14*(2), 440–446. https://doi.org/10.1111/ggi.12126

Choi, J., & Twamley, E. W. (2013). Cognitive rehabilitation therapies for Alzheimer's disease: A review of methods to improve treatment engagement and self-efficacy. *Neuropsychology Review*. NIH Public Access. https://doi.org/10.1007/s11065-013-9227-4

Chu, H., Yang, C.-Y., Yen-Hsueh Liao, L.-I., Chang, C.-H., Lin, C.-C., & Chou, K.-R. (2011). The effects of a support group on dementia caregivers' burden and depression. *Journal of Aging and Health, 23*(2), 228–241. https://doi.org/10.1177/0898264310381522

Clare, L. (2017). Rehabilitation for people living with dementia: A practical framework of positive support. *PLoS Medicine, 14*(3). https://doi.org/10.1371/journal.pmed.1002245

Clare, L., & Woods, R. T. (2004). Cognitive training and cognitive rehabilitation for people with early-stage Alzheimer's disease: A review. *Neuropsychological Rehabilitation*. https://doi.org/10.1080/09602010443000074

Cohen, C. A., Colantonio, A., & Vernich, L. (2002). Positive aspects of caregiving: Rounding out the caregiver experience. *International Journal of Geriatric Psychiatry, 17*(2), 184–188. https://doi.org/10.1002/gps.561

Courtney, C., Farrell, D., Gray, R., et al. (2004). Long-term donepezil treatment in 565 patients with Alzheimer's disease (AD2000): Randomised double-blind trial. *Lancet, 363*(9427), 2105–2115. https://doi.org/10.1016/S0140-6736(04)16499

Delavande, A., Hurd, M. D., Martorell, P., & Langa, K. M. (2013). Dementia and out-of-pocket spending on health care services. *Alzheimer's and Dementia, 9*(1), 19–29. https://doi.org/10.1016/j.jalz.2011.11.003

Dementia and Human Rights—Cahill, Suzanne—Google Books. (n.d.) Accessed on June 30, 2020. https://books.google.co.in/books?id=5b1VDwAAQBAJ&pg=PT291&lpg=PT291&dq=cahill+and+dooley+historical+context&source=bl&ots=JU5vKkBzWa&sig=ACfU3U3Uf6ItI0-aS2BbcoXczF7yB6e8bg&hl=en&sa=X&ved=2ahUKEwiYyZPc8ajqAhXXfn0KHYTVCWUQ6AEwAHoECAkQAQ#v=onepage&q=cahillanddooley historical context&f=false

Dementia, World Health Organization (World Health Organization). (2020). Accessed on June 30, 2020, https://www.who.int/news-room/fact-sheets/detail/dementia

Doraiswamy, P. M., Leon, J., Cummings, J. L., Marin, D., & Neumann, P.J. (2002). Prevalence and impact of medical comorbidity in Alzheimer's disease. *The Journals of Gerontology Series A: Biological Sciences and Medical Sciences, 57*(3), M173–M177.

Ernst, E. (2002). The risk—benefit profile of commonly used herbal therapies: ginkgo, St.Johns ward, ginseng, Echinacea, Saw palmetto, and Kava. *Annals of Internal Medicine, 136*(1), 42–53.

Ernst, E., & Pittler, M. H. (1999). Ginkgo Biloba for dementia: A systematic review of double-blind, placebo-controlled trials. *Clinical Drug Investigation, 17*, 301–308.

Gill, S. S., Bronskill, S. E., Normand, S. L., et al. (2007). Antipsychotic drugs use and mortality in older adults with dementia. *Annals of Internal Medicine, 146*(11), 775–786.

Gill, S. S., Anderson, G. M., Fischer, H. D., et al. (2009). Syncope and its consequences in patients with dementia receiving cholinesterase inhibitors: A population based cohort study. *Archives of Internal Medicine, 169*(9), 867–873.

Gill, S. S., Bronskill, S. E., Sharon, L. T., Normand, G. M., Anderson, K. S., Lam, K., Bell, C. M., et al. (2007b). Antipsychotic drug use and mortality in older adults with dementia. *Annals of Internal Medicine, 146*(11), 775–786.https://doi.org/10.7326/0003-4819-146-11-200706050-00006

Gitlin, L. N., Hodgson, N., Jutkowitz, E., & Pizzi, L. (2010). The cost-effectiveness of a nonpharmacologic intervention for individuals with dementia and family caregivers: The tailored activity program. *American Journal of Geriatric Psychiatry, 18*(6), 510–519. https://doi.org/10.1097/JGP.0b013e3181c37d13

Graff, M. J. L., Adang, E. M. M., Vernooij-Dassen, M. J. M., Dekker, J., Jönsson, L., Thijssen, M., Hoefnagels, W. H. L., & Olde Rikkert, M. G. M. O. (2008). Community occupational therapy for older patients with dementia and their care givers: Cost effectiveness study. *BMJ, 336*(7636), 134–138.https://doi.org/10.1136/bmj.39408.481898.BE

Gustafsson, M., Sjölander, M., Pfister, B., Jonsson, J., Schneede, J., & Lövheim, H. (2016). Drug-related hospital admissions among old people with dementia. *European Journal of Clinical Pharmacology, 72*(9), 1143–1153.

Haas, C. (2012). Strategies, development, and pitfalls of therapeutic options for Alzheimer's disease. *Journal of Alzheimer's Disease, 28*, 241.

Howard, R., Mc Shane, R., Lindesey, J., et al. (2012). Donepezil and memantine for moderate-to-severe Alzheimer's disease. *N Engl J Med, 366*(10), 893–903.

Hepburn, K. W., Tornatore, J., Center, B., & Ostwald, S. W. (2001). Dementia family caregiver training: Affecting beliefs about caregiving and caregiver outcomes. *Journal of the American Geriatrics Society, 49*(4), 450–457. https://doi.org/10.1046/j.1532-5415.2001.49090.x

Hertzog, C., Kramer, A.F., Wilson, R.S., Lindenberger, U. (2008). Enrichment effects on adult cognitive development: Can the functional capacity of older adults be preserved and enhanced? Psychological Science in the Public Interest, *9*, 1–65. https://doi.org/10.1111/j.1539-6053.2009.01034.x

Hitch, S. (1994). Cognitive therapy as a tool for caring for the elderly confused person. *Journal of Clinical Nursing, 3*(1), 49–55.https://doi.org/10.1111/j.1365-2702.1994.tb00358.x

Kelly, F. (2010). Recognising and supporting self in dementia: A new way to facilitate a person-centred approach to dementia care. *Ageing and Society, 30*(1), 103–124. https://doi.org/10.1017/S0144686X09008708

Kim, D. H., Brown, R. T., Ding, E. L., et al. (2011). Dementia medications and risk of falls, syncope, and related adverse events: meta-analysis of randomized controlled trials. *Journal of the American Geriatrics Society, 59*(6), 1019–1031.

Kirkley, C., Bamford, C., Poole, M., Arksey, H., Hughes, J., & Bond, J. (2011). The impact of organisational culture on the delivery of person-centred care in services providing respite care and short breaks for people with dementia. *Health and Social Care in the Community, 19*(4), 438–448. https://doi.org/10.1111/j.1365-2524.2011.00998.x

Kitwood, T. (1997a). The experience of dementia. *Aging and Mental Health, 1*(1), 13–22.https://doi.org/10.1080/13607869757344

Kitwood, T. (1997b). The experience of dementia. *Aging and Mental Health, 1*(1), 43. https://doi.org/10.1080/13607869757344

Kudlicka, A., Martyr, A., Bahar-Fuchs, A., Woods, B., & Clare, L. (2019). Cognitive rehabilitation for people with mild to moderate dementia. *Cochrane Database of Systematic Reviews,* (8). https://doi.org/10.1002/14651858.CD013388

Lau, D. T., Mercaldo, N. D., Harris, A. T., Trittschuh, E., Shega, J., & Weintraub, S. (2010). Polypharmacy and potentially inappropriate medication use among community-dwelling elders with dementia. *Alzheimer Disease & Associated Disorders, 24*(1), 56–63.

Lee, J., Sun, F., & Chima, E. (2018). Risks for abuse and neglect of elders with cognitive impairment or dementia: Insights from three groups. *Innovation in Aging, 2*(Suppl 1), 960.https://doi.org/10.1093/GERONI/IGY031.3558

Loraine, J., Taylor, S., & McAllister, M. (2014). Cognitive and physical stimulation therapy. *Journal of the American Medical Directors Association.* https://doi.org/10.1016/j.jamda.2013.11.016

Mallet, L., Spinewine, A., & Huang, A. (2007). The challenge of managing drug interactions in elderly people. *Lancet, 370*(9582), 185–191.

Martínez, T., Suárez-Álvarez, J., & Yanguas, J. (2016). Instruments for assessing person centered care in gerontology. *Psicothema, 28*(2), 114–121. https://doi.org/10.7334/psicothema2015.263

Martini De Oliveira, A., Radanovic, M., Cotting, P., De Mello, H., Buchain, P. C., Dias, A., Vizzotto, B., et al. (2015). Nonpharmacological interventions to reduce behavioral and psychological symptoms of dementia: A systematic review. https://doi.org/10.1155/2015/218980

Mendez, M. F.,Shapiran, J. S., McMurtray, A., et al. (2007). Preliminary finding: Behavioural worsening on donepezil in patients with frontotemporal dementia. *The American Journal of Geriatric Psychiatry,15*(1), 84–87.

McKeith, I., Del Ser, T., Spano, P., et al. (2000). Efficacy of rivastigmine in dementia with Lewy bodies: A randomised, double-blind, placebo-controlled international study. *Lancet, 356*, 2031–2036.

Mittelman, M. S., Roth, D. L., Coon, D. W., & Haley, W. E. (2004). Sustained benefit of supportive intervention for depressive symptoms in caregivers of patients with Alzheimer's disease. *American Journal of Psychiatry, 161*(5), 850–856. https://doi.org/10.1176/appi.ajp.161.5.850

National Institute for Health and Clinical Excellence. (2001).Guidance on the use of donepezil, rivastigmine and galantamine for the treatment of Alzheimer's disease. NICE, https://www.nice.org.uk/guidance/ta217/resources/donepezil-galantamine-rivastigmine-and-memantine-for-the-treatment-of-alzheimers-disease-pdf-82600254699973

Park-wyllie, L. Y., Mamdani, M. M., Li, P., et al. (2009). Cholinesterase inhibitors and hospitalization for bradycardia: A population based study. *PLoS Med* (9), e1000157.

Patel, L., & Grossberg, G. T. (2011). Combination therapy for Alzheimer's disease. *Drugs and Aging, 28*, 539–546.

Petersen, R. C., Thomas, R. G., Grundman, M., et al. (2005). Vitamin E and donepezil for the treatment of mild cognitive impairment. *New England Journal of Medicine, 352*, 2379–2388.

Prince, M. J., Fan, Wu., Guo, Y., Gutierrez, L. M., Robledo, M. O., Sullivan, R., & Yusuf, S. (2015). The burden of disease in older people and implications for health policy and practice. *The Lancet, 385*(9967), 549–562. https://doi.org/10.1016/S0140-6736(14)61347-7

Prizer, L. P., & Zimmerman, S. (2018). Progressive support for activities of daily living for persons living with dementia. *The Gerontologist, 58*(suppl_1). https://doi.org/10.1093/geront/gnx103

Raina, P., Santaguida, P., Ismaila, A., et al. (2008). Effectiveness of cholinesterase inhibitors and memantine for treating dementia, evidence review for a clinical practice guideline. *Annals of Internal Medicine, 148*, 379–397.

Rolinski, M., Fox, C., Maidment, I., et al. (2004). Cholinesterase inhibition for dementia with Lewy bodies, Parkinson's disease dementia and cognitive impairment. *Cochrane Database Systematic Review* (1):CD004395.

Røsvik, J., Kirkevold, M., Engedal, K., Brooker, D., & Kirkevold, Ø. (2011). A model for using the VIPS framework for person-centred care for persons with dementia in nursing homes: A qualitative evaluative study. *International Journal of Older People Nursing, 6*, 227–236. https://doi.org/10.1111/j.1748-3743.2011.00290.x

Sabat, S. R. (2002). Surviving manifestationsof selfhood in Alzheimer's disease. *Dementia, 1*(1), 25–36. https://doi.org/10.1177/147130120200100101

Sabat†, S.R., Harré†, R. (1992). The construction and deconstruction of self in Aazheimer's disease. Ageing and Society, *12*, 443–461. https://doi.org/10.1017/S0144686X00005262

Seelye, A. M., Schmitter-Edgecombe, M., Das, B., & Cook, D. J. (2012). Application of cognitive rehabilitation theory to the development of smart prompting technologies. *IEEE Reviews in Biomedical Engineering*. https://doi.org/10.1109/RBME.2012.2196691

Shaji, K. S., Smitha, K., Praveen Lal, K., 7 Prince, M. J. (2003). Caregivers of people with Alzheimer's disease: A qualitative study from the Indian 10/66 dementia research network. *International Journal of Geriatric Psychiatry, 18*(1), 1–6.https://doi.org/10.1002/gps.649

Steel, R. M., & Lawrie, S. M. (2018). Medical psychiatry. dementia. In S. H. Ralstan, I. D., Penman, M. W. J., Strachan, & R. P., Hobson (Eds.), *Davidson's principles and practice of medicine* (23rd ed, pp. 1191–1194). Elsevier.

Stockholm County Council, the Health and Medical Care Administration. (2018) Janusmed interactions. https://janusmed.sll.se/interaktioner. Accessed on June 2, 2020.

Taulbee, L. R., & Folsom, J. C. (1966). Reality orientation for geriatric patients. *Psychiatric Services, 17*(5), 133–135. https://doi.org/10.1176/ps.17.5.133

VandeWeerd, C., & Paveza, G. J. (2006). Verbal Mistreatment in older adults: A look at persons with Alzheimer's Disease and their caregivers in the state of Florida. *Journal of Elder Abuse and Neglect, 17*(4), 11–30. https://doi.org/10.1300/J084v17n04_02

Vandeweerd, C., Paveza, G. J., Walsh, M., & Corvin, J. (2013). Physical mistreatment in persons with Alzheimer's disease. *Journal of Aging Research*.https://doi.org/10.1155/2013/920324

Volkert, D., Chourdakis, M., Faxen-Irving, G., Frühwald, T., Landi, F., Suominen,M. H., Vandewoude, M., Wirth, R., & Schneider, S. M.(2015). ESPEN guidelines on nutrition in dementia. *Clinical Nutrition, 34*(6), 1052–1073.

Wilkinson, D. (2012). A review of the effects of memantine on clinical progression in Alzheimer's disease. *International Journal of Geriatric Psychiatry, 27*, 769–776.

Wilson, B. A., Evans, J. J., & Keohane, C. (2002). Cognitive rehabilitation: A goal-planning approach. *Journal of Head Trauma Rehabilitation, 17*(6), 542–555. https://doi.org/10.1097/00001199-200212000-00006

Woods, B., O'Philbin, L., Farrell, E. M., Spector, A. E., & Orrell, M. (2018). *Reminiscence therapy for dementia.* John. https://doi.org/10.1002/14651858.CD001120.pub3

Chapter 12
Psychological Health of Dementia Caregivers: An Overview

Ipsita Basu and Susmita Mukhopadhyay

Abstract **Background**: Worldwide, dementia is one of the foremost health challenging factors among elderly. Behavioural and psychological symptom of dementia (BPSD) has been found to be related with cognitive impairment, and it deteriorates with the progression of the disease. A majority of the persons with dementia depend partially or totally on their families for daily activities and care. Moreover, providing care for a family member with dementia has been found to be associated with caregiver's increased level of burden and poor psychological health. **Purpose of the review**: This article offers a rationalized review of the dementia caregiver's psychological health based on care demands, demographic factors and effects of different behavioural problems of the care recipients. It also offers an overview of dementia and its symptoms. **Methods**: Literature search was carried out using PubMed, Research Gate and Google Scholar which were published between 1990 and 2019. Key words that were mainly used for searching relevant literatures are as follows—dementia, dementia caregiver, carer, caregiving, psychological health (i.e. stress, depression, anxiety, burden and distress), demographic factors and so on. **Findings**: Some common disruptive behaviours such as wandering, delusion, hallucination, apathy and motor disturbances are observed in a person with dementia. Prolong exposure to these types of disrupting behaviour affects caregivers' psychological health enormously. The other responsible factors for caregivers' deteriorating health conditions are number of care hours/day and years, type of help provided towards care recipient's daily activities and so on. Additionally, studies found that sociodemographic factors, less family support, lack of social activities and social interactions are some common causes of adverse psychological health conditions among family caregivers. **Summary**: Due to dearth of proper institutional dementia care services, family caregivers have to play the main supporting role to look after their demented family members. In order to carry out several such responsibilities, they develop many adverse health conditions, specially the deteriorating mental health condition. Literature reviews suggest that coping strategies may help to reduce the adverse psychological health condition among dementia caregivers.

I. Basu (✉) · S. Mukhopadhyay
Biological Anthropology Unit, Indian Statistical Institute, 203 Barrackpore Trunk Road, Kolkata 700108, India

M. K. Shankardass (ed.), *Dementia Care*,
https://doi.org/10.1007/978-981-16-3864-0_12

Keywords Dementia · Family caregiving · Caregiver's psychological health

Introduction

The biggest concern in demography at global level is the growth of the ageing population. Human life expectancy towards longer life has increased in the past century, giving rise to elderly population. It is occurred because of socio-economic developments; improvements in healthcare systems make control over infectious diseases, at the same time, risk of certain non-communicable chronic diseases is emerging rapidly, and dementia is one of them (World Alzheimer's Disease Report, 2013). Worldwide, dementia has become a health challenging factor among elderly. With the ageing of the population, numbers of people with dementia will also increase substantially in the next few decades. According to World Alzheimer Report (2016), approximately 46.8 million people are living with dementia around the world. It is predicted to increase by about 131.5 million by the year 2050. Studies have suggested that prevalence of dementia is higher in developed countries than developing. Overall, it has been accepted that lower level of awareness about dementia is more commonly found in developing countries including India (Boise et al., 1999). With enhanced health quality and increased life span, the proportion of elderly individuals in India is projected to rise dramatically in the future decades. As a result, there will be upsurge of older people with dementia. The number of persons with dementia will be doubled with each increment of 5 years of age. Consequently, India will home the largest populations of elderly individuals with dementia. It was reported that approximately 3.7 million Indian people are living with dementia. By 2050 that number is expected to increase to about 14.3 million. It was also found that dementia hold the second position among different chronic diseases (shown in the pie chart) to the total chronic disease burden among people aged 60 years and above (Bharath & Dias, 2010; Shaji et al., 2010). Prevalence of demented people varies from region to region as well as from urban to rural areas. The prevalence of dementia in Eastern India is 0.8–1.28%, 2.44–4.1% in Western India, 1.83% in Northern India and 3.6% in Southern India (Das et al., 2012). In case of rural areas, it varies from 0.6 to 10.6% while 0.9–7.5% in urban areas (Das et al., 2006; Rajkumar et al., 1997; Rodriguez et al., 2008).

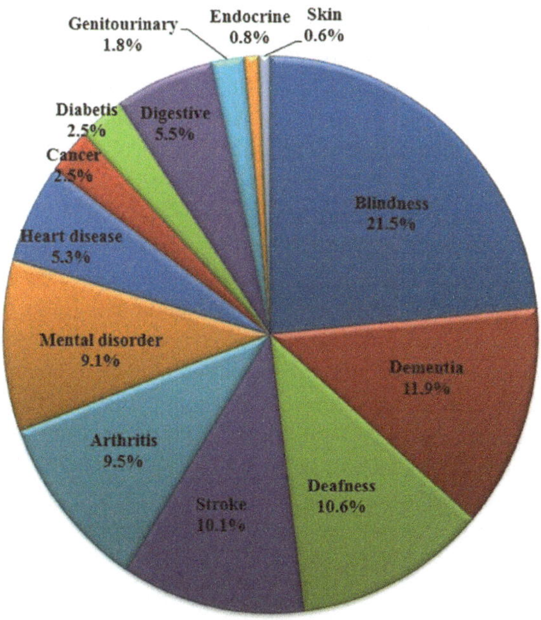

[Ref: The Dementia India Report, 2010]

As reported, with improving lifespan, degenerative diseases such as dementia emerging as a new health problem among elderly people. It is not an inevitable occurrence of normal ageing but is more common among the elderly. Dementia is a set of conditions that causes number of cognitive impairments. As a result person having dementia experience physical, emotional and functional deterioration as well as changes in behaviour and personality are also observed (Alzheimer's Disease International, 2009). Generally, dementia is described as a decline memory ability. It comprises a wide range of neurological disorders and condition is severe enough that can interfere performance of daily activity of living. Impairments also observed in case of language, visuospatial ability, calculation, judgement, problem-solving and so on. The most common types of dementia are Alzheimer's dementia (AD), vascular dementia (VD), frontotemporal dementia (FTD) and Lewy body dementia (LBD) (Harada et al., 2013).

Alzheimer's type of dementia is the most common types found in 60–80% of the cases. In Alzheimer's disease, some changes take place in the brain which can damage and destroy nerve cells. As a result, it prevents nerve cells to send signals to each other and eventually stop working. Therefore, the brain cells die, and brain shrinks in size. The symptoms of the disease are usually gradual, and though almost undetectable, decline in many areas of intellectual abilities is accompanied by physical decline. Memory failure, personality change and problems related to daily activities are the direct result of this types of dementia (Buckner, 2004). The second most common type is vascular dementia (VD). In this dementia, there are number of dead areas are observed in brain, which are outcomes of sequence of small strokes. The brain cells in a small area of the brain are destroyed, either because of the less blood supply to that blocked area or because of blood cells bursts. Each stroke or infarct may be small, but together the dead area of brain results in memory and reasoning difficulties for the person experiencing it (Khan et al., 2016). Frontotemporal dementia (FTD) has the early onset even before 65 years of age. Symptoms may arrive even in between ages 50 and 60 year. As the name suggests, FTD basically occurs in either frontal lobe (behind forehead) or temporal lobe (behind eyes). In this type of dementia, two different types of protein accumulate in the brain cells of these two lobes and damage the brain cells. Person with FTD will experience problems in language, behaviour, personality and movement (Rosness et al., 2016). Lewy body dementia (DLB) includes both Parkinson's disease and dementia with Lewy bodies. Physical slowness, confusion and rambling can be observed in these types of dementia (Walker et al., 2015). Overall, it was observed that persons with severe dementia fail to perform their basic daily activities such as eating, bathing, maintaining own hygiene and so on. Demented person gradually loses the capability of walking as well as speaking, and with the progression of dementia, they require complete supervision. Behavioural and psychological symptoms are essential part of dementia. These are the source of caregivers' morbidity, burden and adverse quality of life (Chiu et. al., 2006).

	Alzheimer's dementia	Vascular dementia	Frontotemporal dementia	Dementia with Lewy bodies
History	Memory loss, spatial disorientation, language failure	Mental and physical decline	Early personality change, stereotyped behaviour	Confusion, physical slowness
Memory	Severe amnesia	Variable loss	Variable loss	Variable loss/amnesia
Language	Aphasia	Dysarthria	A dynamic speech	Incoherent, rambling
Perception	Primary recognition failure	Preserved	Preserved	Misperception
Motor skill	Impaired spatial configuration	Impaired sequencing	Impaired sequencing	Impaired sequencing spatial configuration

Reference: Shaik and Varma (2012)

In Asian countries, cultural beliefs and traditional practices impose that grown-up children or other family members mostly females (as majority of them were unemployed and mostly stayed at home) expected to take responsibilities of caring for elderly people. However, demographic transitions (i.e. high fertility rate to low fertility rate) and urbanization dramatically affect its social structure and system, forcing people to live in smaller families and left fewer younger people to care for elderly members. At the same time, better education and increased workforce mobility are responsible for changes in role and status of women within family. So, certainly there are scarcity of people left at home to care for elderly as large number of women also join out of home employment (World Alzheimer report, 2013). Moreover, due to lack of proper institutional healthcare services, people with dementia live in their home and received daily assistance and care by their family members. The largest proportion of these caregivers are spouses followed by children and children-in-law, who are mostly female (Martin & Sörensen, 2011; Prince, 2004). The time of onset of caregiving is hard to identify. Needs and/or level of care increase with time; it mostly depends on the severity conditions of dementia care recipients (Gaugler et al., 2003). The caregivers provide support towards care recipients' daily activities as well as instrumental activities, and sometime, support extends to complete supervision and surveillance for care recipients (Schulz & Martire, 2004).

It is well acknowledged that providing care for people with dementia is a widely acclaimed social phenomena that influence the life of caregivers (Pinquart & Sorensen, 2007). Family caregivers are sometimes labelled as the invisible second patients (Brodaty & Donkin, 2009). Experience of being a family caregiver with a person of dementia, however, is sometimes encouraging but generally negative with elevated rates of psychological morbidity and adverse mental health conditions (Cheng, 2017). Caregiving is known to be accompanied by the burden among dementia caregivers. They described the experience of caregiving for an individual with dementia as stressful, frustrating and demanding (Butcher et al., 2001). Caregiving is basically a multidimensional response to psychological, physical, emotional and social factors associated with their caregiving experiences. These kind of experiences may lead to the development of physical and psychological illness and even increasing morbidity and mortality among family caregivers (Kasuya et al., 2000, Kiecolt-Glaser et al., 2003; Vitaliano et al., 2004). Caregiving can be defined as a duty or responsibility to care for the person with dementia, and most importantly, they did not financially compensate for the tasks they are performing towards their demented recipients. These kind of tasks may be physically as well as mentally demanding (Schulz & Sherwood, 2008).

Caring for the person with dementia creates chronic stress and psychological distress among family caregivers which affect their quality of life (Akpınar et al., 2011; Laks et al., 2016). Literatures have shown that caring for a person with dementia is more burdensome than caring for an adult with other disabilities (Pinquart & Sorensen, 2007). Caregivers may face a variety of challenges as they simultaneously balance caregiving with other demands such as a job, social relationships and so on. Sometimes, it has also found that caregivers may force to leave their job to carry out their caregiving role. It is evident that because of these kind of responsibilities, they are at risk of developing adverse health-related problems (Kang et al., 2014). Some studies showed that the total burden has two dimensions, subjective and objective, associated with dementia caregiving (Brodaty & Donkin, 2009). The subjective burden refers to physical and emotional impact of caregiving and psychological conditions of caregivers. Whereas, objective burden results from the degree of dependency of the demented person, behavioural problems related to dementia, time spent in caregiving and so on (Brodaty & Donkin, 2009; Clare et al., 2002; Sanders,

2005). Studies showed that dementia caregivers are at risk of developing cardiovascular diseases also (Patterson & Grant, 2003; Roepke et al., 2012; Richardson et al., 2013).

To prevent caregivers' deteriorating health conditions and improve their quality of life, it is important to implement coping or intervention therapies from the early stage of their caregiving role (Salamizadeh et al., 2017). Overall, family support in the early or moderate stages of dementia may improve the perceived health of the caregivers and could avoid the new onset of mental health problems among caregivers (Stansfeld et al., 2017).

This article provides an update and rationalized review of the dementia caregiver's psychological health with focus on care demands, demographic factors, effects of different behavioural problems of the care recipients and other health related problems that are associated with caregivers' adverse psychological health. Additionally, it also discusses about some coping and intervention strategies which are helpful to maintain caregivers' health status.

Methodology

Literatures have been studied to get better understand the psychological well-being of caregivers. To better understand the caregivers' situation, one need to know about dementia and its symptoms. Literature search was carried out using PubMed, Research Gate and Google Scholar which were published in between 1990 and 2019. Search items included "caregivers", "dementia caregiver", "family caregivers", "family caregiving", "career", "caregiving", "caregivers' psychological health", "stress", "depression", "anxiety", "burden", "distress", "demographic factors", "care hours", "care years", "dementia", "dementia types", "dementia symptoms", "family culture of caregiving", "behavioural problems in dementia", "cognitive impairments in dementia", "coping strategies" and "intervention". The articles' title, abstract and reference lists have also been searched and reviewed. The articles were selected based on the following criterion: (i) focus on family caregiving (ii) caregivers of people with dementia (iii) focused in dementia, its types and symptoms (iv) coping strategies (v) intervention (vi) published from 1990 to 2019 (vii) published in English. Selected studies were quantitative, qualitative or mixed in nature.

The researchers carried out the article selection and analysis for cross-checking. At first, we selected the research articles which were published in English language only. After that, title and abstract were checked for inclusion. Removal of duplicates were performed carefully. Reading of the full article allowed for greater evaluation of the scientific studies and then did the final selection. The following article presents a narrative review of the existing literatures.

Discussion

Effects of Dementia on Informal Caregivers

A caregiver is someone who extend their help to disabled individuals or those having difficulty to perform their daily activities as well as they try to ensure the safety of the care recipients. Informal caregivers are mainly the family members, friends or neighbour who has significant relationship with care recipients and provide care without exchange of money. Moreover, primary family caregivers are those who are directly involved in care for disabled family member's daily need in non-professional manner (Schulz & Martire, 2004). Most of the studies showed that family caregivers experience negative effect on caregivers' health, but few studies have shown that caring for people with dementia can have positive effect on caregivers. Family members generally accomplish the role of caregiver out of love, commitment, respect and/or a sense of duty for their demented care recipient. Therefore, by providing care, they may feel sense of mastery and accomplishment that leads to personal growth and sense of reward and fulfilment. Studies reported that providing care to the person with dementia has been identified as the source of satisfaction (Deeken et al., 2003; Savage & Bailey, 2004). Satisfaction mainly depends on interpersonal relationship between caregiver and recipient, intrapsychic orientation of caregiver and a strong desire to provide a good level of care (Grant & Nolan, 1993). A study showed that the source of great pleasure for any caregiver who can retain the dignity and self-esteem of their care recipients (Nolan & Grant, 1992). Another study pointed out that even after extensive care, caregivers find satisfaction if they knew the way of coping with the situation (Kristensson, 2004).

The whole process of dementia caregiving involves many changes related with the progression of the disease, and thus, it is an intricate process (World Alzheimer Report, 2013). In Asia, elderly care has traditionally been family-based and regulated within joint family system. Likewise in Indian society, cultural values and traditional practices impose that adult children or family members expected to take responsibilities of caring for their elderly family members (Jathanna et al., 2010; Mattoo et al., 2013). Demented individual required high level of attention majority of which was given by the family member (Brodaty & Donkin, 2009). In India, demographic transitions and urbanization dramatically affect its social structure and social system; the gradual breakdown of the joint or extended family into nuclear ones has adversely affected the improvised system of care for the elderly within family. Better education and increased workforce mobility are responsible for changes in role and status of women within family. As a result, employment of women has further aggravated the problem of providing care to the elderly as women were the main caregivers of the family (Holroyd, 2001; Huang et al., 2006; Ohaeri, 2002; World Alzheimer Report 2013). In the absence of formal care services in most developing countries, dementia care recipients mostly stayed at home and mostly dependent on the informal care, in particular family care (Prince, 1997). Caregivers are mostly spouses, children and children-in-laws (Ng, 2008; Nolan, 2001; Prince, 2004). As a result, caregivers have to compel their work to maintain caregiving task properly. Studies have reported that responsibility of care for the people with dementia can increase distress levels among the caregivers (Ng, 2008; Nolan, 2001; Prince, 2004).

Caregivers' Mental Health Condition

In low and middle income country, usually family caregivers received little help from the family, community as well as from the governmental body. As caregivers try to balance their job, family, and caregiving responsibilities, they often neglect their own physical and emotional well-being. Caregivers frequently experience stress and depression because of inadequate personal, social, emotional and financial resources (Cassie & Sanders, 2007). Caregiving has a wide-ranging and complex impact on caregivers, and there are numerous other factors that can accelerate how caregivers feel and respond as a result of their caregiving role. According to studies, caring for those with dementia is more difficult and time-consuming than caring for someone with another disability (Mohide et al., 1998; Ory et al., 1999; Schulz et al., 1990). A study pointed out that caregivers' burden is a multidimensional response. There are four major divisions of these responses: (i) the context of care that includes characteristics of caregivers, types of caregiver–recipient relationship along with availability of social support (ii) the behavioural problems related to person with dementia create the stressful condition among the caregivers (iii) elements that mitigate stress perception, i.e. coping strategies and management of caregiving situation (iv) acquire the knowledge of consequences of caregiving for the betterment of the caregivers' health condition (Papastavrou et al., 2007). One psychosocial result of caregiving is perceived burden. Burden includes awkwardness, overload, feeling of entrapment, isolated from society, less social activities, poor communication, work pressure and so on. In this situation, caregivers report more distress than any other non-caregivers (Stephens et. al., 1991).

A study reported that physical health decline, psychological morbidity and poor quality of life are observed among dementia caregivers who experienced chronic stress (Given et al., 1999). Stress may lead to poor care conditions, negligence or even harassment of an individual with dementia (Mahoney et al., 2005). Women caregivers were more inclined to higher level of burden and depression. As a result, they suffer from social or relational deprivation (Papastavrou et al., 2007). Another study found that dementia caregivers in Hong Kong have experienced significant amount of psychological, financial, and physical burdens (Petrus & Chung, 2006). A Taiwan study found that even after institutionalization of the demented people, caregiver still suffers from stress (Lau et al., 2008). In India it was found that half of the dementia caregivers experienced higher level of stress and more than two-third of the caregivers experienced distress due to their caregiving activities (Basu & Mukhopadhyay, 2019).

Behavioural and Psychological Symptoms of Dementia (BPSD) Care Recipients and Caregivers' Mental Health

Behavioural symptoms of dementia are a well-known cause of caregivers' burden, depression and other complication in connection with mental well-being. Behavioural and psychological symptoms of dementia (BPSD) is generally a heterogeneous range of psychological reactions, psychiatric symptoms and typical behaviour occurring throughout the duration of dementia (Finkel et al., 1996). BPSD is defined as "symptoms of disturbed perception, thought content, mood or behaviour that frequently occur in patients with dementia" (Finkel et al., 1998). Behavioural and personality changes are the first observed phenomenon among the person with dementia even before cognitive decline. The most common BPSD includes anxiety, self-centeredness, agitation, wandering and so on. It was reported that BPSD can be related to lower quality of life for person with dementia as well as their family members and caregivers. Unlike cognitive symptoms, BPSD did not show any linear pattern of deterioration. These symptoms are more distressing because of their unpredictable and disruptive nature and also difficult to manage. Therefore, unpredictable nature of these symptoms may create high levels of stress, depression and anxiety among caregivers (McCarty et al., 2000). In addition, some longitudinal studies showed that symptoms in early stage of dementia as well as remarkable increased of symptoms are the predictors of caregivers' burden over time (Gaugler et al., 2005, 2010; Shim et al., 2016). Researches have indicated that mental health of the dementia caregivers is more affected by BPSD than cognitive impairments (Schulz & Gali, 1991; Victoroff et al., 1997). Another study showed that rate of change in behavioural problems among demented person was a significant predictor of caregivers' adverse mental health condition (Gaugler et al., 2000). A study by Hooker et al. (2002) revealed that the deterioration on both mental and physical health of the caregivers depends on the amount of increase in problematic behaviour among the care recipients. On the basis of the findings, authors are also pointing out that if caregiving phases get longer than expected, level of stress will be increased and would have negative implication on caregivers' mental health. Another study showed that among many BPSD, wandering is the most common symptoms found among person with dementia. Wandering behaviour is considered to be the most problematic and dangerous behaviour found among demented person (Müller-Spahn, 2003; Siders et al., 2004). It is associated with falls, injuries and getting lost (Buchner & Larson, 1987; Teri et. al., 1988). These kind of situations are the source of caregivers' stress and anxiety. A study in rural India showed high prevalence of BPSD which was common in both developed and developing countries. Article also revealed symptoms such as activity disturbance, assertiveness and delusion in specific were very annoying and stressful to the caregivers (Shaji et al., 2009). According to studies, behavioural difficulties associated with dementia have a greater impact on caregivers' daily life than cognitive or functional impairment of care receivers (Annerstedt et al., 2000; Papastavrou et al., 2007). A study in Singapore found that caregivers' adverse health condition was positively correlated with presence of behavioural symptoms of dementia such as delusion, hallucination, depression, insomnia, incontinence and so on (Heok & Li, 1997). Another study pointed out that BSPD is normally unrecognized as caused by underlying disease process. Over time long dealing with these symptoms could create stress among caregivers (Hooker et al., 2002).

Results from a study by Katherina and associates (2019) indicated that cognitive impairment and behavioural problems of care recipients were the two important primary stressors to increase the stress level among caregivers. Caregivers who provided care to recipients with more cognitive impairment associated with increased level of burden. The progression of cognitive impairment observed among recipients leads to increased reliance on their caregivers. When recipients are cognitively challenged, caregivers are more involved in supporting them with functional, emotional, and physical requirements. So, these kind of activities may escalate caregivers' quality of life.

Effect of Socio-demographic and Socio-cultural Factors on Caregivers' Health

Numerous studies have found that sociodemographic factors are also responsible for dementia caregivers' adverse mental health condition. Caregivers' mental health status is determined by many factors, viz. age, employment status, education, income, relationship, social support and so on. A study found that caregivers' age and education were strongly associated with caregivers' level of stress and anxiety. This study also revealed that age of the caregivers may be the indirect indicators of poor mental health conditions of caregivers as older caregivers experience age-related physical vulnerability. They also pointed out that being a spouse, it may be husband or wife, has significant effect on their level of anxiety (Basu & Mukhopadhyay, 2019). Studies also reported that the women caregivers are likely to experience 6 times more depressive and anxiety symptoms than those had no caregiving responsibilities. A study conducted in Brazil found that women caregivers under 60 years of age, with high school and college degree, expressed highest level of burden (de Moraes & da Silva, 2009). Higher level of education has significant positive effect on caregivers' health (Cassie & Sanders, 2007; Regiane & Almeida, 1999). Many studies also revealed that gender plays a significant role to determine caregivers' health status. Approximately, 80% of the caregivers were females globally, and the relationship with the recipients could be mother, spouse, daughter or daughter-in-law (Given et al., 1999; Ng, 2008; Nolan, 2001; Prince, 2004). Studies have found that female caregivers are likely to experience higher level of stress, depression and burden than male caregivers (Harries, 2000; Campbell et al., 2008; Ormel et al., 1994; Regiane & Menezes, 2004). Study by Mohamed et.al. (2010) found that spouse caregivers showed higher level of burden. A comparative study conducted in India, China and South-East Asia found that caregivers live in a large family household with demented member and living with many family members helps to reduce caregivers level of burden (Prince, 2004).

Family relationship between caregiver and care recipient may influence caregivers' mental health condition (Campbell et al., 2008). Study showed that close relationship with care recipient has significant influence on caregivers' burden (Annerstedt et al., 2000). Decreased quality of life has adverse effect on caregivers' health. Studies pointed out that insufficient social interaction, less participation in social activities, getting minimum time to maintain own interests are the significant predictor of increased level of burden and depression (Cummings et al., 2004; Mohamed et al., 2010). Lack of social support is also seen as an essential component for caregivers' increased level of burden. It also showed that support from family positively related with caregivers' health, and in case of family dysfunction, the result was reverse (Tremont et al., 2006). A study of Chinese caregivers revealed that informal and emotional support of family members has an important connection with life satisfaction and reduce depressive symptoms (Au et al., 2009). Caregivers who did not get support from their family members enunciate their feeling of anger, frustration and unhappiness. Moreover, lack of support can directly affect physical and mental health of caregivers (Kerpershoek et al., 2018; Landmark et al., 2013).

Culture plays an important role in the demand of help-seeking behaviour of the family caregivers. While Latin-American caregivers express their feelings, needs, family crisis openly (Caudle, 1993), caregivers in Asia believe in maintaining secrecy in divulging the truth. The Chinese think that the family is the main and most important source of life. They clearly differentiate insiders from outsiders. Family matters can be discuss openly to the insiders but not to the outsiders. They were worried that demented members might be considered as "mentally ill" which is socially defamed. So, the family caregivers do not share any unfavourable happenings with others outside the family. This restricts them from seeking help outside the family (Li et. al., 1995). Thus they usually restrained themselves to use mental health facilities. Better understanding of caregivers' problems and beliefs that are rooted in socio-cultural back ground will help healthcare professionals to develop better coping strategies for betterment of caregivers' health condition (Gilbert et al., 2007).

Caregiving Related Factors and Caregivers' Health

The number of elderly living with dementia is increasing progressively and it becomes one of the prominent health problem worldwide. Dementia not only affect the person who have it but also create growing burden for family members too (Das et al., 2010). Unlike other diseases, people with dementia need care from its early stage and become dependent upon the caregivers throughout the course of the disease. Studies revealed that right from the beginning of dementia, person need assistance to do their basic activities (such as bathing, dressing, feeding, taking own medicine, managing finance, etc.) as well as instrumental activities.

Providing continuous assistance for all these activities for an extensive period of time give rise to excessive mental pressure, which help develop poor mental health conditions (Arango-Lasprilla et al., 2010). Impairment in basic activities of care recipients' (eating, bathing, dressing, maintaining personal hygiene, etc.) were frequently mentioned as the source of caregivers' burden. Ravio and his associates (2007) found in their study that only one third of the people with dementia completely independent. Besides that, majority of the caregivers have to do at least one type of activities of daily living for their demented care recipients'. A significant numbers of caregiver had to take the responsibility of providing medicine on proper time, take care of family finance, prepare meals and so on. Even most of the care recipients were incapable to go alone anywhere, either inside the house or outside. So these kind of activities are time-consuming and inconvenient for caregivers. As a result, caregivers get inadequate time to satisfy their personal need.

Study found that dementia care recipients' did lack control over urination or defecation; it was considered as the most distressing problem faced by the caregivers. Caregivers try their best to keep care recipient and the house clean, but often they failed to do so. Some caregivers reported that they felt difficulty to have food at home because of the bad smell that prevails around the house. For the caregivers this condition remains unbearable when the living place is small and had to share it with care recipients on daily basis (Shaji et al., 2003). It was found that in the early stage of dementia, BPSD are more prominent than functional impairment. But with advancement of dementia, BPSD eventually decreased while functional disability slowly grabs the attention of caregivers (Garre-Olmo et al., 2016).

A strong correlation with caregivers' level of mental health condition and care recipients' incompetence for daily activities has been reported widely. Caregivers' support for recipients' every day activities and instrumental activities could be physically exhausting, emotionally stressful and time-taking. In a cross sectional study, Basu and Mukhopadhyay (2019) tried to investigate the mental health condition of dementia caregivers and caregiving related factors. Mental health was measured in terms of stress, depression and anxiety. Based on support given to the care recipients' daily activity, caregivers were divided into two groups. One group of caregivers provided maximum support and other provided minimum support. More than 35% of the caregivers who provided maximum level of support experienced higher level of stress and anxiety. Study revealed that difficulty of providing care for recipient's daily activities may develop mental fatigue for the caregivers.

Caregiving experience can be well-defined both in terms of duration and intensity, i.e. the number of hours spent daily, weekly, or monthly and the episode of caregiving could be continued for months or even years. Additionally, they perform all the required tasks which may be physically and mentally arduous (Schulz & Sherwood, 2008). The main cause of deleterious mental health condition is the considerable involvement of caregivers regarding caring times because of recipients' inability to perform daily and instrumental activities (Haro et al., 2014). According to a study, caregivers are usually provided care to their demented care recipients on an average of 1.6 h daily and caring hours increases with inclusion of care related to instrumental activities (average 3.7 h/day) and complete supervision (average 7.4 h/day) (Wimo et al., 2007). A large survey on dementia caregivers was carried out in five European countries including Germany, France, Poland, Scotland and Spain. A strong relationship was found between dementia severity and caring hours. In this study it was found that caregivers provided more than 10 h of care per day for care recipients' with moderate-to-severe level of dementia. It was reported by the caregivers, functional impairment was the most problematic than behavioural and cognitive impairment (Georges et al., 2008). Another study in US revealed the similar result. Functional disability and dependency was more associated with numbers of care hours. For these reason caregivers who are employed might cut off their working hours or even force to leave their job (Hughes et. al., 2014). The caring hours in daily basis is another steady predictors of carers' burden and depression (Cheng et al., 2013; D'Onofrio et al., 2015; Park et al., 2015). The caregivers who received help from other reported less stress than those did not get any kind of help. Study showed that sharing work responsibility, other helpful caregivers may reduce the pressure from primary caregivers (Shaji et al., 2003).

A study by McCurry and his colleagues (2007) reported that worldwide about 10 million people are providing care for their demented family members and two-third of them suffered from sleep disturbances. Caregivers' sleep troubles have been connected to deprive immune system, elevated stress hormones, increased cardio-vascular disease and mortality (von Kanel et al., 2010). Studies also revealed that the absence of support from other family members creates additional pressure on the person who actually takes full time care. He/she gets very little insufficient time for sleeping, personal care and even time to participate in social activities (Gratão et al., 2010; Landmark et al., 2013). A study reported that care recipients' gender and severity of dementia significantly influence the sleep efficiency of caregivers. The severe types of dementia might have severe neuropsychiatric symptoms of dementia which would have impact on caregivers' sleep (Mills et al., 2009). Similarly, a study showed that care recipients' behavioural abnormality such as agitation, anxiety, depression may increase the caregivers level of burden, stress and affecting their sleep quality (Andrieu et al., 2007). McCurry et al., (2008) found in their study that depression of demented person and caregivers' management ability were related to caregivers' poor quality of sleep. A study (dos Santos et al., 2009) stated that family caregivers experienced both mental and physical exhaustion to cope with various demands of caregiving which could lead to sleep disturbance. Sleep disturbances has adverse effect on caregivers' performance (Chiu et al., 2013).

Conceptualisation of Dementia Among Caregivers

Although dementia is a medical condition, but cannot be understood in a medical framework alone, because the meaning of illness or any kind of disease embedded in the cultural contexts and differ from culture to culture (Zeilig, 2014). Recognition and response to illnesses, both physical and mental, being mostly determined by social norms and acceptance (Dixit, 2005). However, sometimes these believes may can prevent caregivers to practice in help-seeking behaviour and also keep the actual condition from being diagnosed. As a consequence, indirectly it also affects caregivers own health (Hinton et al., 2005; Sethabouppha & Kane, 2005).

Studies have found the dissimilarities between Eastern and Western thought of dementia which have impact on caregivers' perceived burden. Culture gives a structural idea how people define and interpret the meaning of illness. Likewise, family gives the perception to perceive a disease condition (Gilbert et al., 2007). A study found that an Indian Punjabi family perceived dementia as due to absence of effort by the person themselves and sometimes absence of family attention. So, eventually they believe that the affected person with dementia could only overcome their own problems as well as attention of family members might be a part of the way out from the situation (Fontaine et al., 2007). Other Asian countries like Japan, China, are strongly influenced by their religious belief. A study among Chinese people found that they belief whatever has happened is because of heaven's will and without any argument they are ready to accept it. They still believe that their fate is linked to the origin of dementia and their well-being is sent from heaven. Eventually, they embrace their disease as they admit other responsibilities (Mok et al., 2007). It was also reported that Chinese caregivers usually do not distinguish their caregiving responsibility from their other daily activities; they comprehend caregiving as just part of life. Furthermore, Chinese caregivers of family member with dementia had reported higher depression and burden compared with dementia caregivers from Western societies (Torti et al., 2004). Another study in multi-ethnic group of Asian-American caregivers showed that caregivers followed various models in conceptualization of dementia. These models are biomedical model, a folk model and a mixed folk model. The biomedical model frames dementia as an abnormal disorder in brain. In this model, they consider dementia as a disease. In folk model, dementia perceived as "normal" process of being old. A study found that Vietnamese caregivers refer the term "confused" and accepted dementia as a normal process of ageing. Another study showed that they accept dementia as "memory deterioration" and it is happening because of ageing, mental disorder, perceived a "difficult personality" and not being an open-minded individual. In the mixed folk model, caregivers conceptualized disease as like biomedical model but explain the disease as folk model. As example, they belief dementia is a disease and caused by some psychological stress or the process of normal ageing (Hinton et al., 2005).

Caregiver's Coping and Intervention Strategies

In majority of countries, it has been found that family caregivers often get stigmatized due to lack of understanding about dementia (WHO, 2018). Caring for a loved one with dementia can be onerous, and numerous caregivers experienced deprived quality of life (Basu et al., 2020; Takai et al., 2011; Tomomitsu et al., 2014). Therefore, caregivers adopt different strategies to cope up with the adverse situation and try to make caregiving task as well as the situation easier for them. Avoidance in coping strategies may result in higher level of stress and burden among family caregivers (Valentina et al., 2008). It is the responsibility of government and society to realize that better health will enable caregivers to provide good quality of care. Many a times, it has been found that caregivers may not have time to look after their own health problems and has to ignore it. They may feel overwhelmed with the demand of caregiving responsibility (Ronda & Crew, 2007).

Coping is not only a response to the adverse situation. It' only depends on the people, how they understand and interpret a particular event. They may cope with the situation or became stressed by it (Cooper et al., 2006). In some culture, suffering is considered as a part of life. The caregivers willingly accept the suffering of their family members and did not get afraid by it as well (Campbell et al., 2008). Two independent studies in Singapore and China have revealed that religion and spirituality have very positive impact on dealing with dementia. Their culture and religion led them to take the responsibility of look after their demented member as a natural phenomenon of life or destiny (Campbell et al., 2008; Poulshock & Deimling, 1984). "Sense of coherence" (SOC) is a useful coping strategy. It helped people to cope with stressful situation and has positive effect on health, including mental health (Vincent, 1999). Another study pointed out that higher level of SOC correlated with better quality of life and less depressive symptoms (Cooper et al., 2007). A comparative study revealed that the caregivers who experienced lower level of stress, burden scored higher sense of coherence than their counterparts (Livingston et al., 2005). A study in Taiwan revealed that the caregivers with higher SOC associated with better quality of life (Tang & Li, 2008).

Two studies also found that in developing countries, dementia often perceived as a normal process of ageing, and families often ignore the situation which in many cases resist caregivers to meet their own needs (Dias et al., 2008; Patel & Prince, 2001). With the help of local human resources, a randomized controlled intervention was carried out in Goa, India. In this programme, 41 caregivers received the intervention, and 40 remained as control. Proper knowledge was provided to understand the dementia and the related common behavioural problems. They made caregivers aware of, how to manage problematic actions, support caregivers in activities of daily living. They also provide information about forming network to create support group and give information on governmental provisions available for elderly. These interventions lead to significant improvement in caregivers' mental health conditions. Interventions reduced behavioural problems and improved the functional ability of care recipients (Dias et al., 2008).

Interventions intended to show that enhancing caregiver coping skills helps to improve caregiver quality of life by reducing the stress associated with caregiving activities. Structured interventions on caregivers' skill training have shown promise for caregivers' better quality of life. Information and education are the most important factors in caregivers' interventions. The caregiver and the care recipient should have the easy access to information regarding dementia and its management. It is the responsibility of local health professionals to increase awareness of the caregivers about issues related to dementia caregiving and its consequence to their own health. It is also important to provide opportunities to caregivers to discuss their problems with the health professionals. Providing social and emotional support to the caregivers as well as to the care recipients is another important factor for caregivers' interventions (Shaji & Reddy, 2012). Community Outreached Services is one of the influential aspects in identifying and supporting families engaged in home-based care. In this process, family caregivers from each family have to be identified, and their care-related problems should be addressed. The intervention procedure needs to be culturally accepted as well (Dias et al., 2008; Guerra et al. 2011). The intervention should be generic in nature which could be helpful for family caregivers, even if they did not aware of dementia. This community-based intervention will help to reduce caregivers' burden.

A review of studies is based on a different useful strategies such as psycho-educational intervention, social support, use of exercise and taking proper nutrition, case management and so on. Psycho-education intervention includes behavioural therapy, stress-related coping skills and 1–9 months of follow-up. It was found that psycho-educational intervention has significant effect to reduce depression and improvement in caregivers' quality of life. Other interventions were also improved self-efficacy, well-being and perception of problems related to caregiving (Parker et al., 2008). Another study found that positive coping strategies had a negative correlation with caregivers' burden (Papastavrou et al., 2007). Interestingly, it was observed that the caregivers who experienced lower level of burden were capable to use positive approach and problem-solving in a greater degree than those experienced more burden (McKee et al., 2007). Several studies (Powers et al., 2002; McClendon et al., 2004) elicited that caregivers' with higher burden used emotional coping strategies such as "praying and seeking God's help", "hoping for miracle to happen", "day dreaming" and so on. It was also found that wishful thinking and avoidance of coping strategies lead to adverse quality of life of caregivers.

A study compares coping strategies between two groups of spousal caregiver based on onset of dementia. Person with early onset dementia experienced quicker cognitive and functional deterioration than person with late onset dementia. Caregivers' of early onset of dementia reported that they feel more difficulties to accept the disease than their counterparts (Wawrziczny et al., 2017). They also found difficulties to manage behavioural problems of dementia and were not ready to play a role as caregiver (Arai et al., 2007). The study found that the caregivers adopt many strategies to cope with the adverse situation regardless of onset of dementia. They used mainly the "acceptance and avoidance" strategies. "Acceptance" strategies include positive evaluation, gathered helpful information and seeking social support (Papastavrou et al., 2011), while "Avoidance" comprises not registering the proper problems they usually encountered and of course diverting from the future consequences. Moreover, they cut off their emotional and personal involvement in the caregiving (Lockeridge & Simpson, 2013; Molyneaux et al., 2012; Wawrziczny et al., 2016). Papastavrou and his associates (2011), noticed that the caregivers who used "Acceptance" strategies were associated with the lower level of burden and depression. Tschanz and his fellow colleagues (2013) confirmed that use of acceptance approaches by caregivers helps to decline the rate of cognitive and functional ability of care recipients. Increasing level of satisfaction was correlated with decline feelings of depression, burden and help to perceive better health, give strength to confront adverse situation as well as success (Seligman, 2008). In a study, it was found that women caregivers used emotional coping strategies more than men (Papastavrou et al., 2007). Similarly, Lutzky and

Knight (1994) noticed that high level of depression was more common in women because they use avoidance and escape strategies. Besides that, it may be possible that male caregivers' strong personality and characteristics help them to experience lower level of mental health problems than their counterparts (Thomson et al., 2004). It may be also probable that male caregivers developed a stronger coping management because of their experience in long time formal exposure of employment and apply the known strategies to cope with adverse situation (Wallsten, 2000). Collines and Jones (1997) found in their study that male caregivers maintain a psychological distance from their demented family members to reduce their level of stress and distress.

A study reported that diet and physical activities can moderate caregivers' adverse mental health condition. Physical activities can also decrease the cardiovascular risk of caregivers (von Kanel et al., 2010). The relationship closeness of caregivers and recipients may increase the level of stress among caregivers. Greater closeness associated with better performance and lower depression in the early stage of dementia (Fauth et al., 2012).

Conclusion

This article offers a rationalized review of the dementia caregiver's psychological health based on care demands, demographic factors and effects of different behavioural problems of the care recipients. It also offers an overview of dementia and the coping strategies, and the caregivers perceived for better health condition.

Dementia is basically a common disability, mostly found among the elderly. To determine the onset of the disease is very difficult to understand. Therefore, symptoms of dementia at the early stage often remain overlooked. Forgetfulness, particularly loss of recent memory is the most common symptoms in dementia. People with dementia faced the difficulties in day-to-day living and became dependent upon others for support and care. Due to lack of proper institutional dementia care services, family caregivers have to play the main supporting role to look after their demented family members. As a result, demented people will be able to live in their own homes and communities. Family caregivers' voluntary and familiar care is an important source of assistance for demented individual. Care for an older person with dementia can, however, be distressing and onerous for many family caregivers and also lead to different types of morbidities. More than one third of family caregivers are 65 years or older, making the situation more complicated. As the disease progresses, the caregivers have to sacrifice their job, and eventually, they did not get time for any social activities or gathering. Ultimately, their lives only revolve around their demented care recipient. Caregivers may be feel ignored, abandoned and isolated. Thus, caregivers become exhausted or act out improperly. Caregivers' stress could result in an increased risk of disease, depressive symptoms and chronic illness such as diabetes, heart disease and so on.

The phenomenon of care is not restricted to a specific community or culture. Searching out the cultural factors is most important to extend help towards caregivers. These factors may help to promote healthy caregiving. Healthcare practitioners should direct and educate caregivers on available support services and tools to help them in this extensive position in an adequate and efficient way. It is more important to educate and spread awareness among family caregivers about dementia and its consequence of family caregiving. More intervention research is also needed to provide best support to caregivers in managing problem behaviours at every stage of the dementia and how to deal with their own feelings of loss. Caregiver research also needs to focus on health related issues as long-term care process.

References

Akpınar, B., Küçükgüçlü, Ö., & Yener, G. (2011). Effects of gender on burden among caregivers of Alzheimer's patients. *Journal of Nursing Scholarship: An official publication of Sigma Theta Tau International Honor Society of Nursing, 43*(3), 248–54. https://doi.org/10.1111/j.1547-5069.2011.01402.x

Alice Lau, W.-Y., Lotus Shyu, Y.-I., Lin, L.-C., & Yang, P.-S. (2008). Institutionalized elders with dementia: Collaboration between family caregivers and nursing home staff in Taiwan. *Journal of Clinical Nursing, 17*(4), 482–490. https://doi.org/10.1111/j.1365-2702.2007.01955.x

Alzheimer's Disease International. (2009). World Alzheimer report: The global prevalence of dementia. Available at: https://www.alz.co.uk/research/world-report-2009. Accessed on November 05, 2018.

Andrieu, S., Rive, B., Guilhaume, C., Kurz, X., Scuvée-Moreau, J., Grand, A., & Dresse, A. (2007). New assessment of dependency in demented patients: Impact on the quality of life in informal caregivers. *Psychiatry and Clinical Neurosciences, 61*(3), 234–242. https://doi.org/10.1111/j.1440-1819.2007.01660.x

Annerstedt, L., Elmståhl, S., Ingvad, B., & Samuelsson, S. M. (2000). Family caregiving in dementia: An analysis of the caregiver's burden and the 'breaking-point' when home care becomes inadequate. *Scandinavian Journal of Public Health, 28*(1), 23–31. https://doi.org/10.1177/140349480002800106

Arai, A., Matsumoto, T., Ikeda, M., & Arai, Y. (2007). Do family caregivers perceive more difficulty when they look after patients with early onset dementia compared to those with late onset dementia? *International Journal of Geriatric Psychiatry, 22*(12), 1255–1261. https://doi.org/10.1002/gps.1935

Arango-Lasprilla, C. J., Lehan, T., Drew, A., Moreno, A., Deng, X., & Lemos, M. (2010). Health-related quality of life in caregivers of individuals with dementia from Colombia. *American Journal of Alzheimer's Disease and Other Dementias, 25*(7), 556–561. https://doi.org/10.1177/1533317510382287

Au, A., Lai, M.-K., Lau, K.-M., Pan, P.-C., Lam, L., Thompson, L., & Gallagher-Thompson, D. (2009). Social support and well-being in dementia family caregivers: The mediating role of self-efficacy. *Aging and Mental Health, 13*(5), 761–768. https://doi.org/10.1080/13607860902918223

Basu, I., & Mukhopadhyay, S. (2019). Factors related to adverse mental health condition of demented family caregivers: A study in West Bengal. *India Anthropological Review, 82*(4), 373–388. https://doi.org/10.2478/anre-2019-0028

Basu, I., Mukhopadhyay, S., & Maulik, N. (2020). Family caregiving in dementia: Caregivers' mental health and nature of caregiving. *Antrocom Journal of Anthropology, 16*(2), 201–212.

Bharath, S., & Dias, A. (2010). Prevalence, impact, costs and services for dementia A report prepared for the Alzheimer's and Related Disorders Society of India. https://www.semanticscholar.org/paper/THE-DEMENTIA-INDIA-REPORT-2010-Prevalence-%2C-impactBharathDias/4293238f0f89c5b16c43697e51a3475759c8829e. Accessed March, 2020.

Boise, L., Camicioli, R., Morgan, D. L., Rose, J. H., & Congleton, L. (1999). Diagnosing dementia: Perspectives of primary care physicians. *The Gerontologist, 39*(4), 457–464. https://doi.org/10.1093/geront/39.4.457

Brodaty, H., & Donkin, M. (2009). Family caregivers of people with dementia. *Dialogues in Clinical Neuroscience, 11*(1), 217–228. PMID: 19585957; PMCID: PMC3181916.

Buchner, M. D., & Larson, E. B. (1987). Falls and fractures in patients with Alzheimer-type dementia. *The Journal of the American Medical Association, 257*(11), 1492–1495. https://doi.org/10.1001/jama.257.11.1492

Buckner, L. R. (2004). Memory and executive function in aging and AD: Multiple factors that cause decline and reserve factors that compensate. *Neuron, 44*(1), 195–208. https://doi.org/10.1016/j.neuron.2004.09.006

Butcher, K. H., Holkup, P. A., & Buckwalter, K. C. (2001). The experience of caring for a family member with Alzheimer's disease. *Western Journal of Nursing Research, 23*(1), 33–55. https://doi.org/10.1177/019394590102300104

Campbell, P., Wright, J., Oyebode, J., Job, D., Crome, P., Bentham, P., Jones, L., & Lendon, C. (2008). Determinants of burden in those who care for someone with dementia. *International Journal of Geriatric Psychiatry, 23*(10), 1078–1085. https://doi.org/10.1002/gps.2071

Cassie, M. M., & Sanders, S. (2007). Familial caregivers of older adults. *Journal of Gerontological Social Work, 50*(1), 293–320. https://doi.org/10.1080/01634370802137975

Caudle, P. (1993). Providing culturally sensitive health care to Hispanic clients. *The Nurse Practitioner, 18*(12), 40–51. https://doi.org/10.1097/00006205-199312000-00013

Cheng, S.-T. (2017). Dementia caregiver burden: A research update and critical analysis. *Current Psychiatry Report, 19*(9), 64. https://doi.org/10.1007/s11920-017-0818-2

Cheng, S.-T., Lam, L. C. M., & Kwok, T. (2013). Neuropsychiatric symptom clusters of Alzheimer's disease in Hong Kong Chinese: Correlates with caregiver burden and depression. *The American Journal of Geriatric Psychiatry: Official Journal of the American Association for Geriatric Psychiatry, 21*(10), 1029–1037. https://doi.org/10.1016/j.jagp.2013.01.041

Chiu, M.-J., Chen, T.-F., Yip, P.-K., Hua, M.-S., & Tang, L.-Y. (2006). Behavioral and psychologic symptoms in different types of dementia. *Journal of the Formosan Medical Association, 105*(7), 556–562. https://doi.org/10.1016/S0929-6646(09)60150-9

Chiu, Y.-C., Lee, Y.-N., Wang, P.-C., Chang, T.-H., Li, C.-L., Hsu, W.-C., & Lee, S.-H. (2013). Family caregivers' sleep disturbance and its associations with multilevel stressors when caring for patients with dementia. *Aging and Mental Health, 18*(1), 92–101. https://doi.org/10.1080/13607863.2013.837141

Clare, L., Wilson, B. A., Carter, G., Breen, K., Berrios, G. E., & Hodges, J. R. (2002). Depression and anxiety in memory clinic attenders and their carers: Implications for evaluating the effectiveness of cognitive rehabilitation interventions. *International Journal of Geriatric Psychiatry, 17*(10), 962–967. https://doi.org/10.1002/gps.735

Collins, C., & Jones, R. (1997). Emotional distress and morbidity in dementia carers: A matched comparison of husbands and wives. *International Journal of Geriatric Psychiatry, 12*(12), 1168–1173. https://doi.org/10.1002/(SICI)1099-1166(199712)12:12%3c1168::AID-GPS711%3e3.0.CO;2-F

Cooper, C., Katona, C., Orrell, M., & Livingston, G. (2006). Coping strategies and anxiety in caregivers of people with Alzheimer's disease: The LASER-AD study. *Journal of Affective Disorders, 90*(1), 15–20. https://doi.org/10.1016/j.jad.2005.08.017

Cooper, C., Balamurali, T. B. S., & Livingston, G. (2007). A systematic review of the prevalence and covariates of anxiety in caregivers of people with dementia. *International Psychogeriatrics, 19*(2), 175–195. https://doi.org/10.1017/S1041610206004297

Cummings, L. J., Schneider, L., Tariot, P. N., Kershaw, P. R., & Yuan, W. (2004). Reduction of behavioral disturbances and caregiver distress by galantamine in patients with Alzheimer's disease. *American Journal of Psychiatry, 161*(3), 532–538. https://doi.org/10.1176/appi.ajp.161.3.532

D'Onofrio, G., Sancarlo, D., Addante, F., Ciccone, F., Cascavilla, L., Paris, F., Picoco, M., Nuzzaci, C., Elia, A. C., Greco, A., Chiarini, R., Panza, F., & Pilotto, A. (2015). Caregiver burden characterization in patients with Alzheimer's disease or vascular dementia. *International Journal of Geriatric Psychiatry, 30*(9), 891–899. https://doi.org/10.1002/gps.4232

Das, S. K., Biswas, A., Roy, T., Banerjee, T. K., Mukherjee, C. S., Raut, D. K., & Chaudhuri, A. (2006). A random sample survey for prevalence of major neurological disorders in Kolkata". *Indian Journal of Medical Research, 124*(2), 163–172. PMID: 17015930.

Das, S., Hazra, A., Ray, B. K., Ghosal, M., Banerjee, T. K., Roy, T., Chaudhuri, A., Raut, D. K., & Das, S. K. (2010). Burden among stroke caregivers: results of a community-based study from Kolkata, India. *Stroke, 41*(12), 2965–2968. https://doi.org/10.1161/STROKEAHA.110.589598

Das, K. S., Pal, S., & Ghosal, M. K. (2012). Dementia: Indian scenario. *Neurol India, 60*(6), 618–24. https://doi.org/10.4103/0028-3886.105197

de Moraes, S. R. P., & da Silva, L. S. T. (2009). An evaluation of the burden of Alzheimer patients on family caregivers. *Cadernos de saúde pública, 25*(8), 1807–1815. https://doi.org/10.1590/s0102-311x2009000800017

Deeken, F. J., Taylor, K. L., Mangan, P., Yabroff, K. R., & Ingham, J. M. (2003). Care for the caregivers: A review of self-report instruments developed to measure the burden, needs, and quality of life of informal caregivers. *Journal of Pain and Symptom Management, 26*(4), 922–953. https://doi.org/10.1016/s0885-3924(03)00327-0

Di Mattei, V. E., Prunas, A., Novella, L., Marcone, A., Cappa, S. F., & Sarno, L. (2008). The burden of distress in caregivers of elderly demented patients and its relationship with coping strategies. *Neurological Sciences: Official Journal of the Italian Neurological Society and of the Italian Society of Clinical Neurophysiology, 29*(6), 383–389. https://doi.org/10.1007/s10072-008-1047-6

Dias, A., Dewey, M. E., D'Souza, M., Dhume, R., Motghare, D. D., Shaji, K. S., Menon, R., Prince, M., & Patel, V. (2008). The effectiveness of a home care program for supporting caregivers of persons with dementia in developing countries: A randomised controlled trial from Goa, India. *PLoS One, 3*(6), E2333. https://doi.org/10.1371/journal.pone.0002333

Dixit, S. (2005). Meaning and explanations of mental illness: A social representations approach. *Psychology and Developing Societies, 17*(1), 1–18. https://doi.org/10.1177/097133360501700101

dos Santos, J., Mota, D. D. C. F., & Pimenta, C. A. M. (2009). Comorbidities between fatigue and depression in patients with colorectal cancer. *Revista da Escola de Enfermagem da U S P,* 43(4), 904–909. https://doi.org/10.1590/s0080-62342009000400024

Ekvall, K. A. (2004). *Informal caregiving at old age: Content, coping, difficulties and satisfaction.* Doctoral Dissertation. Lund University, Lund, Sweden.

Fauth, E., Hess, K., Piercy, K., Norton, M., Corcoran, C., Rabins, P., Lyketsos, C., & Tschanz, JoAnn. (2012). Caregivers' relationship closeness with the person with dementia predicts both positive and negative outcomes for caregivers' physical health and psychological well-being. *Aging and Mental Health, 16*(6), 699–711. https://doi.org/10.1080/13607863.2012.678482

Finkel, I. S., Silva, J. A. S., Cohen, G. D., Miller, S., & Sartorius, N. (1996). Behavioral and psychological symptoms of dementia: A consensus statement on current knowledge and implications for research and treatment. *International Psychogeriatrics, 3*(11), 497–500. https://doi.org/10.1017/S1041610297003943

Finkel, I. S., Silva, J. A. S., Cohen, G. D., Miller, S., & Sartorius, N. (1998). Behavioral and psychological symptoms of dementia: A consensus statement on current knowledge and implications for research and treatment. *The American Journal of Geriatric Psychiatry, 6*(2), 97–100. https://doi.org/10.1097/00019442-199805000-00002

Fontaine, L. J., Ahuja, J., Bradbury, N. M., Phillips, S., & Oyebode, J. R. (2007). Understanding dementia amongst people in minority ethnic and cultural groups. *Journal of advanced nursing, 60*(6), 605–614. https://doi.org/10.1111/j.1365-2648.2007.04444.x

Garre-Olmo, J., Vilalta-Franch, J., Calvó-Perxas, L., Turró-Garriga, O., Conde-Sala, J. L., & López-Pousa, S. (2016). A path analysis of patient dependence and caregiver burden in Alzheimer's disease. *International Psychogeriatrics, 28*(7), 1133–1141. https://doi.org/10.1017/S1041610216000223

Gaugler, J. E., Davey, A., Pearlin, L. I., & Zarit, S. H. (2000). Modeling caregiver adaptation over time: The longitudinal impact of behavior problems. *Psychology and Aging, 15*(3), 437–450. https://doi.org/10.1037//0882-7974.15.3.437

Gaugler, J. E., Zarit, S. H., & Pearlin, L. I. (2003). The onset of dementia caregiving and its longitudinal implications. *Psychology and Aging, 18*(2), 171–180. https://doi.org/10.1037/0882-7974.18.2.171

Gaugler, J. E., Kane, R. L., Kane, R. A., & Newcomer, R. (2005). The longitudinal effects of early behavior problems in the dementia caregiving career. *Psychological Aging, 20*(1), 100–116. https://doi.org/10.1037/0882-7974.20.1.100

Gaugler, J. E., Wall, M. M., Kane, R. L., Menk, J. S., Sarsour, K., Johnston, J. A., Beusching, D., & Newcomer, R. (2010). The effects of incident and persistent behavioral problems on change in caregiver burden and nursing home admission of persons with dementia. *Medical Care, 48*(10), 875–883. https://doi.org/10.1097/MLR.0b013e3181ec557b

Georges, J., Jansen, S., Jackson, J., Meyrieux, A., Sadowska, A., & Selmes, M. (2008). Alzheimer's disease in real life—the dementia carer's survey. *International Journal of Geriatric Psychiatry, 23*(5), 546–551. https://doi.org/10.1002/gps.1984

Gilbert, P., Bhundia, R., Mitra, R., McEwan, K., Iron, C., & Sanghera, J. (2007). Cultural differences in shame-focused attitudes towards mental health problems in Asian and non-Asian student women. *Mental Health Religion & Culture Religion & Culture, 2*, 127–141. https://doi.org/10.1080/13694670500415124

Given, C. W., Given, B., Stommel, M., & Azzouz, F. (1999). The impact of new demands for assistance on caregiver depression: tests using an inception cohort. *The Gerontologist, 39*(1), 76–85. https://doi.org/10.1093/geront/39.1.76

Grant, G., & Nolan, M. (1993). Informal carers: sources and concomitants of satisfaction. *Health and Social Care, 1*(3), 147–159. https://doi.org/10.1111/j.1365-2524.1993.tb00210.x

Gratao A. C. M., Vale, F. D. A. C. D., Roriz-Cruz, M., Haas, V. J., Lange, C., Talmelli, L. F. S., & Rodrigues, R. A. P. (2010). The demands of family caregivers of elderly individuals with dementia. *Revista da Escola de Enfermagem da U S, 44*(4), 873–880. https://doi.org/10.1590/S0080-62342010000400003

Guerra, M., Ferri, C. P., Fonseca, M., Banerjee, S., & Prince, M. (2011). Helping carers to care: The 10/66 dementia research group's randomized control trial of a caregiver intervention in Peru. *Revista brasileira de psiquiatria orgão oficial da Associação Brasileira de Psiquiatria, Asociación Psiquiátrica de la América Latina, 33*(1), 47–54. https://doi.org/10.1590/S1516-44462010005000017

Harada, N. C., Love, M. C. N., & Triebel, K. L. (2013). Normal cognitive aging. *Clinical Geriatric Medicine, 29*(4), 737–752. https://doi.org/10.1016/j.cger.2013.07002

Haro, M. J., Kahle-Wrobleski, K., Bruno, G., Belger, M., Dell'Agnello, G., Dodel, R., Jones, R. W., Reed, C. C., Vellas, B., Wimo, A., & Argimon, J. M. (2014). Analysis of burden in caregivers of people with Alzheimer's disease using self-report and supervision hours. *The Journal of Nutrition, Health & Aging, 18*(7), 677–684. https://doi.org/10.1007/s12603-014-0500-x

Harries, T. (2000). Where inner and outer worlds meet: psychological research in the tradition of Gorge W. Brown. *The Journal of Child Psychology and Psychiatry, 43*(3), 421–413. https://doi.org/10.1111/1469-7610.0032a

Hinton, L., Franz, C. E., Yeo, G., & Levkoff, S. E. (2005). Conceptions of dementia in a multiethnic sample of family caregivers. *Journal of the American Geriatrics Society, 54*(8), 1405–1410. https://doi.org/10.1111/j.1532-5415.2005.53409.x

Holroyd, E. (2001). Hong Kong Chinese daughters' intergenerational caregiving obligations: A cultural model approach. *Social Science & Medicine, 53*(9), 1125–1134. https://doi.org/10.1016/S0277-9536(00)00406-8

Hooker, K., Bowman, S. R., Coehlo, D. P., Lim, S. R., Kaye, J., Guariglia, R., Li, F. (2002). Behavioral change in persons with dementia: relationships with mental and physical health of caregivers. *The Journals of Gerontology. Series B, Psychological Sciences and Social Sciences, 57*(5), 453–460. https://doi.org/10.1093/geronb/57.5.p453

Huang, C.-Y., Musil, C. M., Zauszniewski, J. A., & Wykle, M. L. (2006). Effects of social support and coping of family caregivers of older adults with dementia in Taiwan. *International Journal of Aging & Human Development, 63*(1), 1–25. https://doi.org/10.2190/72JU-ABQA-6L6F-G98Q

Hughes, B. T., Black, B. S., Albert, M., Gitlin, L. N., Johnson, D. M., Lyketsos, C. G., & Samus, Q. M. (2014). Correlates of objective and subjective measures of caregiver burden among dementia caregivers: Influence of unmet patient and caregiver dementia-related care needs. *International Psychogeriatric, 26*(11), 1875–1883. https://doi.org/10.1017/S1041610214001240

Jathanna, R. P. N., Latha, K. S., & Bhandary, P. V. (2010). Burden and coping in informal caregivers of persons with dementia: A cross sectional study. *Journal of Health and Allied Sciences, 9*(4), 1–6. http://www.ojhas.org/issue36/2010-4-7.htm

Kang, S. H., Myung, W., Na, D. L., Kim, S. Y., Lee, J.-H., Han, S.-H., Choi, S. H., Kim, S. Y., Kim, S., & Kim, D. K. (2014). Factors associated with caregiver burden in patients with Alzheimer's disease. *Psychiatry Investigation, 11*(2), 152–159. https://doi.org/10.4306/pi.2014.11.2.152

Katherina, N. T., Gaugler E. J., & Jacobs-Lawson, J. (2019). Dementia caregiving outcomes: the impact of caregiving onset, cognitive impairment and behavioral problems. *Journal of Gerontological Social Work, 62*(5), 543–563. https://doi.org/10.1080/01634372.2019.1625993.

Kasuya, R., Polgar-Bailey, P., & Takeuchi, R. (2000). Caregiver burden and burnout: A guide for primary care physicians. *Postgraduate Medicine, 108*(7), 119–123. https://doi.org/10.3810/pgm.2000.12.1324

Kerpershoek, L., de Vugt, M., Wolfs, C., Woods, B., Jelley, H., Orrell, M., Stephan, A., Bieber, A., Meyer, G., Selbaek, G., Handels, R., Wimo, A., Hopper, L., Irving, K., Marques, M., Gonçalves-Pereira, M., Portolani, E., Zanetti, O., Verhey, F., & Consortium, A. (2018). Needs and quality of life of people with middle-stage dementia and their family carers from the European Actifcare study: When informal care alone may not suffice. *Aging & Mental Health, 20*(7), 897–902. https://doi.org/10.1080/13607863.2017.1390732

Khan, A., Kalaria, R. N., Corbett, A., & Ballard, C. (2016). Update on vascular dementia. *Journal of Geriatric Psychiatry and Neurology, 29*(5), 281–301. https://doi.org/10.1177/0891988716654987

Kiecolt-Glaser, J. K., Preacher, K. J., MacCallum, R. C., Atkinson, C., Malarkey, W. B., & Glaser, R. (2003). Chronic stress and age-related increases in the proinflammatory cytokine IL-6. *Proceedings of the National Academy of Sciences of the United States of America, 100*(15), 9090–9095. https://doi.org/10.1073/pnas.1531903100

Heok, K. E. E., & Li, T. S. (1997). Stress of caregivers of dementia patients in the Singapore Chinese family. *International Journal of Geriatric Psychiatry, 12*(4), 466–469. https://doi.org/10.1002/(SICI)1099-1166(199704)12:4<466::AID-GPS517>3.0.CO;2-U

La Jenny, F., Ahuja, J., Bradbury, N. M., Phillips, S., & Oyebode, J. R. (2007). Understanding dementia amongst people in minority ethnic and cultural groups. *Journal of Advanced Nursing, 60*(6), 605–614. https://doi.org/10.1111/j.1365-2648.2007.04444.x

Laks, J., Goren, A., Dueñas, H., Novick, D., & Kahle-Wrobleski, K. (2016). Caregiving for patients with Alzheimer's disease or dementia and its association with psychiatric and clinical comorbidities and other health outcomes in Brazil. *International Journal of Geriatric Psychiatry, 31*(2), 176–185. https://doi.org/10.1002/gps.4309

Landmark, B. T., Aasgaard, H. S., Fagerström, L. (2013). "To be stuck in it—I can't just leave": A qualitative study of relatives' experiences of dementia suffers living at home and need for support". *Home Health Care Management & Practice, 25*(5), 217–223. https://doi.org/10.1177/1084822313487984

Lau, W. Y. A., Shyu, Y. I. L., Lin, L. C., & Yang, P. S. (2008). Institutionalized elders with dementia: collaboration between family caregivers and nursing home staff in Taiwan. *Journal of clinical nursing, 17*(4), 482–490. https://doi.org/10.1111/j.1365-2702.2007.01955.x.

Lin, T.-Y., Tseng, W.-S., & Eng-Kung, Y. (Eds.). (1995). *Chinese societies and mental health.* Oxford University Press.

Livingston, G., Regan, R. M. C. R., & Katona, C. (2005). The caregivers for Alzheimer's disease Problems Scale (CAPS): Development of a new scale within the LASER-AD study. *Age and Ageing, 34*(3), 287–290. https://doi.org/10.1093/ageing/afi103

Lockeridge, S., & Simpson, J. (2013). The experience of caring for a partner with young onset dementia: How younger carers cope. *Dementia, 12*(5), 635–651. https://doi.org/10.1177/1471301212440873

Lutzky, S. M., & Knight, B. G. (1994). Explaining gender differences in caregiving distress: The roles of emotional attentiveness and coping styles. *Psychology Aging, 9*(4), 513–519. https://doi.org/10.1037/0882-7974.9.4.513

Mahoney, R., Regan, C., Katona, C., & Livingston, G. (2005). Anxiety and depression in family caregivers of people with Alzheimer disease: The LASER-AD study. *American Journal of Geriatric Psychiatry, 13*(9), 795–801. https://doi.org/10.1176/appi.ajgp.13.9.795

Martin, P., & Sörensen, S. (2011). Spouses, adult children, and children-in-law as caregivers of older adults: A meta-analytic comparison. *Psychology and Aging, 26*(1), 1–14. https://doi.org/10.1037/a0021863

Mattoo, K. S., Nebhinani, N., Anil Kumar, B. N., Basu, D., & Kulhara, P. (2013). Family burden with substance dependence: A study from India. *The Indian Journal of Medical Research, 137*(4), 704–711. PMCID: PMC3724250.

McCarty, H. J., Roth, D. L., Goode, K. T., Owen, J. E., Harrell, L., Donovan, K., & Haley, W. E. (2000). Longitudinal course of behavioral problems during Alzheimer's disease: Linear versus curvilinear patterns of decline. *Journal of Gerontology: Medical Sciences, 55*(4), M200–M206. https://doi.org/10.1093/gerona/55.4.m200

McClendon, J. M., Smyth, K. A., & Neundorfer, M. M. (2007). Survival of persons with Alzheimer's disease: caregiver coping matters. *Gerontologist, 44*(4), 508–519. https://doi.org/10.1093/geront/44.4.508

McCurry, M. S., Logsdon, R. G., Teri, L., & Vitiell, M. V. (2007). Sleep disturbances in caregivers of persons with dementia: Contributing factors and treatment implications. *Sleep Medicine Review, 11*(2), 143–153. https://doi.org/10.1016/j.smrv.2006.09.002

McCurry, S. M., Pike, K. C., Vitiello, M. V., Logsdon, R. G., & Teri, L. (2008). Factors associated with concordance and variability of sleep quality in persons with Alzheimer's disease and their caregivers. *Sleep, 31*(5), 741–748. https://doi.org/10.1093/sleep/31.5.741

McKee, J. K., Whitlack, J. E., Ballinger, B. B., Gillhooly, M., Gordon, D. S., Mutch, W. J., & Philip, I. (2007). Coping in family supporters of elderly people with dementia. *British Journal of Clinical Psychology, 36*(3), 323–340.

Mills, J. P., Ancoli-Israel, S., von Känel, R., Mausbach, B. T., Aschbacher, K., Patterson, T. L., Ziegler, M. G., Dimsdale, J. E., & Grant, I. (2009). Effects of gender and dementia severity on Alzheimer's disease caregiver's sleep and biomarkers of coagulation and inflammation. *Brain, Behavior, and Immunity, 23*(5), 605–610. https://doi.org/10.1016/j.bbi.2008.09.014

Mohamed, S., Rosenheck, R., Lyketsos, C. G., & Schneider, L. S. (2010). Caregiver burden in Alzheimer's disease: Cross sectional and longitudinal patient correlates. *American Journal of Geriatric Psychiatry, 18*(10), 917–927. https://doi.org/10.1097/JGP.0b013e3181d5745d

Mohide, E.A., Torrance, G. W., Streiner, D. L., Pringle, D. M., & Gilbert, R. (1998). Measuring the wellbeing of family caregivers using the time trade-off technique. *Journal of Clinical Epidemiology, 41*(5), 475–482. https://doi.org/10.1016/0895-4356(88)90049-2

Mok, E., Lai, C. K. Y., Wong, F. L. F., & Wan, P. (2007). Living with early-stage dementia: the perspective of older Chinese people. *Journal of Advanced Nursing, 59*(6), 591–600. https://doi.org/10.1111/j.1365-2648.2007.04368.x

Molyneaux, V. J., Butchard, S., Simpson, J., & Murray, C. D. (2012). The co-construction of couplehood in dementia. *Dementia, 11*(4), 483–502. https://doi.org/10.1177/1471301211421070
Müller-Spahn, F. (2003). Behavioral disturbances in dementia. *Dialogues in Clinical Neuroscience, 5*(1), 49–59. PMCID: PMC3181717.
Ng, T. G. (2008). Singapore: Characteristics of family caregivers and care recipients and their caregiving experiences. *Hallym International Journal of Aging, 10*(1), 41–62. https://doi.org/10.2190/HA.10.1.c
Nolan, M. (2001). Supporting family carers in the UK: Overview of issues and challenges. *British Journal of Nursing, 10*(9), 608–613. https://doi.org/10.12968/bjon.2001.10.9.9380
Nolan, M., & Grant, G. (1992). Regular respite: An evaluation of a hospital rota bed scheme for elderly people. Age Concern Institute of Gerontology Research Papers. Ace Books.
Ohaeri, U. J. (2002). The burden of caregiving in families with a mental illness: A review. *Current Opinion in Psychiatry, 16*(4), 457–465. https://doi.org/10.1097/01.yco.0000079212.36371.c0
Ormel, J., Vonkroff, M., Ustun, T. B., Pini, S., Korten, A., & Oldehinkel, A. J. (1994). Common mental disorder and disability across culture. Results from the WHO Collaborative study on Psychological Problems in general health care. *Journal of the American Medical Association, 272*(22), 1741–1748. https://doi.org/10.1001/jama.272.22.1741
Ory, G. M., Hoffman, R. R., Yee, J. L., Tennstedt, S., & Schulz, R. (1999). Prevalence and impact of caregiving: A detailed comparison between dementia and nondementia caregivers. *Gerontologist, 39*(2), 177–185. https://doi.org/10.1093/geront/39.2.177
Papastavrou, E., Kalokerinou, A., Papacostas, S. S., Tsangari, H., & Sourtzi, P. (2007). Caring for a relative with dementia: Family caregiver burden. *Journal of Advanced Nursing, 58*(5), 446–457. https://doi.org/10.1111/j.1365-2648.2007.04250.x
Papastavrou, E., Tsangari, H., Karayiannis, G., Papacostas, S., Efstathiou, G., & Sourtzi, P. (2011). Caring and coping: The dementia caregivers. *Aging & Mental Health, 15*(6), 702–711. https://doi.org/10.1080/13607863.2011.562178
Park, M., Sung, M., Kim, S. K., Kim, S., & Lee, D. Y. (2015). Multidimensional determinants of family caregiver burden in Alzheimer's disease. *International Psychogeriatric, 27*(8), 1355–1364. https://doi.org/10.1017/S1041610215000460
Parker, D., Mills, S., & Abbey, J. (2008). Effectiveness of interventions that assist caregivers to support people with dementia living in the community: A systematic review. *International Journal of Evidence-Based Healthcare, 6*(2), 137–172. https://doi.org/10.1111/j.1744-1609.2008.00090.x
Patel, V., & Prince, M. (2001). Ageing and mental health in a developing country: Who cares? Qualitative studies from Goa, India. *Psychological Medicine, 31*(1), 29–38. https://doi.org/10.1017/s0033291799003098
Patterson, L. T., & Grant, I. (2003) Interventions for caregiving in dementia: Physical outcomes. *Current Opinion in Psychiatry, 16*(6), 629–633. https://doi.org/10.1097/00001504-200311000-00005
Petrus, N. G., & Wing-Chung, H. O. (2006). Experience in coping with Alzheimer's disease at home: A study of Chinese family caregivers. *Journal of Social Work in Disability & Rehabilitation, 4*(4), 1–14. https://doi.org/10.1300/J198v04n04_01
Pinquart, M., & Sorensen, S. (2007). Correlates of physical health of informal caregivers: A met-analysis. *Journal of Gerontology, 62*(2), 126–137. https://doi.org/10.1093/geronb/62.2.p126
Poulshock, W. S., & Deimling, G. (1984). Families caring for elders in residence: issues in the measurement of burden. *Journal of Gerontology, 39*(2), 230–239. https://doi.org/10.1093/geronj/39.2.230
Powers, D. V., Gallagher-Thompson, D., & Kraemer, H. C. (2002). Coping and depression in Alzheimer's caregivers: Longitudinal evidence of stability. *The journals of gerontology. Series B, Psychological sciences and social sciences, 57*(3), 205–211. https://doi.org/10.1093/geronb/57.3.p205
Prince, M. (1997). The need for research on dementia in developing countries. *Tropical Medicine and Health, 2*(10), 993–1000. https://doi.org/10.1046/j.1365-3156.1997.d01-156.x

Prince, M., Prina, M., & Guerchet, M. (2013). World Alzheimer Report 2013. Journey of Caring. An Analysis of Long-Term Care for Dementia. Alzheimer's Disease International.

Prince, M., Comas-Herrera, A., Knapp, M., Guerchet, M., & Karagiannidou, M. (2016). World Alzheimer Report 2016: Improving Healthcare for People Living with Dementia: Coverage, Quality and Costs Now and in the Future. Alzheimer's Disease International (ADI).

Prince, M., & 10/66 dementia Research Group. (2004). Care arrangements for people with dementia in developing countries. *International Journal of Geriatric Psychiatry, 19*(2), 170–177. https://doi.org/10.1002/gps.1046

Rajkumar, S., Kumar, S., & Thara, R. (1997). Prevalence of dementia in a rural setting: A report from India. *International Journal of Geriatric Psychiatry, 12*(7), 702–707. https://doi.org/10.1002/(sici)1099-1166(199707)12:7<702::aid-gps489>3.0.co;2-h

Ravio, M., Eloniemi-Sulkava, U., Laakkonen, M.-L., Saarenheimo, M., Pietilä, M., Tilvis, R., & Pitkälä, K. (2007). How do officially organized services meet the needs of elderly caregivers and their spouses with Alzheimer's disease? *American Journal of Alzheimer's Disease and Other Dementias, 22*(5), 360–368. https://doi.org/10.1177/1533317507305178

Regiane, G., & Almeida, O. P. (1999). Behavioural problems in patients with dementia and their burden on carers. *Arquivos de Neuro-Psiquiatria, 57*(2B), 427–434. https://doi.org/10.1590/S0004-282X1999000300014

Regiane, G., & Menezes, P. R. (2004). Impact on caregivers of elderly patients with dementia treated at a psychogeriatric service. *Revista de Saúde Pública, 38*(6), 835–841. https://doi.org/10.1590/S0034-89102004000600012

Rodriguez, J. J. L., Ferri, C. P., Acosta, D., Guerra, M., Huang, Y., & Jacobl, K. S. (2008). Prevalence of dementia in Latin America, India, and China: A population based cross-sectional survey. *Lancet, 372*(9637), 464–474. https://doi.org/10.1016/S0140-6736(08)61002-8

Roepke, K. S., Allison, M., Von Känel, R., Mausbach, B. T., Chattillion, E. A., Harmell, A. L., Patterson, T. L., Dimsdale, J. E., Mills, P. J., Ziegler, M. G., Ancoli-Israel, & S., Grant, I. (2012). Relationship between chronic stress and carotid intima-media thickness (IMT) in elderly Alzheimer's disease caregivers. *Stress, 15*(2), 121–129. https://doi.org/10.3109/10253890.2011.596866

Ronda, C. T., & Crews, J. E. (2007). Framing the public health of caregiving. *American Journal of Public Health, 97*(2), 224–228. https://doi.org/10.2105/AJPH.2004.059337

Rosness, A. T., Engedal, K., & Chemali, Z. (2016). Frontotemporal dementia: An updated clinician's guide. *Journal of Geriatric Psychiatry and Neurology, 29*(5), 271–280. https://doi.org/10.1177/0891988716654986

Salamizadeh, A., Mirzaei, T., & Ravari, A. (2017). The impact of spiritual care education on the self-efficacy of the family caregivers of elderly people with Alzheimer's disease. *International Journal of Community Based Nursing and Midwifery, 5*(3), 231–238. PMID: 28670585.

Sanders, S. (2005). Is the glass half empty or full? Reflections on strain and gain in caregivers of individuals with Alzheimer's disease. *Social Work in Health Care, 40*(3), 57–73https://doi.org/10.1300/J010v40n03_04

Savage, S., & Bailey, S. (2004). The impact of caring on caregivers' mental health: A review of the literature. *Australian Health Review: A Publication of the Australian Hospital Association, 27*(1), 111–117. https://doi.org/10.1071/ah042710111

Schulz, R., & Gali, W. M. (1991). A 2-year longitudinal study of depression among Alzheimer's caregivers. *Psychology and Aging, 6*(4), 569–578. https://doi.org/10.1037/0882-7974.6.4.569

Schulz, R., & Martire, L. M. (2004). Family caregiving of persons with dementia: Prevalence, health effects, and support strategies. *American Journal of Geriatric Psychiatry, 12*(3), 240–249. PMID: 15126224

Schulz, R., & Sherwood, P. R. (2008). Physical and mental health effects of family caregiving. *American Journal of Nursing, 108*(9 Suppl), 23–27. https://doi.org/10.1097/01.NAJ.0000336406.45248.4c

Schulz, R., Vistainer, P., & Williamson, G. M. (1990). Psychiatric and physical morbidity effects of caregiving. *Journal of Gerontology, 45*(5), 181–191. https://doi.org/10.1093/geronj/45.5.P181

Seligman, M. E. P. (2008). Positive health. *Applied Psychology, 57*(s1), 3–18. https://doi.org/10.1111/j.1464-0597.2008.00351.x
Sethabouppha, H., & Kane, C. (2005). Caring for the seriously mentally ill in Thailand: Buddhist family caregiving. *Archives of Psychiatric Nursing, 19*(2), 44–57. https://doi.org/10.1016/j.apnu.2005.02.004
Shaik, S. S., & Varma, A. R. (2012). Differentiating the dementias: a neurological approach. *Progress in Neurology and Phychiatry, 16*(1), 11–18. https://doi.org/10.1002/pnp.224
Shaji, K. S., & Reddy, M. S. (2012). Caregiving: A public health priority. *Indian Journal of Psychological Medicine, 34*(4), 303–305. https://doi.org/10.4103/0253-7176.108191
Shaji K.S., Smitha, K., Praveen Lal, K., & Prince, M. (2003). Caregivers of people with Alzheimer's disease: A qualitative study from the Indian 10/66 Dementia Research Network. *International Journal of Geriatric Psychiatry, 18*(1), 1–6. https://doi.org/10.1002/gps.649
Shaji, K. S., George, R. K., Prince, M., & K. S. Jacob. (2009). Behavioral symptoms and caregiver burden in dementia. *Indian Journal of Psychiatry, 51*(1): 45–49. https://doi.org/10.4103/0019-5545.44905
Shaji, K. S., Jotheeswaran, A. T., & Girish, N. (Eds.). (2010). *The dementia India report: Prevalence, impact, costs and services for dementia.* Alzheimer's & Related Disorders Society of India.
Shim, S. H., Kang, H. S., Kim, J. H., & Kim, D. K. (2016). Factors associated with caregiver burden in dementia: 1-year follow-up study. *Psychiatry Investigation, 13*(1), 43–49. https://doi.org/10.4306/pi.2016.13.1.43
Siders, C., Nelson, A., Brown, L. M., Joseph, I., Algase, D. L., Beattie, E., & Cadena, S. (2004). Evidence for implementing nonpharmacological interventions for wandering. *Rehabilitation Nursing: The Official Journal of the Association of Rehabilitation, 29*(6), 195–206. PMID: 15597998
Stansfeld, J., Stoner, C. R., Wenborn, J., Vernooij-Dassen, M., Moniz-Cook, E., & Orrell, M. (2017). Positive psychology outcome measures for family caregivers of people living with dementia: Asystematic review. *International Psychogeriatrics, 29*(8), 1281–1296. https://doi.org/10.1017/S1041610217000655
Stephens, M. A. P., Kinney, J. M., & Ogrocki, P. K. (1991) Stressors and well-being among caregivers to older adults with dementia: The in-home versus nursing home experience. *Gerontologist, 31*(2), 217–223. https://doi.org/10.1093/geront/31.2.217
Takai, M., Takahashi, M., Iwamitsu, Y., Oishi, S., & Miyaoka, H. (2011). Subjective experiences of family caregivers of patients with dementia as predictive factors of quality of life. *Psychogeriatrics, 11*(2), 98–104. https://doi.org/10.1111/j.1479-8301.2011.00354.x
Tang, T. A., & Li, C.-Y. (2008). The important role of coherence in relation to depressive symptoms for Taiwanese family caregivers of cancer patients at the end of life. *Journal of Psychosomatic Research, 64*(2), 195–203. https://doi.org/10.1016/j.jpsychores.2007.08.013
Teri, L., Larson, E. B., & Reifler, B. V. (1988). Behavioral disturbance in dementia of the Alzheimer's type. *Journal of the American Geriatrics Society, 36*(1), 1–6. https://doi.org/10.1111/j.1532-5415.1988.tb03426.x
Thomson, L. R., Lewis, S. L., Murphy, M. R., Hale, J. M., Blackwell, P. H., Acton, G. J., Clough, D. H., Patrick, G. J., & Bonner, P. N. (2004). Are there sex differences in emotional and biological responses in spousal caregivers of patients with Alzheimer's disease? *Biological Research for Nursing, 5*(4), 319–330. https://doi.org/10.1177/1099800404263288
Todd, R. J., Lee, S. J., Berg-Weger, M., & Grossberg, G. T. (2013). Caregiver health: Health of caregivers of Alzheimer's and other dementia patients. *Current Psychiatry Reports, 15*(7), 367. https://doi.org/10.1007/s11920-013-0367-2
Tomomitsu, M. R. S. V., Perracini, M. R., & Neri, A. L. (2014). Factors associated with life satisfaction among elderly people who are caregivers and non-caregivers. *Ciência & Saúde Coletiva, 19*(8), 3429–3440. https://doi.org/10.1590/1413-81232014198.13952013
Torti, Jr, F. M., Gwyther, L. P., Reed, S. D., Friedman, J. Y., & Schulman, K. A. (2004). A multinational review of recent trends and reports in dementia caregiver burden. *Alzheimer Disease & Associated Disorders, 18*(2), 99–109. PMID: 15249854.

Tremont, G., Davis, J. D., & Bishop, D. S. (2006). Unique contribution of family functioning in caregivers of patients with mild to moderate dementia. *Dementia and Geriatric Cognitive Disorders, 21*(3), 170–174. https://doi.org/10.1159/000090699

Tschanz, T. J., Piercy, K., Corcoran, C. D., Fauth, E., Norton, M. C., Rabins, P. V., Tschanz, B. T., Deberard, M. S., Christine, S., Smith, C., Lee, L., & Lyketsos, C. G. (2013). Caregiver coping strategies predict cognitive and functional decline in dementia: The Cache County Dementia Progression Study. *The American Journal of Geriatric Psychiatry, 21*(1), 57–66. https://doi.org/10.1016/j.jagp.2012.10.005

Victoroff, J., Mack, W., & Nielson, K. A. (1997). Psychiatric complications of dementia: Impact on caregivers. *Dementia and Geriatric Cognitive Disorders, 9*(1), 50–55. https://doi.org/10.1159/000017022

Vitaliano, P. P., Young, H. M., & Zhang, J. (2004). Is caregiving a risk factor for illness? *Current Directions in Psychological Science, 13*(1), 13–16. https://doi.org/10.1111/j.0963-7214.2004.01301004.x

Vincent K. A. (1999). Treatment of depressive disorders of spousal caregivers of persons with Alzheimer's disease: a review. *American Journal of Alzheimer's Disease, 14*(5), 289–293. https://doi.org/10.1177/153331759901400501

von Kanel, R., Brent Mausbach, J. E., Dimsdale, P. J. Mills, Patterson, T. L., Ancoli-Israel, S., Ziegler, M. G., Roepke, S. K., Allison, M. A., & Grant, I. (2010). Problem behavior of dementia patients predicts low-grade hypercoagulability in spousal caregivers. *Journal of Gerontology: Biological Science & Medical Sciences, 65A*(9), 1004–1011. https://doi.org/10.1093/gerona/glq073

Walker, Z., Possin, K. L., Boeve, B. F., & Aarsland, D. (2015). Lewy body dementias. *The Lancet, 386*(10004), 1683–1697. https://doi.org/10.1016/S0140-6736(15)00462-6

Wallsten, S. S. (2000). Effects of caregiving, gender and race on the health, mutuality and social supports of older couples. *Journal of Ageing and Health, 12*(1), 90–111. https://doi.org/10.1177/089826430001200105

Wawrziczny, E., Antoine, P., Ducharme, F., Kergoat, M.-J., & Pasquier, F. (2016). Couples' experiences with early-onset dementia: An interpretative phenomenological analysis of dyadic dynamics. *Dementia, 15*(5), 1082–1099. https://doi.org/10.1177/1471301214554720

Wawrziczny, E., Pasquier, F., Ducharme, F., Kergoat, M.-J., & Antoine, P. (2017). Do spouse caregivers of persons with early-and late-onset dementia cope differently? A comparative study. *Archives of Gerontology and Geriatrics, 69*, 162–168. https://doi.org/10.1016/j.archger.2016.12.002

Wimo, A., Winblad, B., & Jonsson, L. (2007). An estimate of the total worldwide societal costs of dementia in 2005. *Alzheimer's & Dementia: THe Journal of the Alzheimer's Association, 3*(2), 81–91. https://doi.org/10.1016/j.jalz.2007.02.001

World Health Organization (WHO). (2018). Dementia: mhGAP training of healthcare providers training manual and supporting material. WHO Document Production Services.

Zeilig, H. (2014). Dementia as a cultural metaphor. *The Gerontologist, 54*(2), 258–267. https://doi.org/10.1093/geront/gns203

Ipsita Basu Currently pursuing her Ph.D. in Anthropology from Indian Statistical Institute, Kolkata, India. She did her both bachelor and master degree in Anthropology from University of Calcutta. After completing her master degree, she got engaged at Anthropological Survey of India as Junior Research Fellow. Later she got awarded with Junior Research Fellowship by University Grant Commission and joined Biological Anthropology Unit at ISI, Kolkata as UGC-Research Fellow. Presently, she has been working with the issues related with health aspect of dementia family caregivers.

Susmita Mukhopadhyay Ph.D., is a Professor of Anthropology at Indian Statistical Institute, Kolkata, India. She has specialization in Biological Anthropology. Her research addresses the health issues of women of different age groups. Presently, she is supervising Ph.D. students working on the interrelationship of health and its bio-cultural concomitants among various socio-economic set ups.

Chapter 13
Dementia Care in Malta: Policy, Experiences and Narratives

Marvin Formosa

Abstract This chapter embeds global trends in dementia policy and care in a unique national context, that of Malta, which is a European Union Member State. In 2015, the Maltese government launched the *National Dementia Strategy* so that Malta became the 21st country to launch a plan aimed at enhancing the quality of life of persons with dementia and family carers. The government followed this strategy by implementing a range of policy initiatives—most importantly, a Dementia Intervention Team. The most recent national statistics on caregivers of persons with dementia in Malta reported that the main carer for persons with dementia living in the community was the spouse, followed by the daughter. Informal caregiving with persons with dementia is such an onerous experience that many caregivers require assistance with a range of care work activities and are in receipt of a range of public service. The planning and writing of this chapter also aimed at exploring the subjective experience of dementia. A case study found that persons with dementia tend to experience "narrative dispossession" and spent most of their daily living balancing their private (the "self") and public worlds (how others view them). This chapter advocates a life course approach to dementia, rather than treating it as an end-of-life phenomenon, while also calling for a relational approach to dementia policy that addresses the needs of all those involved in supporting people with dementia, whether in an unpaid or paid capacity.

Introduction

Following the new millennium, a watershed feature in global gerontological and geriatric circles was certainly the increase of public awareness on dementia. Though there is certainly still much to be done, in less than two decades, dementia transformed itself from a group of symptoms spoken about only in professional circles frequented by geriatricians, psychiatrics and neurologists to featuring as an integral part of

---

M. Formosa (✉)
Department of Gerontology and Dementia Studies, Faculty for Social Wellbeing, University of Malta, Msida, Malta
e-mail: marvin.formosa@um.edu.mt

M. K. Shankardass (ed.), *Dementia Care*,
https://doi.org/10.1007/978-981-16-3864-0_13

popular culture. This occurred as widespread television, radio, drama and social media campaigns left a huge impact on both political cadres and the general public. Indeed, it is promising that while as much as 28 countries have already launched national strategies for dementia, another 30 countries have such national plans in development (Alzheimer's Disease International, 2019a). On the education front, it is noteworthy that many learning providers in higher education are providing continuous professional development on dementia care to social and healthcare workers (Karlin et al., 2017). At the same time, one notices the establishment of an increasing number of postgraduate degree programmes on dementia studies (Surr & Gates, 2017). Such courses target professionals already working in the field of gerontological and geriatric care, as well as other specialists seeking specialised knowledge in the multidisciplinary processes of dementia care. The intricacies of dementia care are no longer the sole prerogative of the healthcare profession and can now be found in disparate fields ranging from architecture to information technology to counselling (Burns & Robert, 2019).

One also observes a mushrooming of non-governmental organizations dedicating their energies to progress the well-being and quality of life of both older persons with dementia and their family carers. Many of these organizations are presently grouped under the Global Alzheimer's and Dementia Action Alliance which is the first global body to bring together caregiver networks, professional associations, and public entities to tackle the multifaceted challenges concerning dementia care work. Supra-organizations such as the World Bank and the United Nations have also entered the fray, and in what was perhaps the culmination of the travails of international policy on dementia care the World Health Organization (WHO) (2017a) established a *Global Dementia Observatory*. Seeking to move beyond mere policy rhetoric to real policy implementation, this observatory adopted the *Global action plan on the public health response to dementia 2017–2025* (WHO, 2017b) which called upon policymakers to initiative policy that meets the following responsibilities:

> Protecting the rights, freedoms and wishes of people with dementia and their carers; Actively involving people with dementia, their carers and families in developing policies, legislation, strategies and plans; Engaging all relevant stakeholders including public and private sector, and civil society to implement a coordinated policy response to dementia; Creating equitable access to person-centered, gender-sensitive, culturally-appropriate and human rights-oriented care and support for people with dementia and their carers; Ensuring the social and financial protection of people with dementia and their carers.
>
> WHO (2017b): 1

This international focus on dementia is hardly surprising when one considers that someone in the world develops dementia every three seconds. The global number of people living with dementia worldwide in 2015 was estimated at 46.8 million and projected to rise to 75 and 131.5 million in 2030 and 2050, respectively (Alzheimer's Disease International, 2019a). An estimated 7.7 million new cases of dementia are presently anticipated in each year, meaning that one anticipates and average of 3.6 million (46%) new cases per year in Asia, 2.3 million (31%) in Europe, 1.2 million (16%) in the Americas, and 0.5 million (7%) in Africa in the foreseeable future (WHO, 2015). As expected, studies demonstrate that the incidence of dementia increased exponentially with age, so that incidence doubles itself with every 5.9-year increase in age, from 3.1/1000 person years at age 60–64, to 175.0 every 1000 person years at age 95-plus (ibid.). Such rates of prevalence imply that the economic impact of dementia should not be taken lightly, with Alzheimer's Disease International underlining how the

> ...total estimated worldwide cost of dementia is US$818 billion in 2015, which represents 1.09% of global GDP. By 2018, the global cost of dementia will rise above a US$ trillion. This figure includes costs attributed to informal care (unpaid care provided by family and others), direct costs of social care (provided by community care professionals, and in residential home settings) and the direct costs of medical care (the costs of treating dementia and other conditions in primary and secondary care).
>
> Alzheimer's Disease International (2019b): online version

This chapter embeds the above global trends in dementia policy and care in a particular national context, that of Malta, which is a European Union Member State, and consists of three islands—Comino, Gozo and Malta—93 kms south of Sicily and 290 kms north of Libya. As Comino is uninhabited, and with Gozo having a mere population of 31,592 persons, Malta constitutes the major island of this archipelago state with as much as 397,752 residents (2014 Figures) (National Statistics Office, 2016). The chapter's objectives are threefold. The first objective is to present an overview of the development of dementia care policy in Malta by delineating key and effective milestones. The second objective is to report on qualitative research carried out with caregivers of persons with dementia by drawing out their experiencing of learning and reacting to such a diagnosis. The third and final objective is to focus on the impact of this condition on person diagnosed with dementia, especially with respect to his/her embodiment, selfhood and dementia. While a diagnosis of dementia changes the ways people engage with each other—between those living with dementia, caregivers and relatives, and the wider community—medical understandings, necessary as they are, provide no insights into the lived experiences of dementia. The final section concludes this chapter by taking stock of the emergent data arising from these three objectives to recommend the need to democratize dementia care. The democratization of dementia care does not only advocate good quality care for persons with dementia and caregivers but also ensures that no one is left out in any dementia care roadmap.

Dementia Policy in Malta

At the end of 2016, Malta's total population reached 460,297 persons. While 25.2% (116,009 persons) were aged 60-plus, the 80-plus cohort reached 4.1% (19,019 persons) (National Statistics Office, 2018). The largest share of older persons is made up of women, 53.5% of the total. In fact, the sex ratios (the number of males per 100 females) for cohorts aged 65-plus and 80-plus numbered 82 and 59 respectively (ibid.). Malta is no exception to the global increase in the prevalence of dementia.

Table 1 Persons with dementia by gender (2018, 2020, 2030, 2040)

2018		2020		2030		2040	
Males	Females	Males	Females	Males	Females	Males	Females
2,032	5049	2227	5465	3236	7657	4130	9807
Total		Total		Total		Total	
7,081		7,692		10,893		13,937	

Source Grant Thornton Malta (2018)

Initially, research studies projected that there were 4,072 individuals aged 60-plus with dementia in Malta in the year 2005 (Abela et al., 2007). A subsequent study by Scerri and Scerri (2012) raised this estimation to 5198 for the year 2010, while also projecting the number of Maltese people over 60 years of age who will experience dementia in 2030 will reach some 10,000 persons or 2.3% of the total Maltese population. A few years later, Alzheimer Europe (2013) estimated that the number of people with dementia in Malta in 2012 was 5,301 which was equivalent to 1.26% of the total population. A recent study by Grant Thorton Malta (2018) updated these figures for the year 2018, while also fine-tuning the projections of persons with dementia for the years 2020, 2030 and 2040 (see Table 1).

The emergence of public policy on dementia care can be traced to May 2009 when the Department of Health launched the National Dementia Strategy Group with the aim of devising a series of recommendations that would provide a strategic framework in order to deliver quality improvements in local dementia services and address any shortfalls in dementia care (Scerri, 2015a). In a retroactive view of dementia care policy in Malta, Scerri claimed that

> This group was composed of an interdisciplinary team coming from the medical and allied health care professions together with representatives of the society. The work undertaken was divided into three main categories: (1) detailed analysis of the current services available to individuals with dementia and their caregivers, (2) a wide consultation process with stakeholders working in the field of dementia management and care including professional bodies, and (3), a questionnaire designed for the public in order to obtain information regarding the various aspects of informal dementia care.
>
> Scerri (2015b): 176-7

Unfortunately, the emergent results were not encouraging. While field analysis on the services available for individuals with dementia showed considerable lack of support for them and for their carers, the available staff lacked training in person-centred dementia care. Similarly, the feedback received from wide consultation and survey processes was relative negative, finding that "knowledge on dementia at an undergraduate level is variable, fragmented and lacks focus on the social issues involved in disease progression…[and that] there is a lack of educational methods that prepare students to work in interdisciplinary teams or collaboration settings"

(Scerri, 2015c: 8). The good news, however, was that "results from the public questionnaire showed that dementia was not considered to be a taboo subject and therefore the stigma associated with these individuals and their families was diminishing… especially evident among the carers, relatives and healthcare professionals" (ibid., 2015b: 179). Other key inferences included a general lack of awareness on the meaning of dementia as not being a normal part of ageing, reluctance to seek medical advice after the first symptoms have appeared, and the healthcare profession being highly convinced that they do not have enough training in dementia care. As a result, the then National Dementia Strategy Group advocated the government to tackle dementia care through a two-pronged approach by making antidementia drugs freely available to persons with dementia since their cost is putting significant financial strain on the affected families who cannot afford to purchase such medication, as well as by embedding community and residential care for persons with dementia in a person-centred strategic approach as such a combination can positively contribute to reducing behavioural symptoms and improve the quality of life of individuals with dementia (Scerri & Scerri, 2012).

The Maltese government reacted immediately and began implementing policy initiatives that raised awareness on dementia, made one antidementia medication free available, commenced the collection of data on the number of dementia cases, and published two information booklets *In-Nannu Nesa x'Jisimini* (Grandpa forgot my name) (Zahra, 2009) and *X'hin Hu? Fatti Dwar id-Dimensja* (What time is it? Facts about Dementia) (Scerri & Zahra, 2013). However, the key milestone was, when in 2015, the government launched the much-anticipated *National Dementia Strategy* (Parliamentary Secretariat for Rights of Persons with Disability and Active Ageing, 2015) so that Malta became the 21st country to launch a plan aimed at enhancing the quality of life of persons with dementia and their family carers (Formosa & Galea, 2020; Formosa & Scerri, 2020). The strategy included the following key objectives and actions:

Increasing awareness and understanding of dementia. One fundamental aspect of this strategy is to increase awareness and understanding of dementia among the general public and healthcare professionals in order to reduce stigma and misconceptions about the condition.

- Reduce stigma
- Provide information on risk factors for dementia
- Provide guidance for persons with dementia and caregivers.

Timely diagnosis and intervention. Early symptom recognition and interventions through appropriate referral pathways together with the necessary pharmacological, psychological and psychosocial support offer the best possible management and care for individuals with dementia. This strategy also encourages the development of advanced care directives.

- Promote early diagnosis and timely care

Workforce development. Good quality care will be ensured through the provision of training and educational programmes for staff in direct contact with individuals with dementia giving particular importance to challenging behaviour and palliative care. Caregivers and family members also be provided with adequate.

- Ensure healthcare professionals have required qualifications to work with persons with dementia
- Recruit trained staff to work in various areas of dementia management and care

Improving dementia management and care. A holistic approach in service provision for individuals with dementia, their caregivers and family members will be adopted. Apart from providing all pharmacotherapeutic options to persons with dementia,, individuals receiving a diagnosis will have care plans developed by a multidisciplinary team.

- Provide pharmacological and non-pharmacological treatments
- Develop individualized care pathways
- Provide easy access to community services
- Provide long-term and palliative care.

Ethical approach to dementia management and care. This strategy aims to promote an ethical approach to dementia management and care and provide individuals with dementia and their caregivers with the necessary psychological support needed in making important decisions regarding their health and welfare.

- Provide resources to help persons with dementia and their families with decision-making

Research. Information regarding epidemiology of dementia in Malta, patterns of detection and diagnosis and delivery of care are needed for proper planning and allocation of health and social care funds and for outcome evaluation. Since the delivery of care is context specific, the strategy aims to promote and support epidemiological research in the field of dementia.

- Provide opportunities for undergraduate and graduate studies in dementia care

In keeping up with the above key recommendations and action, the Parliamentary Secretariat for Persons with Disability and Active Ageing followed the launch of the *National Dementia Strategy* by implementing a range of policy initiatives. Most importantly, a free-for-use National Dementia Helpline, open 24 hours every day, was inaugurated in order to open up a direct channel between community-dwelling individuals with dementia and their caregivers with state-run support services. Furthermore, the government works in partnership with the Malta Dementia Society, the latter being the sole non-government organization representing people with dementia and their caregivers in the Maltese Islands. Other initiatives included the publication and distribution of information booklets on dementia and how to care for a relative with dementia; producing a travelling street-theatre production about dementia in a number of villages and towns across the Maltese Islands; providing all healthcare staff working with people with dementia in state-run residential/nursing care homes with intensive training in dementia management and care; producing a 13-episode TV programme on various aspects of dementia care; a 10-week training programme for relatives of persons with dementia living in the community to train them in promoting independence, dementia-friendly environment, assisting in activities of daily living, physical activities and the role of caregivers in providing quality care; opening a new dementia activity centre; establishing Dementia Awards where individuals and organizations working in the field of dementia will be awarded for their effort; and providing scholarships for public sector employees to read a University-run Master degree programme in Ageing and Dementia Studies. However, the key culmination of the national strategic was the establishment of a Dementia Intervention Team which includes a Dementia Care Coordinator, a Practise Nurse and an Occupational Therapist to provide care for individuals with dementia living at home (Formosa, 2017). The team's composition makes sure that there was a varied mix of skills involved in each case's interventions to achieve a holistic assessment and care planning. Its mission to enable persons with dementia to continue actively ageing in the community, with the scope of delaying institutionalization through person-centred care, and includes four key objectives:

> …(a) providing disease-related education and information, (b) offering emotional support and coaching, (c) linking families to medical and non-medical services and resources, and (d) mobilising and organising the informal care network. The intervention protocols consist of Assessment of Care Needs, Development of Care Goals, Development of Action Steps, and Ongoing Monitoring of Action Steps.
>
> Gobey (2018): 155

One also notes a number of initiatives that the government is seeking to implement in the foreseeable future. These include the launch of an online training platform for dementia caregivers in collaboration with the WHO's *iSupport Programme* and the appointment of Dementia Ambassadors whose role is to brainstorm initiatives that foster dementia-friendly communities.

A reflective look at the implications of *National Dementia Strategy* and policy implementations makes it evident that the past decade was a real turning point for dementia care policy in Malta (Formosa & Cassar, 2019). Of course, there is always room for improvement. For instance, Wright and O'Connor (2018: 161) pointed out that the Malta's policy document includes "very little mention of the contribution that technology could make in easing the burden which dementia poses, such as assistive technologies to enhance quality of life of those living with dementia, or augment and improve the way in which health and social care workers deliver care and provide support". It is thus important that any future local policy on dementia care acknowledges "the prospective potential of using robotics to support healthcare delivery, supplement the dementia workforce and change care environments…and in general, consideration of technological advances and their contribution to dementia care" (ibid.: 161). One can also add that one finds "no direct reference to environmental factors…so implications of the policies environmental impact on pollution, land ethics, biodiversity and local/national ecologies are not considered" (ibid.). Moreover, there is also an absence of attention towards the needs of, and implications for, ethnic minorities, LGBTQI + people or other marginalized groups, such as those living with learning and physical disability, those living with mental ill-health and those, especially women, living in poverty. This is a crucial lacuna because, as Wright (ibid.) underlines, "the failure of policy strategies, plans and programmes to explicitly make reference to the needs of the most disadvantaged and vulnerable across societies means their very particular needs remain marginal rather than central". Indeed, policies, plans and programmes which do not advocate a critical social justice perspective are unable to bring about progressive and enduring modifications in the quest of meeting the challenges of dementia care (Hulko, 2009).

A Juggling Act! Experiences of Informal Carers of Persons with Dementia

Dementia is one of the most challenging age-related illnesses, not only for those just diagnosed, but also for their family caregivers. The most recent national statistics on caregivers of persons with dementia in Malta reported that the main carer for persons with dementia living in the community was the spouse (wives, 18.9%; husbands 17.4%), followed by the daughter (24.7%) (Directorate for Health, Information and Research, 2016). Only a minority of non-relatives were carers (1.5%), and caregivers were predominantly female (64.0%) compared to 36.0% who were male (ibid.). Informal caregiving with persons with dementia is such an onerous experience that many informal caregivers require assistance with a range of care work activities. A recent survey targeting informal carers of persons with dementia found that almost all respondents required assistance in bathing (94.7%), handling of medication (94.7%) and transportation (92.1%) (Spiteri & Scerri, 2019). Other activities indicated by caregivers as being difficult to provide on their own included helping their relative with dementia with feeding and to walk/stand up (60.5% and 65.8%, respectively). Moreover, and as expected considering the psychological burden associated with caring for person living with dementia, helping them deal with emotional difficulties was also considered to be in extraordinary in terms of requirement (89.5%). Many of the surveyed informal caregivers professed to make use of a range of public services operated by the Junior Ministry for Active Ageing and included enrolling their relatives in active ageing and dementia activity centres, utilizing respite care and subscribing to telecare services in that respective order. Rather unexpectedly, Spiteri and Scerri (2019) found that informal caregivers were not using a number of dementia-specific services that were aimed at supporting them in their caregiving roles. These included the organization of memory classes and clinics, and the Dementia Intervention Team. It was, however, positive to note that most informal caregivers had knowledge about the availability of the Dementia Helpline. Cross-tabulation analysis found that sociodemographic data that showed a significant effect on service utilization included the caregivers' civil status and the age of the informal carer. On the one hand, caregivers who experienced a breakdown of their marriage tended to be the most frequent users of available services:

...strong significant differences were found in Night Shelters...with informal caregivers who had a separated status making use of the shelters more often than those who indicated that they were single ($p=0.001$) or in a married/domestic relationship ($p<0.001$)...

Spiteri and Scerri (2019): 11

On the other hand, most informal caregivers of community-dwelling persons living with dementia in the Maltese Islands are located in the 70–79 age cohort, a result that was also reported in two previous studies (Gobey, 2013; Muscat, 2015). Citing again Spiteri and Scerri,

The age of persons living with dementia was found to play a significant role in a number of services including the Carer at Home Scheme, Night Shelters, Memory Classes, the Dementia Rehabilitation Programme and the Dementia Information Booklet. As the age of the individual with dementia increased, so did the use of the Carer at Home Scheme. This was expected as age is positively related to dependency in dementia and this can have a negative impact on the caregiver.

Spiteri and Scerri (2019): 11

Reflecting international trends documenting the feminisation of care work in later life, the majority of informal caregivers participating in Spiteri and Scerri's (2019) study were women. While men caregivers approach care in a task-oriented and problem-solving style, women take on a more emotional/relational approach. In comparison, men tend to worry less than women, handle their concern for partner's feeling less negatively, use outdoor activities or travel as coping strategies rarely have any confidants and generally act in a harsh manner to get tasks done (Roberto et al., 2019). Although Spiteri and Scerri's (2019) study provided sparse information in gender disparities in dementia care work, as they found that gender was not related to the Telephone Rent Rebate and the Telecare Plus services as men were found to be less likely to use these services compared to females, another study by Innes and colleagues (2011) found that Malta is no exception to international gender trends. This study presented the ensuing example to illustrate the gendered expectations to provide care imposed on daughters and daughters-in-law:

I want that we all look after her, but it is not going to happen because my brothers don't want to...they are saying that I am obliged to do it because I am the girl, and this is the 'illness' of the family that says that the girl (cares) and the boys follow this understanding [daughter].

informal caregiver, as cited in Innes et al. (2011): 173

Although many caregivers experience a range of positive and satisfying experiences of feeling needed and valuable in their role (McCabe et al., 2016), the burdens associated with caring for persons with dementia—especially in terms of psychological, social and financial aspects—have long been exposed and deliberated upon internationally (Gaugler et al., 2018). As informal caregivers in Innes and colleagues' study claimed,

I feel obliged. I feel so guilty. I can't get away from that. It's terrible. I can't even live my own life...I really wish I can run away but I can't. I can't even leave. She has trapped me really. [daughter]

informal caregiver, as cited in Innes et al. (2011): 173

Another flagged issue in local research constituted the relationship between communication problems and the caregiving experience. Indeed, many international studies have investigated that interface between language difficulties and dementia, as even at the early stage of the disease persons "exhibit anomia, literal and neologistic errors, reduction in phrase length, difficulties in sentence repetition, and impairment in comprehension, especially for written sentences" (Rousseaux et al., 2010: 3884). In this respect, Gobey (2013) found that as the severity of language and gesture expression and comprehension worsens with the progression of dementia, the person with dementia's communication difficulties are related to caregivers' burdens. However, if the communication improves, the caregiver burden decreases significantly. This occurs as communication breakdowns are highly problematic and often lead to declines in the quality of their interaction and relationship between the caregiver and the persons cared for. Citing caregivers again,

> I cannot share my feelings with him anymore, he isn't there to hear/understand what I am going through…I cannot confide with her as I used to. She is just not there anymore. [wife]
>
> Communication is a constant battle for us - it has become more of an inconvenience rather as a means to interaction. [wife]
>
> I simply cannot communicate with my wife, she just doesn't respond, I find that showing her with my hands and in the environment to work better at this point. Sometimes it's impossible to get her to understand me, even the most basic of things! husband
>
> informal caregivers, as cited in Gobey (2013): 98

In addition to communication, a vast range of personal and social abilities deteriorate with the progression of dementia, to the extent that it becomes extremely problematic for persons with dementia to remain independent. In Briffa Farrugia's (2018) study, one male caregiver recounted how the most unpleasant part of the caring role is observing his wife becoming progressively unable to perform the activities of daily living. In the words of one husband: "I'm her husband so I do not feel it right that I go with her when she uses the bathroom. But then, it is a nightmare. She hides "it" in any place she finds. I shout at her…Then chaos!". [husband]" (cited in Briffa Farrugia, 2018: 47). The management of aggressive behaviour is another problematic (and fearful) area for caregivers. Many caregivers related their frustration that their relatives, who always had calm and discreet personalities, became very aggressive with the passing of time, or became prone to wandering during the night:

> The memory loss did not bother me much as I knew what was happening. But when she became very angry, saying bad words that I have never heard her use before that became something I could not handle…It is like you are talking with the wall. I try to keep myself calm but when she gets very aggressive I sometimes end up being angry myself'. [son]
>
> I do not trust going out with him anymore because his behaviour is unpredictable. Before we used to go out to do the shopping together and he was sociable to the people. But now he becomes aggressive and angry - I'm shy to be with him in front of people like that because his behaviour is uncontrollable. [wife]
>
> I lock all the doors before we go to sleep because she wanders at night. But then sometimes I worry that if something happens to us both, they find us all locked in. When she slept during the night it was much better because then during the day I have the strength to take care of

> her, but when I wake up about four times in the night with her, I am already tired before I start. [husband]
>
> informal caregivers, as cited in Briffa Farrugia (2018): 47, 48

The experience of such adversities, in tandem with the actual care work, can become so stressful that many caregivers experience mental health issues during their caregiving role. In Muscat's (2015) study, 32% of surveyed caregivers were diagnosed with depression, a result that corresponded with the results in Mahoney and colleagues' (2005) study. Muscat argued that

> …those caregivers who were not diagnosed with depression might still suffer from depression at some stage. They might experience emotional distress soon after their relative had been diagnosed with dementia, while others might experience depression as dementia progresses…
>
> Muscat (2015): 110

In a nutshell, one can never overstate the extent that dementia caregivers experience unrelenting and changing psychosocial and physical challenges during the course of their care work. There is also no doubt about the importance to understand caregivers' needs from their own perspectives as "this information will assist in the development of intervention programmes or models of care to ensure caregivers' perceived needs are met" (Tretteteig et al., 2016: e71).

Experiencing Dementia: A Case Study Narrative

The subjective experience of dementia for older Maltese persons is largely absent from the extant literature. One reason for this omission is that persons "with dementia suffer from 'narrative dispossession', a phenomenon in which they stop elaborating stories or even influence their narrative comprehension of daily experiences" (Villar et al., 2019: 5). In Baldwin's words, the

> …onset and progression of dementia threatens narrative abilities through its effect on language and cognitive function. Opportunities for narrativity may also be limited by a decrease in social interaction or the response of others who unnecessarily take over the narrative function or who recuperate attempts at narrative expression (such as 'wandering', 'challenging behaviour' and fidgeting) as symptomatic of the dementia. But the *narrative dispossession* of people living with dementia can also be seen as a function of how we conceptualise and operationalise narrative and narrativity…a one-sided approach to narrative agency may also serve to dispossess people living with dementia from their (potential) narratives.
>
> Baldwin (2006): 107 - italics added

Evidence documenting "narrative dispossession" is found in the vast range of studies which record how "the stories of the people with dementia are hijacked by their caregivers, who assume not just their care, but also the narrative identity of the person" they are caring for (Villar et al., 2019: 5). Such happenings imply that the "opportunities for people with dementia to express themselves narratively are dramatically reduced" as they have few to no opportunity to participate in interactions that allow them to construct their life history or let their voices be heard. It largely follows that disempowerment leads to depersonalization, marginalization and social exclusion, which all together are tantamount to a "social death" (Westwood, 2018). This context is no exception, especially since caregivers (both in institutional and community settings) many time "address people with dementia in 'elderspeak', which consists of speaking more slowly and using inappropriate terms of endearment, exaggerated intonation, simplified syntax, collective pronoun substitution, and questions that indicate a desired response". Such a situation does not arise by coincidence or chance but is the direct result of, to draw on Foucault's (1998) philosophy, the hegemony of biopower, which precludes the embodiment of dementia in a humanist dimension. In fact, once the person has been diagnosed.

> Dementia becomes the defining characteristic that gives meaning to their behavior, regardless of their life stories, preferences, needs, expectations, or the actual circumstances that could account for such a behavior from the point of view of the person with dementia.
>
> Villar et al. (2019): 4

The planning and writing of this section was aimed at exploring the subjective experience of dementia and how persons with dementia tend to maintain their personhood in this process. Hence, the research question was as follows: *What is the experience of dementia for diagnosed Maltese individuals*? This research question was opted for on the basis that it promises much potential to help us understand better how a person with dementia loses his/her own voice, lose significant social relationships and cease participating actively in the community, and become labelled and stereotyped by the disease or, to be more precise, by the biomedical meanings that applied health professionals and popular culture attach to the disease. The scope of the research question demanded an interpretative-qualitative methodology as only such a philosophical stance allows us to "study of human behaviour that relies on the analysis of narrative data to create an interpretation of the meaning of these behaviours from the perspective of the participants themselves, within their own social context" (Kuckelman Cobb & Forbes, 2002: 197). Indeed, qualitative research is particularly useful when the topic under investigation is insufficiently understood or when it is best communicated through detailed examples and life histories. In choosing among the different types of qualitative research, this study opted for a narrative gerontological inquiry on the basis that this line of approach "enables us to explore the complex dynamic between our inner personal meaning and the constraints that are imposed on us from the outside" (Kenyon & Randall, 1999: 2). The suitability of narrative inquiry attitude to research the life histories of persons with dementia has been championed by many a research study (Patterson et al., 2018). However, the planning of an effective design proved more challenging than expected. Although when exploring lived experiences through in-depth interviews, a sample size of between four and ten is considered appropriate (Smith et al., 2009), even such a limited number proved to be highly challenging, even when gatekeeping and snowballing sampling were adopted. As a result, a decision was taken for this qualitative-narrative inquiry to opt for a case study research design. The latter is not uncommon in dementia studies since many researchers either also find it difficult to recruit persons with dementia in their research study or do not find more than one person that matches their inclusion criteria (Gauthier & Rosa-Neto, 2011). Definitions of case studies vary but all promote the notion that the researcher aims at knowing a single entity. This study followed Stake (1995: xi) in deeming case study research as "the study of the particularity and complexity of a single case, coming to understand its activity within important circumstances," and viewed such research as an example of an "instrumental case study" since the case is of secondary interest but only in that it plays a supportive role to facilitate our understanding of something else (ibid.).

The interviewee was a male, in his late 70s, and who will be referred to under the pseudonym of Richard. He was interviewed twice, both in his daughter's home (with whom he lives and who fulfils the role of an informal caregiver), once in her presence, and a subsequent time alone. Each interview lasted one hour, and followed a semi-structured interview schedule containing open-ended questions. It is not possible to provide a thorough background sketch of Richard because since Malta is a microstate, and there are only about 2000 male persons with dementia, and hence, relatively easy for informers to be identified. Richard is a widower, was diagnosed with dementia 17 months prior to the interview, and although he seemed to functioned relatively independently, his daughter did everything for him, surveilled him constantly or made sure that he was never alone in the house, and just gave him enough privacy to wash himself while she waited behind a plastic curtain. It is not the intention here to elaborate upon his daughter's modus operandi as a caregiver, and it suffices to say that she thought that she was being a dutiful daughter and that her relentless shadowing of her father was, in her opinion, confirmation of her altruistic love. Richard opened up the interview by describing how, initially, he passed off memory problems as a part of his normal ageing journey. However, as memory problems become more severe, and his attempts to control and conceal growing them through strategies of vigilance and avoidance were becoming increasingly unsuccessful, and even causing increasing stress and strained relationships with significant others, he started to suspect that there was something wrong. Richard described such a prediagnostic period as follows:

> My day oscillated from being caught out that I am in such fault, playing it down when this happens, and being super vigilant that I do not mess things up. [Interviewer: Caught up in what way?] In various ways, I remember not being able to connect sentences, not being able to solve minor DIY issues, and of course, needing more time than before to do anything, and yes, even finding it difficult to recall names of persons and localities. The newer the situation I found myself in, the more difficult it was for me to know what I had to do or what was expected from me (Richard)

Richard recounted how what bothered him most was not the above experiences as such, but the feeling that he was losing control over his life, and how the future looked as an unfamiliar and unpredictable terrain. As expected, this resulted in a perceived loss of self-esteem and locus of control and even threatened his sense of security, autonomy and being a meaningful member of society. Richard eventually decided to disclose such symptoms to his doctor who, after a number of tests, disqualified the possibility of other ailments and diagnosed him as having dementia. Much literature reported how, following a diagnosis of dementia, persons generally exhibited three key emotional responses: "(a) responses suggesting a lack of insight and/or an active denial of the diagnosis, (b) grief reactions/emotional crisis related to the experience of actual or anticipated losses associated with dementia, and (c) positive coping responses to maximize the disease outcome" (Aminzadeh et al., 2007: 281).

In Richard's case, the latter was the case as the diagnosis disclosure resulted in a number of psychological benefits such as confirmation of suspicion, sense of relief, opportunity to develop positive coping skills and an opportunity to make most of time. He also claimed that his diagnoses served him with the possibility of an improved understanding of the problems associated with dementia and potential solutions, better decision-making and future planning, and increased access to available treatment and support services. As Robinson and colleagues (2011: 1035) concluded, "although people with dementia may experience short-term distress following the disclosure of their diagnosis, a small number of studies exploring the longer term effects of disclosure show that the majority of people do not develop significant psychological ill-health".

At the time of the interview, Richard highlighted the dilemma between balancing his private world (the "self") and his public world (how others view them). On one hand, when he is with his family relatives, and his closes friends and acquaintances, who all know about his condition, Richard approaches his condition with a sense of humour. Indeed, humour is common coping mechanism against the adversities that later life may bring:

> Humor can do more than offer the elderly opportunities for maintaining self-worth. It can offer respite from pain and chronic disability. There is a tendency to denigrate oneself when illness, disability, accidents, or trauma occurs. The elder may question, *why me*? Self-blame, which erodes self-esteem and self-worth, ensues. Chronic pain or disability is often seen as the price for imagined or wrongful past actions. Humor can help to relieve the nagging pain or discomfort through blaming the incident rather than oneself as humour mitigates the negativity.
>
> Lurie and Monahan (2015): 85

On the other hand, when Richard is with people he is either relatively or totally unfamiliar with, he finds himself anxious that he will be "caught out", and as a result, he engages in a series of self-protective strategies in the hope of protecting his self-identity and to maintain that normality and continuity which he harboured prior the onset of dementia. Self-protective strategies included minimizing problems, sticking to or setting a routine, self-monitoring, relying on support from children or friends and avoiding most but especially new activities. However, Richard dwelled at length how such strategies may fail and how he sometimes wishes to isolate himself from others:

> My belief is that although I have dementia, I should not be put in a corner and forgotten for ever. I used to be very active up until a few years ago, or since I started feeling different, but this is no longer possible. The constant advice is to take life slowly and take it easy, and if make some blunder, such as mixing names, people, even my daughter, expect me to be supervised all the time, chaperoned I say to myself, as if I am a little child. Facing such

humiliation, I prefer to stay at home and isolate myself from everyone and everything. Isn't this what everybody wants for me? (Richard)

It is thus apparent that from the perspective of a person with dementia, and as argued in more detail in Robinson and colleagues' (2011) study, Richard's greatest difficulty was to have to deal with the prospect of losses which included loss of self-esteem, decline of social status and roles and forfeiture of function and activities of daily living, especially driving. This led him to experience feelings of exclusion and stigmatization or, in Beard and Fox's terms (2008), a kind of "social demotion". Indeed, Richard underlined many times his desire to be treated as "normal" rather than "disabled":

> When I enter a room people either go silent or else show an exaggerated sense of familiarity. They emphasise in asking me how I am, there are no longer questions or comments about football or politics but simply how I am. This makes me feel like a patient of some disease or as if I am suffering from disability. Of course, I am to an extent afflicted with a disease, but I am frustrated that I cannot hold a normal conversation with anybody, all people are interested is how I am coping with dementia and how I am feeling. The Richard of today is no longer *me* but a patient with a dementia label stuck on his forehead on his wat to forgetfulness and senility. (Richard)

Richard's assertions validate what some authors, such as Sabat and Steven (2006), have termed as "excess disability", which refers to the loss of social abilities due to factors other than the disease. Indeed, the progressive and irreversible deterioration of personal autonomy in dementia leads to a parallel restriction of the opportunities to sustain and reinforce the remaining competences that a person with dementia still has. As Villar, Serrat and Bravo-Segal pointed out,

> From this point of view, labeling someone as 'a person with dementia' involves losing hope or expectation about the remaining competences of the individual, or about any kind of improvement or learning, quickly leading to a pattern of dependency in which the remaining competences of the person with dementia are ignored.
>
> Villar et al. (2019): 5

Such circumstances generally lead to a self-fulfilling prophecy in which a radically negative narrative of dementia leads people to behave in a way that confirms that such a narrative is the true state of dementia living (Stone & Byrant, 2012).

Conclusion

This chapter calls for a life course approach to dementia rather than as an end-of-life phenomenon. Despite the fact that research and policy work on dementia has been ongoing for a number of years now, MacDonald et al. (2019: 1) were spot on when they claimed that "the shifts that we are looking for are not happening fast enough". Dementia is still projected in political, professional and cultural discourse in medical terms and as a key cog wheel of the care industry. However, dementia is more than a medical diagnosis, it is ultimately a social and relational experience.

As Kate Swaffer (2019), diagnosed with dementia at the age of 49, and now a key inspirational advocate in dementia care, argued:

> Before dementia, people have good and bad days; life with dementia is no different. It is possible to live a positive and productive life, some of the time, even with dementia, if supported well. I suggest, with a lot more proactive disability support from the day of diagnosis, and moving away from the out-dated dysfunctional and medicalised attitudes about dementia, many more will be living positively with dementia, for much longer, either confounding the experts even further, or teaching them it is possible. With or without dementia, we are all born with a death sentence. There is no need for a diagnosis of dementia to be seen as the catalyst to give up living, and meekly go home and wait for that to happen.
>
> Swaffer (2019) - online version

Of course, this is not the same as saying that we need to reject the biomedical model in a lock, stock and barrel manner, since the neurology of dementia has led to many a preventive measure. Rather, as MacDonald et al. (2019) argued, it is to say that this is not enough:

> Healthy older adults, as well as those living with dementia symptoms, are being discriminated against and there is a need for society to acknowledge the damage caused by ageist stereotyping and how that enables exploitation…It is time to raise public awareness of covert abuse and neglect of older adults, especially those with cognitive deficits, so that they can be welcomed and accepted by society rather than being disenfranchised on the basis of 'dementia' and alienated from their families and communities.
>
> MacDonald (2019): 2

The question is, therefore, how it is best to approach dementia as a social and cultural experience. Following MacDonald et al. (2019), there are two fronts of action namely, the need to confront and change social attitudes, and improvements in the understandings and practices brought to experience living with dementia. On the one hand, there requires a strong current of advocacy that mitigates against the dominant social conceptualisations of dementia as a tragedy such as a heart breaking catastrophe, an identity-absorbing subculture, and stigmatizing. As McDonald concluded,

> Until and unless dementia can be prevented and/or cured, we need to learn to live with it. The side-lining of a social model contributes to fear and stigma; these are exacerbated when the medical model is foregrounded. Each diagnosis of dementia impacts on many people. As the disease progresses so, too, do the social demands. We live in societies ill-equipped to deal with these demands, whether for the person with the diagnosis, or those who will care for them as they become progressively dependent, whether at home or in residential care. We need to challenge people who hear a diagnosis of dementia and say, 'how tragic', and encourage them to say, 'how can I support this person and her carers?'
>
> McDonald (2019): 31

This is especially warranted because, as Mears (2019) argued, while there have been various strategies for raising awareness of ageism and its consequences, translating this wisdom into outcome for older people is yet to occur. One key reason why this transference has not occurred is due to the fact that care work of persons with dementia is, essentially, a female phenomenon, so that understanding dementia care requires an investigation of that intersectionality between ageism and sexism. On the other hand, one needs to strengthen the ongoing work on humanizing lives which experience dementia. Here, we do not have to re-invent the wheel and may go back to one pioneering work in dementia studies when Kitwood (1997) drew attention to the way in which sociopsychological factors contribute to undermine the personhood of persons with dementia. His exhaustive descriptions of the factors which contribute to the dehumanization of both community-dwelling and institutionalized persons with dementia is also followed by much practical guidance on how to prevent these processes from occurring. The way forward is to balance the "first-person perspectival givenness" and the "social personae" of a person of dementia, so as to highlight how personal lives experiencing dementia emerge from relational connections (Bitenc, 2020). Indeed, a holistic approach to dementia policy "cannot focus on the needs of any one individual or group of individuals but rather must address the needs of all those involved in supporting people with dementia, whether in an unpaid or paid capacity" (Clarke & Wolverson, 2016, as cited in MacDonald, 2019: 211). It is hoped that this chapter acts as a catalyst for a more informed, holistic and relational approach to dementia care policy in Malta.

References

Abela, S., Mamo, J., Aquilina, C., & Scerri, C. (2007). Estimated prevalence of dementia in the Maltese Islands. *Malta Medical Journal, 19*(2), 23–26.

Alzheimer's Disease International. (2019a). *From plan to impact II. The urgent need for action.* Alzheimer's Disease International. https://www.alz.co.uk/adi/pdf/from-plan-to-impact-2018.pdf. Accessed on October 2, 2019.

Alzheimer's Disease International. (2019b). *Dementia statistics.* Alzheimer's Disease International. https://www.alz.co.uk/research/statistics. Accessed on October 2, 2019.

Aminzadeh, F., Byszewski, A., Molnar, F., & Eisner, M. (2007). Emotional impact of dementia diagnosis: Exploring persons with dementia and caregivers' perspectives. *Ageing & Mental Health, 11*(3), 281–290.

Baldwin, C. (2006). The narrative dispossession of people living with dementia: Thinking about the theory and method of narrative. In K. Milnes, C. Horrocks, N. Kelly, B. Roberts, & D. Robinson (Eds.), *Narrative, memory and knowledge: Representations, aesthetics and contexts* (pp. 101–109). University of Huddersfield.

Beard, R. L., & Fox, P. J. (2008). Resisting social disenfranchisement: Negotiating collective identities and everyday life with memory loss. *Social Science & Medicine, 66*(7), 1509–1520.

Bitenc, R., (2020). *Reconsidering dementia narratives.* Routledge.

Briffa, F., & Natalie. (2018). *Community services for people with dementia and their caregivers: An analysis of the support services provided in Malta.* Unpublished dissertation, Master of Gerontology and Geriatrics, Department of Gerontology and Dementia Studies, University of Malta.

Burns, A., & Robert, P. (Eds.). (2019). *Dementia care: International perspectives.* Oxford University Press.

Directorate for Health, Inforsmation and Research. (2016). *Dementia Report 2014/2015.* Ministry for Health. http://health.gov.mt/en/dhir/Pages/Registries/dementia.aspx. Accessed on October 2, 2019.

Alzheimer Europe. (2013). *Dementia in Europe Yearbook 2013*. Alzheimer's Disease International. http://alzheimer-europe.org/content/download/79291/491583. Accessed on 12 July 2021.
Formosa, M. (2017). Responding to the active ageing index: Innovations in active ageing policies in Malta. *Journal of Population Ageing, 10*(1), 87–99.
Formosa, M., & Cassar, P. (2019). Visual arts dialogues with older persons in long-term care facilities. *Educational Gerontology, 5*(1), 23–42.
Formosa, M., & Galea, R. (2020). Critical educational gerontology at a senior center in Malta: Possibilities and limitations for critical consciousness. *Educational Gerontology, 46*(2), 59–71.
Formosa, M., & Scerri, C. (2020). Punching above its weight: Current and future aging policy in Malta. *The Gerontologist, 60*(8), 1384–1391.
Foucault, M. (1998). *The history of sexuality Volume 1: The will to knowledge*. Penguin.
Gaugler, J., Reese, M., & Mittelman, M. (2018). The effects of a comprehensive psychosocial intervention on secondary stressors and social support for adult child caregivers of persons with dementia. *Innovation in Aging, 2*(2), 1–10.
Gauthier, S., & Rosa-Neto, P. (Eds.). (2011). *Case studies in dementia: Common and uncommon presentations*. Cambridge University Press.
Gobey, A. (2013). *Investigating the effects of communication difficulties on the burden of Maltese caregivers of persons with dementia*. Unpublished dissertation, Master of Gerontology and Geriatrics, Department of Gerontology and Dementia Studies, University of Malta.
Gobey, A. (2018). The dementia intervention team: Spearheading dementia and care management in the community. In M. Formosa (Ed.), *Active and healthy ageing in Malta: Gerontological and geriatric inquiries in Malta* (pp. 149–166). Book Distributors Limited.
Grant, Thornton Malta. (2018). *Dementia in Malta: Estimating the number of persons with dementia and the demand for dementia specific residential care*. Grant Thorton Malta (unpublished document).
Hulko, W. (2009). "From 'not big deal' to 'hellish': Experiences of older people with dementia. *Journal of Aging Studies, 23*(3), 131–144.
Innes, A., Abela, S., & Scerri, C. (2011). The organisation of dementia care by families in Malta: The experiences of family caregivers. *Dementia, 10*(2), 165–184.
Karlin, B., Young, D., & Dash, K. (2017). Empowering the dementia care workforce to manage behavioral symptoms of dementia: Development and training outcomes from the VOICE Dementia Care Program. *Educational Gerontology, 38*(4), 375–391.
Kenyon, M. G., & Randall, L. W. (1999). Introduction: Narrative gerontology. *Journal of Aging Studies, 13*(1), 1–5.
Kitwood, T. (1997). *Dementia reconsidered: The person comes first*. Open University Press.
Kuckelman Cobb, A., & Forbes, S. (2002). Qualitative research: What does it have to offer to the gerontologist? *Journal of Gerontology: Medical Sciences, 57A*(4), M197 M202.
Lurie, A., & Monahan, K. (2015). Humor, aging, and life review: Survival through the use of humor. *Social Work in Mental Health, 13*(1), 82–91.
MacDonald, G. (2019). Why person-centred' is not enough: A relational approach to dementia. In G. Macdonald & J. Mears (Eds.), *Dementia as a social experience: Valuing life and care* (pp. 195–214). Routledge.
MacDonald, G., Mears, J., & Naderbagi, A. (2019). Reframing dementia: The social imperative. In G. Macdonald & J. Mears (Eds.), *Dementia as a social experience: Valuing life and care* (pp. 1–19). Routledge.
Mahoney, R., Regan, C., Katona, C., & Livingstone, G. (2005). Anxiety and depression in family caregivers of people with Alzheimer's disease—the LASER-AD study. *The American Journal of Geriatric Psychiatry, 13*(9), 795–801.
McCabe, M., You, E., & Tatangelo, G. (2016). Hearing their voice—A systematic review of dementia caregivers' needs. *The Gerontologist, 56*(5), e70–e88.

McDonald, T. (2019). Negative positioning of 'dementia' in an environment of completion for resources. In G. Macdonald & J. Mears (Eds.), *Dementia as a social experience: Valuing life and care* (pp. 23–33). Routledge.
Mears, J. (2019). 'Nobody cares about me': Older women, caring and dementia. In G. Macdonald & J. Mears (Eds.), *Dementia as a social experience: Valuing life and care* (pp. 34–45). Routledge.
Muscat, M. (2015). *Depression, anxiety and quality of life of caregivers of individuals with dementia living in the community in Malta.* Unpublished dissertation, Master of Gerontology and Geriatrics, Department of Gerontology and Dementia Studies, University of Malta.
National Statistics Office. (2016). *Malta in figures 2016.* National Statistics Office.
National Statistics Office. (2018). *World population day: 11 July 2018.* National Statistics Office.
Parliamentary Secretariat for Rights of Persons with Disability and Active Ageing. (2015). *Empowering change: A national strategy for Dementia in the Maltese Islands 2015–2023.* Parliamentary Secretariat for Rights of Persons with Disability and Active Ageing.
Patterson, K., Clarke, C., Wolverson, E., & Moniz-Cook, E. (2018). Through the eyes of others—the social experiences of persons with dementia: A systemic literature review and synthesis. *International Psyhcogeriatrics, 30*(6), 791–805.
Roberto, K., McCann, B. R., Blieszner, R., & Savla, J. (2019). A long and winding road: Dementia caregiving with grit and grace. *Innovation in Aging, 3*(3), 1–12.
Robinson, L., Gemski, A., Abley, C., Bond, J., Keady, J., Campbell, S., Samsi, K., & Manthorpe, J. (2011). The transition to dementia—individual and family experiences of receiving a diagnosis: A review. *International Psychogeriatrics, 23*(7), 1026–1043.
Rousseaux, M., Séve, A., Vallet, M., Pasquier, F., & Mackowiak-Cordoliani, M. A. (2010). An analysis of communication in conversation in patients with dementia. *Neuropsycholgia, 48*, 3884–3890.
Sabat, R., & Steven. (2006). Mind, meaning, and personhood in dementia: The effects of positioning. In J. Hughes, S. Louw, & S. R. Sabat, (Eds.), *Dementia: Mind, meaning, and the person* (pp. 287–302). Oxford University Press.
Scerri, A., & Scerri, C. (2012). Dementia in Malta: New prevalence estimates and projected trends. *Malta Medical Journal, 24*(3), 21–24.
Scerri, C., & Zahra, T. (2013). *X'ħin hu? Fatti dwar id-dimensja.* Parliamentary Secretariat for the Elderly and Community Care.
Scerri, C. (2015a). Malta's vision for a national strategic policy. *International Journal on Ageing in Developing Countries, 2*(1), 113–142.
Scerri, C. (2015b). Dementia in Malta: Trends, policy and professional development. In M. Formosa & C. Scerri (Eds.), *Population ageing in Malta: Multidisciplinary perspectives* (pp. 169–191). Malta University Press.
Scerri, C. (2015c). The travails of dementia policy development in a small island state: Perspectives from Malta Hellenic. *Journal of Nuclear Medicine, 18*(1), 2–12.
Smith, A. J., Flowers, P., & Larkin, M. (2009). *Interpretative phenomenological analysis.* Sage.
Spiteri, C., & Scerri, C. (2019). Utilisation of services by informal caregivers of community-dwelling persons living with dementia making use of the dementia activity centres in the Maltese Islands. *Malta Journal of Health Sciences, 6*(1), 5–13.
Stake, E. R. (1995). *The art of case study research.* Sage.
Stone, R., & Bryant, N. (2012). The impact of health care reform on the workforce caring for older adults. *Journal of Aging and Social Policy, 24*(2), 188–205.
Swaffer, K. (2019). The reliability and meaning of a dementia diagnosis. *Australian Journal of Dementia Care.* https://journalofdementiacare.com/the-reliability-and-meaning-of-a-dementia-diagnosis/. Accessed on October 2, 2019.
Surr, C., & Gates, C. (2017). Effective dementia education and training for the health and social care workforce: A systematic review of the literature. *Review of Educational Research, 87*(5), 966–1002.

Tretteteig, S., Vatne, S., & Rokstad, A. M. (2016). The influence of day care centres for people with dementia on family caregivers: an integrative review of the literature. *Aging & mental health, 20*(5), 450–462. https://doi.org/10.1080/13607863.2015.1023765

Villar, F., Serrat, R., & Bravo-Segal, S. (2019). Giving them a voice: Challenges to narrative agency in persons with dementia. *Geriatrics, 4*(20), 1–9.

Westwood, S. (Ed.). (2018). *Ageing, diversity and equality: Social justice perspectives*. Routledge.

World Health Organization. (2015). *The epidemiology and impact of dementia: Current state and future trends*. World Health Organization. https://www.who.int/mental_health/neurology/dementia/dementia_thematicbrief_epidemiology.pdf. Accessed on October 2, 2019.

World Health Organization. (2017a). *Global action plan on the public response to dementia 2017–2025*. World Health Organization.

World Health Organization (2017b). *Global action plan on the public response to dementia 2017–2025 [Infosheet]*. World Health Organization: https://www.medbox.org/dementia/global-action-plan-on-the-public-health-response-to-dementia-infosheet/toolboxes/preview? Accessed on October 2, 2019.

Wright, T., & O'Connor, S. (2018). Reviewing challenges and gaps in European and global dementia policy. *Journal of Public Mental Health, 17*(4), 157–167.

Zahra, T. (2009). *In-nannu nesa x'jisimni*. Parliamentary Secretariat for the Elderly and Community Care.

Chapter 14
Walking a Legal Tightrope: Assessing the Legal Capacity of Persons Experiencing Dementia to Make and Revoke Enduring Powers of Attorney in Australia

Eileen Webb

> *The law does not prescribe any fixed standard of sanity as requisite for the validity of all transactions. It requires, in relation to each particular matter or piece of business transacted, that each party shall have such soundness of mind as to be capable of understanding the general nature of what he is doing by his participation. Gibbons v Wright (1954) 91 CLR 423, 437–438.*
> (Dixon CJ, Kitto and Taylor JJ)

Abstract This chapter considers the legal conundrum arising where an older person has entered into an enduring power of attorney (EPOA) but was, at the time of entering into, or later revoking, the EPOA, experiencing dementia. If it can be established that the older person did not have legal capacity to enter into or revoke the EPOA, the instrument is rendered *void* ab initio and therefore subsequent dealings made pursuant to the EPOA can be nullified. The chapter examines the legal consequences of such a finding of incapacity and suggests that, with the rising incidence of dementia in Australia, there should be an enhanced focus on properly and consistently assessing legal capacity where a person seeks to enter into or revoke an EPOA.

Introduction

A cornerstone of Australian law is the presumption that an individual has capacity to administer their own affairs and has the right to do so (Fougere, 2017; Schyvens, 2017). This presumption prevails in the absence of evidence indicating incapacity.

E. Webb (✉)
Discipline of Law, Justice, and Society, University of South Australia, North Terrace, Adelaide 5000, Australia
e-mail: Eileen.webb@unisa.edu.au

M. K. Shankardass (ed.), *Dementia Care*,
https://doi.org/10.1007/978-981-16-3864-0_14

Indeed, even where a person may be experiencing loss of capacity through, for example, the onset of dementia, a person will still be regarded as capable of making certain decisions depending on the nature of the person's condition and the complexity of the legal decisions they are making (*Re D (J)* 243–4); *Hoffman v Waters* 505). Furthermore, where decision-making capacity is diminished, supported and substitute decision-making may assist (Lewis et al., 2020b).

An enduring power of attorney (EPOA) is a statutory substitute decision-making tool that enables another person or persons to administer an individual's legal and financial affairs (Purser, 2017). An EPOA can operate both before *and after* a loss of capacity. This can be compared to general powers that are effective only until the grantor loses capacity (Dal Pont, 2019). In this way, a person appointed under an EPOA can deal with the individual's legal and financial affairs even in circumstances where the individual who granted the EPOA has lost capacity (Bagshaw, 2013; Britt, 2016). This arrangement places a considerable degree of power in the hands of the person acting as EPOA, and it is perhaps not surprising that misuse of EPOA is a common source of financial abuse of older people in Australia (Somes & Webb, 2018, p. 225). Indeed, recent inquires considering abuse of older persons have emphasized the need for reform of laws relating to EPOA including harmonization of state and territory laws, a national register of EPOAs, and harsher penalties for misuse (Reis, 2018, p. 9).

However, the focus on the errant activities of some EPOAs has left other pertinent issues affecting EPOA neglected. One such issue involves circumstances where a person seeking to enter into, or revoke, an EPOA does not have the requisite legal capacity to do so. This is critical because entry into an EPOA provides a license for the attorney to enter into a variety of subsequent legal and financial transactions. If a finding of legal incapacity at the time of execution is ultimately made, the law provides that the EPOA is *void* ab initio—from the beginning—and will therefore cast doubt on the validity of subsequent dealings made pursuant to the EPOA (*Szozda v Szozda*, [28]). This will also be the case where a person has purportedly revoked an existing EPOA but lacks the legal capacity to do so. In this case, a newly appointed attorney may act with the authority of the EPOA, but it is subsequently established that, at the time of the revocation and the new appointment, the grantor lacked legal capacity to do so.

Perhaps surprisingly, Australian legal decisions and commentary rarely address these circumstances. Most cases involving the disputed exercise of an EPOA do not question the initial validity of the instrument itself but ponder the attorney's subsequent misuse of the EPOA through misappropriation of the older person's property and assets. Therefore, it is simply assumed that the EPOA is valid, and the focus is on the attorney's conduct in exercising of the powers under the instrument.

However, where the EPOA is void from the outset because of a grantor's lack of legal capacity when the EPOA was signed, *all* transactions whether legitimate or otherwise entered into pursuant to the EPOA are rendered vulnerable. The consequences are potentially far-reaching, especially where a series of transactions has taken place. The prospect of "unpicking" these concluded agreements is, to say the least, fraught with uncertainty (*Scott v Scott,* [236]–[237]).

The rising incidence of dementia in Australia and a lack of understanding of how to assess the requisite legal capacity even among lawyers, is, and is likely to become more, problematic in the future. The trigger for entering into an EPOA is often the concern about a perceived loss of cognitive function in an individual identified by the person concerned or their family (Van Wijnagaarden et al., 2018, p. 2). Such loss may be noticed at an early stage or over a much longer period where episodes may be explained away by explanations such as tiredness or illness. Therefore, by the time a person may seek to enter into an EPOA, there may have been significant deterioration of mental acumen (NSW Law Society, 2016). If a person lacks legal capacity to enter into, or revoke, an EPOA, and this is not discovered prior to the instrument being entered into, the EPOA will be regarded as having "never legally got off the ground" (McCullagh, 2018 [11.170]). State and territory law societies and professional disciplinary bodies are well aware of this nullifying effect and have moved to raise awareness among lawyers as to their professional obligations towards potentially vulnerable clients (NSW Law Society, 2018). However, the pressures of legal practice, uncertainty about assessment of legal capacity and the increasing popularity of "do it yourself" EPOAs mean that this issue will be difficult to identify initially and later, in the event of a dispute, address.

The chapter is comprised of five parts. Part 1 considers the nature of an EPOA and distinguishes such instruments from other powers of attorney and agency relationships while Part 2 discusses the rising incidence of dementia in Australia. Part 3 commences with a discussion of the nature of legal capacity together with an analysis of the legal tests to be applied and the consequences of the EPOA being regarded as *void* ab initio. Several recent Australian cases will be discussed highlighting the sometimes perplexing consequences where a grantor lacked legal capacity upon entering into, and in this case continually revoking, EPOAs. Part 4 considers the role and responsibilities of lawyers in assessing legal capacity and examines recent case law. Part 5 concludes the chapter with a series of recommendations to assist grantors, attorneys and lawyers and provides suggestions for law reform.

The Nature of an Enduring Power of Attorney

At this stage, it is instructive to compare EPOAs with general powers of attorney.

General Powers of Attorney

A general power of attorney provides for a grantor (the person on whose behalf the POA is being entered into) to appoint an attorney to act on behalf of the grantor with respect to legal and financial affairs. A power of attorney has been described as a formal grant of agency powers (*Vella v Permanent Mortgages Pty Limited* [2008] NSWSC 505; *Scott v Scott* [2012] NSWSC 1541).

Like an agency relationship, a general power of attorney is an authority to act in accordance with the grantor's instructions being conferred on the attorney (Dal Pont, 2019). Like other contracts, a general power of attorney will lapse upon supervening incapacity of the principal (*Smith v Smith* [19]). In Australia, powers of attorney are regulated pursuant to statute in each state and territory. The legislation in each jurisdiction is not uniform; however, all provide for necessary formalities that must be complied with.

Enduring Powers of Attorney

General powers of attorney become ineffective upon the grantor losing capacity. In contrast, an EPOA enables another person to administer the principal's legal and financial affairs after the loss of capacity. Each state and territory jurisdiction in Australia has introduced legislation regulating EPOAs ((South Australian Law Reform Institute (SALRI), 2020, 441–443). In summary, an enduring attorney is authorized to do on behalf of the principal who can lawfully authorize an attorney to do (McCullagh [11.50]). Examples include paying debts and bills, recurrent charges, operate bank accounts, sell property, manage investments and execute deeds, contracts or other legal documents.

Although some EPOAs can take effect upon execution—should the grantor want this to be the case—its pivotal role arises and continues after the grantor's loss of capacity.

Entry into an EPOA

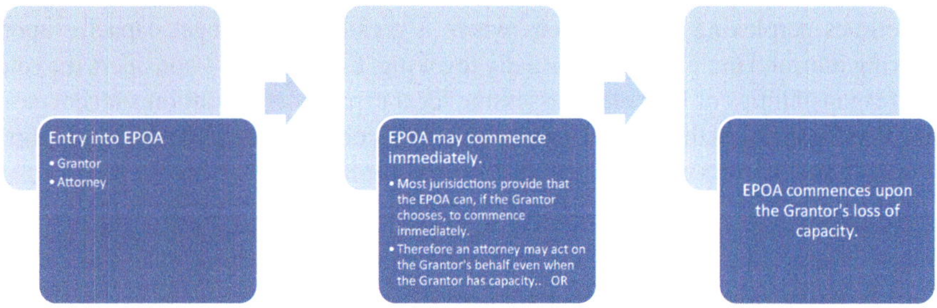

Revoking an EPOA

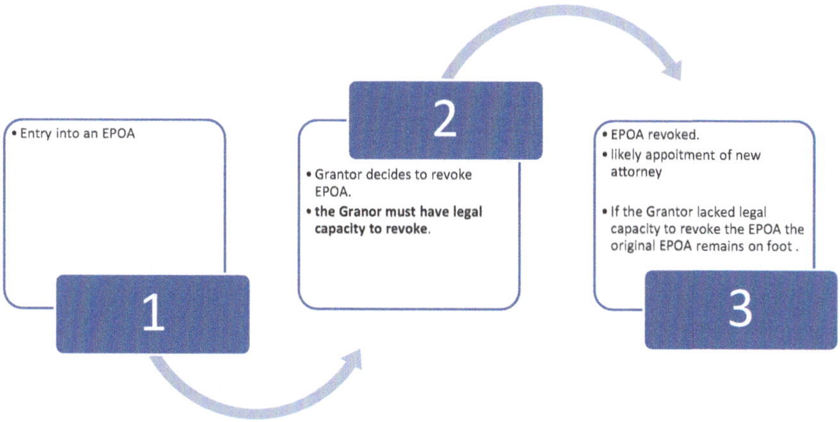

In Australia, EPOAs can be entered into through a lawyer or a designated government body such as, depending on the jurisdiction, the public advocate or the public trustee. Also, EPOAs can be entered into through more informal means such as via downloadable forms or a "do it yourself" kit that can be purchased from a newsagent or post office. In such cases, an EPOA can be witnessed by a range of people who may not necessarily have met the older person before.

The exercise can be problematic in several ways. As will be discussed later in this chapter, a lack of capacity is not always apparent, even to an experienced lawyer (Law Society of South Australia, 2017) although state and territory legal bodies are providing more guidance regarding assessment of capacity (NSW Law Society, 2016). More informal arrangements may pay little heed to capacity, a concern that has seen some recent enhancement to legislation regarding the number, qualification and independence of witnesses in several jurisdictions (Attorney General's Department, 2020, p. 5). Furthermore, it is fair to say that most discussions regarding EPOAs are in relation to their misuse at the hands of a dishonest attorney (Wuth, 2013). While this is certainly a critical issue, particularly in relation to elder financial abuse, this neglects very real concerns regarding the validity of the initial EPOA brought about by issues such as undue influence, mistake and, of course, a lack of capacity.

Revocation is a pertinent issue because a grantor has the right to revoke and enter into a new EPOA at any time while the grantor has capacity. Often, as a person's cognitive function becomes compromised, they may be vulnerable to persuasion of exploitation or, in some cases, develop out of character perceptions of persons who were formerly close to the grantor. Indeed, an indication of encroaching symptoms of dementia can be regular revocation and subsequent entry into EPOAs with other attorneys (*Szozda v Szozda* [36]).

The Incidence of People Experiencing Dementia in Australia

Statistics and Prevalence

"Dementia" describes a collection of symptoms caused by degenerative disorders that gradually cause impairment to the brain (Dementia Australia, 2020; Wojt et al., 2020). The condition may impact upon memory, speech, cognition and understanding, mobility and personality. As time goes on, dementia may cause detriment to a person's mental health and therefore impact upon cognition and decision-making ability (Australian Institute of Health and Welfare, 2019).

Dementia is most commonly experienced by people over the age of 65. In Australia, around one in ten people, over 65 experience dementia (Dementia Australia 2020), but this increases to 25% by the late 80s (Productivity Commission, 2011, 44).

Dementia is the leading cause of death in Australian women and the third leading cause of death of men (Dementia Australia, 2018). It is estimated that around 459,000 Australians are living with dementia, and it is anticipated that this will increase to 590,000 by 2028 and 1,076,000 by 2058 (Dementia Australia, 2018).

International and Domestic Obligations

According to Article 12 of the UN Convention on the Rights of Persons with Disabilities, people with disability "have the right to enjoy legal capacity on an equal basis with others in all aspects of life" (UNCRPD). In ratifying this convention, Australia is deemed to accept "the obligation to recognize that people with disability enjoy legal capacity on an equal basis with others and that appropriate measures should be implemented to provide people with disability access to the support required to exercise legal capacity" (Article 12(4) UNCRPD; ALRC, 2013 [22]).

Australian law, while not specifically adopting the convention within its domestic law, reflects like principles in the common law and, more recently, within relevant state and territory legislation. Indeed, at common law, there is a presumption that all persons have capacity to make decisions (*Borthwick v Carruthers*). There are some recognized exceptions, for example in the case of most, but not all, contracts entered into by minors (*Steinberg v Scala (Leeds) Pty Ltd*) but compare *Chapple v Cooper.* A person is assumed by law to have capacity to make their own decisions unless it can be established that this is not the case (See for example, *Murphy v Doman* [34]–[36]; *Scott v Scott*, [233]; *Coffey v Coffey* (No. 2) [15]–[19]).

Identifying Decision-Making Capacity

In Australia, the population is ageing, and a significant amount of wealth is in the hands of this demographic (Klineberg, 2021). And, as noted above, the incidence of dementia is expected to rise (Dementia Australia, 2018). Therefore, when dealing with an individual's financial and legal affairs, it becomes critical to identify (so far as is possible) the point where an individual's cognitive function "no longer represents competent decision-making" (Moye & Marson, 2007).

This is a difficult task that can confound even experienced clinicians (Gregory, 2007). Assessing decision-making capacity can be arduous because of the manifestation and progression of dementia. The condition is often slow to reveal itself and the individual concerned, and their family may be oblivious to its progress or behaviour can be "explained away" by other reasoning. Furthermore, capacity is a fluid concept. Therefore, at different times, the same person may appear lucid and in charge of their affairs—and in all likelihood will be capable of entering into certain transactions—but can regress and be incapable at a different time (Trachsel et al., 2015). Varying degrees of affliction can be observed at different times of the day and night, for example the concept of "sundowning", where the person becomes disorientated at the end of the day (Sollitto, 2020). Dementia can be impacted upon by factors such as illness, bereavement and stress. Conversely, in some cases capacity may be enhanced through medication or lifestyle changes.

Finally, a diagnosis of dementia may expose an older person to manipulation in relation to their financial and other personal affairs through becoming more dependent on relatives or friends. In these circumstances, the older person may be less likely to question the nature of transactions they enter into at the request of another person.

What is Legal Capacity?

Individuals must make a variety of decisions in relation to, *inter alia,* personal, health legal and financial issues. To do so, an individual must be able to understand the significance—and potential consequences—of a transaction (NSW Health, 2015). This is particularly the case in relation to legal transactions where a higher degree of capacity is required because of the complexity or the significance of the decision. In *Hunter and New England Area Health Service v A*, the court noted that legal capacity is "task and decision specific" stating at [24]:

> It cannot be said that a person completely "lacks capacity" because, even in the direst circumstances, an individual may be capable of making some decisions, for example an individual may have the capacity required to make a contract to buy a cup of coffee but not "the capacity to decide to give away one's fortune" (*Scott v Scott* [240]).

In relation to individuals with dementia, capacity is assessed according to the nature of the particular decision, whether straightforward or situated at a particular point on an increasing scale of complexity. Therefore, particularly in the earlier stages of a dementia diagnosis, a person will be able to continue to make their own decisions. Indeed, an early diagnosis is often the catalyst for seeking advice in relation to the perpetration of an EPOA, and arrangements can be made before the disease progresses too far. In circumstances of a more severe onset, or a later diagnosis, capacity in relation to each decision should be carefully assessed. Obviously, the more complex the proposed transaction, the higher standard of cognitive functioning would need to be demonstrated (Lewis et al., 2020a, b). Capacity in each instance must be assessed and not extrapolated from one decision to another: (O'Neill & Peisah, 2018, [10.3.1]).

Making or Revoking an EPOA—How Do We Assess Legal Capacity?

Commentators have often debated the character of an EPOA in relation to the requisite standard of capacity required and drawn comparisons with other interests such as a contract, a trust and a will (Dal Pont, 2019). The comparison to a contract is logical in that an EPOA demonstrates characteristics of other forms of contractual agency (*Rawlings v John Handcock Mutual Like Insurance Co*, 297). However, like wills and some trusts, EPOAs are voluntary transactions that do not require consideration (Dal Pont, 2019). Decided Australian cases seem to have reached a conclusion that an EPOA requires a high degree of capacity, equal to or even exceeding, on some views, than that required to make a will (*Szozda v Szozda*; Carney et al., 2019).

Identifying the Standard—The Common Law and Statute

Under the Australian Constitution, powers are divided between the commonwealth and state/territory governments. Jurisdiction in relation to EPOAs is the responsibility of each Australian state or territory. As noted in the Diagram 14.1, some jurisdictions utilize common law authorities while others have adopted discrete tests to determine capacity.

All jurisdictions share the same starting point, that is, the presumption of capacity (South Australian Law Reform Institute, 2020, 4.4.83). However, the differences appear through the jurisdictions that adhere to the common law position while other jurisdictions address the elements of capacity, specifically in the relevant statute. Nevertheless, even those jurisdictions with specific provisions tend to look at common law authorities which is appropriate to interpret the statutory provisions. The onus of proving incapacity is on the person alleging that the grantor was incapable at the time of entering into or revoking the EPOA, a position consistent with the presumption: (*Re Caldwell; Re MD*). Table 14.1 summarizes the various state and territory responses.

Diagram 14.1 Sources of law for determining capacity to enter into an EPOA

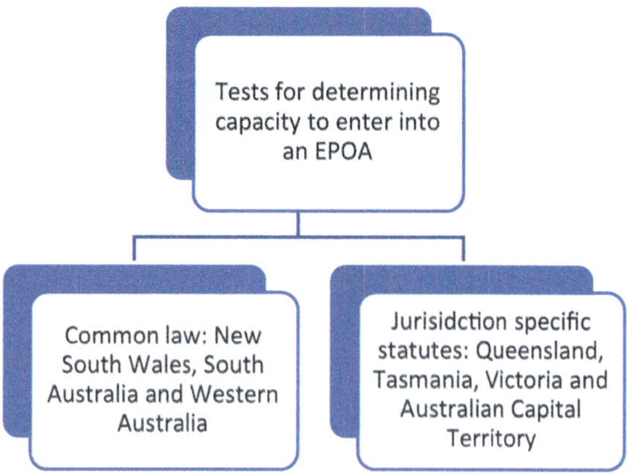

Table 14.1 Tests applicable to determining capacity to enter or revoke and EPOA in each Australian state or territory

Jurisdiction	Legislation	Applicable test
Queensland	*Powers of Attorney Act 1998*	Understanding the nature and the effect of the power
New South Wales	*Powers of Attorney Act 2003*	Common law
Australian Capital Territory	*Powers of Attorney Act* 2006	Presumption of understanding
Victoria	*Powers of Attorney Act 2014*	The statutory test reflects the common law
Tasmania	*Powers of Attorney Act 2000*	Understanding the nature and the effect of the power
Northern Territory	*Advance Personal Planning Act*	"Planning capacity" Decision-making capacity"
South Australia	*Powers of Attorney and Agency Act 1984*	Common law
Western Australia	*Guardianship and Administration Act* 1990	Common law

The Common Law Test

The central question in assessing the validity of an EPOA is to determine whether the grantor had the requisite capacity at the time of entering into the instrument. A pivotal case in relation to assessment of legal capacity at common law is the Australian High Court decision in *Gibbons v Wright.* Although it should be noted that the case itself did not involve an EPOA, the case is the authority for the approach taken by Australian courts regarding assessment of legal capacity generally. The decision has been discussed and applied in the context of entering into and revoking EPOAs. At pages 437–438, the High Court noted:

> The law does not prescribe any fixed standard of sanity as requisite for the validity of all transactions. It requires, in relation to each particular matter or piece of business transacted, that each party shall have such soundness of mind as to be capable of understanding the general nature of what he is doing by his participation.

Therefore, a person experiencing dementia can enter into a valid EPOA so long as the person had the "*soundness of mind to understand the general nature of the agreement*", in this case an EPOA Gibbons v Wright, 437; *Banks v Goodfellow*).

The High Court considered what an assessment of that "soundness" may entail and the standard of mental acuity that must be demonstrated when entering into or revoking an EPOA. The court noted:

> …the mental capacity required by the law in respect of any instrument is relative to the particular transaction, which is being effected by means of the instrument, and may be described as the capacity to understand the nature of what is being explained. (at 438).

In summary, the High Court was of the view that,

> to establish capacity, the person concerned must be provided with relevant information and explanation pertaining to the transaction and must understand the nature and the consequences of the particular transaction: (O'Neill & Peisah, 2018 10.3)

English and Australian courts have considered the degree of capacity necessary for a person to enter into a valid EPOA. In *Re K (Enduring Powers of Attorney),* Hoffman J noted that the grantor must understand the nature of an EPOA and the delegation of responsibility involved, but that the Grantor did not have to be independently capable of managing their financial or legal affairs. At [316], His Honour listed four factors to consider when assessing legal capacity in these circumstances. First, the grantor must appreciate that the EPOA results in the attorney obtaining complete authority over their affairs; second, that the attorney can do anything that the grantor can do themselves with their property; third, that the attorney's role continues after the grantor becomes mentally incapable; and fourth, that if grantor becomes mentally incapable, the power will be irrevocable without confirmation by the court.

In comparison, the decision in *Ranclaud v Cabban* seems to require a higher degree of understanding on the part of the grantor. Noting the criteria in *Gibbons v Wright*, Young J of the New South Wales Supreme Court concluded that a grantor had to demonstrate an appreciation of the nature and scope of the rights being bestowed upon the attorney. In His Honour's view, it was necessary to understand not just that the attorney was authorized to look after the grantor's affairs, but appreciate the tasks the attorney could do without further reference to the grantor. In this case, the grantor had entered into several powers of attorney benefitting two individual attorneys within a nine-month period. Young J considered that this inconstancy demonstrated a lack of the necessary mental capacity. Indeed, it was stated at 57,548 that:

> The plaintiff is living in a supportive environment in that at the nursing home she has a lot of tasks done for her without which she would not be able to cope. In such an environment the plaintiff would be far more likely to be able to manage her personal affairs than elsewhere. However, even taking those matters into account she is still unable to make decisions and to instruct people as to the details of what she wants. She is unable to be consistent about those matters. She has poor memory and assumes that persons now dead are alive which affects her judgment. I take into account that if one spent an hour or so without someone experienced in dealing with geriatrics it may be possible to ascertain what the plaintiff's wish is. I am fairly sure that this is so on very basic levels. I am not at all sure, however, that she would be able to make a more complicated decision involving a greater evaluation of the rights and wrongs of alternatives.

His Honour noted the scope of the broad scope of an EPOA noting:

> Such a power permits the donee (attorney) to exercise any function which the donor may lawfully authorise an attorney to do. When considering whether a person is capable of giving that sort of power one would have to be sure not only that she understood that she was authorising someone to look after her affairs but also what sort of things the attorney could do without further reference to her.

Recent Australian Cases

Both these decisions and *Gibbons v Wright* have been referred to in a series of recent Australian cases addressing the validity of an EPOA. The facts of *Ghosn v Principle*

Focus (No 2) involved an elderly man who entered into an EPOA upon returning to his native Lebanon. He purported to appoint his nephew, Ghosn, as attorney. Ghosn travelled to Australia and purported to transfer to himself all the shares in his uncle's companies and sold properties for less than their market value. At issue was whether the elderly man possessed the requisite legal capacity to appreciate the potential consequences of empowering his nephew in this way. Forrest J held that Mr. Ghosn did not have the requisite capacity noting at [101]:

> I need to be satisfied that when Mr Moussi signed the Powers of Attorney, he understood not only the nature of the instrument but also the practical consequences in a general sense. He need to understand that Mr Abi Ghosn had an enduring power to transfer the shares and control of the trustee companies to others including himself. He had to understand that the assets held by those companies could be transferred by Mr Abi Ghosn to anyone including Mr Ghosn himself at any price he saw fit.

In *Szozda v Szozda*, a family had substantial assets through a family property development business. The business was conducted through a complex structure of companies and trusts. The matriarch of the family, who at the relevant times was well into her 90s and experiencing dementia, continued to have a controlling stake in the business. Tensions had arisen in the family, and several family members had fallen out. Over a period of 3 years, the matriarch purported to grant and revoke a series of EPOAs to members of both sides of the family. Through the ever-changing EPOAs, a variety of simple and complex transactions were entered into. Of particular importance was whether a proxy given to a third person—pursuant to the matriarch's EPOA—to cast a vote on behalf of the matriarch at a meeting of a board of directors of the family company was effective.

It was evident that the matriarch's mental health had declined in the preceding few years and lacked legal capacity at the time of the purported revocations and new appointments. It was held that the proxies purportedly appointed under it, and the resolutions purportedly passed under those proxies were all void. Interestingly, the only transaction disputed was the proxy issue although, should they have decided to do so, all the other transactions made under the impugned EPOAs would be similarly *void* ab initio. The status of an EPOA was noted by Barrett J to affect the grantor, the attorney and all those who have upon its purported exercise.

Barrett J made an assessment of the standard of legal capacity required to enter into (or revoke) an EPOA. His Honour noted at [34] that the grantor should ask two pivotal questions:

> First, is it to my benefit and in my interests to allow another person to have control over the whole of my affairs so that they can act in those affairs in any way in which I could myself act - but with no duty to seek my permission in advance or to tell me after the event, so that they can, if they so decide, do things in my affairs that I would myself wish to do…
>
> Second, Is it to my benefit and in my interests that all these things – indeed everything I can myself lawfully do - can be done by the particular person who is to be my attorney?

In His Honour's view, "on the scale of complexity of decision-making", the creation of an EPOA stood at a very high point in that it "strikes at every conceivable aspect of the donor's property and affairs". (Per Justice Barrett (2012) at [33];

O'Neill and Peisah 2018) It was only if the grantor has some "reasonably informed appreciation of what the property and affairs amount to and what the appointee will be able to do can the decision be said to be a properly informed decision" (Per Justice Barrett (2012) [33]).

Later, in *Scott v Scott,* Lindsay J was of the view that entering into or revoking an EPOA required a high degree of understanding, a standard akin to that required when entering into a will. At [200], His Honour noted that, in such circumstances, the court could consider:

> An exploration of all the circumstances of the case will, not uncommonly, call for consideration of events leading up to, and beyond, the time of execution of the disputed power of attorney, as well as on the focal point of the time of execution itself. A longitudinal assessment of mental capacity, along a time line extending either side of the focal point, may be necessary, or at least permissible, in order to examine the subject's mental capacity in context. Medicos and lawyers, alike, tend to embrace that approach. It is difficult to do otherwise. Context has a temporal as well as spatial and relational dimensions…Where an Enduring Power of Attorney confers on an attorney power to dispose of the principal's property to or for the benefit of the attorney or third parties, the nature and degree of mental capacity required to grant such a power may approximate that required for the making of a valid will.

In *Antov v Bokan*, an elderly woman experiencing dementia passed away in 2014. She had left a will by which she appointed one of her daughters, Antov, as executor and made Bokan the sole beneficiary. Subsequently, Antov claimed she was entitled to a property owned by the deceased because of, *inter alia,* the content of a power of attorney and "Contract for a Gift" allegedly executed by the deceased in 2009. It was noted at [523] that, at the time of entering into those instruments:

> …Ljubica could not see, was immobile and was suffering from dementia such that it could not be said she had capacity to understand the import of a document such as the Macedonian Power of Attorney. It is noted that there is no suggestion that a draft of any such document was prepared and read to Ljubica at Radmila's unit and prior to its execution; and that, insofar as there is any evidence that suggests the document was read to Ljubica (in particular, referring ot the evidence of Mr Danilov), there is no evidence of the kind of questioning of Ljubica that might suggest there was any comprehension by her of what was happening.

The medical evidence as a whole supported the conclusion that, at the time of entering into the transactions, the older woman's capacity was such that she could not have understood the nature, implications and far-reaching ramifications of the power of attorney and what it authorized the attorney to do. Furthermore, there is no evidence that shows that the older woman was properly informed of the power of attorney's significance and consequences. Indeed, the court referred at [119] to the Szozda decision in holding that "she would not have understood them even if they had been spelled out to her by a lawyer in the most comprehensive and punctilious way".

Therefore, the instruments were vitiated by incapacity and each was void and of no effect.

Comments

It can be seen from these cases that the Australian courts require a high degree of understanding by the prospective grantor regarding the nature and effect of entering into or revoking an EPOA. It seems the standard is higher in Australia than in the UK and requires that the prospective grantor understands that the grant bestows extremely wide powers on the attorney for an indefinite period. Therefore, a grantor cannot simply appreciate that the attorney enters into transaction on the grantors behalf. The grantor is required to cognizance of discrete categories of transactions the attorney has authority to pursue. This is a significant standard, akin to, or in some cases regarded as greater than the capacity required to enter into a will (*Szozda v Szozda cf Scott v Scott).* Therefore, it seems that the issue of lack of capacity upon entering into or revoking an EPOA is a "sleeping giant". It has to be pondered how many EPOAs are being acted upon throughout Australia where the grantor did not demonstrate the understanding required in decided cases.

The Obligations of Lawyers

Lawyers are subject to ethical and professional duties to protect the interests of their client. Rule 8 of the *Legal Profession Uniform Law Australian Solicitors' Conduct Rules 2015* states that lawyers must follow a client's "lawful, proper and competent instructions". Because of specific references to competency, a lawyer must pay heed to considerations of capacity. Relying on the presumption of mental competency is insufficient if there are warning signs that capacity may be lacking. Therefore, a lawyer should take care when assessing whether a client's capacity to enter into a particular type of transaction before taking instructions or proceeding with the advice (McCullagh, 2018, 11.230–11.280). Law societies and disciplinary bodies throughout Australia are acutely aware of pitfalls regarding legal capacity when entering into or revoking EPOAs and have taken steps to enhance training and awareness of lawyers through continuing legal education, workshops and regular bulletins. This is especially so when assessing the demeanour of the older person and the circumstances through which they attend the lawyer (Purser, 2017; Lewis et al., 2020b).

How Can Lawyers Assess Capacity?

Assessing legal capacity is an inexact process, and, if there is any doubt, the lawyer should refer the client to a medical professional. Generally, the lawyer should consider factors in three stages. First, the lawyer should consider the demeanour of the older person; second, the circumstances under which the prospective grantor is seeing the lawyer; and third, assess the true extent of the older persons understanding of the nature and effect of the EPOA (NSW Law Society, 2018).

In relation to the demeanour of the older person, lawyers should be mindful of "warning signs" that could indicate a lack of legal capacity. The New South Wales Law Society Guidelines suggest that such signs include difficulty with recall or have memory loss; ongoing difficulty with communications or is disoriented (NSW Law Society, 2018). This can be difficult where a lawyer is seeing a client for the first time. Indeed, another warning sign can be the older person simply "showing up, usually with a relative or associate seeking an EPOA. Alternatively, where a client is known to the lawyer, the lawyer should be mindful of changes in the client's attitude or demeanour including deterioration in personal presentation, mood or social withdrawal (NSW Law Society, McCullagh, 2018 [11.206]). Several excellent checklists have been developed by law societies in most Australian jurisdictions, and it is important that lawyers make careful use of the considerations therein (Queensland Law Society, 2014). It goes without saying that if the lawyer is in any doubt, a medical assessment should be recommended before proceeding further.

Meeting with a Lawyer

The circumstances of entering into or revoking an EPOA should be reviewed when, for example, an older person is receiving medical treatment, is in hospital or has entered an aged care facility.

When an older person attends a lawyer's office, it is essential that the lawyer sees the client alone, in the absence of anyone who may benefit through the EPOA, including the attorney. The lawyer should also be alert to any indications of undue influence or the older person looking uncomfortable in the presence of the persons accompanying them (Reis, 2019).

The discussion of the cases above demonstrates that the older person should appreciate the consequences of entering into an EPOA. Again, the lawyer should carefully ascertain the level of understanding. In straightforward cases, this may not be difficult although in the case of more complex estates, a prospective grantor would, it seems, need to demonstrate a higher degree of comprehension (Lewis et al., 2020a, b).

The lawyer should also ensure they take detailed notes. If an issue arises and questions are raised as to the capacity of the older person at the time the EPOA was entered into or revoked, such notes will be an invaluable evidentiary resource. It will also ensure that the lawyer's professional duties are upheld.

The lawyer should ensure that the meeting is recorded (if permission is granted), a scribe is in attendance (again with permission) or that the lawyer takes extensive and detailed file notes. If in any doubt regarding capacity, the lawyer should seek permission for a medical assessment. It is also a good practice to take instructions, take a few days to prepare the EPOA and then revisit issues raised in the first meeting with the older person when they attend to execute the document. This can allow the lawyer to assess the demeanour of the older person on another day and time and ask essential questions and confirm details.

Conclusions

The title of this chapter refers to the "legal tightrope" that must be negotiated when a person experiencing cognitive decline is entering into or revoking an EPOA. The "balance" that must be sought is between the utility of entering into a practical and effective substitute decision-making instrument to manage the grantor's legal and financial affairs with the risk of misuse inherent in embarking on such agreements. Those risks include the often-discussed possibility of wayward attorneys using fraud and undue influence to exploit the grantor's assets and the subject of this chapter, a lack of capacity by the grantor at the time the EPOA is entered into or revoked. Of course, both instances are not mutually exclusive. For example, an attorney may be aware that the grantor is struggling with cognition through advanced dementia but, for the attorney's personal gain, persuades the grantor to enter into or revoke an EPOA (*Anton v Bokan).*

The crucial difference is that, if the EPOA was originally valid, that is, that the grantor had capacity at the time the EPOA was entered into, the transactions entered into by the attorney will be regarded as valid. Although there may be attempts to "claw back" some funds through personal actions against the attorney, for the most part money and property distributed under the EPOA will stand. However, when the grantor did not have capacity at the time the EPOA was entered into or revoked and a new EPOA entered into, the instrument is of no legal effect. As drastic as it sounds, a purported appointment of an EPOA by an incapable principal is void ab initio—it never got off the ground. As noted above, in *Gibbons v Wright,* the High Court contrasted an EPOA with a contract, which is voidable by the incapable person if it can be shown that the other party ought reasonably have been aware of the incapacity. In contrast, an EPOA is an instrument that does not affect title but confers authority from the grantor to the attorney. Therefore, incapacity will render the appointment of an attorney void. As the purported transfer of authority to the attorney is void from the outset, all transactions entered into on behalf of the grantor by the attorney are ineffective and, it would seem, could be challenged. In *Szodza v* Szozda, the question before the court was limited to the proxy issues. Although it was acknowledged that many other transactions had occurred too and that they would have similarly been void, the prospect of "unpicking" a myriad of complex transactions seems to have been a bridge too far. Nevertheless, it would seem only a matter of time that the estate of a person who lacked legal capacity when entering into or revoking an EPOA seeks to establish the lack of capacity as evidence that transactions undertaken by the attorney were done so without valid authority.

Recommendations

This chapter has considered circumstances where a grantor of an EPOA lacks legal capacity at the time that person entered into, or revoked, the instrument. An ageing population, and the encouragement of individuals (and their families) to plan for the future by putting in place measures enabling supported and substituted decision-making will see an increase in the number of EPOAs. Most of these instruments will be prepared and witnessed without the involvement of a legal professional. This is a significant gap regarding the validity of an essential legal instrument with complex and costly ramification likely to follow in the event of a finding of incapacity

Therefore, how should we approach this quandary? This chapter will conclude with six recommendations that could assist in addressing this issue.

1. **Enhance community education regarding EPOAs and encourage people to enter into such instrument before experiencing cognitive decline.**

It is essential that people start to consider substitute decision-making at an earlier stage in their lives and not as a "last resort" when there is, for example, a diagnosis of dementia. It is trite to state that persons can lose capacity at any age, but it seems that many people tend to "put off" entry into an EPOA until a crisis occurs or a grave medical diagnosis is received. Enhanced community education could see people better understand the nature and effect of an EPOA—factors that must be established when determining if there was legal capacity to enter into or revoke an EPOA. It could also ensure that a trusted, appropriate individual or individuals are nominated as the attorney rather than a relative or "new friend" when negotiating illness or a life crisis (Purser, 2017).

Therefore, a person would better appreciate the effect and potential ramifications of the transaction—that it involves authorization of the attorney to take responsibility for the older person's legal and financial affairs and that the attorney can act without further recourse to the grantor. (Zii v ZIJ)

2. **Enhance education programmes for Justices of the Peace and other professionals who may witness an EPOA.**

A potential in several Australian jurisdictions is the ease with which an EPOA can be entered into or revoked privately. In other words, although it is desirable to seek a lawyer's advice, people often do not do this as they see lawyers as complicating the situation and being too costly. Indeed, on the one hand, older people are encouraged to enter into "do it yourself" or online EPOAs. Government departments encourage the uptake of people entering into informal EPOAs, thus avoiding the need for input from organizations such as the public trustee in the event of incapacity. However, the lack of legal oversight can result in misunderstanding of the nature and effect of the instrument, the powers granted to the attorney and gives rise to the possibility of undue influence and fraud.

At present, Justices of the Peace are commonly asked to witness EPOAs, and, in most cases, it will be the first time the Justice of the Peace has met the grantor. Justices of the Peace take a short course, but there is little in the way of dedicated training in relation to capacity. There have been recommendations for greater responsibilities to be placed on Justices of the Peace, for example being attuned to suspected elder abuse and the obligation to refuse to proceed where elder abuse or a lack of capacity is suspected. Other suggestions include imposing penalties on Justices of the Peace who are negligent, and that legislation should provide for an opportunity for the older person to receive compensation where a Justice of the Peace has acted negligently or fraudulently. (Blundell et al., 2017) Other professionals who are empowered to witness documents do not receive training at all. Given the crucial nature of EPOAs and the ramifications in the event of incapacity or abuse, a higher standard of training is essential where a lawyer is not involved in witnessing the EPOA.

3. **Lawyers—or witnesses with thorough professional training—should have sole responsibility for the carriage and witnessing of EPOAs.**

Although this suggestion will, no doubt, be unpopular due to the perception of the cost and inconvenience, the high incidence of undue influence, fraud and other conduct resulting in elder financial losses, the extent of abuse and the increasing number of older people who are likely to experience dementia would make such a safeguard desirable.

4. **Ensure that lawyers are aware of potential incapacity.**

Although lawyers have become more aware of the possibility of financial abuse through misuse of EPOAs, there has been less focus on the mental faculties of the grantor at the time of giving instructions and entering into the EPOA.

As noted above, the various law societies throughout the states and territories are mindful of the pitfalls of misuse of EPOAs and have also awakened to the concerns regarding incapacity at the time the EPOA was entered into or revoked. Several jurisdictions have produced comprehensive continuing education materials and checklists that can be utilized by lawyers to ensure that the prospective grantor has the legal capacity to understand the nature and effect of the EPOA. If in doubt, the lawyer should refer the older person for an assessment by a medical professional.

In this respect, there should be more emphasis on lawyers' professional conduct duties. For too long, many lawyers have tended to rush transactions involving EPOAs and have not been as thorough as they should with regard to the possibility of cognitive decline and potential elder abuse. Disciplinary proceedings against lawyers who do not fulfil their duties of care in relation to the entry into or revocation of EPOAs are becoming more commonplace, and it is essential that higher standards are adopted throughout the legal industry in this respect.

5. **Consider whether capacity assessments should be an essential step prior to entry into or revocation of an EPOA.**

Again, this suggestion may be unpopular due to cost, inconvenience and, on some views, ageism. It is not suggested for a moment that, because a person is ageing, that the person automatically lacks capacity. Indeed, the presumption in the law is the opposite. However, the consequences of entering to an EPOA without understanding the consequences, or when under a cognitive disadvantage, may serve to provide greater protection for many older people. It may also serve to protect them from potential abuses and the consequences to third parties who enter into transactions under the EPOA that are subsequently revealed to have been invalid.

6. **Consider the ramifications of the proposed national EPOA register.**

In 2017, the Australian Law Reform Commission's report released its report, Elder Abuse—A National Legal Response (ALRC, 2017). The ALRC inquiry provided a comprehensive examination of the laws relevant to elder abuse, and the ensuing report made several recommendations including the introduction of a mandatory national register of EPOA's.

This recommendation was made because of the concerning levels of elder abuse and fraud associated with EPOAs. Under the National Plan to address Elder Abuse 2019–2023 (COAG, 2019), the Council of Australian Governments has advanced a proposal for a national register with a view to introduction in 2022 (Jacobsen, 2020). Nevertheless, this prospect has been the subject of some reservations because of concerns regarding formality, cost and inconvenience that may deter grantors from registering EPOAs (Law Council of Australia 2020). This formality and the probability of legal oversight may assist in diluting the instances of elder abuse and the unexpected consequences of a void EPOA where a person lacked legal capacity upon execution.

Cases

Antov v Bokan [2018] NSWSC 1474.

Banks v Goodfellow (1870) LR 5 QB 549.

Chapple v Cooper (1844) 13M&W 252; 153 ER 105.

Coffey v Coffey (No. 2) [2015] NSWSC 338.

Gibbons v Wright (1954) 91 CLR 423, 437–438 (Dixon CJ, Kitto and Taylor JJ):

Ghosn v Principle Focus (No 2) [2008] VSC 774.

Hoffman v Waters [2007] SASC 273.

Hunter and New England Area health Service v A [2009] NSWSC 761.

Murphy v Doman (2003) 58 NSWLR 51.

Scott v Scott [2012] NSWSC 1541,

Steinberg v Scala (Leeds) Pty Ltd [1923] 2 Ch 452.

Ranclaud v Cabban (1988) NSW ConvR 55–385.

Rawlings v John Handcock Mutual Like Insurance Co (2001) 78 SW 3d 291.

Re Caldwell [1999] QSC 182.

Re D(J) [1982] Ch 237.

Re K (Enduring Powers of Attorney) [1988] Ch 310.

Re MD [2005] QGAAT 29.

Scott v Scott [2012] NSWSC 1541.

Szoda v Szoda [2010] NSWSC 804.

Zii v ZIJ (2018) NSWCATAP 255.

References

Attorney-General's Department. (2020). Enhancing protections relating to the use of enduring power of attorney instruments: Consultation regulation impact statement. https://www.ag.gov.au/Consultations/Documents/consultation-regulation-impact-statement/enhancing-protections-relating-use-enduring-power-of-attorney-instruments.pdf

Australian Institute of Health and Welfare. (2019). Dementia overview 2019 Australian Government Canberra. https://www.aihw.gov.au/reports-data/health-conditions-disability-deaths/dementia/overview

Australian Law Reform Commission (ALRC). (2013). Equality, capacity and disability in commonwealth laws IP 44 2013 Australian Government Canberra. https://www.alrc.gov.au/publication/equality-capacity-and-disability-in-commonwealth-laws-ip-44/equality-capacity-and-disability-in-commonwealth-laws/legislative-and-regulatory-framework/#:~:text=By%20ratifying%20the%20CRPD%2C%20Australia,in%20exercising%20their%20legal%20capacity

Australian Law Reform Commission (ALRC). (2017). Elder abuse—A national legal response, ALRC report 131, Australian Government Canberra. https://www.alrc.gov.au/wp-content/uploads/2019/08/elder_abuse_131_final_report_31_may_2017.pdf

Bagshaw, D. (2013). Financial abuse of older people by family members: Views and experiences of older Australians and their family members. *Australian Social Work, 66*(1), 86.

Barrett, J. Law society of New South Wales elder law and succession committee "listen to the judges" series 11 September 2012 Justice R I Barrett.

Blundell, B., Webb, E., Clare, J., Moir, E., & Clare, M. (2017). *Review into the prevalence and characteristics of elder abuse in Queensland*, Curtin University, Australia.

Britt, S. L. (2016). The intergenerational transference of money attitudes and behaviors. *Journal of Consumer Affairs, 50*(3), 539.

Carney, T., Then, S. N., Bigby, C., Wiesel, I., Douglas, J., & Smith, E. (2019). Realising 'will, preferences and rights': Reconciling differences on best practice support for decision-making? *Griffith Law Review, 28*(4), 357.

Council of Attorneys General (COAG). (2019). National plan to respond to the abuse of older Australians (Elder Abuse) 2019–2023, Canberra. https://www.ag.gov.au/RightsAndProtections/protecting-the-rights-of-older-australians/Documents/National-plan-to-respond-to-the-abuse-of-older-australians-elder.pdf

Dal Pont, G. (2019). *Powers of attorney*, 3rd edn. LexisNexis, Sydney.

Dementia Australia. (2018). Dementia prevalence data 2018–2058, commissioned research undertaken by NATSEM, University of Canberra.

Dementia Australia. (2019). Dementia statistics. https://www.dementia.org.au/about-dementia/what-is-dementia

Fougere, C. Guardianship, financial management and the NDIS: NCAT's experience. http://www.ncat.nsw.gov.au/Documents/speeches_and_presentations/20170323_paper_fougere_agac_hobart.pdf

Gregory, R., Roked, F., Jones, L., & Patel, A. (2007). Is the degree of cognitive impairment in patients with Alzheimer's disease related to their capacity to appoint an enduring power of attorney? *Age and Ageing, 36*(5), 527.
Jacobsen, D. (2020) Enduring power of attorney national register options bright law Sydney. https://www.brightlaw.com.au/enduring-power-of-attorney-national-register-options/
Klineberg, A. (2021). Superannuation and economic security for older Australians: A human rights perspective. *Australian Journal of Human Rights, 1*.
Law Society of South Australia. (2017). Client capacity guidelines (Law Society Document, 6 November 2017) https://www.lawsocietysa.asn.au/Public/Publications/Guidelines/ClientCapacityGuidelines.aspx
Lewis, B., Purser, K., & Mackie, K. (2020a). A human rights-based approach to elder law. In *The human rights of older persons* (pp. 67–81). Singapore: Springer.
Lewis, B., Purser, K., & Mackie, K. (2020b). Legal capacity and decision-making. In *The human rights of older persons* (pp. 139–173). Singapore: Springer.
McCullagh, R. (2018). *Australian elder law—Accommodation, agency and remedies*. Thomson Reuters, Australia.
Moye, J., & Marson, D. (2007). Assessment of decision-making capacity in older adults: An emerging area of practice and research. *The Journals of Gerontology: Series B, 62*(1), 3. https://doi.org/10.1093/geronb/62.1.P3
NSW Health. (2015). Exploitation of people with disabilities—Capacity & dementia a guide for NSW health care, ageing and community workers. Mini-legal kit Series 1.3.
NSW Law Society. (2016). When a client's mental capacity is in doubt: a practical guide for solicitor. https://www.lawsociety.com.au/sites/default/files/2018-03/Clients%20mental%20capacity.pdf; www.lawsociety.com.au/practising-law-in-NSW/ethics-and-compliance/ethics/committee-guidance
NSW Law Society. (2018). Guidelines for solicitors preparing and enduring power of attorney Sydney. https://www.lawsociety.com.au/sites/default/files/2018-04/Enduring%20POA.pdf
O'Neill, N., & Peisah, C. (2018). *Capacity and the Law Sydney University Press*. http://www.austlii.edu.au/au/journals/SydUPLawBk/2011/1.html
Productivity Commission. (2011). Caring for Older Australians (AG Canberra 2011 Ch 3 44).
Purser, K. (2017). Substitute decision-making. In *Capacity assessment and the Law* (pp. 129–148). Springer, Cham.
Queensland Law Society, Allens Linklaters and Queensland Advocacy Incorporated. (2014). Queensland handbook for practitioners on legal capacity. https://www.qls.com.au/Knowledge_centre/Ethics/Resources/Client_instructions_and_capacity/Queensland_Handbook_for_Practitioners_on_Legal_Capacity
Ries, N. M. (2018). When powers of attorney go wrong: Preventing financial abuse 148 Precedent 9.
Ries, N. M. (2019). Enduring powers of attorney and financial exploitation of older people: A conceptual analysis and strategies for prevention. *Journal of Aging & Social Policy*, 1.
Schyvens, M. (2017). The Australian guardianship tribunal system: Lessons to share with Canada (2017 Canadian Elder Law Conference Vancouver, 2 November 2017).
South Australian Law Reform Institute (SALRI). (2020). *Valuable instrument or the single most abused legal document in our judicial system? A Review of the role and operation of enduring powers of attorney in South Australia*. University of Adelaide.
Sollitto, M. (2020). Understanding and minimizing sundowners syndrome, 2020 ageing care. https://www.agingcare.com/articles/sundowners-syndrome-133187.htm
Somes, T., & Webb, E. (2018). Financial abuse. In S. Field (Ed.), *Elder law in Australia* (p. 225). Federation Press.
Van Wijngaarden, E., Van der Wedden, H., Henning, Z., Komen, R., & The, A. M. (2018). Entangled in uncertainty: The experience of living with dementia from the perspective of family caregivers. *PLoS ONE, 13*(6), 1.

Wojt, I., Cairns, R., & Tan, E. (2020). Poisoning exposures in people with dementia (2014–2019): A retrospective study. *Journal of the American Medical Directors Association*
Wuth, N. (2013). Enduring powers of attorney: With limited remedies—It's time to face the facts! *Elder Law Review, 7*, 1.

Chapter 15
Cognitive Impairment and Dementia: A Risk Factor for Elder Abuse and Neglect

Sara Alon

Abstract Elder abuse and neglect are social and health problem. The WHO (Missing voices, views of older persons on elder abuse, 2002; Elder abuse, 2018) defined abuse as a single or repeated act or lack of appropriate action, occurring within trust relationship, which causes harm or distress to an older person. Elder abuse can manifest itself in physical harm; sexual abuse; emotional or psychological abuse; financial exploitation; and neglect—failure to provide basic needs either by commission or omission. Elder abuse may result in physical disability, psychological distress and increased risks of morbidity and mortality. Cognitive impairment and dementia are ranked as the twentieth-century leading disease. Dementia, dependency and stress experienced by the caregiver were found as predictors to the occurrence of elder abuse (Borenstein in Am Psychol 74:719, 2019; DeLiema et al in J Gerontol Soc Sci 73(5):e49–e58, 2018). The present article will focus on the relationship between cognitive impairment and dementia. A definition of elder abuse and neglect will be presented followed by description of the various dimensions of the phenomenon and its consequences. Due to the fact that abuse and neglect are under identified and reported, it is essential to discuss detection and methods of intervention in this field.

Keywords Cognitive impairment · Dementia · Risk factor · Elder abuse and neglect · Intervention

> Rina is 79 years old. She lives with her husband, who is 85 years old. Rina suffers from dementia. She is entitled by law to nursing care at home. Rina has occasionally been found by neighbors wandering the streets, unsuitably dressed for the weather. At meetings with the husband, the social worker explained to him about dementia. She also recommended hiring a home career and bringing Rina to a day care center suited to her condition. Over time, Rina's cognitive condition and health deteriorated. It seemed that her husband did not really understand the severity of her condition, and did not provide her with the necessary care. At a home visit, the social worker saw Rina sitting alone. She sat there, staring, and had difficulty understanding what was being said. Her husband says he does not need any

S. Alon (✉)
The Bob Shapell School of Social Work, Tel Aviv University, Tel Aviv-Yafo, Israel

Present Address:
16 Rimalt St, Ramat Gan, Israel

M. K. Shankardass (ed.), *Dementia Care*,
https://doi.org/10.1007/978-981-16-3864-0_15

help. It is his duty and obligation to care for his wife. When asked how he is getting along with shopping and errands, he said that twice a week he locks the door and goes out.

Is Rina suffering from inadequate treatment? Is she neglected?

What is Elder Abuse and Neglect?

Elder abuse is defined as a single or repeated act or lack of appropriate action, occurring within a trust relationship, which causes harm or distress to an older person. Elder abuse can take various forms such as:

Physical abuse is the infliction of pain and injury, the inappropriate use of drugs, restraint; **sexual abuse** is any sexual behaviour directed towards an older adult without his/her consent; **emotional** or **psychological abuse** is the infliction of mental anguish by intimidation, humiliation, threats and terrorizing; **financial exploitation** is the illegal or improper use of funds or resources, deliberate undue influence, misuse of an older adult's finances; and **violation of rights** is deprivation of liberty, the right to make choices, social exclusion (WHO, 2002).

Neglect is the failure to provide basic needs either by *commission* (denial of food, denial of medical treatment or therapeutic services), or by *omission* (inadequate treatment, treatment which is not in accordance with the person's needs) (Storey & Perka, 2018; WHO, 2018).

Elder abuse and neglect are either by commission or omission, result in negative health outcomes including physical symptoms such as high blood pressure, heart problems, chronic pain, morbidity and mortality (Fisher & Regan, 2006; Dong et al., 2011; Schofield et al., 2013). Elder abuse can result in psychological distress including PTSD symptoms, anxiety, depression and somatic complaints (Cloitre et al., 2009; Ramsey-Klawsnik, 2017).

As illustrated above, Rina is physically and mentally frail. She is unable to take care of herself; her husband, who is the primary caregiver, is not sufficiently aware of her changing needs. Rina's needs are not being met. She suffers from inadequate care and neglect. As a result, her well-being, health and safety are compromised.

Elder abuse is an alarming public health and human rights issue. A study, based on the best available evidence from 52 studies in 28 countries from diverse regions, estimated that over the past year, 15.7% of people aged 60 years and older were subjected to some form of abuse (Yon et al., 2017). With regard to people with dementia, the rates are higher (WHO, 2018).

Dementia

Dementia is a syndrome—usually of a chronic and progressive nature—in which there is deterioration in cognitive function, beyond what might be expected in the normal ageing process. Dementia affects memory, cognition, orientation, comprehension, calculation, learning capacity, language and judgement. The ability to perform everyday activities is reduced (World Alzheimer Report, 2019). According to this report, around 50 million people around the world suffer from dementia. Every year, there are nearly 10 million new cases worldwide. Increase of the ageing population may result in increasing numbers of people with dementia.

Dementia is one of the major causes of disability and dependency among older people worldwide. Dementia has a physical, psychological, social and economic impact—not only on people with dementia, but also on their caregivers, their families and society at large.

Dementia and Elder Abuse and Neglect

Older adults with cognitive impairment are particularly vulnerable to abuse and/or neglect (Dong et al., 2014; Serra et al., 2018).

Prevalence of elder abuse in regard to people with cognitive decline and dementia is difficult to study. Elder abuse and neglect occur behind closed doors, and abusive behaviours are commonly perpetrated against vulnerable people with cognitive impairment, by those on whom they depend. Thus, the phenomenon is underestimated and victims are frequently unable, frightened or embarrassed to report its presence (Selwood & Cooper, 2009). It is estimated that abuse and neglect towards older people with dementia who live in the *community* range from 5.4 to 62.3% (Cooney et al., 2006; Yan, 2014; Yan & Kwok, 2011).

Elder abuse and neglect also exist in *long-term care facilities*, in which some of the residents suffer from dementia. Findings from several studies reported that approximately 80% of nursing home staff said they had observed abusive behaviours towards the residents in the year before the study was conducted (Goergen, ; Wang, 2005). In a large study conducted in Israel, Ben Natan et al. (2009) examined staff members from 22 different nursing homes. Findings showed that slightly more than half (53.1%) of the staff members sampled reported committing an abusive behaviour towards the residents over the past year. About two-thirds (64.3%) of the cases were incidents of neglect. Seventy per cent (70%) of the respondents reported that they had been present during incidents in which another staff member abused a resident, employing one or more types of abuse. A systematic review conducted by Yon et al. (2018) revealed that overall abuse estimates, based on staff reports, suggest that 64.2% of staff admitted abusing residents in the past year. Prevalence estimates for abuse subtypes reported by older residents were highest for psychological abuse (33.4%), followed by physical (14.1%), and financial (13.8%) abuse, neglect (11.6%), and sexual abuse (1.9%).

Barriers to Identifying Elder Abuse and Neglect

Research findings report that millions of older adults are exposed to abuse and neglect; yet, only a small number of these cases are detected, identified and treated. Several types of barriers were described by researchers and professionals in previous studies:

Barriers related to *cultural and social norms* is, for example, maintaining the value of family privacy. Family life is perceived as "sacred", and the behaviours occurring within the family are a private matter. In addition, in certain societies and cultures, violent behaviours are perceived as "educational" means of discipline (Chokkanathan et al., 2014; Hughes & Luo, 2015; WHO, 2002).

Barriers related to the *victim* is lack of awareness about his/her rights, shame and guilt, protecting the abusive family member and fear of the abuser (Alon, 2015; Chokkanathan et al., 2014; Dow et al., 2019). People with dementia often have difficulty communicating with others and would not reveal being exposed to abuse or neglect. Dementia is also associated with a high level of dependency. Thus, older adults may be reluctant to disclose abuse because they fear retaliation or the loss of caregiver support. Behavioural symptoms such as agitation and aggression are associated with dementia. Caregiver–recipient relationship may become stressful and may result in abuse and/or neglect (Dong et al., 2014).

Barriers related to the *perpetrator*: Family members, paid caregivers, friends and healthcare professionals may feel shame, guilt or even deny offensive behaviours towards the older adult with dementia(Dow et al., 2019).

Barriers related to *professionals* are lack of awareness, lack of knowledge, ethical dilemmas and concerns about the outcomes of the intervention (Anetzberger et al., 2000; Schmeidel et al., 2012). There is often reluctance among professionals to ask a family member directly about abuse. This reluctance may stem from the fear that discussing and acknowledging the issue would necessitate the older adult's protection, possibly leading to separation and removing the client from his/her home. This may result in an "all-or-nothing" approach to abuse, where it is ignored until the problem becomes severe.

Organizational barriers are lack of services, overload, organizational priorities, insufficient knowledge, lack of coordination among various services and lack of appropriate protocols for collaborative service delivery (Anetzberger et al., 2000). All or part of the above may result in under-identification and under-detection of cases of elder abuse and neglect. Thus, there is a gap between the existing data available about the extent of the phenomenon and the reality in which only a limited proportion of the victims are identified and treated (Alon, 2015; Schmeidel et al., 2012).

Theoretical Framework for Understanding the Occurrence of Elder Abuse

Due to the complexity of the phenomenon and the multiplicity of situations that can cause elder abuse, researchers and professionals find it difficult to embrace one comprehensive theory that can explain the occurrence of elder abuse and neglect. However, some approaches and perspectives may provide certain theories and explanations.

Ageism and negative attitudes towards older adults: *Ageism* is defined as a systematic process of stereotyping and discrimination against individuals or groups on the basis of their age (Butler, 1975). Negative stereotypes attributed to old age and the ageing process are at the root of these negative attitudes, which often affect older adults' social status. Decreasing a person's value can create circumstances and an atmosphere of legitimacy regarding abusive behaviours towards older adults (Phelan, 2008).

Ageism and Dementia: lack of awareness, lack of understanding about dementia and stigmatization can lead to **negative attitudes** towards a person with dementia (Alzheimer's Disease International, 2019). As a result, individuals with dementia may be at high risk of abuse (Dong et al., 2014). This abuse can be manifested in deprivation of rights, humiliation, isolation and home confinement—as illustrated in Rina's case.

Psychological Models: It examines the occurrence of abuse from the individual (victim and or abuser) perspective. A person with dementia may be a potential victim or potential abuser (Cooper & Livingston, 2014). There is a positive correlation between personality disorders and mental illness and abuse of an older parent (Pillemer et al., 2016; Ramsey-Klawsnik, 2004). Substance abuse can also result in abusive behaviours (Labrum & Solomon, 2018; Pillemer et al., 2016.(Pathological relationships between parents and their adult children and complex dynamics in family relationships can also cause abuse and/or neglect (Band-Winterstein et al., 2016; Labrum & Solomon, 2018; Santos et al., 2018).

Dependency: An individual's dependency on his caregiver (family member or paid caregiver) may be fertile ground for abusive behaviours (Borenstein, 2019; Mysyuk et al., 2016). The dependency (financial and emotional) of an adult child on his parent was also found to be a risk factor associated with abuse and neglect. Adult children who find themselves in this situation may experience a sense of loss of control. In response to this perceived feeling of powerlessness, and in an attempt to regain control, they may act abusively towards the elderly parent (Band-Winterstein et al., 2016).

Stress and burden of care can lead to abusive behaviours towards those who are perceived as the "cause" of the difficulty and frustration (DeLiema et al., 2018; Wang et al., 2019). Caring for a person with dementia can be overwhelming. Physical, emotional and financial pressures can cause great stress to both families and caregivers alike (Alzheimer's Disease International, 2019). As a result, primary caregivers may "take out" their frustration and stress on the care recipient.

Risk Factors Perspective: Given the fact that there is no single comprehensive theory that can provide an explanation for the occurrence of elder abuse and neglect, researchers tend to draw on risk factors that were found to be predictors of elder abuse (Cohen & Alon, 2014; Pillemer et al., 2016; Yon et al., 2017).

Risk factors related to the ***victim's characteristics*** are age 75+, morbidity, functional impairment, **cognitive decline**, dependency and social isolation. Lower levels of cognitive function were also associated with increased risk for elder abuse. Non-compliant older adult with dementia may be restless and aggressive. This may intensify caregiver-care recipient interaction and result in elder abuse (Cohen, 2013; Dong et al., 2014; Rivera-Navarro et al., 2018).

Perpetrator's characteristics: a family member, caregiver, unemployment, coping with difficult life events/crisis situations, psychopathology, dependency on the elderly parent, social isolation, were found as risk factors for elder abuse and neglect (Dong et al., 2014; Pillemer et al., 2016; Rivera-Navarro et al., 2018).

Contextual circumstances: Cohabitation, codependency, poor relationship and/or complicated family dynamics and caregiving relationships are all predictors of elder abuse and neglect (Acerino et al., 2010; DeLiema et al., 2018; Dong et al., 2014; Fulmer et al., 2005; Labrum & Solomon, 2018; Pillemer et al., 2016).

Risk factors associated with abuse and neglect in long-term care facilities include extremely demanding and stressful work, lack of knowledge, lack of skills and experience, low staff-to-patient ratio, job dissatisfaction, lack of continuing education and a high staff turnover rate (Ben Natan & Tabak, 2013; Ben Natan et al., 2009).

Residents with dementia are more vulnerable to abuse and neglect. They are totally dependent on staff members, and many of them have behavioural, neurological and physical limitations. As a result, care provision is more challenging. All of these factors may result in an inadequate level of care, neglect and abuse (Ben Natan et al., 2009; Goergen, 2001; Wangmo et al., 2017).

In sum, risk factors may serve as the best accounting framework for understanding and addressing the multiple dimensions of elder abuse and neglect.

The Risk Factors Model, illustrated below, encompasses aspects related to victim's characteristics; perpetrator's characteristics; mutual factors such as codependency and cohabitation; victim–perpetrator relationship and societal factors. This model can be a useful tool for practitioners in detecting elder abuse and neglect, and in planning interventions with the victim, perpetrator and family members. It can also be useful for service providers in developing programmes to address elder abuse.

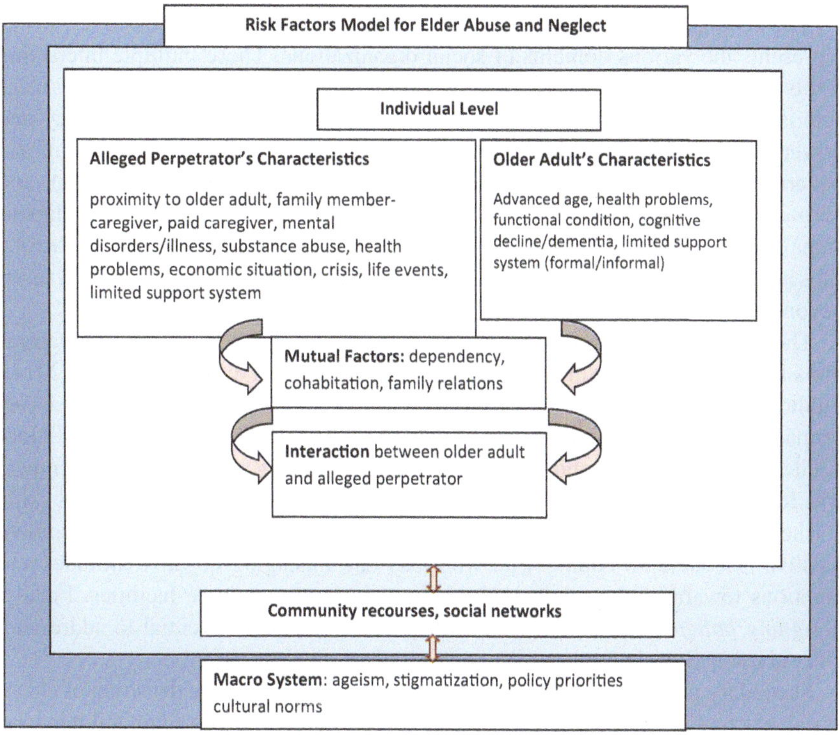

All of the risk factors described in the model are reflected in Rina's case. Both spouses are of advanced age. Rina suffers from cognitive decline and depends on her husband, who is her primary caregiver. The interaction between them led to neglect. Rina's condition may deteriorate due to the lack of social support.

Responding to Elder Abuse and Neglect

Elder abuse and neglect affect millions of older adults throughout the world. Responding and solving the problem of elder abuse requires multiple strategies based on different approaches and solutions aimed at reducing the risks associated with elder abuse and neglect.

Over the years, several paradigms addressing elder abuse were discussed such as the ecological approach; public health models; domestic violence models, which focus on empowerment and protecting the victim; protective services that provide information; legal services (guardianship and reporting cases of elder abuse and neglect); and the criminal justice perspective, which focuses on persecuting offenders. Each paradigm has its own ideology, strategies and interventions methods. Three perspectives will be addressed in the current chapter.

The ecological approach is one of the most comprehensive perspectives, which represents the various domains of social organizations. These multiple interacting levels include the *individual level* (the victim's characteristics; the perpetrator's characteristics); the *victim–perpetrator relationship* (cohabitation, codependency and power and control relationships); the *family system* (solidarity, intergenerational patterns); *community and environment* (service providers, social networks); and *sociocultural* processes. This perspective provides practitioners with a multidimensional assessment framework with which to understand the full scope of risk factors. This approach focuses on the need to target specific populations and broader systems beyond the individual level (Burnes, 2017; Schiamberg & Gans, 1999).

The **public health approach** can help policymakers, professionals and organizations in their dealings with elder abuse and neglect (Alon & Berg-Warman, 2014). Public health models are aimed at prevention and treatment in the field of elder abuse: primary prevention includes *raising awareness* through campaigns among older adults, family members and the general public. *Training*—training staff members was found to be effective in raising awareness and acquiring knowledge and skills (Fisher et al., 2016; Ben Natan & Tabak, 2013). Courses and workshops may promote positive discourse on coping with workload and managing negative attitudes and emotions towards older adults and residents in long-term care facilities. Finally, *designing policies, legislation* and *developing services* are essential to addressing elder abuse and neglect.

Secondary prevention focuses on *pro-active screening* and the detection of elders at risk (such as a person with dementia, an older adult living with adult children who engage in substance abuse, isolated older adults, etc.); *helping primary caregivers* through the provision of information, helping them to utilize available resources, support groups for caregivers and family therapy (Cooper et al., 2016). Providing supportive services such as day care centres, a paid caregiver, housekeeping and hot meals were found to be effective in reducing neglect (Alon & Berg-Warman, 2014). *Strengthening formal and informal networks* and strengthening *collaboration* between professionals and organizations such as welfare services, NGOs and health-care services are important factors in implementing an effective response to abuse and neglect.

Tertiary prevention includes assessing victim's needs, case work, psycho-educational groups, legal consultation and supplying supportive services (home care, hot meals, etc.). Intervention with perpetrators includes referral to a mental health clinic, referral to detox services and group therapy—all aimed at reducing risk and harm (Alon & Berg-Warman, 2014).

Towards a Holistic Approach to Address Elder Abuse and Neglect

Based on the cumulative knowledge and practices developed, it is recommended to employ diverse strategies and approaches in order to deal effectively with elder abuse and neglect. A combination of the ecological approach and the public health approach may yield successful responses in dealing with elder abuse and neglect.

Due to the fact that abuse and neglect usually occur "behind closed doors", it is essential to take active measures in order to detect and identify cases of abuse and neglect. In this context, healthcare professionals, social workers and other practitioners play a key role in detecting, treating and stopping abusive behaviours towards older adults. Their expertise and accessibility to the elderly population places them in a central position to address the problem. The encounter between professionals and older adults may be the only chance to detect and identify abuse and/or neglect as illustrated in Rina's case. Using validated tools can improve the identification of elder abuse and neglect (Alon & Berg-Warman, 2014; Alon et al., 2019; Cohen, 2013).

Detecting abuse and neglect might be challenging regarding people with dementia, especially when it is difficult for the client to provide information. Therefore, the professional should establish trust relations with the client and his caregiver. For example, the social worker who visited Rina and her husband should introduce herself as follows: "My name is.... we met the other day and talked about ...". In order to avoid embarrassment, she should begin with general questions and ask Rina about herself and her daily life routines. This communication may provide preliminary information and general impressions about Rina's cognitive state. At the same time, the social worker should try to establish a trust relationship with Rina's husband.

Home visits can provide additional information. A social worker or a nurse should be alert and observe behavioural signs such as discomfort, fear, distress or indifference. Detecting neglect is likely to be evident as described in Rina's case. Poor personal hygiene, unsuitable clothing and evidence of malnourishment or dehydration may confirm the suspicion of neglect.

In order to obtain more information and substantiate the identification of abuse or neglect, it is recommended to ask directly: "Do you feel safe at home?"; "How are you getting along?"; "Do you need help?" If the answer is positive, ask "Who helps you?"; "Does anyone refuse to help you when needed?".

The professional has to be alert to signs of physical abuse and/or sexual abuse such as unexplained bruises, scars and broken bones. He/she should ask the client: "Has anyone hurt you?"; "Has anyone touched you without your consent?" If the answer is positive, he/she should ask: "Who did this?"; "How did it happen?"; "How often does this happen?".

If financial exploitation is suspected, social workers should ask: "Who manages your money?"; "Who makes the financial decisions?"; "Do you keep track of your own bank account?"; "Has anyone taken your belongings or money without your permission?"; "Did you sign a document that you didn't understand?".

Intervention: Following the initial intake, the individual's needs must be defined. The older adult and his family member–caregiver should be provided with information about his/their rights and information about available services in the community. Victims and perpetrators can be offered and/or referred to either individual or group therapy. If needed, supportive services should be offered (day care centre, seniors club, meals on wheels, home care, a volunteer, legal counsel, etc.). A geriatric assessment is recommended in cases of cognitive impairment.

In Rina's case, the social worker should provide the husband with information about dementia and the existing services that can help both Rina and himself as a caregiver. For example, a day care centre could provide adequate care for Rina and a break for the husband.

Stopping abuse and neglect might require legal interventions such as ongoing mandatory reporting, separating victims and abusers, and court orders to restrain the perpetrator. In cases of cognitive decline and dementia, appointing a legal guardian should be considered.

In complex situations, it is recommended to consult with an **Advisory Multidisciplinary Team**. Professionals from diverse disciplines and fields of expertise (social workers, attorneys, medical doctors, mental healthcare practitioners and law enforcement representatives) complement one another and collaborate to provide holistic, reliable and effective interventions (Burnes, 2017; Cafaro Schneider et al., 2010).

It is also important to raise awareness among older adults and family caregivers; conducting educational workshops in day care centres and clubs for seniors; and conducting seminars and conferences for professionals.

Professionals should also be involved in recruiting partners from healthcare services, welfare and social services, home care agencies and non-profit organizations. Collaborations between professionals from different organization may also improve coping with elder abuse and neglect (Alon & Berg-Warman, 2014).

Such a holistic model was developed in Israel in 2005 and has been implemented ever since. The model includes the establishment of a Specialized Unit for the Prevention and Treatment of Elder Abuse, with trained social workers and a multidisciplinary advisory team, which accompanies the unit's staff. Social workers screen and detect older adults that are exposed to abuse and neglect. They employ diverse techniques such as counselling, group therapy, mediation and providing supportive services to the victim, perpetrator and family members. They also employ legal means such as filing complaints with the police, obtaining protection orders and appealing to the court for the appointing of a legal guardian. Complex cases are brought up for discussion in the multidisciplinary team for developing intervention plans. On the community level, the Specialized Unit's staff raises awareness and conducts seminars and conferences for older adults, family members and professional partners.

Findings from the evaluation study that accompanied the implementation of the programme showed an increase in the number of identified cases. In 20% of the cases, abusive behaviours had completely stopped. In two-thirds of the cases, provision of supportive services reduced neglectful behaviours and improved their living conditions. At the organizational level, working procedures became systematic; social workers employed a variety of interventions; and collaboration between and among professionals/services were strengthened (Alon & Berg-Warman, 2014).

Conclusion

Dementia and elder abuse and neglect have been recognized internationally as extensive and serious problems, requiring the attention of healthcare systems, social welfare organizations, policymakers and the general public.

The current article discusses dementia as a risk factor associated with elder abuse and neglect. Over the years, studies have established the relationship between dementia and elder abuse; yet, most of the cases are not identified and treated properly.

Coping effectively with the phenomenon of abuse and neglect of people with dementia requires understanding the causes and circumstances that may lead to abusive behaviours. In the current article, several theories and perspectives explaining the occurrence of elder abuse and neglect were introduced, among them the ecological perspective and risk factors model, which may be useful for both researchers and professionals. Based on the ecological approach, a multi-system activity is required in order to deal with the phenomenon.

At the social level: Awareness must be raised regarding dementia and elder abuse and their consequences for the older adult, his family, service providers and society at large.

At the organizational level: Education and training of professionals is imperative to increase awareness about dementia as a risk factor for abuse and neglect. Professionals must be made aware of the phenomenon and acquire knowledge and skills to detect and identify abuse and neglect, especially regarding clients with cognitive decline and dementia. Professionals who deal with the elderly have the best chance to detect, identify, and expose cases of elder abuse and neglect. Therefore, systematic screening and the detecting of abuse and neglect among older adults, in general, and people with dementia, in particular, may promote the implementation of successful interventions and help put a stop to abusive behaviours.

At the individual level: Providing information about dementia can help the family cope more effectively as primary caregivers. Providing information about existing services will enable the family and the older adult to utilize existing services, such as day care centres, support groups, volunteers and placement in a nursing home, if necessary.

In sum, effective dealing with elder abuse and neglect consists of formulating policy; allocating resources; raising awareness; and providing education for older adults and caregivers, by developing specialized services, supplying supportive services and training professionals. Only multi-systemic and inclusive efforts may succeed in reducing the damages of elder abuse and neglect and improving the quality of life of millions of older adults.

References

Acierno, R., Hernandez, M. A., Amstadter, A. B., Resnick, H. S., Steve, K., Muzzy, W., et al. (2010). Prevalence and correlates of emotional, physical, sexual, and financial abuse and potential neglect in the united states: The national elder mistreatment study. *American Journal of Public Health, 100*(2), 292.

Alon, S. (2015). Elder abuse and neglect: Dimensions and coping strategies. In: M. Cohen, & D. Perlotzki (Eds). *Practical gerontology* (pp 445–494). JDC Israel Eshel [in Hebrew].

Alon, S., Shiendler, M., Doron, I., & Yuz, F. (2019). *Elders at risk: Legal, therapeutic and ethical issues*. JDC-ESHEL [In Hebrew].

Anetzberger, G. J., Palmisano, B. R., Sanders, M., Bass, D., Dayton, C., Eckert, S., & Schimer, M. R. (2000). A model intervention for elder abuse and dementia. *The Gerontologist, 40*(4), 492–497.

Band-Winterstein, T., Semloy, Y., & Avieli, H. (2016). Shared reality of the abusive and vulnerable: The experience of aging for parents living with abusive adult children coping with mental disorder. *International Psychiatric, 26*(11), 1917–1927.

Ben Natan, M., Lowenstein, A., & Eisikovits, Z. (2009). Psycho-social factors affecting elders' maltreatment in long-term care facilities. *International Nursing Review, 57,* 113–120.

Ben Natan, M., & Tabak, N. (2013). Combating the maltreatment of older persons by staff in long term care nursing homes: Legal aspects. *Medicine Law, 32*, 53–64.

Borenstein, R. F. (2019). Synergetic dependency in partner and elder abuse. *American Psychologist, 74*, 713. https://doi.org/10.1037/amp0000456.

Burnes, D. (2017). Community elder mistreatment intervention with capable older adults: Toward a conceptual practice model. *The Gerontologist, 57*(3), 409–416.

Butler, R. N. (1975). *Why survive: Being old in America*. John Hopkins University Press.

Cafaro Schnieder, D., Mosqueda, L., Falf, E., & Huba, G. J. (2010). Elder Abuse Forensic Center. *Journal of Elder Abuse and Neglect, 22*(3–4), 255–274.

Chokkanathan, S., Natarajan, A., & Mohanty, J. (2014). Elder abuse and barriers to help seeking in Chennai India: A qualitative study. *Journal of Elder Abuse and Neglect, 26*, 60–79.

Cloitre, M., Stolbach, B. C., Herman, J. L., Kolk, B. V., Pynoos, R., Wang, J., & Petkova, E. (2009). A developmental approach to complex PTSD: Childhood and adult accumulative trauma as predictors of symptom complexity. *Journal of Traumatic Stress, 22*(5), 399–408.

Cohen, M. (2013). The process of validation of a three-dimensional model for the identification of abuse in older adults. *Archives of Gerontology and Geriatrics, 57*, 243–249.

Cohen, M., & Alon, S. (2014). Development and dissemination of a tool for identification of elder abuse and Neglect. *Gerontology, 41*(2), 75–98. [Hebrew].

Cooney, C., Howard, R., & Lawlor, B. (2006). Abuse of vulnerable people with dementia by their carers: Can we identify those most at risk? *International Journal Geriatric Psychiatry, 21*, 564–571.

Cooper, C., Barber, J., Griffin, M., Rappaport, P., & Livngstone, G. (2016). Effectiveness of START psychological intervention in reducing abuse by dementia family carers: Randomized controlled trial. *International Psychogeriatric, 28*(6), 881–887.

DeLiema, M., Yonashiro-Cho, J., Gassoumis, Z. D., Yon, Y., & Conrad, K. J. (2018). Using latent class analysis to identify profiles of elder abuse perpetrators. *Journals of Gerontology: Social Sciences, 73*(5), e49–e58.

Dong, X., Chen, R., & Simon, M. A. (2014). Elder abuse and Dementia: A review of the research and health policy. *Health Affairs, 33*(4), 642–649.

Dong, X. Q., Simon, M. A., Beck, T. T., Farran, C., McCann, J. J., De Leon, C. M., Laumann, E., & Evans, D. A. (2011). Elder abuse and mortality: The role of psychological and social wellbeing. *Gerontology, 57*(6), 549–558.

Dow, B., Gahan, L., Gaffy, E.,Joosten, M., Vrantsidis, F., & Jarred, M. (2019). Barriers to disclosing elder abuse and taking action in Australia. *Journal of Family Violence*. https://doi.org/10.1007/s10896-019-00084-w.

Fisher, B. S., & Regan, S. L. (2006). The extent and frequency of abuse in the lives of older women and their relationship with health outcomes. *The Gerontologist, 46*(2), 200–209.
Fulmer, T., Paveza, G., VaneWeerd, C., Fairchild, S., Guadagno, L., Bolton-Blatt, M., & Norman, R. (2005). Dyadic Vulnerability and risk profiling for elder neglect. *The Gerontologist, 45*(4), 525–534.
Goergen, T. (2001). Stress, conflict, elder abuse and neglect in German nursing homes: A pilot study among professional caregivers. *Journal Elder Abuse and Neglect, 13*, 1–26.
Goergen, T. (2004). A multi-method study on elder abuse and neglect in nursing homes. *Journal of Adult Protection, 6*, 15–25.
Hughes, J., & Luo, H. (2015). Abuse of older adults in culturally minorized groups within Canada. In K. Jagielska, J. Malgorzata, & N.G. Pikula (Eds.). *Violence against the elderly challanges-research-action.* Nova.
Labrum, T., & Solomon, P. L. (2018). Elder mistreatment perpetrators with substance and/or mental health conditions: Results from the National Elder Mistreatment Study. *Psychiatry Quarterly, 89*, 117–128. https://doi.org/10.1007/s11126-017-9513-z.
Mysyuk, Y., Westendorp, R. G. J., & Lindenberg, J. (2016). Perspectives on the Etiology of violence in later life. *Journal of Interpersonal Violence, 31*(18), 3039–3062.
Phelan, A. (2008). Elder abuse, ageism, human rights and citizenship: Implication for nursing discourse. *Nursing Inquiry, 15*(4), 320–329.
Pillemer, K., Burnes, D., Riffin, C., & Lachs, M. S. (2016). Elder abuse: Global situation, risk factors, and prevention strategies. *The Gerontologist, 56*(Suppl 2), S194–S205.
Ramsey-Klawsnik, H. (2017). Older adults affected by Polyvictimization: A review of early research. *Journal of Elder Abuse & Neglect, 29*(5), 299–312.
Rivera-Navarro, J., Sepulveda, R., Contador, I., Fernandez-Calvo, B., Ramos, F., Tola-Arribas, M. A., & Goni, M. (2018). Detection of maltreatment of people with dementia in Spain: Usefulness of the Caregiver Abuse Screen (CASE). *European Journal of Ageing, 15*, 87–99.
Santos, A. J., Nunes, B., Kislaya, I., Gil, A. P., & Ribeiro, O. (2018). Older adults' emotional reactions to elder abuse: Individual and victimization determinants. *Health Social Care Community*, pp. 1-–12. https://doi.org/10.1111/hsc.12673.
Schiamberg, L. B., & Gans, D. (1999). An ecological framework for contextual risk factors in elder abuse by adult children. *Journal of Elder Abuse and Neglect, 11* (1), 79–103.
Schmeidel, A., Daly, J. M., Rosenbaum, M. E., Schmuch, G. A., & Jogerst, G. J. (2012). Health care professionals' perspectives on barriers to elder abuse detection and reporting in primary care settings. *Journal of Elder Abuse and Neglect, 24*(17), 17–36.
Schofield, M., Powers, J. R., & Loxton, D. (2013). Mortality and disability outcomes of self reported elder abuse: A 12-year prospective investigation. *Journal of American Geriatric Society, 61*(5), 679–685.
Selwood, A., & Cooper, C. (2009). Abuse of people with Dementia. *Reviews in Clinical Gerontology, 19*, 35–43.
Serra, L., Contador, I., Fernandez-Calvo, B., Ruisoto, P., Jenaro, C., Flores, N., Ramos, F., & Rivera-Navarro, J. (2018). Resilience and social support as protective factors against abuse of patients with dementia; A study on family caregivers. *International Journal Geriatric Psychiatry, 33*, 1132–1138.
Storey, J. E., & Perka, M. R. (2018). Reaching out for help; recommendations for practice based on an in-depth analysis of an elder abuse intervention program. *British Journal of Social Work, 48*, 1052–1070.
Wang, J. J. (2005). Psychological abuse behavior exhibited by caregivers in the care of the elderly and correlated factors in long-term care facilities in Taiwan. *Journal of Nursing, 13*, 271–280.
Wang, M., Huimin, S., Junjian, Z., & Juan, R. (2019). Prevalence and associated factors of elder abuse in family caregivers of older people with Dementia in central China, Cross-Sectional Study. *International Journal of Geriatric Psychiatry, 34*(2), 299–307.
Wangmo, T., Nordström, K., & Kressig, R. W. (2017). Preventing elder abuse and neglect in geriatrics institutions: Solutions from nursing care providers. *Geriatric Nursing, 38*, 385–392.

WHO. (2002). *Missing voices, views of older persons on elder abuse*. World Health Organization.
World Alzheimer Report, Attitudes to dementia. (2019). Disease International (ADI).
World Health Organization. (2018). *Elder abuse*. WHO.
Yan, E. (2014). Abuse of older persons with dementia by family caregivers: Results of a 6- month prospective study in Hong Kong. *International Journal Geriatric Psychiatry, 29*, 1018–1027.
Yan, E., & Kwok, T. (2011). Abuse of older Chinese with Dementia by family caregivers: An inquiry unto the role of caregiver burden. *International Journal Geriatric Psychiatry, 26*, 527–535.
Yon, Y., Mikton, C. R., Gassoumis, Z. D., & Wilber, K. H. (2017). Elder abuse prevalence in community settings: A systematic review and meta analysis. *Lancet Global Health, 5*, 147–156.
Yon, Y., Ramiro-Gonzalez, M., Mikton, C. R., Huber, M., & Sethi, D. (2018). The prevalence of elder abuse in institutional settings: A systematic review and meta-analysis. *The European Journal of Public Health, 29*(1), 58–67.

Chapter 16
Ethical Issues in Dementia—Global Challenges

Avinash De Sousa and Russell D'Souza

Abstract As the world population ages, there is a large number of people with dementia that shall dwell in our nations with the years to come. The growing number of individuals suffering from dementia shall escalate the ethical issues related to dementia that are relevant for clinical practice and research. Some of these issues and overview of them shall be addressed in the current chapter. These include issues early detection and diagnosis, issues in dementia related to genetic testing, the role of neuroscience to enhance cognition in these subjects, ethical issues with the use of medications in mild cognitive impairment. There are also other ethical dilemmas related to proper diagnostic disclosure, apt informed consent procedures, proper conflict of interest declarations, research study issues in dementia care trials, issues with the use of placebo and the medical treatments in these subjects. This moves on to issues in the late stages of dementia, framing of therapeutic goals and end-of-life care in these patients. Issues related to old-age home care, testamentary capacity, certification, human rights, elder abuse and many such factors are discussed.

Keywords Dementia · Ethical issues · Ethics · Dilemmas

Introduction

As the number of people ageing in the world increases, there is a surge in people that suffer from dementia which is ever increasing. As the population of old people with dementia shall increase, the complex treatment and ethical issues concerning them shall also increase (Brannelly, 2006). Ethics is a general term that means "*exploring*

A. De Sousa (✉)
Desousa Foundation, Mumbai, India

Department of Psychiatry, Lokmanya Tilak Municipal Medical College, Mumbai, India

Present Address:
Carmel, 18, St. Francis Road, Off SV Road, Santacruz West, Mumbai 400054, India

R. D'Souza
UNESCO Chair in Bioethics—Haifa, Asia Pacific Region and Head of the UNESCO National Training Programme in Bioethics, Haifa, India

M. K. Shankardass (ed.), *Dementia Care*,
https://doi.org/10.1007/978-981-16-3864-0_16

human values and understanding what constitutes a good and moral life". Ethics has an approach that looks at what is considered right or good and what is considered bad or unjust in any specific cultural context (Miles, 2005). The ethics of dementia and dementia care is complex and intertwined with multiple personal, psychological, social and environmental or familial issues that concern the patient. Dementia affects rational judgement and the independence of the patient and raises concerns with regard to the quality of life as many decisions need to be taken which cannot be done by the patient (Foster et al., 2014). Changes in healthcare systems in many nations, which are in keeping with the ageing of our populations, will generate new ethical dilemmas for doctors, nurses and caregivers in dementia care (Hughes et al., 2002). In this chapter, we shall discuss the global ethical challenges that emerge in the clinical care, research, treatment and practice involving persons affected by dementia. These challenges shall be described at various levels and the tortuosity concerning each issue shall be deliberated upon. There is no sound solution that shall provide but the person involved in dementia care shall have to look at every aspect and every patient as an individual and take decisions in the interest of the patient.

Challenge 1: Ethical Issues Related to the Diagnosis of Dementia

Multiple ethical issues concern a diagnosis of dementia. There are many people who are brought early for an assessment, and they may have age associated cognitive decline rather than dementia. Should we warn such patients about a diagnosis of dementia that may ensue or follow a wait and watch policy. We must consider those who are asymptomatic, those with mild degrees of intellectual disability who might be at risk for developing dementia in the long run and how we would assess such patients (Post, 2000). The differentiation of those who have dementia and age associated cognitive impairment is established by assessment of scores on a number of quantitative assessment rating scales. The persons administering these batteries must be equipped and well aware of the scoring and administration before making a diagnosis (Laakkonen et al., 2008). The capacity of the patient and insight into his condition needs to be ascertained before you decide whether a diagnosis of dementia is made or not as that would determine whether we disclose the same to the patient or not (Cornett & Hall, 2008).

Understanding the patient and his decision-making capacity as well as his level of insight provides the basis for the use of ethical principles in the disclosure of a diagnosis of dementia. One of the basic ethics principles is the respect for the autonomy of the patient and all diagnosis disclosures are based on a total consideration for the patient's autonomy (Lepeleire et al., 2004). All patients have the right to make their own autonomous choices concerning their medical, psychological and mental health but those that lack a decision-making capacity are considered incapable of doing so and are considered incapable of consenting or withholding it for treatment in the case of dementia (Whitehouse et al., 2004). A patient not understanding the gravity and nature of his diagnosis will not be able to make treatment decisions, and his decisions then shall have to be made by his family. Hence, the disclosure of diagnosis has many confounding variables that need to be assessed before a decision is made (Bamford et al., 2004). The diagnosis in medical care may often be made by a general physician, gerontologist, psychiatrist, neurologist or geriatric psychiatrist. It is imperative that the treating doctor is aware of dementia assessment and diagnosis before the diagnosis is confirmed and conveyed to either the patient or the relatives.

Challenge 2: Ethical Issues Related to Driving and Patients with Dementia

One that drives and suffers from dementia is at an increased risk of being involved in and causing traffic accidents. The prevalence rate of road traffic accidents among persons that suffer from dementia compared to normal old people without dementia has always been on the higher side. There is a 5–tenfold greater risk of road accidents for patients with mild to moderate dementia that drive cars when compared to people of the same age with no dementia (Brown & Ott, 2004). Dementia is a research established risk factor for risky driving in old age. Studies have not yet been able to accurately point out whether the risky driving is attributable to cognitive or functional factors or environmental sensory issues that are affected in dementia (Carmody et al., 2013).

Evidence bases suggest that impaired performance on tests of visuospatial ability, executive functioning and dexterity are pivotal factors to assess when granting fitness to drive in patients that suffer with dementia. These findings while true must not be the only yardstick as driving is a complex sensorimotor activity. It involved multiple complex cognitive abilities, like visual judgement, visuospatial adjustments. Left–right orientation, moto and sensory skills and to combine all these together which is a vital task in driving. Thus, sound neuropsychological assessment and even simulated driving tests as well as on-road driving tests are essential before fitness for driving is certified (Snyder, 2005). Some other ethical and situational dilemmas emerge. Even naturalistic evaluations as on-road driving tests and assessments may not adequately reflect a person's true capability and capacity to drive over periods of time. Patients with dementia may be anxious and take more precautions during their driving test and also may be uncomfortable in an electronic driving simulator while they may perform adequately while driving their own car (Carter et al., 2015).

Driving sometimes may be the only source of joy or recreation for the patient and his means of transportation and contact with the outside world. It is preferable to desist from removal of driving privileges early. Individual needs, personal dignity and promoting a better quality of life must be considered while making fitness to drive decisions. The reasons for not allowing driving need to be explained to the patient that suffers from dementia and his family and other forms of transport for his daily life must be discussed and planned for (Carr & O'Neill, 2015).

The greatest challenge in driving assessment and taking a fitness to drive decision is in patients that suffer from mild dementia and there is no way to ascertain how the dementia may progress. The dementia severity, history from the family, driving ability and neuropsychological assessment reports are factors that should serve as screening factors, and this must be supplemented by a driving assessment from a professional motor school and driver (Herrmann et al., 2006). The lack of clear guidelines and research in this area is also a clear problem that raises the ethical dilemmas in this area.

Challenge 3: Ethical Issues Related to Testamentary Capacity in Patients with Dementia

Testamentary capacity (TC) is defined as *"the capability to execute a will and is determinate upon both the ability to understand relevant facts and an appreciation of the reasonably foreseeable consequences of taking specific actions regarding the formation of a will"* (Kennedy, 2012). There are huge ramifications ethically when someone is certified being fit to make a will. The doctor has to first assess whether the patient has the cognitive capacity to make the will and also assess whether he is under no duress by any of the relatives to do so.

Many patients with dementia may have been assessed for TC at a time when they were cognitively aware and sound and the will is challenged in the court of the law after dementia ensues and the doctor is accused of false certification and supporting a particular relative of the patient (Jacoby & Steer, 2007). It is vital that all relatives of the patient (rather the heirs) be present when the patient is being assessed and the assessment be done in video recording so that no challenge to the assessment be made in court. It is also prudent that a will be made and registered in 24–48 h of the TC assessment as many a times the TC fitness certificate is obtained and the will is made many months after the patient may have deteriorated (Shulman et al., 2007). Many cases of property disputes based on wills made by patients who were fine, later developed dementia and even died come up years of the TC assessment and maintaining all papers and copies of the documents are essential for any legal and ethical issues that may arise (Shulman et al., 2005).

Challenge 4: Ethical Issues in Dementia Clinical Trials and Informed Consent

There is a huge debate on patients with dementia participating in clinical studies and drug trials. Overall, the ability of patients with dementia to understand information and consent has been inadequately studied in ethical and psychiatric research. There is a need to study informed consent capacity clinically and using proper standardized instruments in order to establish the capacity of decision-making and informed consent in cases like dementia (Slaughter et al., 2007). Notwithstanding that studies on informed consent in geriatric psychiatry are sparse, it has been in such studies that there is no association between decision-making capacity and any specific diagnosis like dementia. Certain mental health conditions may lead to higher rates of diminished capacity than others. Impairments in memory, executive function, cognition and thought processes seem to diminish decision related capacities in patients that suffer from dementia (Dewing, 2007).

Another ethical issue that plagues research in dementia is the notion of surrogate consent given by the relative of the patient. There is no clear guideline, and no proper regulation that addresses the issue of surrogate consent in studies that involve patients with dementia (Cubit, 2010). Surrogate consent also has ethical dilemmas, because relatives have to decide on behalf of another person whose capacity is impaired and this would involve third parties making a risk benefit assessment for someone else whose preferences and choices are oblivious to them. It is noteworthy to appreciate that surrogate consent is looked at favourably in patients that suffer from dementia and have cognitive impairment and poor decision-making capacities (Vries et al., 2013). This is an area that is debatable and arguable. Yet for research in dementia to continue and to enable treatment-related decisions to be made in these populations many institutional ethics committees and hospitals permit surrogate consent in these populations (Cohen-Mansfield, 2003).

The ideal surrogate for consent is one who understands all the healthcare needs of the patient concerned. A close friend who knows the patient for years may at times be a better person as a decision-maker than a close family member. The surrogate decision-maker must be one who understands the patient's healthcare values, beliefs and goals. Clinicians may have to determine at times the best surrogate decision-makers when multiple relatives are present. In many cases, direct blood relatives like the kin are chosen over close friends and family in the absence of the spouse. If the spouse is alive, then he or she is the best decision-maker for the patient (Kim et al., 2011).

In consent with dementia patients that do understand have fair cognitive capacities, the researcher shall have to go slow and explain details clearly and even repeat information in different ways to make the patient understand. From a psychological standpoint, the psychology of ageing has its own effects on the informed consent process (Warner & Nomani, 2008). Ageing may make a patient feel submissive and develop a passive standpoint in relation to the environment with more introversion and belief on the physician including a dependency and reliability on what he suggests rather than taking one based on one's own beliefs and decision-making powers. Many a times there may be a difference of opinion between the clinician's choice and the patient's choice (Hellström et al., 2007). Thus, research in dementia where informed consent needs to be obtained is a huge ethical dilemma for many.

Challenge 5: Ethical Issues in Old-Age Home Residents with Dementia

The staff of an old-age home that caters to patents that suffer from dementia experience ethical dilemmas wherein there may be problems ascertaining right from wrong and taking a decision may be cumbersome. They also face ethical issues associated with colleagues and the interactions with dementia patients and their caregivers (Jakobsen & Sørlie, 2010). Research on caregivers and ethically difficult situations has been conducted widely, and it has been documented that caregivers of patients that suffer from dementia go through ethically vexing situations daily that cause frustration and anguish on a daily basis. Dementia care in old-age homes faces greater challenges in the west due to an increased ageing population and scarcity of caregivers for the same (Edberg et al., 2008).

Despite their training, the caregivers in an old-age home experienced difficulty when faced by such complex ethical situations. Sometimes they may have to take surrogate decisions as the patient with dementia is not capable of making a decision. One has to be careful when things are done against the patient's will as the dignity of the dementia patient is paramount in all decisions that are made (Brodaty et al., 2003). The caregivers of dementia patients feel confused when the balance between autonomy and dignity is challenged. Two factors in an old age are important: one, the use of weak restraint to preserve a patient's dignity, and the caregiver's demands for the use of physical restraint when needed (Rabins et al., 2006). Thus, multiple issues may worry staff when it comes to the ethical decisions they make for patients with dementia. It is also prudent to mention that the basic ethos and value of the organization that runs the old-age home also plays a vital role in the ethical issues and how they are negotiated.

Challenge 6: Therapeutic Lying in Dementia

Research base indicates therapeutic lying does in fact rise up in the dementia care process. The family of the patient with dementia engages in deception that is both routine and contextual. Deception in the form of lying or making up stories by the caregiver is fuelled by the patient's failing awareness and reduced cognitive capacity (Culley et al., 2013). Nursing home and old-age home care staff admit to lying or deceiving patients that suffer from dementia in variety of situations that will help the dementia patient eventually. Therapeutic lying helps ease patient distress, family stress, improves medical compliance and helps in better care (James & Caiazza, 2018). Many staff and caregivers believe that when open disclosure or the truth would lead to aggression by the patient suffering from dementia, the lie is much more forthcoming. When it comes to telling the truth, the intention of the person is very important. Beneficence and non-maleficence may promote the caregiver being deceptive as the truth may cause the patient with dementia distress, anguish, depression and anger (Tuckett, 2012).

Many people may argue that therapeutic lying is unethical. But the ethical argument is that lying is fine if it meets the need of the patient with dementia, is person-centred and therapeutic. It is important that truth telling and lying be weighed on a situational basis and the ethics requires an appreciation of the contextual or situation to situation basis (Elvish et al., 2010). Therapeutic lying has to be understood under the ethical tenets of compassion, benevolence and non-maleficence. This is essential so that the caregiver acts in the best interest of the patient. Beneficence means doing good and involves a concern for the well-being of the patient that suffers from dementia. Many issues remain unstudied when it comes to therapeutic lying in dementia. The potential threat to the relationship between caregivers, nurses and patients with dementia is an area of concern when therapeutic lying is exerted. Therapeutic lying always requires careful consideration. Further research will help us ascertain and determine its place in the care of patients that suffer from dementia (Cantone et al., 2019).

Challenge 7: Ethical Issues with Respect to Biomarkers in Dementia

There are a large number of biomarkers that are being studied to assess the risk, early diagnosis and disease progression in dementia. The first concern is the economic cost of testing versus the tangible benefits. Equity and equality of distribution of biomarker resources is paramount. There are also issues in connection with the practicality of the screening process. One must have a valid informed consent prior to screening in order to ensure patient autonomy, but dementia is a progressive disorder with cognitive decline and this may challenge the validity of such consent (Wright et al., 2009). At some time during the screening process, one may seek assent from caregivers or family members to provide consent on behalf of the patient. The screening would also raise ethical issues on how a society would cater to individuals with a high dementia risk and often the predictive value of screening tests is usually low (Chiu & Brodaty, 2013). Should existing health insurance systems support screening to ease the financial burden on the patient? Should individuals at high risk still be allowed to hold a driver's license, sign contracts, apply for a bank loan, or make a valid will. There is also a debate that this biomedicalization of ageing could affect routine medical care and override patient choice (Hughes et al., 2017). Dementia is a disorder where there is a lack of its clarity regarding lying on a continuum between a disease, an illness, a disability, an impairment, or an accelerated form of ageing and even a variant of normal ageing. Biomarkers need to be viewed as tools that help us understand the disease causation and risk better with no confirmations and must not be viewed as a tool that can save us from ageing itself (Prvulovic & Hampel, 2011).

Challenge 8: Ethical Issues with Regard to Genetic Testing in Dementia

There has been a large amount of recent progress in unravelling the genetics of dementia and related disorders with a huge surge of research in this area. The development of genetics for dementia has led to misperceptions, confusions and new ethical issues among public health and doctors that care for patients with dementia (Post, 1994). The genetic tests for dementia need to be carefully ordered, as they are expensive and sound genetic counselling for both the patient (if possible) and family members are needed once the results are obtained. Genetic testing is unique due to the diagnostic and predictive implications for caregivers in addition to the patient with dementia (Hedera, 2001). The results may lead to possible job and insurance discrimination based on the results. Another paramount issue is insurance coverage of these tests and privacy issues with the access of the test results to third parties. Genetic testing using susceptibility genes are of limited diagnostic value the predictive value and its effects on first-degree relatives need analysis. Also, while many patients may opt for genetic testing but the testing of first degree relatives and subjects at risk are not recommended as individual risk cannot be quantified and there are no gene based interventions even if the risk for dementia is known (Goldman, 2015). The genetics of dementia is just in its nascent stage and as it evolves, the ethical and societal ramifications shall be huge.

Challenge 9: Surrogate Decision-Making for Patients with Dementia

Often individuals with early-stage dementia may be able to process and give informed consent for clinical research or procedures but as dementia progresses there is loss of decisional capacity. When this happens, the patient depends upon a legally authorized representative (i.e. caregiver or other) to provide this surrogate consent, probably with the lack of objection of the person that suffers from dementia (Fetherstonhaugh et al., 2017). The ethical and legal standards for this substituted decision-making is knowing the patient's wishes, knowing and reflecting the patient's values and always considering his or her best interests (i.e. what is perceived to be best for the person). This caregiver who makes decisions may also serve as a partner for the dementia subject in medical and surgical procedures, clinical trials and research studies. Family and caregivers usually serve this role as study partners in dementia-related clinical research (Black et al., 2013).

The nature and intent of the surrogate decision-maker is vital. Surrogates are most likely to choose only best interests for dementia patients as the ethical standard would suggest. Many surrogates however see from their own perspective and favour substituted judgement over best interests. The interests of the caregivers and other family members who serve as the study partner comes into question. Thus, this informed consent made on behalf of the patient is complex, dynamic, personal and an ethically challenging task (Koppelman, 2002). The most common reasons for giving consent have been cited to be the desire that study or clinical trial participation would benefit the dementia patient in some way and serve for the greater good of dementia patients in the future. Many patients with dementia are unable to give an informed consent but may still may be able to appoint their decision-maker. It is therefore important that we exhort patients with dementia to appoint this decision-maker and express these preferences for the future while their cognitive capabilities are intact (Black et al., 2009).

Challenge 10: Ethical Aspects of Stopping Medication in Dementia

There is current thought that like advanced cancers, dementia is a progressive illness and that dementia patients nearing the end of their lives may not receive adequate end-of-life care as needed. Research in the area of how and why medication must be stopped in patients with dementia is limited. Stoppage of medication in dementia has been understudied while it may reduce the treatment burden on the patient and also serve to improve the quality of life of the patient (Parsons et al., 2010). Most of the evidence speaks against the discontinuation of drugs used in dementia, avoiding antipsychotic agents, stoppage of statins and antihypertensives as well as antihyperglycemic drugs and anticoagulants. There are large ethical dilemmas in accurately estimating life expectancy and the ethical considerations regarding withholding or withdrawing treatment and wishes patient and/or their family have to be considered (Tjia & Givens, 2012). The cultural considerations of the patient and some religious aspects also need to be considered. The physician never takes these decisions and these are left to the family members (Parsons et al., 2014). There is need for doctors and healthcare workers to consider all the facts when speaking to relatives and supervising their decision for medication discontinuation. There is a need for medical and ethical bodies to develop guidelines on how the medicines may be discontinued if needed in these patients while keeping the dignity of the patient in mind at all times (Jiménez et al., 2013).

Challenge 11: End-of-Life Care and Related Decisions in Patients with Dementia

Dementia has been recognized as a terminal illness. There is a scarcity of clinical guidelines with regard to end-of-life care in dementia. Dementia care pathways have not been developed to make sure that palliative care and end-of-life care from various medical settings could be transferred to patients with dementia when needed. The need for developing this pathway for advanced dementia care might improve quality of life, education and also facilitate better integrated care (Caron et al., 2005). A multidisciplinary team approach is needed, with a wide range of specialties, like geriatric psychiatry, palliative care specialists, occupational and physiotherapists, general physicians, specialist nursing care support and specially trained hospital staff (Robinson et al., 2005). We have already addressed the issue of stoppage of medication in the paragraph above. Another ethical dilemma in the end-of-life care of dementia is whether to continue hospital treatment or discharge the patient and let him spend his final days with his family at home.

Negotiating ethical dilemmas like this needs clinical judgement and wisdom on the part of the clinician. These can only be acquired through experience and practice. This wisdom needs weight to be put on the value of caring that enhances human flourishing (Elliott et al., 2009). The end-of-life care of dementia and the acceptance of death in patients with severe and end stage dementia is no way linked to the concept of ending their life by active or passive euthanasia. The ubiquity of these clinical and ethical complexities in end-of-life care in dementia needs specialized knowledge by treating doctor. Training in these skills also needs to be imparted to result sound judgement on the part of the treating physician (Congedo et al., 2010).

Challenge 12: Living Wills and Advance Directives in Patients with Dementia

Advance care documents in dementia care, include advance directives, living wills and valid powers of attorney made when the patient was cognitively intact, have become popular tools, because the patient can have greater control over treatment decisions in case of future incompetence. They are means of conserving patient autonomy in patients who may be incompetent and have an inability to express their needs, desires and preferences with regard to treatments and related decisions and end-of-life care (Boer et al., 2010). Advance directives for dementia treatment are needed patients with dementia are incompetent and do not have the capacity to think about the medical choices and decisions to be made in the future. In cases of dementia, many patients already have cognitive impairment and are unable to execute any kind of detailed advance directive for their care anyways (Berghmans, 1998). Since the directives are meant to apply for the future, the person formulating them must be able to think and anticipate the future well. Formal advance directives for dementia research can be useful and valid as supplements to informed consent by caregivers. However, the major challenge is to get dementia patients to prepare these directives when they are in a sound state of mind and for surrogate decision-makers to follow them when the patient has lost his cognitive capabilities (Mezey et al., 2000).

Challenge 13: Ethical Aspects of Sexual Issues in Patients with Dementia

Intimacy and sexuality expressed by nursing home, old-age home and hospital patients with dementia remains an ethically vexing and highly sensitive issue for these facilities, their nursing staff and caregivers. Dealing with the sexual desire or behaviour of dementia patients is extremely difficult and puts huge burden and shame on caregivers and nursing staff. The caregivers and healthcare staff often do not know how to react when dementia patients express sexuality. It is important to consider how to deal with such expressions of intimacy and sexuality by patients with dementia in an ethically responsible way (Tarzia et al., 2012). The nursing staff and caregivers have to balance the patient's rights and dignity, their duty of care, and the existent ageism and negative attitudes towards sexuality in older adults. In many cases, the sexual expression by patients with dementia is overlooked, ignored, or even discouraged and thought to be not needed (Benbow & Beeston, 2012). Another vital concept is that when a patient with dementia may make sexual overtures, one has to ascertain whether there is a genuine sexual need, or is this a part of the dementia brain damage or even a symptom of the behavioural and psychological symptoms of dementia. One must also know that the dementia patient while expressing sexuality may have little ability to consent for the same. The patient must be protected ethically against sexual misuse or abuse in the light of the same (Kontos et al., 2016).

Sexual activity decreases increasing age but many older people remain sexually active, and there is no age limit to sexual responsiveness as shown in various studies. Sexuality is an important issue in those suffering from dementia. Attitudes and beliefs towards sexuality and dementia are shaped by stereotypical thinking of the common man and healthcare workers or caregivers alike (Dhingra et al., 2016). Dementia care training courses should include sexuality as a module their training curricula. One must assess sexual needs of a patient with dementia based on environmental factors, group living situations, consent for sexuality and genuine desire versus it being a product of ensuing brain damage. Ethical frameworks must investigate how staff would view these issues while providing holistic person-centred care to patients that suffer from dementia (Grigorovich & Kontos, 2018).

Challenge 14: Covert Medication, Restraint and Its Use in Patients with Dementia

The practice of drug administration in food and beverages is well known in psychiatry and has also been used in treatment of dementia but data on its usage is not available. Covert medication and restraint of patients with dementia touches on the legal and ethical issues of patient competence, decision-making, desires, autonomy and insight (Latha, 2010). Medicating patients without their knowledge must not be used to just calm an aggressive patient but must also be used to enhance compliance in many cases. The ethical need is ensuring the well-being of a patient with dementia that lacks the ability to give informed consent (Singh, 2008). This covert/surreptitious administration of medication may be seen as a breach of trust and compensates the doctor patient relationship. Physical restraint may imply the desire to dominate or suppress the patient and violates the basic human rights of the patient with dementia. The practice of covert administration of psychiatric medication is only now covered in the mental health legislation in developing countries (Welsh & Deahl, 2002). There has been increasing importance accorded to respect for autonomy, and this is how the debate of covert medication use and restraint has come into the light. Although some ethicists view covert medication administration as a deprivation of the basic rights of the patient, it is also prudent to state that, at times paradoxically, withholding medication needed to treat mental illnesses can also be viewed as a violation of patient's rights (Haw & Stubbs, 2010). Both restraint and covert medication have to be carried out with the sole premise in ethics that one is protecting the patient and doing no harm in the first place.

Challenge 15: Electronic Tagging of Patients with Dementia

Electronic tagging in patients with dementia though a new technology has sprung up new ethical questions. The first ethical dilemma is, one being drawn between the complexity of safety of the dementia and the civil liberty restriction aspect of human rights by tagging. We may argue that electronic tagging is far better than the rights of dementia patient as tagging would provide for an excellent source to track if there would be any violation of the person (with dementia) who may get robbed or attacked, or were found alone or on a lonely road along with the chances of an accident occurring. Tagging may help reduce the anxiety of the caregivers, both in home-based and nursing home settings, also allowing mobility and freedom to the patient. Electronic tagging is like wearing a waist band and is less tough on the patient with dementia than the use of physical restraint in a hospital to minimize wandering away behaviour (O'Neill, 2013).

When patients with dementia lack decision-making capacity, the caregivers make this decision but ethical issues arise when we have patients with dementia with partial decision-making capacity and whether they could be involved in these decisions. Does tagging entail the patient's interests or rather a peace of mind for the caregiver? It has also been suggested by some authorities that wandering, where the person does not get lost, may help patients with dementia by providing them ease of movement, exercise and a greater space to move about (Bail, 2003).

Many people may feel that electronic tagging patients with dementia is like tagging of criminals. Many patients with dementia in India are kept in locked rooms and a wandering episode may pre-empt a patient's placement into residential care or dementia homes. Technology like electronic tagging may increase the chances that the patient is kept and managed at home. Greater freedom of movement and independence while providing psychological relief for relatives and caregivers. There will be a need for a 24-h call centres for monitoring these devices and tracking them and resources for the same are scarce in our country (Miskelly, 2005).

Patients with dementia need to be mobile, free and engage in social interaction and dependence on electronic tagging may be detrimental to their quality of life. Clinicians and caregivers must note that tracking methods should never be a substitute for quality care and the fostering of meaningful relationships in patients with dementia (Gu, 2015). The last ethical aspect of electronic tagging that remains is testing it in randomized controlled trial. There are small case reports and case series but no randomized clinical trials of its efficacy. Though there are no hazards noted so far, the efficacy is difficult to be reliably established unless such trials are undertaken. Nursing homes and old-age homes where multiple wandering incidents may happen in a day is the best place to try such devices (Lodha & Sousa, 2020). Only with continued usage will we realize the full ethical gamut of electronic tagging in dementia.

Challenge 16: Everyday Ethics in Nursing Homes and Dementia Day Care

Everyday ethics constitutes a whole new set of ethical questions. A working definition of everyday the ethics that concerns not just only questions of life, death and right or wrong but about how one handle life on a daily basis. The ethics of day-to-day life is a part of healthcare ethics as is the ethics of the complex situations one may encounter (Powers, 2001). The dementia care medical team, hospital staff and caregivers must engage in ethical decision-making and ethical deliberation while being open minded to matters of significance. Everyday ethics in dementia care calls for a case-based rather than a guideline-based approach in dementia care (Powers, 2005).

One must have an ethically driven approach to dementia care that promotes values and principles governing our opinions about what "should be" in the care and treatment of patients with dementia. Everyday ethical issues in dementia involve clinical debates on the limits of interventions in dementia, physical control and restraint, informed consent and caregiver preferences, human dignity and community values as these apply to dementia care (Bolmsjö et al., 2006). An ethics committee in old-age homes and nursing homes can help as a platform to examine recurrent ethical issues and can be an ethical decision-making guide to caregivers, treatment teams and the nursing home community as a whole. Homes where residential care for patients with dementia is provided must work towards an ethical framework for managing the ethical issues in everyday care of patients with dementia (Powers, 2003).

Challenge 17: Ethics of Neuroimaging in Dementia

Neuroimaging in dementia may help us ascertain brain changes that predict the onset and risk for the disorder. This is coupled with clinical expertise that helps us determine at risk individuals. The key question is that once neuroimaging helps us determine these at risk groups—should we start intervention in this group to prevent the onset of dementia and will it work. Would neuroimaging as a tool helps us predict whether these drugs would enhance normal function in at risk individuals. Would starting Donepezil in at risk individuals for dementia prevent the onset of the illness and reduce the risk of developing dementia. These are some of the ethical dilemmas that dementia care faces in the wake of neuroimaging (Roskies, 2002). Will a day arrive when neuroimaging may be used as a diagnostic and prognostic tool like we use echocardiography in cardiology. It shall then be imperative to make the cost of cutting edge technology less so that it is available to all. Can neuroimaging substitute the human observation and the mental status examination which we assume is far superior in cases with dementia. Too much of medicalization and biological basis may also ruin the sanctity of the psychological components of psychiatric treatment and affect the doctor patient relationship in general. We are treating human beings and not just brains. This is one of the ethical problems on relying too much on neuroimaging. Another question also is when treatment may ameliorate neuroimaging findings and the patient may still not feel better—where does one then turn to when such phenomena ensue (Levy, 2007).

There is need for clinicians, patients, caregivers, neuroscientists and students to be aware of these ethical dilemmas as they venture into the field of neuroscience. Neuroethics poses interesting crossroads for dementia research of the future and ranges from whether this research is needs to the long-term ramifications of these research projects and how these research projects can be used for the greater good of society and mankind in general (Farah, 2005).

Conclusions

Multiple changes are occurring in the clinical and ethical milieu involving patients with dementia. We will have a large number of individuals affected by dementia because of the large ageing population in the future. As our healthcare systems learn and evolve, a variety of ethical issues in dementia will emerge. The above challenges are the major ethical challenges in dementia and covering the entire ethical conundrum of dementia would entail a book itself. Dementia is a disorder that is challenging and shall keep challenging us ethically in the future. We need to train healthcare workers, dementia care medical personnel and caregivers to be able to rise up to this challenge.

Acknowledgements Nil

Conflict of Interest Nil

Funding Nil

References

Bail, K. D. (2003). Electronic tagging of people with dementia: Devices may be preferable to locked doors. *BMJ, 326*(7383), 281–286.

Bamford, C., Lamont, S., Eccles, M., Robinson, L., May, C., & Bond, J. (2004). Disclosing a diagnosis of dementia: A systematic review. *International Journal of Geriatric Psychiatry, 19*(2), 151–169.

Benbow, S. M., & Beeston, D. (2012). Sexuality, aging, and dementia. *International Psychogeriatrics, 24*(7), 1026–1033.

Berghmans, R. L. (1998). Advance directives for non-therapeutic dementia research: Some ethical and policy considerations. *Journal of Medical Ethics, 24*(1), 32–37.

Black, B. S., Fogarty, L. A., Phillips, H., Finucane, T., Loreck, D. J., Baker, A., Blass, D. M., & Rabins, P. V. (2009). Surrogate decision makers' understanding of dementia patients' prior wishes for end-of-life care. *Journal of Aging and Health, 21*(4), 627–650.

Black, B. S., Wechsler, M., & Fogarty, L. (2013). Decision making for participation in dementia research. *The American Journal of Geriatric Psychiatry, 21*(4), 355–363.

Bolmsjö, I. Å., Sandman, L., & Andersson, E. (2006). Everyday ethics in the care of elderly people. *Nursing Ethics, 13*(3), 249–263.

Brannelly, T. (2006). Negotiating ethics in dementia care: An analysis of an ethic of care in practice. *Dementia, 5*(2), 197–212.

Brodaty, H., Draper, B., & Low, L. F. (2003). Nursing home staff attitudes towards residents with dementia: Strain and satisfaction with work. *Journal of Advanced Nursing, 44*(6), 583–590.

Brown, L. B., & Ott, B. R. (2004). Driving and dementia: A review of the literature. *Journal of Geriatric Psychiatr Neurol, 17*(4), 232–240.

Cantone, D., Attena, F., Cerrone, S., Fabozzi, A., Rossiello, R., Spagnoli, L., & Pelullo, C. P. (2019). Lying to patients with dementia: Attitudes versus behaviours in nurses. *Nursing Ethics, 26*(4), 984–992.

Carmody, J., Traynor, V., Iverson, D., & Marchetti, E. (2013). Driving, dementia and A ustralian physicians: Primum non nocere? *Internal Medicine Journal, 43*(6), 625–630.

Caron, C. D., Griffith, J., & Arcand, M. (2005). End-of-life decision making in dementia: The perspective of family caregivers. *Dementia, 4*(1), 113–136.

Carr, D. B., & O'Neill, D. (2015). Mobility and safety issues in drivers with dementia. *International Psychogeriatrics, 27*(10), 1613–1622.

Carter, K., Monaghan, S., O'Brien, J., Teodorczuk, A., Mosimann, U., & Taylor, J. P. (2015). Driving and dementia: A clinical decision pathway. *International Journal of Geriatric Psychiatry, 30*(2), 210–216.

Chiu, H. F., & Brodaty, H. (2013). Arguments against the biomarker-driven diagnosis of AD. *International Psychogeriatrics, 25*(2), 177–181.

Cohen-Mansfield, J. (2003). Consent and refusal in dementia research: Conceptual and practical considerations. *Alzheimer Disease & Associated Disorders, 17*, S17-25.

Congedo, M., Causarano, R. I., Alberti, F., Bonito, V., Borghi, L., Colombi, L., Defanti, C. A., Marcello, N., Porteri, C., Pucci, E., & Tarquini, D. (2010). Ethical issues in end of life treatments for patients with dementia. *European Journal of Neurology, 17*(6), 774–779.

Cornett, P. F., & Hall, J. R. (2008). Issues in disclosing a diagnosis of dementia. *Archives of Clinical Neuropsychology, 23*(3), 251–256.

Cubit, K. (2010). Informed consent for research involving people with dementia: A grey area. *Contemporary Nurse, 34*(2), 230–236.
Culley, H., Barber, R., Hope, A., & James, I. (2013). Therapeutic lying in dementia care. *Nursing Standard, 28*(1), 35–39.
De Boer, M. E., Hertogh, C. M., Dröes, R. M., Jonker, C., & Eefsting, J. A. (2010). Advance directives in dementia: Issues of validity and effectiveness. *International Psychogeriatrics, 22*(2), 201–208.
De Lepeleire, J., Buntinx, F., & Aertgeerts, B. (2004). Disclosing the diagnosis of dementia: The performance of Flemish general practitioners. *International Psychogeriatrics, 16*(4), 421–428.
De Vries, R., Ryan, K. A., Stanczyk, A., Appelbaum, P. S., Damschroder, L., Knopman, D. S., & Kim, S. Y. (2013). Public's approach to surrogate consent for dementia research: Cautious pragmatism. *The American Journal of Geriatric Psychiatry, 21*(4), 364–372.
Dewing, J. (2007). Participatory research: A method for process consent with persons who have dementia. *Dementia, 6*(1), 11–25.
Dhingra, I., De Sousa, A., & Sonavane, S. (2016). Sexuality in older adults: Clinical and psychosocial dilemmas. *J Geriatr Ment Health, 3*(2), 131–139.
Edberg, A. K., Bird, M., Richards, D. A., Woods, R., Keeley, P., & Davis-Quarrell, V. (2008). Strain in nursing care of people with dementia: Nurses' experience in Australia, Sweden and United Kingdom. *Aging & Mental Health, 12*(2), 236–243.
Elliott, B. A., Gessert, C. E., & Peden-McAlpine, C. (2009). Family decision-making in advanced dementia: Narrative and ethics. *Scandinavian Journal of Caring Sciences, 23*(2), 251–258.
Elvish, R., James, I., & Milne, D. (2010). Lying in dementia care: An example of a culture that deceives in people's best interests. *Aging & Mental Health, 14*(3), 255–262.
Farah, M. J. (2005). Neuroethics: The practical and the philosophical. *Trends in Cognitive Sciences, 9*(1), 34–40.
Fetherstonhaugh, D., McAuliffe, L., Bauer, M., & Shanley, C. (2017). Decision-making on behalf of people living with dementia: How do surrogate decision-makers decide? *Journal of Medical Ethics, 43*(1), 35–40.
Goldman, J. S. (2015). Genetic testing and counseling in the diagnosis and management of young-onset dementias. *Psychiatria Clinica, 38*(2), 295–308.
Grigorovich, A., & Kontos, P. (2018). Advancing an ethic of embodied relational sexuality to guide decision-making in dementia care. *The Gerontologist, 58*(2), 219–225.
Gu, L. (2015). Nursing interventions in managing wandering behavior in patients with dementia: A literature review. *Archives of Psychiatric Nursing, 29*(6), 454–457.
Haw, C., & Stubbs, J. (2010). Administration of medicines in food and drink: A study of older inpatients with severe mental illness. *International Psychogeriatrics, 22*(3), 409–416.
Hedera, P. (2001). Ethical principles and pitfalls of genetic testing for dementia. *Journal of Geriatric Psychiatry and Neurology, 14*(4), 213–221.
Hellström, I., Nolan, M., Nordenfelt, L., & Lundh, U. (2007). Ethical and methodological issues in interviewing persons with dementia. *Nursing Ethics, 14*(5), 608–619.
Herrmann, N., Rapoport, M. J., Sambrook, R., Hébert, R., McCracken, P., & Robillard, A. (2006). Predictors of driving cessation in mild-to-moderate dementia. *Canadian Medical Association Journal, 175*(6), 591–595.
Hughes, J. C., Hope, T., Reader, S., & Rice, D. (2002). Dementia and ethics: The views of informal carers. *Journal of the Royal Society of Medicine, 95*(5), 242–246.
Hughes, J. C., Ingram, T. A., Jarvis, A., Denton, E., Lampshire, Z., & Wernham, C. (2017). Consent for the diagnosis of preclinical dementia states: A review. *Maturitas, 98*, 30–34.
Jacoby, R., & Steer, P. (2007). How to assess capacity to make a will. *BMJ, 335*(7611), 155–157.
Jakobsen, R., & Sørlie, V. (2010). Dignity of older people in a nursing home: Narratives of care providers. *Nursing Ethics, 17*(3), 289–300.
James, I. A., & Caiazza, R. (2018). Therapeutic lies in dementia care: Should psychologists teach others to be person-centered liars? *Behavioural and Cognitive Psychotherapy, 46*(4), 454–462.

Jiménez, M. A., Jaén, M. C., García, M. V., & Barahona-Alvarez, H. (2013). Decision-making in older people with dementia. *Reviews in Clinical Gerontology, 23*(4), 307–316.
Kennedy, K. M. (2012). Testamentary capacity: A practical guide to assessment of ability to make a valid will. *Journal of Forensic and Legal Medicine, 19*(4), 191–195.
Kim, S. Y., Kim, H. M., Knopman, D. S., De Vries, R., Damschroder, L., & Appelbaum, P. S. (2011). Effect of public deliberation on attitudes toward surrogate consent for dementia research. *Neurology, 77*(24), 2097–2104.
Kontos, P., Grigorovich, A., Kontos, A. P., & Miller, K. L. (2016). Citizenship, human rights, and dementia: Towards a new embodied relational ethic of sexuality. *Dementia, 15*(3), 315–329.
Koppelman, E. R. (2002). Dementia and dignity: Towards a new method of surrogate decision making. *Journal of Medicine and Philosophy, 27*(1), 65–85.
Laakkonen, M. L., Raivio, M. M., Eloniemi-Sulkava, U., Saarenheimo, M., Pietilä, M., Tilvis, R. S., & Pitkälä, K. H. (2008). How do elderly spouse care givers of people with Alzheimer disease experience the disclosure of dementia diagnosis and subsequent care? *Journal of Medical Ethics, 34*(6), 427–430.
Latha, K. S. (2010). The noncompliant patient in psychiatry: The case for and against covert/surreptitious medication. *Mens Sana Monographs, 8*(1), 96–129.
Levy, N. (2007). Neuroethics: Challenges for the 21st century. Cambridge University Press.
Lodha, P., & De Sousa, A. (2020). Ethics of electronic tagging of people with dementia. *Indian Journal of Medical Ethics, 5*(1), 83–84.
Mezey, M. D., Mitty, E. L., Bottrell, M. M., Ramsey, G. C., & Fisher, T. (2000). Advance directives: Older adults with dementia. *Clinics in Geriatric Medicine, 16*(2), 255–268.
Miles, S. H. (2005). *The Hippocratic Oath and the ethics of medicine.* Oxford University Press
Miskelly, F. (2005). Electronic tracking of patients with dementia and wandering using mobile phone technology. *Age and Ageing, 34*(5), 497–499.
O'Neill, D. (2013). Should patients with dementia who wander be electronically tagged? *BMJ, 346*, 3606.
Parsons, C., Hughes, C. M., Passmore, A. P., & Lapane, K. L. (2010). Withholding, discontinuing and withdrawing medications in dementia patients at the end of life. *Drugs and Aging, 27*(6), 435–449.
Parsons, C., McCorry, N., Murphy, K., Byrne, S., O'Sullivan, D., O'Mahony, D., Passmore, P., Patterson, S., & Hughes, C. (2014). Assessment of factors that influence physician decision making regarding medication use in patients with dementia at the end of life. *International Journal of Geriatric Psychiatry, 29*(3), 281–290.
Post, S. G. (1994). Genetics, ethics, and Alzheimer disease. *Journal of the American Geriatrics Society, 42*(7), 782–786.
Post, S. G. (2000). Key issues in the ethics of dementia care. *Neurologic Clinics, 18*(4), 1011–1022.
Powers, B. A. (2001). Ethnographic analysis of everyday ethics in the care of nursing home residents with dementia: A taxonomy. *Nursing Research, 50*(6), 332–339.
Powers, B. A. (2003). *Nursing home ethics: Everyday issues affecting residents with dementia.* Springer Publishing Company.
Powers, B. A. (2005). Everyday ethics in assisted living facilities: A framework for assessing resident-focused issues. *Journal of Gerontological Nursing, 31*(1), 31–37.
Prvulovic, D., & Hampel, H. (2011). Ethical considerations of biomarker use in neurodegenerative diseases—a case study of Alzheimer's disease. *Progress Neurobiology, 95*(4), 517–519.
Foster, C., Herring, J., Doron, I., (Eds.) (2014). *The law and ethics of dementia.* Bloomsbury Publishing.
Rabins, P. V., Lyketsos, C. G., & Steele, C. D. (2006). *Practical dementia care.* Oxford Uiniversity Press.
Robinson, L., Hughes, J., Daley, S., Keady, J., Ballard, C., & Volicer, L. (2005). End-of-life care and dementia. *Reviews in Clinical Gerontology, 15*(2), 135–148.
Roskies, A. (2002). Neuroethics for the new millenium. *Neuron, 35*(1), 21–23.

Shulman, K. I., Cohen, C. A., & Hull, I. (2005). Psychiatric issues in retrospective challenges of testamentary capacity. *International Journal of Geriatric Psychiatry, 20*(1), 63–69.
Shulman, K. I., Cohen, C. A., Kirsh, F. C., Hull, I. M., & Champine, P. R. (2007). Assessment of testamentary capacity and vulnerability to undue influence. *American Journal of Psychiatry, 164*(5), 722–727.
Singh, A. R. (2008). Covert treatment in psychiatry: Do no harm, true, but also dare to care. *Mens Sana Monographs, 6*(1), 81–109.
Slaughter, S., Cole, D., Jennings, E., & Reimer, M. A. (2007). Consent and assent to participate in research from people with dementia. *Nursing Ethics, 14*(1), 27–40.
Snyder, C. H. (2005). Dementia and driving: Autonomy versus safety. *Journal of the American Academy of Nurse Practitioners, 17*(10), 393–402.
Tarzia, L., Fetherstonhaugh, D., & Bauer, M. (2012). Dementia, sexuality and consent in residential aged care facilities. *Journal of Medical Ethics, 38*(10), 609–613.
Tjia, J., & Givens, J. (2012). Ethical framework for medication discontinuation in nursing home residents with limited life expectancy. *Clinics in Geriatric Medicine, 28*(2), 255–272.
Tuckett, A. G. (2012). The experience of lying in dementia care: A qualitative study. *Nursing Ethics, 19*(1), 7–20.
Warner, J., & Nomani, E. (2008). Giving consent in dementia research. *Lancet, 372*(9634), 183–185.
Welsh, S., & Deahl, M. (2002). Covert medication–ever ethically justifiable? *Psychiatric Bulletin, 26*(4), 123–126.
Whitehouse, P., Frisoni, G. B., & Post, S. (2004). Breaking the diagnosis of dementia. *Lancet Neurology, 3*(2), 124–128.
Wright, C. F., Hall, A., Matthews, F. E., & Brayne, C. (2009). Biomarkers, dementia, and public health. *Annals New York Academy of Sciences, 1180*(1), 11–19.

Chapter 17
Technology and Dementia Care—Global Challenges

Avinash De Sousa

Abstract There is a huge emphasis on how technology may be useful to persons suffering from dementia. The role of technology in dementia has been supported by governments, policymakers, caregivers and patients. The following chapter provides an overview of the role of technology in dementia care, and in the treatment and management of patients with dementia. Technology has been found to have a role in memory enhancement, treatment support, treatment planning, cognitive assessments, in providing safety and security and enhancing doctor–patient interaction in dementia. Technology may have lesser roles in severe dementia but have a great role to play in mild to moderate dementia. Technology has a role in both outpatient, home-based and nursing home-based care for dementia. The role of safety devices, smoke detectors and electronic tagging of patients with dementia along with emergency alarms have been elucidated in dementia. There is also a trend of the use of GPS devices and voice prompts. The role of technology in reminiscence therapy and therapeutic care in dementia has also been mentioned. The chapter also looks at the ethical issues brought by technology in dementia care.

Keywords Dementia · Technology · Virtual reality · Challenges

Introduction

Dementia is a neurodegenerative disease where there is memory loss and behavioural as well as personality changes with or without deficits in language, executive function and planning leading to social impairment and impaired quality of life (Prince et al., 2015). Dementia is a cluster term for various symptoms as well and presents as several types depending on which area of the brain is damaged and what the cluster of symptoms are. Dementia impacts the personal, social and family life of the individual

A. De Sousa (✉)
Desousa Foundation, Mumbai, India

Department of Psychiatry, Lokmanya Tilak Municipal Medical College, Mumbai, India

Present Address:
Carmel, 18, St. Francis Road, Off SV Road, Santacruz West, Mumbai 400054, India

M. K. Shankardass (ed.), *Dementia Care*,
https://doi.org/10.1007/978-981-16-3864-0_17

affected (Das et al., 2012). Patients with dementia may also develop behavioural and psychological symptoms such as depressive features, general stress, irritability and anger, agitation, anxiousness and disinhibited behaviours that are referred to as behavioural and psychological symptoms of dementia (BPSD) (Das et al., 2014). Till date, there is no cure for dementia, and thus, long-term care provision is inevitable in these patients.

Dementia represents a global pandemic, and many cases have been recorded in the developed world and most ageing countries. It has already been shown that 60–65% of patients suffering from dementia live in low- and middle-income countries (LAMIC) where health care for dementia is scarce. According to the World Health Organization, 50–55 million people suffer from dementia, and nearly 60–65% live in low- and middle-income countries alone (Vugt et al., 2006). Every year, nearly 10–12 million new cases of dementia are detected causing exponential rise in the global burden of dementia. The current healthcare challenge is to develop health and social care services to cope with this growing population of dementia cases and to improve the quality of care for people living with dementia and their caregivers (Lodha et al., 2019).

In a paper on technology and health care, it has been mentioned that "*Electronic health solutions have been described as essential tools to providing cost-effective, quality healthcare services to ageing populations. Technology and assistive devices have the potential to enhance the long-term care needs of people living with dementia and their caregivers. Additionally, technology can enhance knowledge exchange, education and virtual environments, in turn, increasing the standards of the care environments*" (Martínez-Alcalá et al., 2016). The term "technology" in this chapter will refer to the technology and the types of technology used in dementia care only.

Different Types of Technology in Dementia Care

Technology may be of different types, depending on the purpose it serves in care. Technology in its assistive nature is described as assistive devices to support people living with dementia and include prompts, reminders, electronic alarm systems, automatic lights and domestic appliances that switch themselves off at a certain point. There are also elderly friendly remote controls and mobile phones along with monitoring, and therapeutic interventions (Buckley, 2006). Another subset of technological devices is the pervasive telecare technologies which include sensors to detect movement, vibration, pressure, lack of activity, falls and body temperature which sends a beep or message to the caregiver when assistance is needed. The third category of technologies includes surveillance technologies like wearable bands and devices such as electronic tracking and Global Positioning Systems (GPS) locators that can locate a person if they wander away, get lost or try to move out of a particular circumferential area (Øderud et al., 2013).

Traditionally, when at home, people with dementia use the alarm, calendar or maybe prompts from caregiver to enhance memory or for medication compliance enhancement. Assistive technologies in dementia care include touchscreen devices for entertainment and to enhance quality of life along with reminiscence therapy through pictures, music and applications. New technological devices are continually being made and act as research prototypes that mean technology is aimed to support people living with dementia in their environments (Topo, 2009). Technology in dementia care is classified as devices used in ambient-assisted living (AAL): daily task facilitation, mobility assistance, health care and rehabilitation and social inclusion and communication (Li et al., 2015).

Following are some technological assists for dementia care—(Bharucha et al., 2009).

1. **Clocks**: special and digitalized clocks are designed to help patients with dementia easily differentiate between night and morning hours, warding off confusion. It also allows caregivers to tune in a schedule in these clocks.
2. **Communication aid**: adapted telephones, vide chatting forums like Skype and FaceTime along with mobile applications (like Talking Mats) help patients of dementia stay easily connected with their loved ones. There are also picture phones that are enabled with large fonts, preprogrammed with frequently called numbers and sometimes even buttons where the photographs of people can be placed to ease dialling-in and communication.
3. **Monitor for electrical appliance**: this is a novel piece of technology which has been specifically designed for caregivers who stay away from their loved ones who battle dementia. It monitors their use of electrical appliances by the patient with dementia by alerting caregivers if electrical appliances have not been turned off or they remain on.
4. **Home care robots**: designed only to supplement and not replace humane caregiving, these home care robots can help do the general housework and cleaning and they may remind the patient with dementia to take medication or even alert medical professionals when needed by messages.
5. **Reminders**: reminders can help keep properties and messages safe with the caregivers and patients suffering from dementia. These messages are recorded on a device by the caregiver for various reminders and playback out loud at the appropriate time programmed into the device. Reminders can be about medication, day-to-day activities or about upcoming appointments.
6. **Medication manager**: these medical managers can be something as simple as pillboxes that are marked for each day of the week or can sometimes take the form of fancy automated pill dispenser which beeps and opens to remind caregivers and patients to have their medication. Some medication reminders are like vibrating alarms on a watch.

Role of Technology in Dementia Care

Among the diverse symptomatology, cognitive impairment is a common feature of dementia wherein communication, language, understanding, loss of ability to recognize familiar people and things (agnosia), mood, loss of ability to carry out complex purposive actions (apraxia) and interfering with daily functioning are all affected (Cahill et al., 2007). Though psychopharmacological agents are used in the treatment for mild–moderate dementia, little can be done to reverse the condition of manage the underlying causes of the problem. Non-pharmacological management, viz. psychotherapeutic management of dementia is the mainstay in the treatment in dementia care but that also falls short of efficacious long-term outcome especially in all spheres of life. Till technology was recognized as a potent aspect of dementia care, care for patients with dementia was usually a combination of medication, psychotherapy and guidelines for family care. It is well known that dementia care has also risen the global economic burden and primary caregiver (family or personnel taking care of people with dementia) burnout as part of increased burden of dementia care (Cash, 2003).

There has been a need for innovation in dementia care to not only make dementia care long-term, efficacious and cost-effective but also to reduce the load on caregivers in order to improve the quality of life with patients and their caregivers. Yet, given the shrunken cognitive abilities that make remembering and learning new things difficult, it can be challenging to design technology that can be used by the people with dementia. Nevertheless, technology has been developed so as to help people with dementia with day-to-day activities. Such technology is often simple in design developed with an aim to improve intuitiveness and ease of use (Godwin et al., 2013).

Use of technology has several benefits for patients with dementia and they are—(Landau et al., 2010).

1. Replenishes the sense of independence among patients of dementia who are psychologically marred having lost their ability to take care of themselves
2. Improves quality of life
3. Easier risk management and enhanced safety
4. Personalized support catering to individual needs of patients
5. Supportive for patients, caregivers and family.

Technology can be used for wide-ranging purposes in dementia care, right from diagnosis, assessment to delivery care and providing overall support for the patient and the caregiver. What needs to be borne in mind is that the nature of technology use also changes with different stages of dementia. Technology in the form of smartphones, palm devices and tablets have created new avenues on how these devices can be a rising force in our lives. The availability of wearable gadgets, smart home gadgets, Amazon Alexa and Google Home Hub, home robots, virtual reality (VR)-based applications in dementia, the rise of artificial intelligence (AI) and its use in dementia care has opened a new field of gerotechnology; i.e. how technology can enhance the life of the elderly and how these gadgets can be used in the management of dementia (Nygård, 2008).

Though the role of technology in dementia is only recent, it has always been used in dementia care for the assessment of dementia. Two cognitive assessment batteries—Cambridge Neuropsychological Test Automated Battery (Robbins et al., 1994) and Examen Cognitif par Ordinateur (Ritchie et al., 1993)—are available in a digital mode. More recently, technology has also extended to paper–pencil assessment such as the clock drawing test. Additionally, technological use has also extended to make day-to-day activities like tea / coffee making and toast making instrumental.

In the diagnostic phase, technology comes to aid for imaging data such as information gathered through PET scans, MRIs and combined PET and MRI, as well as non-imaging linguistic analysis and gait analysis over time. The computerized analysis of neuroscience data from huge neuroimaging databases that have huge datasets of dementia patients shall pave the way for population precision neuroscience that shall help build better understanding of the role of neurobiology in dementia genesis and progress (Astell et al., 2019).

There are a range of cognitive assistive technologies that carefully deliver the prompts and with a longer-ranging goal to help detect progression of dementia in patients. Lastly, technology has also been used to assist leisure and social activities (Kim et al., 2016). Music and art are activities that are also assisted with technology by using applications like House of Memories and such multiple technological projects. There are also social interactive platforms as well as maintenance of cognitive function with technological platforms like Computer Interactive Reminiscence and Conversation Aid (CIRCA) (Ngandu et al., 2015). Gaming has also been explored for people with dementia with applications such as elder games and living in the Moment. Another aspect of technological assistance covers caregiving. Games designed for use by patients with dementia are more common with the availability of touchscreen devices to dementia patients (Astell et al., 2016). Reminiscence therapy has also been proven to be viable and effective through digital means via applications and with the digitalized use of photograph, video and music along with three-dimensional modelling of the environments (Huldtgren et al., 2017).

Robotics and Dementia

Though robots cannot replace humans, there are certain aspects that robots can take care for patients with dementia. The use of robots and their monitoring using sensors and videoconferencing (via tablets) has been studied. Robotics in dementia expands to areas like eating and feeding support and recreational activities in dementia care homes. Assistive technology has also made caregiving possible from remote distances through dementia technology that promotes care as well as independence. Where at one hand, there may be several caregivers living away from their loved one with dementia; on the other hand, there are also individuals with dementia who'd prefer taking care of themselves with less interference in such instances telecare and technology that allows monitoring of care through technology works best (Gibson et al., 2016).

Technology, to begin with, was considered to improve the quality of life for patients with dementia as well as those of the caregivers and family members. The World Health Organization, in 2017, launched a programme called iSupport which could be accessed online to improve the quality of life for both patients with dementia and their caregivers (Mehta et al., 2018).

Virtual Reality Technology in Dementia Care

When a dementia patient is upset or under stress, they may exhibit various forms of behaviours. Patients with dementia may manifest their frustration via repetitive moaning or may trouble their caregivers by constantly asking them questions. They might also indulge in excessive pacing as they are trying to remember a certain memory or item, which might create discomfort and confusion. Interrupting any of these behaviours may cause frustration or irritation as well as it might lead to increased symptoms of anxiety (Man et al., 2012).

Using the technology of VR has proven to be helpful while dealing with lowering symptoms of anxiety. As these patients may not be exposed to situations in person, virtual reality has the techniques to bring it to them via a headset. Research has found that patients with dementia being in touch with nature reduces stress and improves mood (Blackman et al., 2007). In another study, the effectiveness of virtual reality on engagement, apathy and mood states of people with dementia were studied and the findings revealed that overall, a positive outcome in patients with dementia as well as their family members was noted, thereby improving their overall quality of life (Manera et al., 2016).

Dementia patients need that their locked-in memories to be rekindled via various forms of stimuli that remind them of events of the past. The digitalization of old photograph albums and videos has enhanced the quanta of such stimuli available to patients with dementia. This serves as valuable asset to reminiscence therapy used in the management of dementia (Wijma et al., 2018).

Electronic Tagging in Dementia

Many patients with dementia stay alone without a caregiver and show wandering away behaviour where they must just leave their house and move away. It has been suggested that patients with dementia have an identity card with their details so that in case the patient wanders away, he may be identified and the caregivers may be informed by the police or good people in the community. Though an identity card can be work or carried around, it may not be most effective to inform caregivers about this and may also not be very safe. In recent paper, the author of this chapter has mentioned that "*a medical company in India has recently launched an electronic tagging device for cell phones, to help relatives track the whereabouts of dementia patients beyond a certain perimeter of the home address fed into the device. Weighing the safety of the individual against the restrictions on their liberty, we may argue that electronic tagging may keep the person safe from an accident, or from being robbed or attacked*". Studies are available that show how tagging helps to find wandering patients faster and allows them the freedom of mobility. While it may reduce the anxieties of caregivers, the process of electronic tagging is dignified like wearing a waist band rather than physical restraint to prevent wandering (Lodha & Sousa, 2020; Miskelly, 2004).

Creating Dementia Friendly Technological Environments

In order to provide care and support to the patient and caregiver through assistive technology so as to elevate the quality of life, it is necessary to focus upon the following goals— (Brittain et al., 2010).

1. Look at practical, useful and affordable technology
2. Use technology to improve independence and promote quality of life
3. Assess long-term cost-effectiveness of technology
4. Keep up to date: regularly look at technologies.

Focus must be provided on aspects like what people need to know that it exists and how to find it. Evidence for the impact did the assistive technology made? How to channelize an effective way of technology transfer for both caregivers and patients. One or more approaches can be used to achieve the above-said goals (Mitchell & Burton, 2010).

- Directories
  - Web-based directories.
- Specialist technology providers
  - Telecare available
  - A little assistive technology.
- Equipment catalogues
  - Small dementia sections with stand-alone technology.

Disadvantages of Technology for Dementia Care

Assistive technology in dementia may not help every patient as it needs to be individualized for many patients have different needs and usage as well as choices a "one size does not fit all" in this case. Some patients may revel in technology and some may not like it at all. Failure to meet the individual needs in dementia care would impair quality of life when technology is used and in fact because more stress to patients and caregivers alike. There are situations where technology and telecare may not be of help like—

- The patient may switch off the device
- The patient is distressed by alarm sounds or messages
- Caregivers fail to respond to an alert or are unavailable.

Technology is not a replacement to human care and to caregivers. Many patients with dementia experience loneliness and social isolation. The role of technology is supplementation to human contact and human care (Baldwin, 2005).

Conclusions

Technology has huge applications in dementia care from diagnosis, assessment, care delivery and safety and security. The aim of this chapter was to summarize key areas of technology development in dementia care and its future implications. The main areas where technology provides respite in dementia care are diagnosis, assessment tools and monitoring progress, improving functional ability, recreation for patients, enhanced caregiving and safety management. The growth of technology development in dementia care needs sound rules and policies, research funding and a model that looks to individualize dementia care using technology to best suits the needs of every patient and improve the quality of life of the patient.

Acknowledgements Nil.

Conflict of Interest Nil.

Funding Nil.

References

Astell, A. J., Bouranis, N., Hoey, J., Lindauer, A., Mihailidis, A., Nugent, C., & Robillard, J. M. (2019). Technology and Dementia: The Future is Now. *Dementia and Geriatric Cognitive Disorders, 47*(3), 131–139.

Astell, A. J., Joddrell, P., Groenewoud, H., de Lange, J., Goumans, M., Cordia, A., & Schikhof, Y. (2016). Does familiarity affect the enjoyment of touchscreen games for people with dementia? *International Journal of Medical Informatics, 91*, e1-8.

Baldwin, C. (2005). Technology, dementia and ethics: rethinking the issues. *Disability Studies Quarterly, 25*(3).

Bharucha, A. J., Anand, V., Forlizzi, J., Dew, M. A., Reynolds, C. F., III., Stevens, S., & Wactlar, H. (2009). Intelligent assistive technology applications to dementia care: Current capabilities, limitations, and future challenges. *The American Journal of Geriatric Psychiatry, 17*(2), 88–104.

Blackman, T., Van Schaik, P., & Martyr, A. (2007). Outdoor environments for people with dementia: An exploratory study using virtual reality. *Ageing & Society, 27*(6), 811–825.

Brittain, K., Corner, L., Robinson, L., & Bond, J. (2010). Ageing in place and technologies of place: The lived experience of people with dementia in changing social, physical and technological environments. *Sociol Health Illness, 32*(2), 272–287.

Buckley, J. (2006). The importance of telecare for people with dementia. *Nursing and Residential Care, 8*(5), 212–214.

Cahill, S., Macijauskiene, J., Nygård, A. M., Faulkner, J. P., & Hagen, I. (2007). Technology in dementia care. *Technology and Disability, 19*(2–3), 55–60.

Cash, M. (2003). Assistive technology and people with dementia. *Reviews in Clinical Gerontology, 13*(4), 313–319.

Das, P. C., Kumar, P. R., Amara, A., & Dey, A. B. (2014). Dementia care in India: Issues & prospects. *International Journal of Medicine Pharmaceutical Sciences, 4*, 29–36.

Das, S. K., Pal, S., & Ghosal, M. K. (2012). Dementia: Indian scenario. *Neurology India, 60*(6), 618–624.

De Vugt, M. E., Riedijk, S. R., Aalten, P., Tibben, A., van Swieten, J. C., & Verhey, F. R. (2006). Impact of behavioural problems on spousal caregivers: A comparison between Alzheimer's disease and frontotemporal dementia. *Dementia and Geriatric Cognitive Disorders, 22*(1), 35–41.

Gibson, G., Newton, L., Pritchard, G., Finch, T., Brittain, K., & Robinson, L. (2016). The provision of assistive technology products and services for people with dementia in the United Kingdom. *Dementia, 15*(4), 681–701.

Godwin, K. M., Mills, W. L., Anderson, J. A., & Kunik, M. E. (2013). Technology-driven interventions for caregivers of persons with dementia: A systematic review. *American Journal of Alzheimer's Disease & Other Dementias, 28*(3), 216–222.

Huldtgren, A., Mertl, F., Vormann, A., & Geiger, C. (2017). Reminiscence of people with dementia mediated by multimedia artifacts. *Interact Computers, 29*(5), 679–696.

Kim, C., Wu, B., Tanaka, E., Watanabe, T., Watanabe, K., Chen, W., Ito, S., Okumura, R., Arai, T., & Anme, T. (2016). Association between a change in social interaction and dementia among elderly people. *International Journal of Gerontology, 10*(2), 76–80.

Landau, R., Auslander, G. K., Werner, S., Shoval, N., & Heinik, J. (2010). Families' and professional caregivers' views of using advanced technology to track people with dementia. *Qualitative Health Research, 20*(3), 409–419.

Li, R., Lu, B., & McDonald-Maier, K. D. (2015). Cognitive assisted living ambient system: A survey. *Digital Communications and Networks, 1*(4), 229–252.

Lodha, P., & De Sousa, A. (2020). Ethics of electronic tagging of dementia patients. *Indian Journal of Medical Ethics, 5*(1), 83–84.

Lodha, P., Palsetia, D., & De Sousa, A. (2019). Dementia Research in India—a review of the last 10 years 2008–2018. *Indian Journal of Gerontology, 33*(3), 313–330.

Man, D. W., Chung, J. C., & Lee, G. Y. (2012). Evaluation of a virtual reality-based memory training programme for Hong Kong Chinese older adults with questionable dementia: A pilot study. *International Journal of Geriatric Psychiatry, 27*(5), 513–520.

Manera, V., Chapoulie, E., Bourgeois, J., Guerchouche, R., David, R., Ondrej, J., Drettakis, G., & Robert, P. (2016). A feasibility study with image-based rendered virtual reality in patients with mild cognitive impairment and dementia. *PloS One, 11*(3):e0151487.

Martínez-Alcalá, C. I., Pliego-Pastrana, P., Rosales-Lagarde, A., Lopez-Noguerola, J. S., & Molina-Trinidad, E. M. (2016). Information and communication technologies in the care of the elderly: systematic review of applications aimed at patients with dementia and caregivers. *JMIR Rehabilitation and Assistive Technologies, 3*(1), e5226.

Mehta, K. M., Gallagher-Thompson, D., Varghese, M., Loganathan, S., Baruah, U., Seeher, K., Zandi, D., Dua, T., & Pot, A. M. (2018). iSupport, an online training and support program for caregivers of people with dementia: Study protocol for a randomized controlled trial in India. *Trials, 19*(1), 271.

Miskelly, F. (2004). A novel system of electronic tagging in patients with dementia and wandering. *Age and Ageing, 33*(3), 304–306.

Mitchell, L., & Burton, E. (2010). Designing dementia-friendly neighbourhoods: Helping people with dementia to get out and about. *Journal of Integrated Care, 18*(6), 11.

Ngandu, T., Lehtisalo, J., Solomon, A., Levälahti, E., Ahtiluoto, S., Antikainen, R., Bäckman, L., Hänninen, T., Jula, A., Laatikainen, T., & Lindström, J. (2015). A 2 year multidomain intervention of diet, exercise, cognitive training, and vascular risk monitoring versus control to prevent cognitive decline in at-risk elderly people (FINGER): A randomised controlled trial. *Lancet, 385*(9984), 2255–2263.

Nygård, L. (2008). The meaning of everyday technology as experienced by people with dementia who live alone. *Dementia, 7*(4), 481–502.

Øderud, T., Landmark, B., Eriksen, S., Fossberg, A. B., Brørs, K. F., Mandal, T. B., & Ausen, D. (2013). Exploring the use of GPS for locating persons with dementia. *Assistive Technology Research Series, 33*, 776–783.

Prince, M., Wimo, A., Guerchet, M., Ali, G. C., Wu, Y. T., Prina, M. (2015). The global impact of dementia. *World Alzheimer Report* (pp. 1–82).

Ritchie, K., Allard, M., Huppert, F. A., Nargeot, C., Pinek, B., & Ledesert, B. (1993). Computerized cognitive examination of the elderly (ECO): The development of a neuropsychological examination for clinic and population use. *International Journal of Geriatric Psychiatry, 8*(11), 899–914.

Robbins, T. W., James, M., Owen, A. M., Sahakian, B. J., McInnes, L., & Rabbitt, P. (1994). Cambridge Neuropsychological Test Automated Battery (CANTAB): A factor analytic study of a large sample of normal elderly volunteers. *Dementia and Geriatric Cognitive Disorders, 5*(5), 266–281.

Topo, P. (2009). Technology studies to meet the needs of people with dementia and their caregivers: A literature review. *Journal of Applied Gerontology, 28*(1), 5–37.

Wijma, E. M., Veerbeek, M. A., Prins, M., Pot, A. M., & Willemse, B. M. (2018). A virtual reality intervention to improve the understanding and empathy for people with dementia in informal caregivers: Results of a pilot study. *Aging & Mental Health, 22*(9), 1121–1129.

Chapter 18
Leap into the Unknown: Artistic Collaboration Across Diagnosis of Dementia

Sadie Harmon, Bob Savage, and Erin Partridge

Abstract Direct involvement by people living with dementia on the development of policies, services and research that concerns them is an emerging ethical concern. Equitable collaboration between people with lived experience and professionals is increasingly important in establishing people living with dementia as full citizens retaining personhood until death. Activism by people living with dementia has brought this issue to the fore, but more explorations of these complex collaborations are essential. This chapter will describe an artistic collaboration between Sadie Harmon, Dementia Inclusive Communities Director, and Dr. Erin Partridge, Experiential Researcher in Residence, both at Elder Care Alliance, and Bob Savage, an artist and advocate living well with dementia in a memory care community. The three exchanged two pieces of artwork three times each via mail, discussing over the videoconferencing platform Zoom each time a package arrived. Topics covered in this chapter will include: the use of action research and arts-based inquiry and the role of videoconferencing technology in this project; the impact of cocreation on prescribed hierarchical roles within the provider/recipient binary as well as larger implications for reducing stigma and promoting citizenship; and themes that emerged through the process and how those relate to the lived experience of dementia. This project has implications for community art and art therapy practice, care settings for people living with dementia and the research and inquiry processes involved in better understanding the needs and strengths of people living with dementia.

Keywords Cocreation · Lived experience · Art and dementia · Gerotranscendence · Technology · Risk

S. Harmon (✉) · E. Partridge
Elder Care Alliance, 1301 Marina Village Parkway #210, Alameda, CA 94504, USA
e-mail: sadieharmon@gmail.com

B. Savage
Person With Lived Experience, Plantsville, CT, USA
e-mail: bob-savage@comcast.net

E. Partridge
Department of Art Therapy Psychology, Dominican University of California, San Rafael, CA, USA

M. K. Shankardass (ed.), *Dementia Care*,
https://doi.org/10.1007/978-981-16-3864-0_18

Introduction

This chapter will describe a cocreative arts-based research project between three collaborators: an art therapist and researcher; an artist and dementia care professional; and an artist, advocate and person living well with dementia. The outcomes of this project are explored within the context of process-based creative practice as a direct challenge to conventional organizational structures that place professionals in positions of power over people with lived experience. Three major outcomes emerged from this project. (1) Artistic cocreation repudiated stigmatizing narratives of dementia that centre tragedy and death, revealing an alternative vision grounded in resiliency and transcendence; (2) A shared language of symbolism and meaning developed, becoming increasingly complex and referential through the course of the project; (3) Collaborative artmaking was uniquely effective in disrupting conventional provider/recipient binaries by engendering risk, rewarding diverse contributions and creating a conceptual shared space apart from prescribed roles. Using transcripts from reflective conversations generated by each iteration of our collaborative artwork, this chapter will examine the ways in which creative exchange across dementia diagnosis contributed to the outcomes described above.

A note on terms: We use both cocreative and collaborative to describe our method of working together. Cocreative here denotes specifically the process of working across diagnosis of dementia (Camic et al., 2018; Zeilig et al., 2018), while collaborative distinguishes this project as separate from our individual practices. The terms art and creativity are also used. We use Dissanayake's (1990) definition of art as the human instinct to construct meaning through "making special (p. 92)". Creativity is a method for self-expression that facilitates making special through art.

Background

Sadie Harmon met Bob Savage in October of 2018, when he copresented a session at the LeadingAge National Conference titled "Partnering with People Living with Dementia: Empowering Change", with Stephani Shivers, COO of Livewell, a non-profit care provider and research organization in Connecticut. As the Dementia Inclusive Communities Director at Elder Care Alliance, a non-profit senior living organization, Sadie uses art as an inclusive method of communication and collaboration with people living with dementia. Bob's presentation included a description of "Hello Light", a performance he had organized with other people living well with dementia. "Hello Light" used storytelling to share personal experiences and exemplified the unique capacity of creativity to create meaning around previously unthinkable realities. Sadie was organizing a similar event and introduced herself to Bob. Almost two years later, we still marvel at the string of seemingly random events and choices that led us to that convention centre in Philadelphia and launched a fruitful collaboration and lifelong friendship.

Shortly after that first meeting, Bob and Sadie began collaborating over the video-conferencing platform Zoom. Initially, we talked about art and the ways that we used it in our own lives, especially in relationship to dementia. After several months of bi-weekly Zoom calls, Sadie suggested that her colleague, Dr. Erin Partridge, an art therapist and Experiential Researcher in Residence at Elder Care Alliance, should join a collaborative creative project. The three of us decided to exchange two paintings three times, with each person making an individual contribution to both pieces each time they came to us. We then met over Zoom to discuss every iteration of this work. This shared project gave Bob, Erin and Sadie an opportunity to explore individual creative practice through the lens of a collaboration across diagnosis of dementia.

Research Methods and Use of Technology

For this project, we agreed that we would use an action research framework that employed arts-based inquiry. Our objective was open-ended and exploratory, and we did not initially conceive of a formal research question (Hasemand & Mafe, 2003, as cited in Savin-Baden & Wimpenny, 2014). This project was designed using the Action Research Cauldron, a tool that consists of three overlapping elements: collaborative experiments, conceptual space and relational space. Through the research process, these converge in generative knowledge (Bradbury-Huang, 2010). The collaborative experiment in this case was both the cocreation of the artwork itself and the conversations following each iterative cycle. Although the conversation may seem merely responsive, it provided an experimental ground in which we could playfully test one another's interpretation of the work. These conversations were an integral component of the artwork, responding to and informing new aesthetic additions. The interplay between the verbal and non-verbal is especially loaded in the space of dementia, and artistic practice can provide a bridge for communication (Killick & Craig, 2011; Stallings, 2010). In our first conversation, Bob described the complicated relationship between art and verbal expression:

> Bob: That's why I like a lot of music. It brings you back to where you first heard it and why you appreciated it. Art's the same way. But art is more complex. Because it's much more difficult to verbalize it. It's hard to verbalize the feeling. Like, my feeling right here is pretty simple, but I said it that way and they'll say, well, what do you mean?
>
> Sadie: Well, there's always more, right?
>
> Bob: Well that's what's so fascinating about it.

In our third conversation, after one full cycle of exchange, Bob referenced the relationship of the verbal and non-verbal in the collaboration itself:

> Bob: What's going to be nice, at the very end, is to say, well, what happened here? It's all an intuitional process. Where did it start from, what is it in the middle, and what do we interpret as its results? All nonverbal. Except when each step, we verbalize it, but then we go back and throw more nonverbal in.

A shared conceptual space was generated by the fluid exchange of the verbal and non-verbal. This space is referenced in our conversations and is also apparent in the visual symbolism of the artwork, which began to resemble a landscape. Our conversations formed a relational space through a pattern of shared language, visual metaphor and modes and methods of working. By the end of this process, we had created something totally new that manifested not only as a finished artwork, but also as an updated understanding of our relationships to each other.

Videoconferencing technology was an essential tool in this collaboration. New technology has the potential to combat emotional distress, loneliness and frustration by connecting people living with dementia to friends and family, although its use also raises concerns about privacy and consent (Bennett et al., 2017). Responsive technology, like social robots, also has the potential to itself be a tool for engagement and social connection (Ostrowski et al., 2019). Existing literature examining the use of videoconferencing technology with people living with dementia has focused on its use in telemedicine (Cullum et al., 2006; Martin-Khan et al., 2012; Poon et al., 2005) and caregiver support (Austrum et al., 2015; Marziali & Donahue, 2006; Marziali & Garcia, 2011). Appropriate technology also has incredible potential to move forward creative engagement (HP Development Company, 2016; Mihailidis et al., 2010) and collaborative care for chronic conditions like dementia (Nunes & Fitzpatrick, 2015). We concluded as a group that a videoconferencing platform would not only be appropriate for cocreation, but would introduce a creative constraint that would inform the process in unexpected ways. When we met, Bob had already been using Zoom to attend and host virtual support groups with other people living with dementia. Our early conversations used Zoom in its most basic way, allowing us to watch one another's facial expressions and share the spatial and aesthetic contexts from which we were speaking.

For this project, Zoom became another space in which conventionally defined roles were disrupted. Because of its design as a tool for professional collaboration, Zoom immediately positioned the three of us as equals in a virtual space. All three of us were actively engaged, sidestepping the presumed passivity often conferred onto people living with dementia in care settings. Our Zoom-mediated conversations had the richness of face-to-face interactions, including serious conversations about the realities of life with dementia, as is the case here, in our fourth conversation:

> Bob: I live in Assisted Living. And people just disappear. Right? And I said, what happened? We can't tell you, because of HIPAA. Right?...
>
> Bob: I'm serious about this, right? Now, we've given up all our rights. In order to protect us...
>
> Erin: ...What you're bringing up, one of our assisted livings, they've had multiple conversations about this. About how heartbreaking it is to them when their friend, who they eat lunch with every day is suddenly gone, and no one will tell them anything...
>
> Bob: That's when you hear rumors, right?
>
> Erin: Right, exactly...
>
> Bob: Because one day, I'm going to be—I'll be that empty room. They say, what happened to Bob? I don't know.

We also had many moments of levity and a lot of laughter. Our first conversation started in this way:

Bob: I woke up the box. And there's a warning sign on it. Don't—Let me read it to you.

All [laughter]

Bob: Inside this bag is everything you'll ever need to get started. Please wait to open until you're on Zoom.

All [laughter]

Bob: OK. I'm ready to roll. I have no idea—I'll ask questions as I go along...

Erin: This is like the comedy hour. This is the art comedy hour.

The joke about waiting to open the box containing the artwork until everyone was present on Zoom continued throughout the collaboration, coming up in our third conversation:

Sadie: Can I record you opening it?

Bob: Yes—certainly. I'll make more noises.

Sadie: Oh wait—did you want to open it, open it, or did you just want to open the box?

Bob: I don't know what you're talking about. I opened the box. Now I want to open—open—open it.

Sadie: OK. Yeah. Open, open, open it.

Bob: OK. No chocolate.

Sadie: Oh, you got me.

And in our sixth and final conversation, as Sadie struggled to open the box:

Sadie: I'm looking—usually they have a tab. Oh, here it is.

Bob: Erin, Erin. I think she needs help.

[laughter]

Erin: She does have scissors here.

Sadie: I just ripped off both sides of the tab.

Bob: You're doing well.

[laughter]

Sadie: OK.

Bob: You've got a loooong ways to go.

Erin: How many artists does it take to open a box?

We continued to use Zoom through the process of disseminating the results of this project. It also allowed us to record audio, for which we would collectively establish consent at the beginning of each new conversation.

Interdependence, Stigma and Shared Practice Across Diagnosis of Dementia

As professionals who use the arts in our work with older adults and people living with dementia, both Erin and Sadie have witnessed the power of creativity to unlock new possibilities. For both of us, art has the potential to promote self-expression in a space free from judgement. Older adults may come into the art studio with protestations that they are "bad artists", bringing with them earlier negative experiences with art teachers (Partridge, 2019). A great deal of our time as facilitators is spent creating safe spaces for creative exploration and risk-taking, as well as for failure. Allowing oneself to fail and subsequently learn from those failures is a hallmark of satisfying creative practice (Bayles & Orland, 1993). In his art practice, Bob centralizes risk and failure as necessary for growth. By returning again and again to risk as the primary framework through which he approached this project, Bob pushed Erin and Sadie out of our expected role as facilitators and into that of fellow risk-takers. It also equitably redistributed custodial roles of safe space keepers, with all three of us sharing this emotional labour.

This dynamic can be explored over the course of several exchanges, starting in our initial conversation about the project:

> Erin: So it could be that these pieces get some paint put over some of the busy areas, if it starts feeling too busy, or if you feel that there's a section that needs some open space, you can definitely do that... I imagine that in our final project will be not only these physical pieces, but also the photographs of our process all along the way...
>
> Bob: If we lose the hand, it's OK.

Here, Erin explicitly gives permission to modify the work of other artists and to make space for one's own contribution. In addition, she suggests a method to archive past iterations so that no individual contribution is lost. Bob pushes this further, giving permission to cover up and reimagine the hand. This permission is generous, as Bob had emphatically responded to the hand as the central focal point of the work. This generosity operates strategically as well. By potentially sacrificing the most valuable component of the artwork, Bob moved the collaboration into a new territory with space for experimentation and growth. Later, the roles are reversed, as in this exchange, also in our first conversation:

> Erin: Whatever you are called to put on both of those pages... the only rules we should put around this is that each time, we should do something on both pieces…
>
> Bob: Also, you've got to give room. Right?

Erin reinforces Bob's earlier permission by stating that there are no formal rules around what is preserved and what is altered. By making non-participation against the rules, she creates a constraint that likely will make covering up another artist's work inevitable. In response, Bob becomes the facilitator, checking the comfort level of the group against the stated limit of risk. His words communicate the desire for reciprocal relationships in this collaboration—a give and take of visual imagery, creative expression and "room". In our second conversation, this theme re-emerged, this time as it related to the structure of the collaboration:

> Sadie: As I see these evolve, we're all really thinking about: how is the next person going to find space in this.
>
> Bob: Yeah. I think the fact that we're doing it three times will work pretty well.

The consideration for others continued throughout this process, with any tension in regard to size of contribution resolved in individual studios. A completeness emerged within the artwork that stopped all of us from making drastic new changes. This was not because we felt that we could not, but instead because we were responding to the piece and its layered parts intuitively. Bob described this gradual resolution in our fifth conversation:

> Bob: I did a few minor changes with the hand, because the hand was so beautiful. Actually, my first thought [was] what the hell can I do? Without screwing it up, right? And it took me a while before I figured out what I could do. But there's always a little thing that we can do. But I noticed that as I went from the first to the second, I did a lot. Second to the third it was less.

Throughout these exchanges, interdependence emerges as a theme. We began to accept the impossibility of artistic progress without equally authentic contributions and grew an awareness of the impact that individual choices and actions had on collective work. This rich collaboration counters the tragedy narrative of dementia, which abruptly positions people after a diagnosis as incapable of independence (Reed et al., 2017). The erasure of the complex relationship between doing for oneself and accepting help from others directly contributes to stigma. Wendy Mitchell describes this experience this way:

> Once, we were the strong ones, the leaders, the people who others came to for help, now we are the ones needing help.
>
> I don't want pity; I'm not looking for support. I'm simply having a bad day.
>
> Maybe tomorrow will be a better day……..or next week…… (Mitchell, 2016, "Changing Relationships……." para. 3)

The concept of relational interdependence as it relates to decision-making by people living with dementia and care partners can advance inclusion and citizenship for people living with dementia (Keyes et al., 2019). Interdependence offers a framework through which to appreciate the ways that a dementia diagnosis radically upends the independence/dependence binary. The relational space created by our project was interdependent by design. Because all three collaborators accepted the risks inherent in the cocreative process, a space was created for failure and growth to be encountered on all sides. Within this creative space, we shared the consequences of risk-taking, giving and receiving help in the form of aesthetic and verbalized responses to individual contributions. We became citizens of a new place that had been created by and within the artwork, where risk and reward shifted from collaborator to collaborator.

In one of Sadie and Bob's early conversations, Bob described his experience of being "dementia-tized" by people close to him following his diagnosis of Alzheimer's disease. A person is "dementia-tized", Bob explained, when they are suddenly seen only through the lens of their diagnosis (B. Savage, personal communication, August 10, 2019). For Bob, this experience included losing much of his independence and being regarded primarily as someone who required care and looking after. Prior to his diagnosis, Bob had plans for retirement that built on his successful career in addiction treatment and services. After being "dementia-tized", those plans were all but impossible. Not because Bob lacked the knowledge or skill to carry them out, but because he was suddenly perceived as a person with no active capacity. After his diagnosis, Bob found himself almost totally defined by the overwhelming social stigma he experienced against people living with dementia.

A diagnosis of dementia is often accompanied by both internalized and externalized stigma such that quality of life is immediately negatively impacted (Reed et al., 2017). Stigma can have a significant and detrimental impact on physical as well as emotional health, with people living with dementia receiving a lower standard of care as social regard decreases (Benbow & Jolley, 2012; Brannelly, 2011). The extreme result of stigma is social death or the perception of the person living with a diagnosis of dementia as being as-good-as dead (Sweeting & Gilhooly, 1997). This is a well-established and persistent, if unnamed, fear following a diagnosis (Reed et al., 2017). Bob's personal experience of stigma can be directly connected to the cultural concept of social death as a natural end point after a dementia diagnosis. Králová (2015) writes that to be truly considered a social death, a person must experience three extreme losses: social connection, a social identity and the loss of an embodied social being. Although these losses do not happen immediately, it is instead their understanding as a foregone conclusion that contributes to extreme hopelessness following a diagnosis of dementia. The tragedy narrative of dementia is fuelled by social perception more than by the daily lived experience of people with dementia (Reed et al., 2017; Swaffer, 2014). The fact that a person can live well with dementia for many years and retains their personhood until the end of life has been established by people with lived experience (Mitchell, 2015; Rohra, 2016) as well as through ethical frameworks (Hughes, 2014; Kitwood & Bredin, 1992; Nuffield Council on Bioethics, 2009).

Artistic practice provides a vital outlet to express personhood in the face of the persistent threat of erasure. Cocreative production challenges stigmatizing views of people living with dementia and opens a dialogue about institutional and social change (Gray et al., 2018; Kontos et al., 2020). The discovery of painting and the possibility to communicate through colour presented Bob with an opportunity to re-identify himself as an artist and as a person living well with dementia. Bob's artistic practice deeply connects to his motivating purpose of fighting stigma and supporting others following their own diagnosis of dementia. Artistic collaboration also enables witnessing and being witnessed—a process Allen (1995) formalized in her studio work: "To witness is to see and affirm. It is a skill. It involves putting aside any personal agenda. The witness does not judge, criticise, extol, reject, or evaluate your images. To witness is to be with" (p. 110). The three collaborators in this project were able to be with each other and the cocreated images as they evolved.

At the project's outset, Erin and Sadie were positioned as professional allies to Bob's lived experience of stigma following his diagnosis of Alzheimer's disease. It is difficult for care providers, even those with a creative arts background, to imagine themselves as equal collaborators with people living with dementia. Much more common is the implicit or explicit belief that the arts have a practical benefit towards health and quality of life, with less consideration for the harder to measure goal of creative expression (Zeilig et al., 2018). While art as a healing modality certainly has a valuable place within dementia care and can be a tool for empowerment (Anacleto,

2018; Basting, 2006), it risks being situated by the institution solely as a service or treatment. This orientation places professionals in the provider role with people living with dementia as recipients. Additionally, over-attachment to professional titles and practices can interrupt true collaborative relationships. Our collaboration was guided by arts-related research principles, including knowledge generation through the work itself, accessibility as a focal point and a shared sense of authenticity (Savin-Baden & Major, 2013, as cited in Savin-Baden & Wimpenny, 2014). Through this framework, a relationship outside of the elder care structures within which each of us either work or live began to emerge. This alternative structure fostered a relational process of cocreation, moving away from the provider/patient dynamic and towards what Allen (1995) described as art's "most potent properties, the ability to dissolve boundaries and reveal our interconnectedness with one another, as well as reveal the dignity of our uniqueness" (p. xvi). Relational engagement through the arts is a key component of reclaiming citizenship for people living with dementia (Dupuis et al., 2016; Kontos et al., 2017).

Zeilig et al. (2018) writes that "co-creativity with the arts and with people with dementia is unwaveringly about mutual involvement in an aesthetic process (p. 7)". In order to be successful, this mutual involvement including people living with and without dementia brings together two factors: what Hadley (2020) calls transactional artistic allyship, or equitable access to resources, and transformational aesthetic allyship, where the creative vision and process of disabled artists is prioritized. When this kind of creative collaboration is achieved, it unites people across dementia diagnoses around a common creative project. Large-scale collaborative creative projects in senior living and dementia care settings, like those conceptualized by Anne Basting, often require a tremendous commitment of financial and labour resources, or transactional allyship, alongside transformational aesthetic allyship in the form of comprehensive facilitation that positions the creative contributions of residents, staff and artists in conversation with one another (Basting et al., 2016). Given the tremendous impact of such projects, it is useful to consider employing the same principles on a smaller scale. Our project used accessible shared resources and operated on a modest scale without sacrificing its efficacy as a transformational tool. Through an exchange of creative work and honest feedback, we became three artists collaborating equally, inspired by a finished work of art that none of us could have created alone.

Meaning and Gerotranscendence

Using art, whether in a formal therapeutic capacity or through the development of a personal practice, to process a life-changing diagnosis has a long established precedent in the literature (Waller, 2002). Artists diagnosed with dementia often continue to create (Byers, 2011), developing an evolving aesthetic vocabulary that matches their experience. For those without an artistic practice prior to diagnosis, developing a tool to support meaning-making is essential (Anacleto, 2018; Dissanayake, 1990). When presented in an affirming environment that encourages the development of a deeply personal and unique artistic identity, art can be used as a tool for gerotranscendence (Stephenson, 2013). Maintaining a personal and/or collective art practice after a diagnosis of dementia can also promote resiliency (Newman et al., 2019). In relationship to the stigma many people experience after a diagnosis of dementia, art can be an escape and a medium for self-care as well as a powerful tool for advocacy and identity formation.

When Sadie visited Bob in October of 2019, he took her on a tour of his artwork almost all of which explored his experience of living with dementia. Through an aesthetic vocabulary that includes colour, symbolism and spirituality, Bob connects his past and present experiences of stigma to the vast unknown future after a diagnosis of Alzheimer's disease. Bob speaks about one early piece this way,

> This one is a story of my period over the first nine months. This is when my wife and I were doing great, then we got the diagnosis, and we go down. Almost immediately we were down. And down here is all the end of life stuff... very depressing stuff. And these are colors—black of course is the depressing part. As soon as you get rid of those colors—when I first came down, this was the only color that was there. Then I started thinking of my grandkids, and now we have a lighter color. Then I started thinking—now I'm up to yellow, I'm doing pretty good. And this is where my resilience is, right? (B. Savage, personal communication, October 11, 2019).

In the process of cocreation, visual language that had once been singularly owned by one individual artist soon became the property of the group. Bob's use of colour differs from Sadie's, which differs from Erin's. All three collaborators had different entry-points into art and creative expression. In the course of cocreation, these differences provided a ground for new discoveries and creative risk-taking. This allowed for a variety of shared symbolic language to emerge organically. Recurring themes included the hand, multilayered uses of colour, and the paintings as portals to the unknown.

The Significance of the Hand

The overlap of shared symbols in the artwork complicated the binary between creator and observer. Most notably, the hand became central in both pieces, with each of us contributing a different meaning to its presence. The three contributions coalesced in a shared narrative understanding. The hand was at various times interpreted as an instrument of care, a method of communication and as a guide, beckoning us towards an undefined creative potentiality. Bob described the hand like this in our first conversation:

> Bob: This is the hand of Mother God. That's saying, come with me... And all these other things here [around the hand are]... the electrical messages that I'm now receiving. And I see this as a major, major privilege... Most of the work I do, I have this hand helping me. Most of the art I do, I have this hand. I got to realize that in the last four or five years.

The hand is also a utilitarian tool, used both to create and communicate. Verbal communication in the setting of dementia care can be frustrating and unreliable, leaving caregivers and people living with dementia to resort to other forms of communication. Over the course of a relationship between care partners, a unique language of touch can be established (Huss, 1977). Bob understood the hand as a communicative tool within the context of the painting, whereas Erin externalized this property to the real-life realm of care work:

> Bob: And this is your emotion that you're speaking through your picture... you're communicating through the hand.
>
> Erin: The other thing about hands... in the work that I've been doing over the past eight years, if there's not the ability to communicate verbally, if I can't see what's in front of me, there's still this life, here. In the hands. That seems to transcend almost everything else…

Erin continued to relate a story of a woman with whom she had worked who had been an accomplished oil painter, but had stopped painting as she approached the end of her life. Erin met this woman while working as an art therapist in a skilled nursing environment and, on a whim, tried placing a paintbrush in her hand. Erin described to Bob and Sadie what happened:

> I put a paintbrush in her hand. And I just put it flat in her hand, and she immediately went like this [grips]. Holding a paintbrush, the way you would hold a paintbrush to paint. And so we put a piece of paper down. And she started making these little paint brushing strokes. And her shoulders relaxed down away from her ears. And she could no longer speak, but she was showing us, like, this is where I'm comfortable.

In this anecdote, the hand becomes the physical link to a past artistic practice. The hand in our painting became a symbol of connection to the past and a reassuring invitation to the future. In our second conversation, Erin imparted onto the hand this meaning:

> Erin: I see this hand, emerging out of the yellow and green checkerboard shapes and kind of either reaching out toward something or offering something. And there's a bit of a landscape feel to me. It feels very hopeful, this image. I get a hopeful feeling from it.

In the same conversation, Sadie built on Bob's conception of the hand as coming from an unseen force, this time tying it to our shared construction of a conceptual space:

> Sadie: Something that I'm really interested in in my own work is thinking about art as a way for us to find a different place. So almost like a portal or a door to somewhere other than our everyday reality. And to me, this is what that looks like... like Erin said, a landscape. And I think that's true, but to me it's a kind of a landscape of a place of, like, pure creativity, or something like that… That's kind of what I see in it. And maybe the hand is someone who's welcoming you there. It's kind of your guide in a way.

The hand welcomed the collaborators to a shared space within the artwork, where each individual artist could contribute unique narrative experiences. This created a shared sense of ownership and established each contribution as essential to the whole. Not only does the finished artwork hold a consistent aesthetic vocabulary, but it is also imbued with personal resonance. Again, in our second conversation, Erin introduces a personal anecdote that combines the symbolism of the hand and of colour with the literal action of hand-tracing, which is physically intimate and a form of embodied relationship building:

Erin: When we talked about this piece before we were talking about this idea of energy, and again, this kind of reaching out or connectedness that comes from it. I think that the yellow, for me, makes it clear that this is a good, positive energy and connecting that is happening here in this image. It also reminds me… so much of someone who Sadie and I both knew, who loved all warm colors. She was particularly fond of orange, but she would have loved this piece. And... we did a lot of tracing her hand. So it's bringing to mind a positive memory of someone I knew really well. That's really a loving memory.

As the collaboration continued, the symbolism of the hand evolved as well, taking on meaning from the evolving content and form of the painting. Where the hand at the beginning of the collaboration was portrayed as an omniscient guide, by our third conversation, some of its power had been redistributed through the action of collaborative making. Bob described this process in this way:

Bob: That hand looks loving. Doesn't it? Look how soft it looks. Now, my theory of this is—before this was going on, it was just a hand. And now with all this stuff going on, the hand is softer, and softer, and softer. And more powerful. Because it's non-directed. It's not directed to anywhere, it just—envelops the whole thing.

As the process continues and individual artists give up control and respond more freely to the artwork, the hand gains power and begins to represent the emergence of a shared creative practice. It also builds a bridge from conceptual to relational space. In our fifth conversation, Bob returned again to the significance of the hand, tying it to his own experience communicating with a woman living with the later stages of dementia.

Bob: How do you communicate with a person with different levels of dementia? And I always say, it's an emotional connection... There's some sort of a mysterious bonding. And then you hold the hand, right? You hold the hand because that's the future connection for your communication. There's a woman that I've been involved with now for almost two years... I'm the only one that she'll talk to… She can tell who I am by holding my hand. Instead of seeing me, this is it.

The hand signifies a connection that transcends language and diagnosis of dementia. It becomes a bond and a guide. It is also significant that the hand is the instrument used to both create and engage with the artwork. In our sixth and final conversation, Sadie commented on the depth and texture of the finished artwork:

Sadie: The other thing that I think about this one… it's almost like a window to another place. I think the fact that this one has so much texture—it almost looks like you could put your hand inside of it and you would feel, you know, something else.

From this description, Bob assigns the hand with a metaphysical property, commenting:

Bob: That hand. That hand, as the basis for that, is so mysterious… It's spiritual, and it's scary.

Erin added outlines of hands in the first cycle of collaboration. On both pieces, the hand became a visual and conceptual focal point, with both Bob and Sadie adding embellishments and emphasis. In our conversations, the interpretation of the hand grew increasingly layered, gaining depth and complexity as each artist added their own and responded to others' interpretations. In our first conversation, we conceptualized the hand as an external guide, offering inspiration and the strength to take creative risks. Bob tied the hand to his own art practice, giving it a spiritual significance. Erin related the hand to earlier experiences with art, framing it as a safe and familiar point of reference. Sadie imagined the hand as a method of linkage between multiple real and imagined spaces. Hands were also recognized as a tool for self-expression through artmaking and became a metaphor for inclusive non-verbal communication from the beginning to the end of the lived experience of dementia. As the collaboration progressed, the hand became increasingly essential to our understanding of both the artwork and to our collaboration. The hand became an invisible cocreator, a stand-in for what Barthes (1967) refers to as the process by which the act of creation is no longer tied to the agency and will of the single author. Through this framing, one can interpret the hand as a symbol representing the magic of the process itself.

Colour as Connective Tissue

Another powerful theme that grew more pronounced throughout the collaboration was the use of colour. Colour is an important component in inclusive design for dementia, especially in regard to way finding and engagement with the built environment (Bowes & Dawson, 2019). In art therapy, colour is a valuable expressive tool (Anderson & Gantt, 2009; Kersten & van der Vennet, 2010). Across cultural traditions, colour holds symbolic meaning and spiritual significance (Hutchings, 2015). Artists like Wassily Kandinsky have assigned colour with both a spiritual and aesthetic meaning. In *Concerning the Spiritual in Art*, Kandinsky (1977) wrote of colour,

> Generally speaking, colour is a power which directly influences the soul. Colour is the keyboard, the eyes are the hammers, the soul is the piano with many strings. The artist is the hand which plays, touching one key or another, to cause vibrations in the soul (p. 26).

Bob's art practice has a complex and highly sophisticated relationship to colour. Both Erin and Sadie have active art practices, with well-developed usages of colour. Somewhat surprisingly, the three of us were able to easily negotiate both colour and meaning, with dominant colours shifting in each iteration. Instead of colour use that clashed either aesthetically or symbolically, each individual's colour choices expanded the meaning of the finished piece. In our first conversation, Bob explicitly centralized colour and texture in our shared creative practice:

> Bob: I love colors, right? And I love the light colors… this is friendlier to me. And then there's little fur on there, right? Which is even more gentle. Then you have this medallion down here, or this medal. And again, these beautiful bright colors.

At the beginning of the second cycle of our collaboration, Sadie took a major risk and cut shapes out of the background of the painting, which she glued directly onto the fingers of the hand. Bob commented on the paradox created by this action: the confusion of cutting the hand into pieces that is resolved by the flow of the colours:

> Sadie: I took some risks with cutting [the hand] out.
>
> Bob: I love it. Also, now you have it all over the place. And this is more than colors. It's enhancing the colors. And I like the confusion of how they're put in. Because it's not easy to do this. Right? In order for them to get in line, you have to do it.
>
> Sadie: Right. You just have to go with the flow.
>
> Bob: They'll be coming more, and more, in line. Isn't that beautiful? Wow… And this one. Oh, you put more yellow in the hand.

The use of a sharp tool and the irreversibility of the act of cutting pushed the discussion of risk further and had a poetic connection to the experience of living with an as-of-yet incurable disease. The cutting process also revealed new possibilities in the composition and incited conversation about moving between ease and difficulty.

Colour played a role in our negotiation of the many unknowns inherent in the lived experience of dementia. These unknowns are not all intrinsically negative, but do represent a disquieting possibility for uncertainty and rapid change, which disrupts a sense of security. Bob connected red to the concept of danger, a meaning that in conversation with the colour yellow transformed to represent power and love:

> Bob: I love it. A little mystery here is the red. Up in the sky? And, it's almost like danger… The red is danger. And it's going all different sorts of directions. The danger, as I see it, wants to be re-transformed. And then this circle—That's a mystery. You know, when I look at it, it comes in and out, in and out. Yellow. All the love colors. Power colors. See, they're all there. Right? Wow. And the hand is loving. It's enhanced now.

Neither our understanding nor our use of colour remained static throughout this process. All of us made individual aesthetic choices that responded to the whole, including those of colour and its application. The final piece was a harmonious joining of many colour stories. Colour's role in the artwork was fluid depending on the viewer's or artist's perception of it at the moment of encounter. This flexibility was discussed in our sixth conversation:

> Bob: There's so many different emotions related to one color. So when you're painting, you have to be painting from that one emotion. How green can be happy, it can be sad. Blue can be happy, it can be sad. So if you're painting a bluebird, or if you're painting him blue, right? It's complicated. But you cannot say that in words, right? You can't say it in words. And everybody's going to interpret it differently.
>
> Erin: And even each individual person is going to interpret it differently on different days. Right? I can look at something that's blue one day and it means one thing to me, and then look at that same thing the next day and it's totally different.

Colour was a tool for empowerment as well. In our sixth conversation, Bob tied the ways that art can confirm personhood through the late stages of life with dementia by linking colour with communication and self-expression:

> Bob: This is why colors are so important, when you connect it with your emotions. You're painting from your emotions that you have—you can't put words on them... I see a hand there, and I think, well, I have an emotional connection to it. I don't say, well, now I've got to draw the finger right… you just let it flow.

Erin: Yeah.

Bob: At my age, a lot of my paintings now are that way... You saw the one I sent you last, right? And how it's just that? The project that you're working on, and what you're trying to do with the people with dementia is to connect with that emotion, which they never lose. Right? No matter what the level is.

Colour was a connective tissue in both the artwork's final composition and the process of cocreation. It was its own vocabulary, allowing us to communicate visually and respond playfully to choices made by our collaborators. Certain colours became a signature of a particular artist, as happened with Bob's use of yellow. A rainbow of colours winked through the layers of the completed artwork, reminding us of the process in which we had engaged. From our sixth and final conversation:

Sadie; So, I also wanted to talk—while we're still talking about this one? I really want to talk more about the yellow.

Bob: Yellow is my color.

Sadie: We have almost every color in here. We have purple, we have red, we have this kind of pink, we have orange, we have blue, we have green. And the yellow is such a bright color... To me, the yellow ties it together and gives it this life... You take it for granted that it's there. Like, it has to be there.

Bob: Yeah.

Sadie: I can't even imagine what this would look like without yellow. And we didn't start with any.

Bob: That's right. To me, we had to close it in somehow, or wrap it up.

A New Landscape to Experience the Unknown

Although dementia is not explicitly what this work is "about", it informed our conversation and our understanding of the piece. Artmaking has been posited as a uniquely effective way to facilitate meaning-making for people living with dementia (Basting, 2006) and for affective communication across diagnosis (Jeong, 2019). In her art therapy practice with older adults, Erin has positioned artmaking as a tool for connection and empowerment that directly challenges ageism and stigma (Partridge, 2019). Sadie has used accessible creative practice as a way for people to collaborate in person across diagnosis of dementia (Dementia Inclusive Communities Initiative, 2019). Much of Bob's work addresses his own experience living with dementia. Since creating his first artwork in the first year of his diagnosis of Alzheimer's disease, Bob has developed a complex aesthetic dialogue between what is known and what is yet to come. He described an artwork created shortly after his diagnosis this way (B. Savage, personal communication, October 11, 2019):

What am I going to face? This one is what you're facing. At the same time this came to me… Louisa May Alcott: "I'm not afraid of storms, for I'm learning to sail my ship." And this is her ship. Now, I'm in it... You start to go into the unknown. And that's the storm, right?

By engaging in this collaborative process with Bob, both Erin and Sadie entered into this unknown as co-investigators. Together, we evolved a visual vocabulary that structured our investigation. It is important to note that this vocabulary emerged organically and was not defined by any one of us individually. In our first conversation, Sadie described the artwork we were creating like this:

> Sadie: When I make artwork, what I'm really interested in is… creating other realities. And thinking about the artwork as being, almost like a window or a portal into a different place… So I saw the hand as almost like a citizen of this other place. And then adding the fur was like making a relationship to it. So it was almost like by putting this texture on there, I was making the world more three dimensional. So, it's almost like a place that the three of us can be together that is, like, a purely imagined space.

This "other space" symbolically represented the object of our collaboration, offering a safe and boundaried place to explore shifting roles and power dynamics. The space generated by our shared process was both conceptual and relational. It essentially operated outside of established roles while retaining clear rules and expectations around engagement and interaction. Professional ethics require a boundary between personal life and professional practice, which becomes especially fraught when interaction occurs outside of a physical institution (McGarry, 2009). Boundary setting is imperative in collaborations across dementia diagnosis but presents novel challenges (Bell & Leong, 2019; Kenny et al., 2016). A key consideration is that boundaries and needs may be very different between collaborators and must be flexible and interrogatable. Contributions to collective boundaries are equally important from all parties, and it is crucial to include people living with dementia in considerations of a project's boundaries. Especially for artists working in health care, measures to care for oneself must also be taken (Naismith, 2019). In our collaboration, laughter, trust and regular permission for risk-taking allowed us to enter together into the sometimes frightening unknown inherent in a diagnosis of dementia. A rich lexicon of symbolic imagery populated this void with shared landmarks. We had no deadline or content expectation for individual contributions, which allowed for self-care and reduced anxiety over production. A single mark would have been permissible, as would the reconfiguration of the piece as a whole. We also kept the lines of communication open, sharing questions and clarifications over email, phone calls and text messages.

By establishing a secure structure to hold and contain the unexpected, we were able to collectively enter into rich subject matter such as the complex interplay between what is known and unknown in a dementia diagnosis. Artmaking is a way to reconcile the common stigmatizing assumption that people living with dementia are already dead, lending agency and personhood to people living in all stages of dementia, including later stages (Camic et al., 2018; Treadaway & Kenning, 2016.). In our project, this theme emerged in various ways, allowing each of us to engage with the radical change inherent in the progression of dementia. In our sixth and final conversation, we explored the sun as a powerful metaphor for full personhood that remains intact until death:

> Sadie: We talked about how this one has looked like a landscape. What's interesting to me is that I can still see the landscape there, but it looks like the position of the sun, I always thought was in the upper corner. And now-

Bob: It's in the middle. It's right there. It's amazing…

Sadie: I wasn't thinking about it as a sun when I added the rays. I was just thinking about it as a shape.

Bob: This was the result of the hands. The rays coming out. You know, that one to me is scary. It's scary because they're on a cliff here. I just don't know where the hell to go with it. It's a point now of creativity—Oh, shit. You know, seeing it this way, I see it different now than the last time I saw it. This little thing—this sun on the right there? That is powerful. It's not the human sun, but it's another form of the sun. Not the earth's sun. Maybe, it's a little bit of the sun in all of us.

The Role of Risk

Our creative exploration of the unknown future of a dementia diagnosis was accompanied by a not inconsequential amount of risk. There is an emotional risk inherent in venturing into and sharing personal experience. In order to move forward, it was necessary for us all to respect and honour one another's perspectives on complicated topics including end of life, dementia, art and self-expression, while recognizing that there were some areas in which we differed in our lived experience. Risk also has an embedded role within artmaking, not only as a catalyst for making via acceptance of failure, but also as a political tool, especially in pedagogical contexts (Gale, 2020). The positioning of risk as a method for resistance is particularly applicable for artists living with dementia, whose agency is consistently under threat. A collaborative process across diagnosis of dementia should focus on process (Camic et al., 2018) and embrace failure as generative of an affirmative creative environment (Zeilig et al., 2018). This positions risk as a tool for the collaborative process, which is itself loaded with uncertainty. Each new iteration of our shared artwork was a gift for the artist whose work would follow our own, requiring us to maintain a spirit of acceptance and gratitude.

Because of our structure of rotating gift-giving, the management and distribution of risk shifted throughout the project. In our first conversation, Bob assigned the hand as a metaphor for our collective risk-taking, which Erin subsequently affirmed:

Bob: It's very powerful. I see it as very exciting. In the sense—it's a risk, right? Her hand is shaking a little bit. And she has no idea where it's going to go. All she's doing is putting the seed.

Erin: This is part of the fun and the terror of an interactive project like this, where we're collaborating... You have to be able to be willing to put things on the page that might get covered up.

Later in the same conversation, Bob positioned risk as a potentially intimidating force inherent in the creative process. He identified the distribution of risk equally across three collaborators as a relational strategy within our present collaboration and a future invitation for non-artists to experiment with creativity:

Bob: And the work I'm doing is all new, right? There's a chance for failure every minute. Right? But. As you move, the same power that goes into this art is the same power that goes through to support you taking a risk. And this is a way of communicating about that... You tell people that—they walk away. But the fact that the three of us are going to be doing this, together, we're really expressing that process.

Our conversations positioned risk as a process that engenders the creative state of flow. Killick and Craig (2011) point out that creative flow is especially valuable for people living with dementia, as it is marked by a feeling of integration directly countered by the experience or expectation of dis-integration accompanying a diagnosis of dementia. Bob described in our first conversation the convergence of the integrated process of creative flow with a chaotic nightmare he had experienced, while also, in the last sentence, re-establishing our relationship as equal cocreators:

Bob: But what happens is the thrill—here's what I do, right? I take a painting, and I have a section and I put paint on it, do the next section, do the next section. Now there's one—I had—I was having a terrible nightmare. And this dude scared the hell out of me, right? So I decided to paint him. It was abstract. And I did that. Boom—he showed up. He was kind of weird. But he showed up. Right? I—there's something—I don't know, how do you explain that? I don't know how to explain. But there's a magic there in art that is so hard to explain to anybody. Especially art teachers.

The relationship between flow and risk in creating a cohesive whole was established in our first conversation, which we concluded by discussing our hopes for the future of our collaboration:

Bob: I love the mystery already. I'm going to be blown away when it's over. I have no doubt. I get chills already. It's a strange hope. It's more of an expectation.

Erin: I echo that. I love what you said earlier. You said, "There's a magic there in art that is so hard to explain." And I think that my hope for this project is not that we explain, and surgically take art apart. But rather, we have the experience of being in that mystery, that we get blown away by, or get the goosebumps.

Sadie: I hope that this process will open doors that I didn't even know existed. And I feel like that's already happening, and so I'm just really excited about continuing.

By collectively embracing the mystery of the process, we gained a collective agency outside of our individual goals. The process of artmaking increasingly gained its own power, fortifying for experimentation the safe space we had created through our expectation and boundary setting. In our third conversation, Bob and Sadie discussed the difficult to define experience of the collaborative process establishing its own agency:

Bob: You know my, my—faith is in the mother of God. And this is it, right? This is the power, from the mother of God, falling into the hand. And they're starting to form the symbol, going out the finger. And it's going to really have an impact on whatever's around it.

Sadie: Well that's the amazing thing about art, right? It tells us things that we wouldn't have known.

Bob: Yeah. And, what's so interesting, when you did it, you had no idea how I was going to interpret it. But you understand what I'm saying.

The concept of a separate energy that flowed through the artwork and therefore the process itself was also addressed. In our second conversation, Bob identified ink-marks on the piece as signifiers of this energy:

> Bob: You see the hand? And Sadie, you put those little squigglies out? While, I took those little squigglies and made them major, major events. It's the power. This is the power that goes through the hand, right? That goes through your fingers and takes those squigglies and just explodes. And that's the power of—the mystery of this whole thing.

This theme re-emerged in our third conversation:

> Bob: I go first to the hand... And... all these little icons, or all these little things there? The hand is forcing them to go out through the fingers. And energy is coming in. Do you see up here? The energy that's coming in, and then going out, right? Is so much more forceful as a result of this showing how the force is going through. I love it.

In our sixth and final conversation, Erin and Bob had this exchange:

> Bob: You wonder where the energy comes from in that painting. Right? I think it's very spiritual. I think it is. And that's a lot how art creates… I think if you can create from that then you've got yourself one hell of a powerful piece of art... We just let it go, right? We had no idea. We had no idea when we started out. When you made the hand, right? Did you ever think it would end up this way?
>
> Erin: No. Definitely not. And Bob, you said, "how art creates." I love that. Because it's not how us fancy humans create, it's how the art creates. And the art is what's doing it, not the humans.

By shifting the power to the creative process itself, responsibility for the artwork and the collaborative process no longer rested on any single artist. We were no longer beholden to a process of exchange that relies on a professional to draw out the creative potential of a person living with dementia. Instead, we worked together side-by-side, responding intuitively to what we had created together, no longer concerned with who did what.

Conclusions and Implications

Because of the open nature of this collaborative process, establishing constraints, expectations and shared values was essential. It is not necessary for future collaborations that use our model of cocreation to replicate every one of our conditions. Instead, it is important to understand the qualities of the structure and materials: accessibility, flexibility and ease of use, as well as the values that guided us: mutual respect, acknowledgement of differences in power and lived experience and explicit permission for failure and risk. The following discussion of the details of our collaborative structure is provided only as one example of the ways that these qualities and values might come together to form a successful collaboration.

We limited the scope of our process, agreeing from the beginning that we would each have three opportunities to contribute to the artwork, at which point we would declare it finished. Any material, technique and manipulation were acceptable, but we started out with two heavy sheets of watercolour paper that had already been painted with a purple wash. While we had an expectation of covering up work from previous iterations, we also knew that each step would be photographed, thus preserving each contribution. In our conversations, we adhered to constructive critique and took as valid all aesthetic and symbolic choices. We did not question the legitimacy of any individual interpretation, but were free to contribute a different point of view. When dementia was explicitly addressed in our conversations, or implicitly in the artwork, we differentiated between Bob's lived experience as a person with dementia and Sadie and Erin's experience as professionals who work with people living with dementia. The tools used to facilitate this collaboration were chosen for their affordability and flexibility of use. Zoom was chosen as a videoconferencing platform because all three of us had used it before and had found it relatively easy to use. We mailed the finished artwork between the East and West coasts of the United States, protecting it with two sheets of cardboard. This was also a risk, as the piece might have been damaged or lost in the mail. Erin and Sadie sent Bob a small number of art supplies in the first mailing, in recognition of the fact that Elder Care Alliance reimbursed us for the art supplies that we used. We made it clear, however, that any media was acceptable to use. Watercolour paper proved to be a durable and flexible ground for the artwork, and by the final iteration, we had used a variety of mixed media, including acrylic paint, ink, sewing, wool roving, paper collage, gel medium and glitter.

This project has greatly expanded each of the collaborators' understanding of shared art practice across diagnosis of dementia. The confluence of our individual experiences of art and its uses in personal and professional practice, combined with recent scholarship on diverse applications for a critically rigorous creative exploration in the context of collaborating with people living with dementia, led to a powerfully impactful experience. All three of us recognized stigma and tragedy narratives as major detractors from quality of life for people after a diagnosis of dementia, and each of us had explored the use of art in combating these. Although we did not explicitly set out to create a work of art that was about dementia, Bob's lived experience of stigma and Sadie and Erin's identification as professionals engaged in dementia as a human rights issue deeply informed the work. As much as themes relating to dementia emerged in the work itself, the process is informative of a novel way to disrupt conventional structures of power within the larger field of dementia care and studies. By sharing creative risk and establishing a relational space within the artwork, we formulated a potential path forward for a model of shared citizenship forged via interdependence.

Artmaking and creative flow rest on a densely interwoven give and take of risk and reward, wherein failure yields opportunities for discovery and growth. This structure is valuable in collaborations across diagnosis of dementia, as it privileges diverse skill sets and contributions and intuitively responds to change, subverting some of the "substantial barriers to unlearning" (Rushmer & Davies, 2004, p. ii13) in traditional care providing relationships. Successful cocreation rests on the process as much as if not more than the result and necessitates collaborators who can generously negotiate between risk and growth. When we discussed "making room" for each other on the canvas, this was as much a reference to emotional space as it was to physical space. This sharing of risk broke down socially prescribed roles and gave our relationship within the project a nimbleness and flexibility, directly challenging rigid and hierarchical provider/receiver roles that contribute to feelings of helplessness and loss of agency for people living with dementia.

The outcomes of this project have the potential to impact multiple fields of practice. This project included an art therapist and operated outside of conventional medical models or client/therapist dualities. Because of this, it lends insight to and advances the model of community-based practice for art therapists. This cross-disciplinary and collaborative modality utilizes the unique skills of art therapists in service of working alongside communities to positively transform social environments (Drusedum & Awais, 2016). For artists, it complicates a model of arts in health care that privileges a teacher/student or provider/recipient hierarchy. This conventional structure risks burning out artists without chronic health conditions by positioning them as heroic saviours taking on ever greater responsibility to meet needs created or exacerbated by an inadequate system of social support (Naismith, 2019). At the same time, it impoverishes bodies of creative work created in care settings by disabled artists and artists with chronic, serious or life-limiting health conditions by reducing their meaning to a tokenized representation of successful programming. A collaboration that positions artists living within and outside of long term care as cocreators allows for the creation of artwork that references and makes room for multiple experiences. Our collaboration rigorously contributed to each of our individual practices, an outcome that is beneficial for artists working in many environments and contexts. This outcome also presents a new model for arts in dementia care, conceptualizing the role of providers as that of transactional and transformational allies to resident artists. This project has implications for researchers as well. The use of an action research framework with arts-based methods allowed for us to explore cocreation in a safe environment. The outcomes that emerged from this project were conceptually rigorous and offered a counterpoint to the assumption that methods for providing art in dementia care must be based on quantitative data in order to be effective. This project also contributes to a body of research that is conducted with older adults and people living with dementia as co-investigators. It is our hope that this project will serve as an inspiration to anyone committed to the use of art to create meaning and connection, wherever their practice is located.

We will leave Bob with the last word, as he summarizes the almost impossible to put into words process that we experienced together:

> Bob: This here is going to be the mystery. That will only be determined once we're through.
>
> Sadie: That in between.
>
> Bob: That is called a risk. R-I-S-K. I don't know what the hell it means, but I'm putting it here. Oh, that's beautiful.

Conflicts of Interest We have no conflicts of interest to disclose.

Appendix 1

Paper with the original watercolour wash. Images taken before our collaboration began.

Appendix 2

The two artworks after the first cycle of collaboration. Each artist has contributed once.

Appendix 3

The two artworks after the second cycle of collaboration. Each artist has contributed twice.

Appendix 4

The two artworks after the third cycle of collaboration. Each artist has contributed three times.

References

Allen, P. B. (1995). *Art is a way of knowing*. Shambhala.

Anacleto, S. (2018). Aspirational ethics: Considerations for art therapy practice in a health-care setting for adults with dementia. *Art Therapy, 35*(3), 165–169. https://doi.org/10.1080/07421656.2018.1523653

Anderson, F., & Gantt, L. M. (2009). The formal elements art therapy scale: A measurement system for global variables in art. *Art Therapy, 26*(3), 124–129. https://doi.org/10.1080/07421656.2009.10129372

Arts Access Australia. (2015). *Beyond access: The creative case for inclusive arts practice literature review*. https://www.artsaccess.com.au/assets/Resources/Beyond-Access-Literature-Review-web.pdf

Austrom, M. G., Geros, K. N., Hemmerlein, K., McGuire, S. M., Gao, S., Brown, S. A., Callahan, C. M., & Clark, D. O. (2015). Use of a multiparty web based videoconference support group for family caregivers: Innovative practice. *Dementia, 14*(5), 682–690. https://doi.org/10.1177/1471301214544338

Barthes, R. (1967). The death of the author. *Aspen 5+6*. Retrieved April 29, 2020, from http://www.ubu.com/aspen/aspen5and6/threeEssays.html#barthes

Basting, A. D. (2006). Arts in dementia care: 'this is not the end... it's the end of this chapter'. *Generations, 30*(1), 16–20.

Basting, A., Towey, M., & Rose, E. (Eds.). (2016). *The penelope project: An arts-based odyssey to change elder care*. University of Iowa Press.

Bayles, D., & Orland, T. (1993). *Art and fear: Observations on the perils (and rewards) of artmaking*. Image Continuum Press.

Bell, J., & Leong, T. W. (2019). Collaborative futures: Co-designing research methods for younger people living with dementia. *In Proceedings of the 2019 CHI Conference on Human Factors in Computing Systems (CHI '19)*. Association for Computing Machinery, New York, NY, USA (Paper 352, pp. 1–13). https://doi.org/10.1145/3290605.3300582

Benbow, S. M., & Jolley, D. (2012). Dementia: Stigma and its effects. *Neurodegenerative Disease Management, 2*(2), 165–172. https://doi.org/10.2217/nmt.12.7

Bennett, B., McDonald, F., Beattie, E., Carney, T., Freckelton, I., White, B., & Willmott, L. (2017). Assistive technologies for people with dementia: Ethical considerations. *Bulletin of the World Health Organization, 95*(11), 749–755. https://doi.org/10.2471/BLT.16.187484

Bowes, A., & Dawson, A. (2019). Designing environments for people with dementia: A systematic literature review. *Emerald Publishing Limited*. https://doi.org/10.1108/9781787699717

Bradbury-Huang, H. (2010). Sustainability by collaboration. The SEER case. *Organizational Dynamics, 39*(4), 335–344. https://doi.org/10.1016/j.orgdyn.2010.07.001

Brannelly, T. (2011). Sustaining citizenship: People with dementia and the phenomenon of social death. *Nursing Ethics, 18*(5). https://doi.org/10.1177/0969733011408049

Byers, A. (2011). Visual aesthetics in dementia. *International Journal of Art Therapy, 16*(2), 81–89. https://doi.org/10.1080/17454832.2011.602980

Camic, P. M., Crutch, S. J., Murphy, C., Firth, N. C., Harding, E., Harrison, C. R., Howard, S., Strohmaier, S., Van Leewen, J., West, J., Windle, G., Wray, S., & Zeilig, H. (2018). Conceptualising and understanding artistic creativity in the dementias: Interdisciplinary approaches to research and practise. *Frontiers in Psychology, 9*, 1842. https://doi.org/10.3389/fpsyg.2018.01842

Cullum, C. M., Weiner, M. F., Gehrmann, H. R., & Hynan, L. S. (2006). Feasibility of telecognitive assessment in dementia. *Assessment, 13*(4), 385–390. https://doi.org/10.1177/1073191106289065

Dementia Inclusive Communities Initiative. (2019). *A celebration of dementia inclusive art & community* [Video]. Adobe. https://shared-assets.adobe.com/link/be341f87-3eff-41e7-691f-4e27f158fc30

Dissanayake, E. (1990). *What is art for?* University of Washington Press.

Drusedum, D., & Awais, Y. (2016). A model for art therapists in community-based practice. *Art Therapy: Journal of the American Art Therapy Association, 33*, 144–150. https://doi.org/10.1080/07421656.2016.1199245

Dupuis, S. L., Kontos, P., Mitchell, G., Jonas-Simpson, C., & Gray, J. (2016). Re-claiming citizenship through the arts. *Dementia, 15*(3), 358–380. https://doi.org/10.1177/1471301216637206

Gale, C. (2020). Art school as a transformative locus for risk in an age of uncertainty. *Art, Design & Communication in Higher Education, 19*, 107–118. https://doi.org/10.1386/adch_00016_1

Gray, J., Kontos, P., Dupuis, S., Jonas-Simpson, C., & Mitchell, G. (2018). Dementia (re)performed: Interrogating tensions between relational engagement and regulatory policies in care homes through theatre. In S. Chivers & U. Kriebernegg (Eds.), *Care home stories: Aging, disability, and long-term residential care* (pp. 111–126). Transcript Verlag. https://doi.org/10.14361/9783839438053-012

Hadley, B. (2020). Allyship in disability arts: Roles, relationships, and practices. *Research in Drama Education: The Journal of Applied Theatre and Performance, 25*(2), 178–194. https://doi.org/10.1080/13569783.2020.1729716

HP Development Company. (2016). Dr. Erin Partridge: Sprout by HP Offers New Possibilities. http://h20195.www2.hp.com/v2/GetDocument.aspx?docname=4AA6-8503ENW

Hughes, J. C. (2014). *How we think about dementia, personhood, rights, ethics, the arts and what they mean for care*. Jessica Kingsley.

Huss, A. J. (1977). Touch with care or a caring touch? *American Journal of Occupational Therapy, 31*(1), 11–18.

Hutchings, J. B. (2015). Symbolic use of color in ritual, tradition, and folklore. In A. J. Elliot, M. D. Fairchild, & A. Franklin (Eds.), *Handbook of color psychology* (pp. 340–356). Cambridge University Press.

Jeong, J. (2019). The affective creativity of a couple in dementia care. *Culture, Medicine and Psychiatry*. https://doi.org/10.1007/s11013-019-09662-5

Kandinsky, W. (1977). *Concerning the spiritual in art* (M.T.H. Sadler, Trans.). Dover. (Original work published 1914).

Kenny, J., Asquith, I., Guss, R., Field, E., Slade, L., Bone, A., Oliver, K., Jones, M., Ryan, C., Brooks, M., & Norris, C. (2016). Facilitating an evolving service user involvement group for people with dementia: What can we learn? *The Journal of Mental Health Training, Education and Practice, 11*, 81–90. https://doi.org/10.1108/JMHTEP-09-2015-0046

Kersten, A., & van der Vennet, R. (2010). The impact of anxious and calm emotional states on color usage in pre-drawn mandalas. *Art Therapy, 27*(4), 184–189. https://doi.org/10.1080/07421656.2010.10129387

Keyes, S. E., Clarke, C. L., & Gibb, C. E. (2019). Living with dementia, interdependence and citizenship: Narratives of everyday decision-making. *Disability & Society, 34*(2), 296–319. https://doi.org/10.1080/09687599.2018.1528970

Killick, J., & Craig, C. (2011). *Creativity and communication in persons with dementia*. Jessica Kingsley Publishers.

Kitwood, T., & Bredin, K. (1992). Towards a theory of dementia care: Personhood and well-being. *Ageing and Society, 12*(3), 269–287. https://doi.org/10.1017/S0144686X0000502X

Kontos, P., Grigorovich, A., Dupuis, S., Jonas-Simpson, C., Mitchell, G., & Gray, J. (2020). Raising the curtain on stigma associated with dementia: Fostering a new cultural imaginary for a more inclusive society. *Critical Public Health, 30*(1), 91–102. https://doi.org/10.1080/09581596.2018.1508822

Kontos, P., Miller, K., & Kontos, A. (2017). Relational citizenship: Supporting embodied selfhood and relationality in dementia care. *Sociology of Health & Illness, 1*(1), 182–198. https://doi.org/10.1111/1467-9566.12453

Králová, J. (2015). What is social death? *Contemporary Social Science, 10*(3), 235–248. https://doi.org/10.1080/21582041.2015.1114407

Martin-Khan, M., Flicker, L., Wootton, R., Loh, P., Edwards, H., Varghese, P., Byrne, G., Klein, K., & Gray, L. (2012). The diagnostic accuracy of telegeriatrics for the diagnosis of dementia via video conferencing. *Journal of the American Medical Directors Association, 13*, 487.e19–24. https://doi.org/10.1016/j.jamda.2012.03.004

Marziali, E., & Donahue, P. (2006). Caring for others: Internet video-conferencing group intervention for family caregivers of older adults with neurodegenerative disease. *The Gerontologist, 46*(3), 398–403. https://doi.org/10.1093/geront/46.3.39

Marziali, E., & Garcia, L. J. (2011). Dementia caregivers' responses to 2 internet-based intervention programs. *American Journal of Alzheimer's Disease & Other Dementias*, 36–43. https://doi.org/10.1177/1533317510387586

McGarry, J. (2009). Relationships between nurses and older people within the home: Exploring the boundaries of care. *International Journal of Older People Nursing, 5*, 265-273. https://doi.org/10.1111/j.1748-3743.2009.00192.x

Mihailidis, A., Blunsden, S., Boger, J., Richards, B., Zutis, K., Young, L., & Hoey, J. (2010). Towards the development of a technology for art therapy and dementia: Definition of needs and design constraints. *The Arts in Psychotherapy, 37*, 293–300. https://doi.org/10.1016/j.aip.2010.05.004

Mitchell, W. (2015, November 21). *Learning to live differently*. Which me am I today? https://whichmeamitoday.wordpress.com/2015/03/04/learning-to-live-life-differently/

Mitchell, W. (2016, February 17). *Changing Relationships........* Which me am I today?. https://whichmeamitoday.wordpress.com/2016/02/17/changing-relationships/

Naismith, N. (2019). *Artists practising well*. Aberdeen: Robert Gordon University. https://rgu-repository.worktribe.com/output/235847

Newman, A., Goulding, A., Davenport, B., & Windle, G. (2019). The role of the visual arts in the resilience of people living with dementia in care homes. *Ageing and Society, 39*(11), 2465–2482. https://doi.org/10.1017/S0144686X18000594

Nuffield Council on Bioethics. (2009). *Dementia: Ethical issues*. https://www.nuffieldbioethics.org/assets/pdfs/Dementia-report-for-web.pdf

Nunes, F., & Fitzpatrick, G. (2015). Self-care technologies and collaboration. *International Journal of Human-Computer Interaction, 31*, 150730080814008. https://doi.org/10.1080/10447318.2015.1067498

Ostrowski, A., DiPaola, D., Partridge, E., Park, H. W., & Breazeal, C. (2019). Older adults living with social robots: Promoting social connectedness in long-term communities. *IEEE Robotics & Automation Magazine, 1*(1). https://doi.org/10.1109/MRA.2019.2905234.

Partridge, E. E. (2019). *Art therapy with older adults: Connected and empowered*. Jessica Kingsley.

Poon, P., Hui, E., Dai, D., Kwok, T., & Woo, J. (2005). Cognitive intervention for community-dwelling older persons with memory problems: Telemedicine versus face-to-face treatment. *International Journal of Geriatric Psychiatry, 20*, 285–286. https://doi.org/10.1002/gps.1282

Reed, P., Carson, J., & Gibb, Z. (2017). Transcending the tragedy discourse of dementia: An ethical imperative for promoting selfhood, meaningful relationships, and well-being. *AMA Journal of Ethics, 19*(7), 693–703.

Rohra, H. (2016). *Dementia activist: Fighting for our rights*. Jessica Kingsley Publishers.

Rushmer, R., & Davies, H. T. O. (2004). Unlearning in health care. *Quality and Safety in Health Care, 13*(SUPPL. 2), ii10–ii15. https://doi.org/10.1136/qshc.2003.009506

Savin-Baden, M., & Wimpenny, K. (2014). *A practical guide to arts-related research*. Sense Publishers.

Stallings, J. W. (2010). Collage as a therapeutic modality for reminiscence in patients with dementia. *Art Therapy, 27*(3), 136–140. https://doi.org/10.1080/07421656.2010.10129667

Stephenson, R. C. (2013). Promoting well-being and gerotranscendence in an art therapy program for older adults. *Art Therapy, 30*(4), 151–158. https://doi.org/10.1080/07421656.2014.846206

Swaffer, K. (2014). Dementia: Stigma, language, and dementia-friendly. *Dementia, 13*(6), 709–716. https://doi.org/10.1177/1471301214548143

Sweeting, H., & Gilhooly, M. (1997). Dementia and the phenomenon of social death. *Sociology of Health & Illness, 19*(1), 93–117.

Treadaway, C., & Kenning, G. (2016). Sensor e-textiles: Person centered co-design for people with late stage dementia. *Working with Older People, 20*(2), 76–85. https://doi.org/10.1108/WWOP-09-2015-0022

Waller, D. (2002). *Arts therapies and progressive illness: Nameless dread*. Brunner-Routledge.

Zeilig, H., West, J., & van der Byl Williams, M. (2018). Co-creativity: Possibilities for using the arts with people with a dementia. *Quality in Ageing and Older Adults, 19*(2), 135–145. https://doi.org/10.1108/QAOA-02-2018-0008

Chapter 19
Opening the Window: Art Therapy and Nature as Mindfulness Practices with People Living with Dementia

Michelle C. Olson

Abstract By the year 2030, about 82 million people are projected to be living with dementia and that number is projected to reach 152 million in 2050 (World Health Organization. 2019. "Dementia: Key Facts." https://www.who.int/news-room/fact-sheets/detail/dementia). These astounding figures demonstrate the obligation of healthcare systems and societies to address the issue of improving the lives of those living with forms of dementia and those who care for them. The language used in empirical literature to describe dementia very often focuses on suffering and behaviours and has been argued to do more harm than good because the terminology refers to people by their medicalized conditions. The need to look beyond the diagnosis and symptoms to see each person holistically and empathically is paramount. There has been considerable research on non-pharmacological approaches to dementia care. However, many of these commonly used and predetermined interventions such as the use of stuffed animals, laundry folding or doll therapy often have little or nothing to do with the individual and are not considered to be person-centred. Although these interventions may be beneficial for some people, they cannot be used as a one-size-fits-all approach in dementia care. Great strides have been made in long-term care practices, but they often remain medicalized and task-oriented. This chapter is an exploration of how the creative and expressive arts, nature and mindfulness practices related to people living with dementia have been researched. In addition, my experience using art, nature and mindfulness with clients living with dementia is explored as a meaningful and cohesive intervention. Case examples are presented from over two decades of my experience as a creative arts therapist who has worked with clients living with dementia to share opportunities for mindfulness; encourage exploration of natural materials; and focus on personhood, abilities, strengths, unique perspectives and gratitude.

M. C. Olson (✉)
Evergreen Minds, Inc., Montana State University, New York, USA
e-mail: molson@evergreenminds.org

M. K. Shankardass (ed.), *Dementia Care*,
https://doi.org/10.1007/978-981-16-3864-0_19

Inequities in Dementia Care

People living with dementia are often denied the basic rights and freedoms that are available to other people (World Health Organization, 2019). In eldercare settings, daily choices and agency are often taken from people living with dementia. The use of physical and chemical restraints has declined drastically in the USA since the Omnibus Budget Reconciliation Act of 1987. However, the use of physical and chemical restraints persists in the USA if deemed "medically appropriate" and also continues in other parts of the world (Crutchfield et al., 2019, p. 617; Amira et al., 2017).

Some of these inequities in dementia care are evidenced in language that is often used in eldercare. Outdated words such as *locked units* (dementia care communities), *wandering* (someone desires to walk around) and *elopement* (a person wants to leave an area or explore a new place) are often used in memory care. The italicized words above are problematic because they represent these injustices and do not allow people living with dementia to be seen holistically as individuals with the human rights, needs and desires to which they are entitled.

The term *behavioural and psychological symptoms of dementia* (BPSD) was derived from a meeting of the International Psychogeriatric Association in 1996. The term BPSD was intended to improve the quality of life and the care of those living with dementia and their care partners (Cunningham et al., 2019; Tible et al., 2017). However, this terminology generalizes all people with dementia as the same and has sparked further perpetuation of the institutionalization practices within eldercare. In these settings, terms such as *challenging behaviours* or *inappropriate behaviours* are commonly used by staff. These linguistical choices do not support the unmet needs and desires of individuals living with dementia or address the context of their emotional, physical, or environmental needs or discomfort.

The term BPSD has rightly come under fire by advocate group called GAP in Dementia Care, which is made up of people living with dementia, care partners, academics, eldercare professionals, geriatricians and gerontologists (Swaffer, 2018). Kate Swaffer is a former aged-care nurse who is also living with dementia. She is a founding member of the GAP in Dementia Care group and has helped to generate a movement called #BanBPSD. Members of this movement argue for the removal of negative and stigmatizing labelling such as BPSD and advocate for more compassionate care that respects the human rights of all people. Group members advocate for the removal of "all forms of segregation and restraint by acknowledging everyone has the right to 'freedom of expression'" (Barsness et al., 2018, 8).

There is a prevalence of outdated and ageist language, even in current literature focused on improving quality of life for people living with dementia through the arts. I noted perpetuation of terms such as *patients*, *suffering*, *victims*, *the elderly*. The use of inappropriate language to describe older people living with dementia further marginalizes them and fails to recognize them as a heterogeneous group. It is imperative not to medicalize or victimize people living with dementia. Researchers must recognize their biases and their obligation to reduce ageism and be mindful not to perpetuate negative misperceptions about older people.

Creative Arts in Dementia Care

Likely no one has done more to advance the study and understanding of creative expression as a means of promoting cognitive and physical health in ageing than the late Gene Cohen, a geriatric psychiatrist and pioneer in the gerontological field. Just as the GAP in Dementia Care group is reframing the negative stereotypes and injustices that surround dementia, Cohen (2006a) promoted person-centred approaches that do not focus on decline and despair, but rather, highlight the opportunities and creative abilities that can occur in people living with Alzheimer's disease (Cohen, 2006b).

As cases of dementia continue to increase in the ageing population, the need to move away from pharmaceutical interventions to more person-centred or individual-guided practices within aged care settings is crucial (Deshmukh et al., 2018; Scales et al., 2018; Wang & Li, 2016). Over the last two decades, various cultural arts have been implemented and studied with people living with dementia in eldercare settings. These creative forms of engagement, which include music, visual arts, poetry, dance and storytelling have generated positive psychosocial and behavioural improvements and increased quality of life for people in eldercare settings (de Medeiros & Basting, 2013; Fraser et al., 2015).

There is no shortage of evidenced-based literature on the use of arts with people living with dementia. It should be noted, however, that empirical studies utilizing creative approaches often have limitations in their study design (i.e. not having randomized control groups; small sample sizes; diverse approaches to art therapy, arts-based methods and arts activities; varied measurement scales; diverse competencies, experience and goals of the facilitators delivering these modalities; inconsistent outcome assessment; and the challenge of retaining participants; (Cowl & Gaugler, 2014; Huhtinen-Hildén, 2014; Schneider, 2018; Stuckey & Nobel, 2010; Wang & Li, 2016; Windle et al., 2016).

Additionally, there has been a rise in neuropsychologists seeking to provide evidenced-based practices demonstrating the efficacy of art therapy and arts interventions within dementia care to reduce chronic pain (Rouch et al., 2018), to determine the effect dementia has on art production (Safar & Press, 2011) and to elicit positive increases in psychological well-being via emotional expression and communication (Guseva 2018).

Despite these limitations, researchers in the neurological sciences have grown interested in the creation of art and viewing of art by people living with dementia use has grown in widely recognized community programs such as the I'm Still Here Foundation's ARTZ initiative, Time Slips, Cognitive Dynamics' Bringing Art to Life, the Museum of Modern Art's Meet Me at MoMA, Arts and Minds, The Alzheimer's Association's Memories in the Making arts program and the intergenerational work of Opening Minds Through Art. Anne Basting, the founder of Time

Slips, questions the common one-time activities that "are designed to distract" and asks, "what if programming in care settings provided avenues for meaning-making that were designed to foster learning over time?" and "What if stigmatized health centers became vibrant cultural centers?" (Basting, 2018, 10).

An interesting finding in a review of arts research in dementia care revealed that much of the quantitative research focused on mood symptoms of dementia such as depression, apathy, cognition and agitation (Schneider, 2018). The findings of studies utilizing art therapy for people living with dementia often focus on biomedical conditions that are placating to caregivers by "alleviating 'disruptive' behaviors" versus the holistic and person-centred needs, life enrichment activities and personal expression of the people living with dementia (Beard, 2012, 637).

It is common in long-term care settings to facilitate activities with as many participants as possible, a practice that focuses on quantity over quality. Recreation staff in long-term care homes are often mandated by management to keep residents busy and bring as many elders as possible to daily programs. Creative arts therapists must continue to educate staff in these settings and advocate for providing elders space and opportunities for imaginative process-oriented expressions versus the common product-oriented approach (Beard, 2012). Although art products and even the displaying of artwork may be a part of the process, it is paramount to recognize that to support a therapeutic experience for these elders, the priority cannot and should not be on the facilitator's own agenda and need to fill seats (Partridge, 2019).

So many powerful and important moments and opportunities can be missed if staff are not present for elders during the creative process. This outdated model of programming is unfortunate because it deprives older people with dementia living within these settings of their right to participate in meaningful engagements that honor who they are and provide opportunities for creative expression through symbolic language (Lehikoinen, 2017).

There is no denying that engaging in art can help individuals express thoughts, memories and emotions. However, staff who focus on remaining strengths and opportunities (Cohen, 2006) and use imagination and creative arts as "a mode of expression and an avenue toward meaningfulness" can empower people living with dementia within their community (Basting, 2018, 10).

Creativity: A Gift of Dementia

It may be surprising to some that it is possible to live fully and creatively not despite having a diagnosis of dementia, *but because of it*. Creative expression has been called one of the super-powers of dementia by Dixon (2019) who is an advocate for people living with dementia and who is living with Alzheimer's and vascular dementia himself. As Basting (2009) notes, creativity is not rooted in memory. Creativity can be a means of tapping into memories, expressing oneself in a given moment and connecting with one another in meaningful ways. Having created with older adults with dementia for over two decades, I continue to learn from their unique life experiences and diverse perspectives, and I have witnessed the freedom that can come from expressing one's imagination, thoughts and feelings. I always encourage awareness and connection with others, self-expression and independent choices, and I provide generous materials that draw on the various senses without being child-like. Renowned art therapist, expressive arts therapist and psychologist, Cathy Malchiodi refers to "imagination as an antidote to adversity" (2020, 345). Dementia does not rob a person of their imagination and in many ways, it can enhance it.

Research is inconsistent on how engaging in arts can improve cognition. This is not surprising due to the heterogeneity of the people living with dementia and the difficulty finding methodological approaches that effectively measure these diverse changes and factors. However, it is also important to recognize that preserved and even new-found artistic creativity has been demonstrated in people living with Alzheimer's disease and frontotemporal dementia (Cohen, 2006; Erkkinen et al., 2018). Neurodegenerative diseases do not necessarily restrict artistic expression, but in the case of frontotemporal dementia, the damage to the frontal and temporal lobes can release inhibitions in the right parietal lobe, which can increase creative expression. This phenomenon where suppression or damage in some brain regions leads to stimulation and enhancement in other areas is called *paradoxical functional facilitation* (Erkkinen et al., 2018). We have heard of cases where older people who are living with dementia develop an affinity for creativity that they did not have earlier in their life.

I have very often witnessed this loss of inhibition in the creative process with my clients over the years. I have curated many art exhibits, with the assistance of older adults who express interest in sharing their work. I recall one exhibit that was the culmination of a year-long grant-funded arts program designed specifically for people living with moderate to advanced dementia. In this exhibit, I used the theme "Art is a window to creativity". Loved ones and staff members shared their amazement and enjoyment seeing the emergent creative expressions in people who had not previously demonstrated artistic interest or inclinations. My approach was simple. Get to know each person in the group and learn about their lives, families, work and interests. Get them to connect with one another, share real or imagined stories and provide various materials to explore and use in traditional and innovative new ways. I like to introduce visual prompts to my elder clients to elicit conversations and stories. Then, I string their words together into free verse poetry which is printed out and read back to them. This collaborative approach focuses on abilities and strengths of individuals and emphasizes that there are no right or wrong responses. I often utilize found objects, recycled materials and items from nature to inspire connections with self and others while allowing participants to enjoy the absolute freedom and non-judgement of creative expression. Arts-based, non-verbal approaches can elicit feelings of grief, loss, unresolved trauma to emerge as well as positive life experiences, memories, feelings and playfulness. It is paramount to provide opportunities for people living with dementia to express these important human emotions and experiences in whatever form they arise. Having a safe space where these expressions are acknowledged and validated with empathy can help alleviate distress. Rather than treating the disease and "behaviors", holistic arts-based approaches tend to elders with dementia as capable people, empowering them (Partridge, 2019) and encouraging imagination and engagement in the face of adversity (Cohen, 2006).

Resistance to Art Making

It should be noted that as an art therapist, I often encounter strong resistance to art making from older adults. I have worked with many older people who associate art making as childish and I then introduce the suggestion of the importance of play, exploration and imagination in later stages of life. Malchiodi (2020) prefers the term "expressive rather than creative" in her psychotherapeutic approach because she has found that some people, particularly adults, may find the word *creativity* to be intimidating and thus resistant to art making when introduced this way (2020, 35).

I have experienced resistance especially in people who are artists—meaning people who have spent their lives in a professional or personal artist capacity. I purposely did not use the past tense version (i.e. people who *were* artists). This is because I believe that people maintain their identities throughout their lives regardless of cognitive function. A dementia diagnosis does not take away or diminish a person's innate desire to create. Although the level of ability, or confidence may change, the creativity or desire to be expressive does not simply dissolve, so it must be acknowledged, supported and nourished.

This resistance can often be witnessed in people living in the early and middle stages of dementia. I have held countless conversations with elder artists who are resistant to artmaking because they consciously or subconsciously feel they "can't do it" like they once did, so they refuse to try. This is where honoring the person for who they are now and how they feel is paramount. I have experienced many well-meaning staff try to force this creative process with comments such as, "Come on Mary, I know you like to paint. Make me a pretty picture!" Professional caregivers and family care partners must be cognizant of whose agenda are they are trying to meet. It is important not to use infantilizing and ageist communication with older adults and to utilize an empathetic and person-centered (also called person-directed, individual-centered, client-centered) approach.

A positive and dignified way to engage someone who is living with dementia is to simply offer the person an opportunity for socialization or observation within a group without pressure to participate. An example of this type of resistance occurred with a gentleman named Henry who was living in a lovely assisted living community specifically for people living with dementia. Henry was quiet and cordial and was often a passive observer in the group. He did not consider himself to be an artist, and he was uncertain about how or what to create, so he would politely refuse and agree to just watch. I asked Henry about his career, and he became animated as he spoke of his experience as a junior Rabbi. He was no longer able to give details about what he did in his service, but it was evident that Henry took great pride in his important role. With this information, I was able to then connect Henry in conversations with the other people in the group. This spark also allowed Henry to express something

Fig. 19.1 Sheila and Michelle's collaborative chalk drawing

that interested him in a non-threatening way. I asked what the synagogue looked like and if he would be willing to draw it. Without hesitation, he created a simple drawing that was later turned into a printmaking project. Henry was thrilled to see this important part of his life represented in a new and tangible way that was meaningful for him. When people are honored for who they are and are given autonomy and choice without pressure or judgement, resistant older adults will very often become comfortable enough to join in on their terms and explore the materials (Partridge, 2019).

Another way to engage an elder who is resistant to artmaking is to ask them for their aesthetic opinion or ask for their assistance with an art material or process—empowering them as an instructor. Figure 19.1 is a chalk drawing made by Sheila who was far along in her Alzheimer's journey. Sheila spent her adult life as an accomplished artist and teacher. At every session, Sheila verbalized her accomplishments and those of her mother who she reported to be a very talented artist. Sheila came willingly to the art group on most occasions, but she was extremely resistant to artmaking in the three years I worked with her. Additionally, Sheila was very critical of the work of others and would make cutting comments such as, "This isn't art" or "This is crap!". I often sought her advice, which she seemed to appreciate despite her lack of participation in the artmaking process. It was also important that I was present and attentive to *what she was really saying*. Sheila was aware on some level that things were no longer what they once were. She felt these losses, and she deserved to be heard and empowered. On one occasion, I was able to create a chalk drawing along with Sheila (Fig. 19.1) by asking her opinion about colors, lines and blending techniques. Her verbal words did not match the clarity of what she spontaneously wrote on the paper: "Today is an interesting day". Sheila was heard, and more importantly, she was understood. It was indeed an interesting day.

Another common reason I have found for resistance in artmaking among older adults is the presence of physical challenges, which can include decreased vision, hearing and fine motor dexterity or tremors. I have found ways to help clients work with these challenges and encourage them to view them as opportunities to express themselves in a different way. For instance, if someone has difficulty keeping a brush steady, we may discuss how the broken marks could be turned into something new. Acknowledging frustration is important, and these feelings should be validated and respected.

In the case of Mae, physical limitations were compounded. Mae was legally blind, had a difficult time hearing, and was in the middle stages of Alzheimer's disease. However, she loved to be around people, so like clockwork three times a week, she arrived in the art studio, which was full of windows and light. Routine was important to Mae, and she sat in the same sunny seat each session and asked what we were doing that day. Mae reminded me each time that she could not see and that she was not sure if she could make art. I sat close to Mae's dominant side, and we worked together as a team to ensure that she could create in a way that was comfortable and enjoyable for her. I was usually pleasantly surprised when I would leave Mae, only to come back and find her creating something completely spontaneous. I often left her to explore and imagine on her own and invited her to share her work and her thoughts. Other times, I asked Mae if she would help me mix paint colors because I noticed that she would often become absorbed in the repetition, movement and stirring motion with the paintbrush. This request seemed to please Mae because she was being creative; she was productive, and she was needed. People living with dementia naturally desire and have the right to maintain a sense of purpose and inclusion within their communities and to remain respected and socially connected (Windle et al., 2016).

A common theme in Mae's art revolved around roses. Although her renderings of them were not always recognizable to an observer, they were clearly roses to Mae. This is why it is important not to make assumptions or judgements about what appears to be in a person's artwork. Mae was not able to give details about her roses but spoke about the vibrancy of the colour and how much she loved them. When I asked her daughter about the roses, she informed me that her father had a loving tradition of sending her mother red roses every Friday. This information was invaluable. I was now able to ask Mae about this tradition. Mae lit up as she moved the pastels across the paper and shared memories about receiving these flowers from the "love of her life".

What mattered to Mae was not how the art looked (after all, she could not see it) but for her, it was more the process of expression and how she felt while she engaged in it. Mae felt connected and a part of the community in the art studio, which validated that her feelings and her life were important. Figure 19.2 is a colorful painting that Mae and I created together. With prompts, Mae described vivid recollections of the sunset over the Hudson River. I taped the sides of the paper so Mae could feel the

Fig. 19.2 Mae's sunset over the Hudson painting

paper's edges and asked her what colors she saw in the sky. As she named the colours, I assisted her in dipping into the named colour, then I asked her to move her brush across the paper.

Reframing the common narrative of dementia as something other than the end of personhood where a person experiences only losses and limitations is essential within art therapy and elder health care. Providing opportunities for continued growth can empower and enrich the lives of people who deserve to have opportunities for creative self-expression.

Art can serve as a type of meaning-making which can enable an older adult to "re-author the dominant narrative of their illness or physical disability" (Malchiodi, 2013, 9). Transforming difficult feelings and emotions through creative self-expression with an art therapist can be extremely helpful for older people in re-evaluating, reframing and accepting their physical and cognitive challenges at this moment.

Barbara was a social woman in her eighties with whom I visited for over five years. Barbara's career as a nurse ended after she experienced a series of strokes that caused vascular dementia, limited mobility and some physical pains. Having endured many losses, including the loss of her daughter, her professional identity, her home, and her physical and cognitive agility—art came to be a source of great inspiration and expression for Barbara. In addition, creating art reinforced a sense of community for Barbara, and she would often create around themes of resilience, hope, connections and faith.

Over the years, I have had very insightful conversations with Barbara about her perspectives and her ability to remain hopeful and her constant encouragement of other elders in her care community. Barbara eagerly attended the monthly art group and would completely immerse herself in the creative process in a visibly relaxed and mindful "state of flow" (Chilton, 2013). Figure 19.3 is a fabric banner suspended

Fig. 19.3 Barbara's banner

on driftwood that Barbara painted to represent the themes of gratitude, peace, love of nature, hope and connection with others.

It was important that Barbara continued to serve others, just as she had always done as a nurse. Her wheelchair did not hinder her from helping others. She made a point to offer her peers a friendly visit, a smile, a snack or even a piece of her artwork. Barbara was African American with deeply held religious beliefs. It was important to her to "respect everyone even though we may not like everyone". Barbara was acutely observant of people and contemplative of the world around her. She would vividly describe the beauty in the way the full moon sparkled over the water at night when she could not sleep or the seasonal colours of the treetops. Barbara would clip meaningful articles about mindfulness and images from the newspaper, knowing we could share a conversation about it together. This was her way of processing her losses and finding acceptance, peace and gratitude. Figure 19.4 is a chalk pastel drawing that Barbara created. The words are from a conversation we shared after she made this drawing. We spoke about her theme of forgiveness and resilience and I asked her if I could combine her words with her artwork, to which she was very pleased.

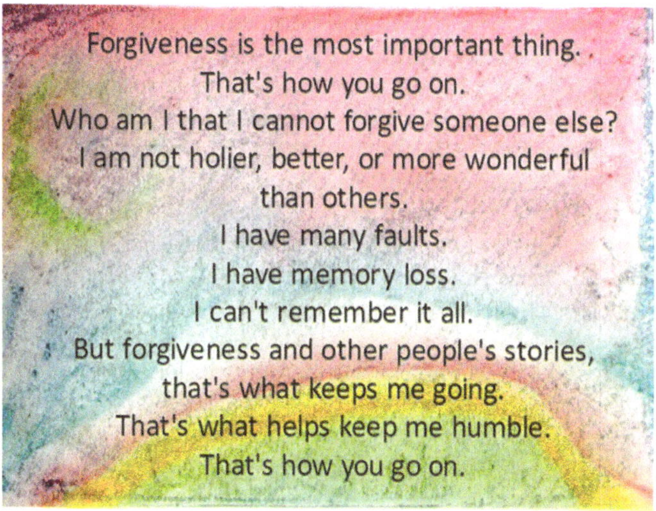

Fig. 19.4 Barbara's chalk pastel drawing with her thoughts overlaid on top

Nature and Health

Humans have been deeply connected to the natural world since the beginning of time. *Biophilia* is a term coined by Harvard University biologist and entomologist, Wilson (1984) to describe human beings' evolution involving an innate connection and attraction to plant and animal life (i.e. nature; Mencagli & Nieri, 2019). Eco-psychology and eco-therapy are informed by systems theory and have emerged since the 1990s. Theorists in these disciplines posit that it is detrimental to our psychological well-being to live disconnected from our ecological surroundings (Roszak et al., 1995).

Eco-psychologist Michael Cohen has been instrumental in the movement of nature and healing. As the founder of the 501c3 organization Project Nature Connect, Cohen established a four-step natural system's thinking process through which he posits that our human disconnection with nature is the root of our psychic suffering, which often occurs without conscious awareness of this separation (Project Nature Connect, n.d.; Sweeney, 2013). Cohen states:

> We suffer our troubles because we do not realize that our planet is our other body. We do not realize that every time we exchange air, water and food with our planet and biosphere. We are creating reciprocity between our individual bodies and the body we share with every other life form around us, Earth. (Project Nature Connect, n.d.)

Mindfulness and awareness of self and self as it relates to the natural world is a major tenet of eco-therapy. More recently, mindfulness-based eco-therapy (Hall, 2016), nature-based expressive arts therapy (Atkins & Snyder, 2018) and eco-art therapy (Sweeney, 2013) have evolved to take on more holistic, sensory, regenerative and mindful ways to achieve psychic balance in reconnecting people to the natural world around them (Sweeney, 2013).

Shinrin-yoku (i.e. forest bathing) was developed in Japan in the 1980s, but the understanding of the deep relationship between humans and nature have ancient origins. The Japanese forest service inaugurated a campaign to extend the medicinal and stress-relieving health benefits of "soaking up" the forest environment (Mencagli & Nieri, 2019, 85). Since then, there has been extensive research conducted on how the forest aromas, sensory experiences and the monoterpenes—which are organic and aromatic substances emitted by plants, fungi and bacteria—have measurable positive physiological effects on humans' immune systems and psychological and cognitive processes (Mencagli & Nieri, 2019, 85).

Theodore Roszak (1992) was a twentieth-century historian who coined the term, *eco-psychology* to refer to the blending of ecology and psychology. The premise of the eco-psychology movement is the interconnectedness between people and the natural environment and the world itself. The aim of eco-psychologists is to raise awareness of how human disconnection from nature can play into pathologies and poor mental health and well-being, something that traditional psychotherapeutic principles do not take into account (Davis & Canty, 2013).

Nature and Eldercare

Little did I realize that my long-time practice and innate desire to engage mindfully with natural materials and the outdoors in all four seasons with elders in long-term care settings had deep roots in arts and psychology. I have been keenly aware of the disconnection of people living with dementia in eldercare homes to one other and to green space. This is not to say that some care homes do not offer the special occasion of an outdoor barbeque, gardening group, or a well-manicured patio for elders to visit with family—but the day to day accessibility to outdoor spaces and the connection with sunshine, fresh air, birdsong and plants is often missing for people living in these homes. This is especially true when people who are living with dementia are paternalized and segregated behind locked doors. Safety must be the highest priority, but a sense of normalcy, humanity, quality of life and well-being should not cease because someone has dementia or lives in an eldercare community.

I have worked in long-term care communities where the views are extraordinarily beautiful, but these views often go unnoticed or unexperienced by people who are no longer unable to move about on their own. In my doctoral research study in long-term care homes, some elder participants expressed sympathy and compassion for their peers who were living with advanced dementia because they witnessed their peers being cut off from others and the beautiful views around them. They even perceived the locked floor where people lived with dementia as punishment and expressed their hope to never experience living there. If long-term care communities are to embrace individual-centred care practices, they must provide the *human right and choice* to experience the outdoors on a regular basis.

Some people living with dementia may feel more safe and secure being inside, but other people living with dementia have expressed the desire to get outdoors, even if they forget where they are and get lost (Førsund et al., 2018). Research has shown that people living with dementia have conveyed feeling a lack of autonomy and control over their lives. They have expressed dissatisfaction with "not being allowed to go outdoors" and have described their experience as "living in a prison without bars" (Førsund et al., 2018, 33). In another small Swedish study, participants with early-stage dementia perceive and reflect on perceived limitations to being outdoors and participating in outdoor activities. One male participant described the experience of being outdoors as, "It's like medicine". Participants also recognized perceived limitations, grief and loss as well as the freedom and life outdoors. One participant reflected, "It [being outdoors] reminds me that there is still a lot of life left to be lived" (Olsson et al., 2013, 796).

It is interesting that almost forty years ago, American researcher, Ulrich (1984) discovered that patients in hospitals who had rooms facing green space experienced less pain and healed more quickly than patients whose windows looked out over urban views. Research has also shown that increased exposure to natural daylight can help decrease depression experienced by people living with dementia (Konis et al., 2018; Sakurai et al., 2017). Despite this evidence of nature's healing qualities, the majority of long-term care communities have still not fully incorporated nature into their environmental designs.

It should be noted that there are innovative organizations such as The Eden Alternative, an international not-for-profit, that centres around the premise of elder-centred care in all aspects of daily life. The Eden Alternative is committed to "creating a Human Habitat where life revolves around close and continuing contact with people of all ages and abilities, as well as plants and animals" (The Eden Alternative, 2016).

Some researchers have called for further clarification and categorization of the terms used in the expansive amount of research on nature's positive effects on human cognitive and psychological functioning (Bratman et al., 2012). These authors argue for more clarity surrounding definitions of terms such as *urban, natural, forest, field* and *urban greenspace*. They have also called for more specific data regarding the frequency and duration of exposure to nature and recommended ways of assessing its behavioural and psychological effects (Bratman et al., 2012, 132).

Research surrounding person-centred opportunities for people living with dementia to have access to and engage with green spaces has been linked with psychosocial well-being, empowerment, positive risk taking and sense of identity and relation to the natural world (de Bruin et al., 2010; Mmako et al., 2020).

Scottish researcher, Mandy Cook, coproduced a unique sensory and social woodland programme for people living dementia. Cook's doctoral research was conducted along with the Forestry Commission Scotland, Alzheimer Scotland and other public-sector organizations. Cook adopted a participatory action research approach that involved male participants who were living with early stage dementia. In this pilot programme, park rangers facilitated urban forest activities with the participants such as woodland walks, making bird and bat houses, planting trees, photographing nature, creating nature-related art, identifying birds and trees, lighting fires and engaging in woodland cooking (Cook, 2015). This programme was influential because it reiterated the need for eldercare across contexts to focus on the strengths and abilities of people who are living with dementia. The study promoted agency and sense of identity and supported rich connections with self and others through nature. In addition, this study provided opportunities for participants to connect with others emotionally while communicating feelings through means other than verbal language (Cook, 2020). This alternative approach can be applied in other eldercare settings where access to nature is limited (Cook, 2020).

Eco-Arts Therapy as Mindfulness Practices in Dementia Care

Roman philosopher, Marcus Tullius Cicero said that art is born of the observation and investigation of nature. Eco-art therapy addresses the "disconnection of our thoughts and feelings from the restorative flow of Earth's wisdom" (Sweeney, 2013, 13). This melding of eco-therapy and art therapy uses the holistic and expressive modalities of art and nature to mitigate conflict resolution; impart self-discovery; and promote personal insight, inner healing and reconnection with the land (Sweeney, 2013). Utilizing natural materials for expression has been a part of the human experience since the earliest days, yet in our busy technological lives, we are often disconnected from the rituals that connect us with one another, ourselves and the rejuvenating properties of our Mother Earth (Speert, 2016; Sweeney, 2013).

There has been a growing body of evidence to support mindfulness-based care practices with people living with dementia (Quintana Hernández & Barrachina, 2015; Russell-Williams et al., 2018; Wells et al., 2013) and mindfulness practices designed to aid in non-pharmacological interventions to prevent cognitive decline (Marciniak et al., 2014; Sapozhnikov, 2015). Despite these findings, religious versus spiritual practices are more commonly offered in Western nursing homes. Meditative and mindfulness practices are relatively unknown in long-term care settings with the exception of some hospice care agencies.

Additionally, elder healthcare professionals are often not exposed to or trained in contemplative or palliative care practices (Ellison & Weingast, 2016). The treatment of disease rather than the person is where the two spectrums diverge. People of Buddhist or secular backgrounds practice mindfulness by using breath as a source of centering and as a way to bring individuals greater intentional awareness of their body, their surroundings and their feelings in the present moment during daily activities (Shonin et al., 2014). The cognitive, emotional and psychological benefits of mindfulness interventions in age-related and cognitive studies have been emerging over the last several years (Berk et al., 2018).

Focusing on the present moment through exploration of nature and art materials can be a meaningful way for groups of individuals to tap into their common humanity and lived experiences. I have led sessions where elders paint their breath, recognizing

Fig. 19.5 Exploration, movement and painting Japanese forest grass

the somatic feelings that arise such as moving like wind in the trees and embodying the grass. These are ways to incorporate simple mindfulness practices with people living with dementia that spark imagination, story-telling and creative inspiration. Figure 19.5 is an example of how mindfulness practice can combine nature and art. The energy in nature and the energy inside of us (i.e. chi energy) is connected. This figure illustrates an art directive of moving the body then moving the paintbrushes to release this energy onto paper.

Exploring and creating with natural materials in a mindful way should include bringing older adults out into nature to awaken the sensory experiences of being outdoors. In my work, there are often barriers to getting people outdoors in these long-term care settings. One way to address these challenges is to observe nature by using windows for inspiration. I often set up the space by having tables near the windows and seating elders so that they can see the views outside. I then ask them to pay close attention to what they see (or imagine) and we will describe details of our observations, the light, shadow, movement, colors, shapes, animals, etc. I often bring an engaging assortment of natural materials that can be used to prompt exploration, mindfulness, thoughts, memories and stories from the elders. Aromatic, interesting and textural objects in particular can inspire meaningful connections between people who may have difficulty expressing thoughts in words. Working in metaphors is a common approach I utilize and something I encourage in my work with older adults including those living with dementia.

Sister Marguerite was someone I knew very well. She was in her 80s and a lifelong artist who never missed an opportunity to laugh and enjoy the company of

Fig. 19.6 Sister Marguerite's oil pastel: self-portrait sitting under the pines

other people. Naturally, some of Sister's favorite things were contemplative, creating art and being in nature. She would speak about how her time in the woods allowed her space to just be alone with her thoughts. The first day I met her, she used oil pastels on dark paper to recreate these vivid experiences. With her unique style of repetitive lines, she depicted herself sitting beneath tall pines where, as she described it, she was "just chillin" (Fig. 19.6). Humour was important to Sister Marguerite and she infused laughter in almost every interaction. In her many decades of being a nun, she had an opportunity to work in a prison, travel and live in other lands, and her artwork often reflected the colors, textures and feelings of these places and the people she befriended.

Her art, although not always discernible to others, was very meaningful for her. She was not only fulfilled and enthralled by the process of artmaking, but by the storytelling and sharing of her thoughts and by her love of the freedom that comes with creative exploration and self-expression. When she was making art, her mind was as vibrant and lucid as her brushstrokes. After several years, her Alzheimer's journey became very advanced and she was no longer able to hold a paint brush alone. Communicating verbally became very challenging for her. I continued to bring Sister Marguerite to the art group as she was able. On our last session together, the group was creating communal prayer flags to celebrate and give thanks for the arrival of autumn. Knowing sister's love of nature, I prompted her with memories of the woodland walks she had shared with me many times before. As we laid shades of green paint onto the fabric using a hand-under-hand technique. I tried to emulate her motions and her brush stroke patterns that I knew so well from our years of creating together. Sister verbalized "the smell of the pines". This was a poignant way

Fig. 19.7 Collaborative prayer flag by Sister Marguerite and Michelle

to attune to sister's important life experiences, with focus on connecting, preserving and honoring personhood (Fig. 19.7).

Another example of using metaphor to express feelings or ideas in my practice occurred in the use of the fleeting beauty of the *sakura*, or cherry blossom, to inspire conversation, painting and poetry. The cherry blossom is a beautiful symbol of the impermanence of life. I believe exploring these transcendent topics with people who have lived long lives and experienced great loss and beauty can been be a powerful outlet for older adults to express these experiences and emotions. One spring, I decided to bring cherry blossoms to the different communities where I worked. I asked the group members to hold, smell and examine the blooms. After a discussion, I then asked them to share their thoughts about the cherry blossom. I typed their exact words into a poem. The participants then painted their interpretation of a cherry blossom, which would be displayed with their poem.

The following poem was created with a small group of women, all of whom were living with advanced dementia. Each person in the group contributed thoughts to the poem. The words are profound and connected us all in that moment.

> To value what we have/Don't fuss with it/Value what is with us for a short time/The Japanese revere this with festivals/Throughout the whole country/It's a sacred thing/Enjoy it before it's gone/When the wind carries the sweet smell of cherry blossoms/You never forget it/Heavenly.

It is misguided to believe that people living with dementia do not have a need or desire to express themselves because they are no longer able to verbalize in a way that we understand. Using familiar materials from nature cultivates rich experiences and promotes connections. In my work with people living with dementia,

I am not seeking psychological healing or treatments, but rather instilling a sense of community, empowerment, emotional well-being, equanimity and exploration (Partridge, 2019). It is imperative as an art therapist for older adults living with dementia to provide deep listening, which may involve much more than verbal cues. It includes paying attention to breath and bodily, hand, and facial gestures. It may include joining in a spontaneous song or rhythm that a participant taps on the table, singing or humming aloud. Being truly present, offering validation and means for self-expression with natural and creative materials encourages connections to self, others and the world. In other words, nature-based art therapists "view nature as the inspiration and the model for our understanding of creative process, and we view creative expression and responding as participatory processes embedded in the ongoing creative processes of the world" (Atkins & Snyder, 2018, 55–56).

Gerontological Lens

As a social gerontologist, my work is informed by the socioemotional selectivity theory and gerotranscendence theory. Socioemotional selectivity theory (Carstensen et al., 1999) is a life course theory that demonstrates the shift in perspective and goals of older people towards the end of their lives. This awareness of mortality in later years creates changes in goal priorities in social and relational contexts. Practice guided by this theory creates what is called the positivity effect, where negative emotions are experienced less often in older adults than younger adults. This is not to downplay the negative thoughts and emotions that older people may experience, and certainly, this includes people living with dementia.

Gerotranscendence theory (Tornstam, 2003) relates to socioemotional selectivity theory in that it demonstrates how older adults' perspectives, life goals and priorities shift as old age approaches. In this theory, Swedish sociologist Lars Tornstam proposes that people are always developing new ways to view and experience their world view. Gerotranscendence theory recognizes that ageing and death are inevitable. However, although pain and sadness are also inevitable at times, meaningful and positive outlooks can and do occur. I have found that it is not uncommon for long-term care staff to project their own fears and discomfort about loss and death onto older residents, inadvertently preventing essential conversations around these topics from occurring.

We cannot and should not deny elders the opportunity to examine their lives and share their inner feelings and perceptions. Although it may be more challenging for people living with dementia to express positive and negative emotions verbally, the creative process can serve as a non-verbal conduit through which they can share these felt experiences, thoughts and emotions. In addition, people living with dementia continue to experience all of the same feelings and emotions as people without dementia. Therefore, it is imperative to recognize the heterogeneity of people with dementia and to not assume one-size-fits all in how they are supported.

Nature and Art as a Window

Utilizing art materials, exploration, imagination, play and connection in conjunction with the generous gifts of nature can comprise a meaningful approach to individual and relationship-centred care. Being with nature and providing people living with dementia opportunities for contemplation can help them express feelings that may be difficult to articulate when words are challenging to find. It is important that eldercare communities recognize that all humans have an innate connection to the natural world around them. A diagnosis of dementia should never take away a person's right to explore the world they continue to live in or their right to express their thoughts and feelings in new, imaginative and mindful ways. Eldercare organizations and architects can imagine for a moment living in a world that is cut off from nature with no ability to open a window or go outdoors and smell the rain, fresh cut grass, spring flowers, fallen leaves, crisp winter air or experience the warmth of the sun or feel the earth. This kind of existence would likely make life quite depressing and grim for most people. That is the reality imposed on people living with dementia in many eldercare communities and it is considered normal and unquestioned.

In my work in advanced dementia care, I have witnessed figurative windows as well. An example that can occur is when someone in a group expresses distressing feelings or recounts a traumatic event, it is not uncommon for the other people in the group to support that person. These experiences do not automatically leave the person once their memory fades. They are still there, and when this window opens, it is important to hold a safe space for these feelings to be expressed—in whatever form they come. Communication is not dependent on verbal language. These empathic responses and connections between people can occur even in later stages of dementia and should not be underestimated.

Social worker Naomi Feil, founder of the validation method (1982), argues that the oldest-old people who are living with dementia, may have unresolved emotions and trauma and struggle with making meaning and expressing these feelings. Feil's validation therapy is a communication method rooted in deep listening, acceptance and empathy. When feelings may not be able to be verbally expressed, they can surface in other ways and often in eldercare, these needs can be misconstrued and bypassed. When working with individuals and small groups, I have witnessed this effective practice of repeating back to my client what they have said and then, share empathy for what they are expressing (this could be verbally, emotionally, bodily movements and artistically).

An example of art as a figurative window occurred with a woman named Bonnie. Bonnie was in her 80 s and was in the mid-to-late stages of Alzheimer's. She always had a bright smile and greeting for everyone. I worked with Bonnie for two years in a small expressive art group that was held in a small private office in a long-term care community. Bonnie was usually very jovial and outgoing. One day we were creating paintings inspired by nature. As Bonnie painted, she described her blue and green painting as "leaves" and then she described it to look like "a baby's leg" and then her words evolved to "my baby". This fluid evolution of art and thoughts is something I often witness with people who are living with dementia. Because of primary progressive aphasia, Bonnie had difficulty verbalizing her thoughts into complete sentences. She spoke in short clusters of words, moving her hand and arm in gestural motions as her eyes filled with tears. I watched, listened and understood that she was remembering her baby that had died decades earlier. It was very important that I did not rush passed these emergent feelings or try to distract her back to her "leaves". She felt safe in this group and needed to express grief. I validated this heavy loss and her sadness as she held hands with another woman in the group. After expressing herself and being heard, Bonnie appeared calm, went back to her usual smile and without prompting, continued painting. Providing support along with opportunities to engage with natural and creative materials for expression can provide a way to empower, self-regulate, soothe and serve as a reminder of our common humanity and the beauty and gratitude of life's moments.

Opening the window (and doors) to the outdoors can promote outward and inward reflection on the human experience throughout life's seasons. This chapter has demonstrated holistic approaches to socioemotional well-being, using nature and art which focuses on *abilities*, connections with one another and with our earth. Individuals living with dementia should be provided opportunities for verbal, kinesthetic and creative self-expression, imagination, play, validation and mindful immersion and exploration of the natural world that *we are all part of*.

Photo: M. Olson

References

Atkins, S., & Snyder, M. (2018). *Nature-based expressive arts therapy*. Jessica Kingsley Publishers.

Barsness, S., Bisiani, L., Greenwood, D., Macaulay, S., Power, Al., & Swaffer, K. (2018). Rethinking dementia care. *Australian Journal of Dementia Care, 7*(4), 8.

Basting, A. (2009). *Forget memory: Creating better lives for people with dementia*. Johns Hopkins University Press.

Basting, A. (2018). How we underestimate the arts in long-term care. *Caring for the Ages, 19*(7), 10–11.

Beard, R. L. (2012). Art therapies and dementia care: A systematic review. *Dementia, 11*(5), 633–656.

Berk, L., Warmenhoven, F., van Os, J. & van Boxtell, M. (2018). Mindfulness training for people with dementia and their caregivers: Rationale, current research, and future directions. *Frontiers in Psychology, 9*(982). https://doi.org/10.3389/fpsyg.2018.00982.

Bratman, G. N., Hamilton, J. P., & Daily, G. C. (2012). The impacts of nature experience on human cognitive function and mental health. *Annals of the New York Academy of Sciences, 1249*(1), 118–136.

Carstensen, L. L., Isaacowitz, D. M., & Charles, S. T. (1999). Taking time seriously: A theory of socioemotional selectivity. *American Psychologist, 54*(3), 165–181.

Chilton, G. (2013). Art therapy and flow: A review of the literature and applications. *Art Therapy, 30*(2), 64–70.

Cohen, G. D. (2006a). The center on aging, health & humanities, The George Washington University. In *The creativity and aging study: The impact of professionally conducted cultural programs on older adults*. https://hsrc.himmelfarb.gwu.edu/cgi/viewcontent.cgi?article=1001&context=son_ncafacpubs.

Cohen, G. D. (2006b). Research on creativity and aging: The positive impact of the arts on health and illness. *Generations, 1*, 7–15.

Cook, M. (2015). *Forests as places of mental wellbeing: The meaning and use of urban forests by people with early-stage dementia*. Economic and Social Research Council/Forestry Commission Scotland Phd Studentship.

Cook, M. (2020). Using urban woodlands and forests as places for improving the mental well-being of people with dementia. *Leisure Studies, 39*(1), 41–55. https://doi.org/10.1080/02614367.2019.1595091

Cowl, A. L., & Gaugler, J. E. (2014). Efficacy of creative arts therapy in treatment of Alzheimer's disease and dementia: A systematic literature review. *Activities, Adaptation & Aging, 38*(4), 281–330.

Crutchfield, P., Gibb, T. S., Redinger, M. J., Ferman, D., & Livingston, J. (2019). The Conditions for ethical application of restraints. *Chest Journal, 155*(3), 617–625.

Cunningham, C. Macfarlane, S., & Brodaty, H. (2019). Language paradigms when behaviour changes with dementia: #BanBPSD. *International Journal of Geriatric Psychiatry, 34*(8) (May 27), 1109–1113. https://doi.org/10.1002/gps.5122.

Davis, J. V., & Canty, J. M. (2013). Ecopsychology and transpersonal psychology. In H. L. Friedman & G. Hartelius (Eds.), *The Wiley-Blackwell handbook of transpersonal psychology* (pp. 595–611). John Wiley & Sons Ltd.

de Medeiros, K., & Basting, A. (2013). "Shall i compare thee to a dose of donepezil?": Cultural arts interventions in dementia care research. *The Gerontologist, 54*(3), 344–353.

de Bruin, S., Oosting, S., van der Zijpp, A., Enders-Slegers, M.-J., & Schols, J. (2010). The concept of green care farms for older people with dementia: An integrative framework. *Dementia, 9*(1), 79–128.

Deshmukh, S. R., John H., & Cardno, A. (2018). Art therapy for people with dementia. *Cochrane Database of Systematic Reviews, 9*, (9). https://doi.org/10.1002/14651858.CD011073.pub2.

Dixon, D. (2019). Caregivers stories: Dallas Dixon living with dementia after being a caregiver to his uncle. Interview with Kimberly D. Scott: *That Kimberly*. Podcast audio, n.d. https://player.

fm/series/kimberly-d-scott-that-kimberly/caregivers-stories-dallas-dixon-living-with-dementia-after-being-a-caregiver-to-his-uncle.
Eden Alternative. (2016). Mission, vision, values, principles. Accessed May 12, 2020. https://www.edenalt.org/about-the-eden-alternative/mission-vision-values/.
Ellison, K. P., & Weingast, M. (2016). *Awake at the bedside: Contemplative teachings on palliative and end-of-life care*. Wisdom Publications.
Eltaliawi, A. G., El-Shinawi, M., Comer, A., Hamazah, S., & Hirshon, J. M. (2017). Restraint use among selected hospitalized elderly patients in Cairo, Egypt. *BMC Research Notes, 10*(633). https://doi.org/10.1186/s13104-017-2978-x.
Erkkinen, M. G., Zúñiga, R. G., Pardo, C. C., Miller, B. L., & Miller, Z. A. (2018). Artistic renaissance in frontotemporal dementia. *JAMA, 319*(13), 1304–1306. https://doi.org/10.1001/jama.2017.19501.
Førsund, L. H., Grov, E. K., Helvik, A. -S., Juvet, L. K., Skovdahl, K., & Eriksen, S. (2018). The experience of lived space in persons with dementia: A systematic meta-synthesis. *BMC Geriatrics, 18*(33). https://doi.org/10.1186/s12877-018-0728-0.
Fraser, K. D., O'Rourke, H. M., Wiens, H., Lai, J., Howell, C., & Brett-MacLean, P. (2015). A scoping review of research on the arts, aging, and quality of life. *The Gerontologist, 55*(4), 719–729.
Guseva, E. (2018). Bridging art therapy and neuroscience: Emotional expression and communication in an individual with late-stage Alzheimer's. *Journal of the American Art Therapy Association, 35*(3), 138–147.
Hall, C. (2016). *Mindfulness-based ecotherapy: Facilitator manual*. N.p.
Huhtinen-Hildén, L. (2014). Perspectives on professional use of arts and arts-based methods in elderly care. *Arts & Health, 6*(3), 223–234.
Konis, K., Mack, W. J., & Schneider, E. L. (2018). Pilot study to examine the effects of indoor daylight exposure on depression and other neuropsychiatric symptoms in people living with dementia in long-term care communities. *Clinical Interventions in Aging, 13*, 1071–1077.
Lehikoinen, K. (2017). Justifying the arts in health and care in Finland: A discourse analytic inquiry. *Cogent Arts & Humanities, 4*, 1–12. https://doi.org/10.1080/23311983.2017.1345048
Malchiodi, C. A. (2013). Art therapy and health care. New York: The Guilford Press.
Malchiodi, C. A. (2020). *Trauma and expressive arts therapy: Brain, body, & imagination in the healing process*. The Guilford Press.
Marciniak, R., Sheardova, K., Čermáková, P., Hudeček, D., Šumec, R., & Hort, J. (2014). Effect of meditation on cognitive functions in context of aging and neurodegenerative diseases. *Frontiers in Behavioral Neuroscience, 8*(17), 1–8.
Mencagli, M., & Nieri, M. (2019). *The secret life of trees: Harness the healing energy of forest bathing and natural landscapes*. Rodale Books.
Mmako, N. J., Courtney-Pratt, H., & Marsh, P. (2020). Green spaces, dementia and a meaningful life in the community: A mixed studies review. *Health & Place, 63*(2020), 102344. https://doi.org/10.1016/j.healthplace.2020.102344.
Olsson, A., Lampic, C., Skovdahl, K., & Engström, M. (2013). Persons with early-stage dementia reflect on being outdoors: A repeated interview study. *Aging & Mental Health, 17*(7), 793–800.
Partridge, E. (2019). *Art therapy with older adults: Connected and empowered*. Jessica Kingsley Publishers.
Project Nature Connect, Inc. n.d., Increase personal and planet wellness. Accessed May 12, 2020. http://projectnatureconnect.org/.
Quintana Hernández, D. J., & Barrachina, M. T. M. (2015). Mindfulness-based stimulation for elderly people with Alzheimer's disease or other types of dementia. *Papeles Del Psicólogo [psychologist Papers], 36*(3), 207–215.
Roszak, T. (1992). *Voice of the earth: An exploration of ecopsychology*. Simon & Schuster.
Roszak, T., Gomes, M. E., & Kanner, A. D., (Eds.). (1995). *Ecopsychology: Restoring the earth, healing the mind*. Sierra Club Books.

Rouch, I., Pongan, E., Leveque, Y., Tillmann, B., Trombert, B., Getenet, J.-C., Auguste, N., Krolak-Salmon, P., Laurent, B., & Dorey, J. M. (2018). Personality modulates the efficacy of art intervention on chronic pain in a population of patients with Alzheimer's disease. *Journal of Alzheimer's Disease, 63*(2), 1–8.

Russell-Williams, J., Jaroudi, W., Perich, T., Hoscheidt, S., El Haj, M., & Moustafa, A. A. (2018). Mindfulness and meditation: Treating cognitive impairment and reducing stress in dementia. *Reviews in the Neurosciences, 29*(7), 791–804.

Safar, L. T., & Press, D. Z. (2011). Art and the brain: Effects of dementia on art production in art therapy. *Journal of the American Art Therapy Association, 29*(3), 96–103.

Sakurai, R., Fujiwara, H., Yasunaga, M., Takeuchi, R., Murayama, Y., Cuya, K. E. K., & Kanosue, K. (2017). Neural basis for the relationship between frequency of going outdoors and depressive mood in older adults. *International Journal of Geriatric Psychiatry, 32*(6), 589–595.

Sapozhnikov, D. (2015). The promise of mindfulness-based interventions as therapies to prevent cognitive decline. *McGill Science Undergraduate Research Journal, 10*(1), 49–54.

Scales, K., Zimmerman, S., & Miller, S. J. (2018). Evidence-based nonpharmacological practices to address behavioral and psychological symptoms of dementia. *The Gerontologist, 58*(1), S88–S102.

Schneider, J. (2018). The arts as a medium for care and self-care in dementia: Arguments and evidence. *International Journal of Environmental Research and Public Health, 15*(6), 1151. https://doi.org/10.3390/ijerph15061151

Shonin, E., VanGordon, W., & Griffiths, M. D. (2014). Changing paradigms: Buddhist insight in western psychological treatments. *Psychology Post Graduate Affairs Group, 92*, 35–39.

Speert, E. (2016). *Eco-art therapy: Deepening Connections with the natural world*. American Art Therapy Association. https://arttherapy.org/eco-art-therapy-deepening-connections-natural-world/.

Stuckey, H. L., & Nobel, J. (2010). The connection between art, healing, and public health: A review of current literature. *American Journal of Public Health, 100*(2), 254–263.

Swaffer, K. (2018). Rethinking dementia care: #BanBPSD. Last modified August 13, 2018. https://kateswaffer.com/2018/08/13/rethinking-dementia-care-banbpsd/.

Sweeney, T. (2013). *Eco-art therapy: Creative activities that let earth teach*. Theresa Sweeney.

The Eden Alternative. (2016). Mission, vision, values, principles. Last modified 2016. https://www.edenalt.org/about-the-eden-alternative/mission-vision-values/.

Tible, O. P., Florian, R., Savaskan, E., & von Gunten, A. (2017). Best practice in the management of behavioural and psychological symptoms of dementia. *Therapeutic Advances in Neurological Disorders, 10*(8), 297–309.

Tornstam, L. (2003). *Gerotranscendence from young old age to old old age*. Uppsala University. http://www.soc.uu.se/digitalAssets/235/235770_3gt-from-young-old-age.pdf.

Ulrich, R. S. (1984). View through a window may influence recovery from surgery. *Science, 224*(4647), 420–424. https://doi.org/10.1126/science.6143402

Wang, Q.-Y., & Li, D.-M. (2016). Advances in art therapy for patients with dementia. *Chinese Nursing Research, 3*, 105–108.

Wells, R. E., Yeh, G. Y., Kerr, C. E., Wolkin, J., Davis, R. B., Tan, Y., Spaeth, R., Wall, R. B., Walsh, J., Kaptchuk, T. J., Pulse, D., Phillips, R. S., & Kong, J. (2013). Meditation's impact on default mode network and hippocampus in mild cognitive impairment: A pilot study. *Neuroscience Letters, 556*, 15–19.

Wilson, E. O. (1984). *Biophilia*. Harvard University Press.

Windle, G., Newman, A., Burholt, V., Woods, B. O'Brien, D., Baber, M., Hounsome, B., Parkinson, C., & Tischler, V. (2016). Dementia and imagination: A mixed-methods protocol for arts and science research. *BMJ Open, 6*, e011634.

World Health Organization. (2019). *Dementia: Key facts*. https://www.who.int/news-room/fact-sheets/detail/dementia.

Chapter 20
Leaving Life Too Early or Leaving the Decision Too Late in Alzheimer-Type Dementias: Choices and Preferences for Medical Assistance in Dying (MAID)

Colin Brewer

Abstract In many Western countries, especially if they have good palliative care services, dementia is now more feared than cancer and surveys consistently show that most indigenous citizens would not want to continue living once dementia had become severe or even moderately severe. If they were to develop a life-threatening condition at this stage of the illness, most would not want to be resuscitated, yet that is what commonly happens. In Britain and Canada, most would like the option of an advance request for medical assistance in dying (MAID) in the form of euthanasia when dementia reaches a certain degree of severity. That is legal in the Benelux countries but even doctors sympathetic to MAID may find the process difficult. The only alternatives, restricted to people with early dementia, are MAID by medically assisted rational suicide (MARS, available in Switzerland) voluntary euthanasia (available in Benelux countries and probably now Canada) or voluntarily stopping eating and drinking (VSED—legal in most countries) before they lose mental capacity. Surveys also show important differences between religious groups in their attitudes to MAID in all diagnostic contexts after discussing some of the ethical, philosophical, religious and diagnostic issues that MAID raises, and this chapter presents some case histories of patients who chose and received MARS in Switzerland.

Keywords Dementia · Alzheimer · Resuscitation · Voluntary euthanasia · Medically assisted suicide · Assisted dying · Autonomy · Palliative · Religion

This chapter is rather different from other contributions to the book, which address the clinical features, diagnosis, epidemiology and management of dementia, as well as its impact on the families who are the informal carers for dementia patients (in developing countries, they are often the only carers). These are all important aspects

Retired psychiatrist. London, UK. Formerly lecturer, department of psychiatry, University of Birmingham and director of the community alcoholism treatment service, Westminster Hospital, London. Currently board member and medical group convener, My Death, My Decision (mydeath-mydecision.org.uk).

C. Brewer (✉)
Bankside, London, UK

M. K. Shankardass (ed.), *Dementia Care*,
https://doi.org/10.1007/978-981-16-3864-0_20

of dementia, but they are of little or no interest to a significant number of people in Western countries (almost the only countries I can speak about from professional experience) who already know enough about dementia to feel that if they were to have the misfortune to suffer from it themselves, then the sooner they died, the better. They would certainly not want to be resuscitated if they were to develop a life-threatening illness, and many would be grateful if the law allowed doctors to speed up their deaths in this increasingly common and increasingly feared end-of-life scenario.

Some very distinguished citizens in Britain and elsewhere have explained why they take this view. The late Baroness Warnock—a leading British ethicist—wrote: “I simply do not want to be remembered as someone wholly dependent on others, especially for the most personally private aspects of my life, nor can I tolerate the thought of outstaying my welcome, an increasing burden on my family, so that no one can be truly sorry when I die and they are free” (Warnock, 2015). Norman Cantor, Emeritus Professor of law at Rutgers University in the USA, has set out very clearly his concern to be remembered as the person he still is and not as the person he would undoubtedly become if he was more than mildly demented. “I care mightily about posthumous recollections of my personality, and I strive to shape my life trajectory (including a dying process) consistently with my personal vision of dignity”. While recognizing that “Some people will confront Alzheimer’s with a measure of resignation … or they may be hoping for a miracle cure to surface, or they might value whatever interactions they can salvage with their loved ones” and that “They are entitled to pursue that resolute path”, his own living will include the following passage. “This wish to hasten my post-competence demise is not based on prospective suffering or distress, but rather on my personal vision of intolerable indignity and degradation associated with cognitive dysfunction … . In addition, it is important to me to avoid being an emotional, physical, or financial burden on my family and friends, even if they would willingly assume such burdens” (Cantor, 2018).

However, it is not only leading academics and philosophers who want to spare their families and friends from having to watch the progressive indignities and personality disintegration of dementia and do not want to be kept alive when they can no longer recognize those families, have meaningful interactions with them or remember conversations that took place only a few minutes previously. I first became aware of just how widespread this attitude is—at least in Britain—in the 1980s, when I briefly had a part-time National Health Service (NHS) job that involved seeing many patients for very short consultations. Often I saw six or eight every hour, and it occurred to me that I could easily carry out some simple and informal patient surveys to satisfy my curiosity and also to provide useful material for the fortnightly column in *General Practitioner* magazine that I wrote for nearly 20 years. One of those surveys asked about the sort of treatment they would want if they ever became demented enough to need institutional care. Nobody asked: “What is dementia?” and although the qualifier “Alzheimer’s” was not so familiar then, everybody knew that dementia implied the progressive and permanent loss of memory and relationships and the gradual destruction of the personality. In the 1980s, institutional care mainly took place in the large NHS mental hospitals that still existed. Most of them had dementia wards where people with advanced—though not necessarily terminal—dementia were cared for (at the State's expense) until they died. Several of my interviewees had personal experience of dementia in their families.

"If you were ever admitted to such a ward", I asked, "I'm interested in how you would want to be treated. Please choose one of three options". The first was: "I would like Matron to slip something into my tea to end my life as soon as possible". The second was: "The Matron option is a bit too radical but at the first sign of pneumonia or other life-threatening illness, I very definitely don't want to be treated and hope I will be heavily sedated to help me to die". Option three was: "I would like to be treated just like any other patient and be resuscitated if I have a serious illness". Not one chose option three. They were about equally divided between options one and two.

Several post-2000 studies show that the results of my small informal survey in the early 1980s (mainly in the 18–40 age group) were very similar to those of larger, more sophisticated and more representative studies today. One, involving 725 members of the general public in South-East England, concluded: "In the face of severe dementia, less than 40% of respondents would wish to be resuscitated after a heart attack, nearly three-quarters wanted to be allowed to die passively and almost 60% agreed with physician-assisted suicide ... . Our survey suggests that a large proportion of the UK general public do not wish for life-sustaining treatments if they were to become demented and the majority agreed with various forms of euthanasia" (Williams et al., 2007). That as many as 40% *would* want to be resuscitated may reflect the intriguing additional finding that: "White respondents were significantly more likely to refuse life-sustaining treatment and to agree to euthanasia compared with Black and Asian respondents". I will return to that difference later.

A more recent Dutch survey described the responses of three groups to two hypothetical end-of-life scenarios, one involving cancer and the other dementia. In both cases, the patients have reached the stage where rational communication is no longer possible. In the dementia scenario, "you ... no longer recognise your family or friends. You refuse to eat and drink and you retreat more and more into yourself. To communicate with you about medical treatments is not possible any more". Clinicians will recognize this situation as one that often arises well before the truly terminal stages of dementia, when patients are usually bed bound, very frail and likely to die within weeks rather than months or years. Respondents were asked to state their preferences in this situation about artificial feeding and hydration, antibiotics in the event of pneumonia, resuscitation in the event of cardiac arrest and artificial respiration. The preferences of 5661 respondents who were members of the Dutch equivalent of British Right-to-Die societies such as My Death, My Decision, Friends at the End (Scotland) or Dignity in Dying and had completed living wills—also known as Advance Decisions—were very similar to those of a representative sample of 1402 Dutch citizens who were not members of Right-to-Die societies and had not completed a living will. In both of those groups, a large majority "want to forgo all four [resuscitation procedures]" in both cancer and dementia (Wijmen et al., 2014). A study that compared British and US attitudes to end-of-life decisions as mental capacity gradually disappeared found "a high prevalence of preference for 'measures to end my life peacefully' when decision-making capacity was compromised, which increased as dementia progressed" (Clarke et al., 2017).

Another Dutch survey—this time, involving only elderly citizens—found that a preference for avoiding life-prolonging treatment in advanced disease with a poor prognosis "ranged from 61 to 79% in cancer and 75–88% in dementia scenarios" (Evans et al., 2014). The authors concluded that "The more frequent desire to forgo treatments in case of dementia than cancer *suggests that physical deterioration is more acceptable than cognitive decline*". [my italics].

The comparatively recent arrival of social media networks has made it easier to do research into public attitudes to medical assistance in dying (MAID) in dementia. A team from Auckland University in New Zealand used the opportunity provided by "netnography" ("a specialized form of ethnography adapted to the unique computer-mediated contingencies of today's social worlds") to obtain the views of "five Facebook communities (UK, Australia, Canada, USA and NZ)" whose followers varied in number from 3247 to 322,961 and "posted news daily, mainly related to [MAID] in contexts: such as court cases, family assisted deaths, law changes, etc.". There were 1007 comments on MAID and dementia in response to 316 posts. Noting an apparent "increase in public awareness and concerns/fears in relation to a dementia diagnosis", they conclude that "There is a profound fear of developing dementia and its ensuing physical and cognitive decline, which has made our online members (as public representative) consider alternative end-of-life options such as expressing their preferences in an advance directive, and considering an assisted death" (Dehkhoda et al., 2019).

Yet doctors—even though they increasingly repeat mantras about patient autonomy and choice—may ignore reasonable assumptions about their patients' wishes and choices when dementia is involved. In a 2002 survey with particularly puzzling and troubling findings, over 500 Finnish doctors were asked how they would deal with two clinical scenarios involving terminally ill patients who developed a life-threatening additional illness, one group suffering from cancer and the other from advanced dementia. 83% said that they would not resuscitate the cancer patient, but only 57% of them would not resuscitate in the case of dementia (Hinkka et al., 2002). Finnish doctors seem to have become less interventionist since that study—but only somewhat less (Piili et al., 2018). In any case, it seems there is nothing very unusual about the enthusiasm of Finnish doctors for the active treatment of patients in the last stages of dementia.

A 2006 US study looked at 154 patients with advanced dementia who "experienced 229 suspected pneumonia episodes during the last 6 months of life". Over half of the episodes were in the final month, which implies that many of the patients recovered from pneumonia before dying from another episode, or from something else. Over 90% received active rather than palliative treatment. "Antibiotic treatment for the 229 episodes was as follows: none, 9%; oral only, 37%; intramuscular, 25%; and intravenous, 29%" (Chen et al., 2006). In other words, over half of these uncomprehending patients, many of them likely to die anyway in the next few months or weeks, were given intensive antibiotic treatment, often involving painful intramuscular injections and the side effects, such as diarrhoea, nausea and rashes, that are common when antibiotics are given in large doses to frail, elderly patients. Very few of them had made living wills when they still had the capacity to do so and "care plans" were often made by the medical and nursing staff rather than by (or in consultation with) the family—if any. "For most institutionalized persons with advanced dementia, a decision to forgo hospitalization"—where aggressive treatment tends to be routine—"is not made until death is imminent. Thus, hospital transfers are common near the end of life" (Lamberg et al., 2005). Only a third had "do not hospitalize" (DNH) orders six months before they died, and this had only risen to half in the last month of life.

As with advanced cancer, doctors often overestimate the likely survival time in severe dementia, failing to accept that it is, in its own way, a terminal illness. "At nursing home admission, only 1.1% of residents with advanced dementia were perceived to have a life expectancy of less than 6 months; however, *71.0% died within that period*". [my italics] Half had a do not resuscitate (DNR) order, but only 1.4% had a DNH order. "Non–palliative [i.e. active treatment] interventions were common among residents dying with advanced dementia: tube feeding, 25.0%; laboratory tests, 49.2%; restraints, 11.2%; and intravenous therapy, 10.1%. Residents with dementia were less likely than those with cancer to have directives limiting care but were *more likely to experience burdensome interventions*".[my italics]. Common and distressing conditions in advanced dementia included "pressure ulcers [i.e. bedsores]...constipation...pain... and shortness of breath" (Mitchell et al., 2004). The ambivalence, confusion and indecisiveness—both ethical and therapeutic—and the high levels of both avoidable and unavoidable discomfort that it reveals are evidently still very common. Perhaps "normal" or "predictable" would be more accurate descriptions.

"Terminal" has two meanings in this context. It can mean: "we think death is very likely in a few days or weeks", but it can also mean: "This patient's dementia is now very severe. Conversation is impossible and nothing seems to give her pleasure. She may live for another six months or a year, or even more, but she will get steadily worse, whatever we do". A very recent review criticized the frequency with which "incurably dying patients receive antibiotics in their last days or weeks of life, even in patients with 'do not resuscitate' (DNR) or 'comfort measures only' orders" (Macedo et al., 2018). So common have pointless, painful interventions and investigations become in all sorts of manifestly terminal illnesses that the term "non-beneficial treatment" (NBT) has been coined to describe them. One recent definition is as follows: "Any treatments, procedures or tests administered to [elderly] patients who are naturally dying and which will not make a difference to their survival, will probably impair their remaining quality of life and can substantially cause them pain or prolonged suffering, or leave them in a worse state of health than they were before admission". Unfortunately—and especially when private, fee-per-item medical systems combine with misplaced therapeutic enthusiasm, inexperience and fear of criticism for not "doing something"—NBTs are common. A 2016 review found that "evidence from 38 studies indicates that on average 33–38% of patients near the [end of life] received NBTs". In some of the reviewed studies and for some NBTs, the incidence was as much as 90% (Cardona-Morrell et al., 2016).

The terminal nature of the more advanced stages of dementia (particularly in its commonest Alzheimer manifestation) is routinely ignored, one leading US geriatrician noting that "There is a pervasive failure—by both physicians and the public—to view advanced dementia as a terminal illness, and there is a strong conviction that technology can be used to delay death. … Although the use of feeding tubes is not unequivocally futile in all cases, balancing the risks and benefits leads to the conclusion that they are seldom warranted for patients in the final stage of dementia" (Gillick & Volandes, 2008). Despite that "conclusion", long-term tube feeding remains far from uncommon in advanced dementia. In Japan—the country with the world's longest life expectancy—a 2014 study found 1199 hospital patients and 2160 long-term care home patients aged 65 years or older with percutaneous endoscopic gastrostomy (PEG). 62.9% of the hospital patients had "advanced dementia". Only 4.2% of the care home patients were said to be "enjoying their own lives" (Nakanishi & Hattori, 2014).

Dutch doctors appear to be less interventionist. In one study, even though only 4.9% of patients had made living wills, "in almost half of the residents … decisions were made not to start potentially life-prolonging treatment such as hospital transfer and artificial nutrition and hydration. In … 53.7%, decisions were made to withdraw potentially life–prolonging treatment such as artificial nutrition and hydration and medication". Nevertheless, a quarter received antibiotics in the last week, mainly for pneumonia (Hendriks et al., 2017). Furthermore, even if the main purpose of prescribing antibiotics is to reduce the severe discomforts that pneumonia can cause (shortness of breath, painful and rapid breathing, and high fever, often leading to hallucinations or increased confusion) rather than to prolong life, they do not seem to have much effect in frail patients whose respiratory muscles are as weak as their other muscles and who therefore have difficulty in clearing their lungs by coughing or cooperating with chest physiotherapy. In the days before organized palliative care—which is generally effective when pain is the major symptom—pneumonia was widely regarded as "the old man's friend", but modern medicine and a reluctance to let nature take its course may have reduced pneumonia's ability to shorten an inevitable dying process and thereby reduce the period of terminal distress. "Nursing home residents with dementia and pneumonia experience severe discomfort which increases in the six days preceding death. Discomfort occurs regardless of treatment with antibiotics or not. Residents dying from pneumonia experienced more discomfort than residents dying of other causes. Moreover, death from respiratory infections has been associated with the largest symptom burden before death. For example, 78% experienced respiratory distress compared to 40% for residents dying from a cardiovascular disorder", and "because residents with dementia are often unable to express their complaints and wishes about treatment, relieving the symptoms of pneumonia in these residents is particularly challenging". (Maaden et al., 2016) Not surprisingly, "discomfort and symptoms were higher for residents who died within 20 days from the diagnosis of pneumonia and lower for residents who were observed asleep", which implies that patients who died from pneumonia often failed to get any sleep during their last miserable days and hours on earth.

In the last decade or so, I have had the opportunity to examine several people with early dementia—usually Alzheimer-type—who, like Prof Cantor and Baroness Warner, were so concerned about the likely effects of more advanced dementia on themselves and their families that they had applied for MAID in Switzerland, where it has been legal for about a century. Unlike Dutch, Belgian and—since 2016—Canadian law, Swiss law does not permit doctors to administer a lethal dose of medication, but it does allow them to prescribe a lethal dose of medication (usually a barbiturate) for a patient to self-administer, either by swallowing or by opening the valve of an intravenous infusion that a doctor has set up. This is subject to a number of requirements and safeguards, one of which is that the patient must have mental capacity when the medication is prescribed and administered. For patients who are not resident in Switzerland, the preliminaries are very bureaucratic and often require several weeks or even months of correspondence.

Most of these patients had no previous history of psychiatric illness. If there is even a hint of such a history—including cases where amitriptyline has been prescribed as an adjunct to pain medication rather as an antidepressant—the Swiss organizations also require a full psychiatric assessment as well as a mental capacity assessment, to exclude the possibility that a request for MAID stems from a depressive illness or a delusional system. The assessment can be complicated by the inappropriate prescribing of antidepressants to patients who have had their worst fears confirmed by a diagnosis of progressive Alzheimer-type dementia and are understandably distressed. This is surely a normal and predictable reaction to bad news and most patients adjust to it. There is no evidence that this kind of understandable misery is reduced in degree or duration by antidepressants, other than by their placebo and non-specific effects and persuasive evidence that it is not (Orgeta et al., 2017). To say this is not to minimize the size of those placebo effects, which can be considerable, but it is just one aspect of the "medicalization of misery" that has led to large and steady increases in antidepressant prescribing in many countries with remarkably little to show for it in terms of reduced suicide rates, or a reduction in demand for non-pharmacological therapies. It also comes at the cost of what is now recognized to be a high incidence of withdrawal syndromes when antidepressants are discontinued or doses are reduced. The tendency to prescribe antidepressants is increased if patients mention that MAID is among the options they are considering. When surveys consistently show that a majority of people in a particular country would rather be dead than significantly demented, such feelings and statements should not be pathologized, though they might usefully be explored in some cases. However, many UK doctors are frightened that even discussing a patient's desire for MAID might expose them to accusations of assisting a suicide—a new offence created when suicide was decriminalized in 1961.

Even when there is no such background, I always take a normal, semi-structured psychiatric history because that generally reveals the patient's existential and philosophical views on MAID and related issues. Similarly, a family history of dementia might have made the patient particularly aware of the nature of dementia and even more keen than the average British citizen to consider ways of limiting its impact and duration. The tests of mental capacity are spelled out in the Mental Capacity Act of 2005 and are so simple and straightforward that the courts have accepted capacity assessments from non-medical professionals—sometimes in preference to assessments by psychiatrists or clinical psychologists. All the patients I see have already been thoroughly assessed by NHS dementia or memory clinics and after detailed neuropsychological testing and CT or MRI scans, and the diagnosis is not usually in doubt. I usually administer the Mini Mental State Examination (MMSE) when I examine the patient, partly to see whether significant deterioration had occurred since the last testing and partly because it is easy to repeat if a second examination is needed. This happens quite often because the bureaucratic process for foreigners seeking MAID in Switzerland can take months if vital identity documents cannot easily be located. This means that a situation might arise where a patient is given a date some weeks ahead and travels to Switzerland but by the time the Swiss doctor does the final assessment before writing the prescription, progression of the dementia causes capacity to be lost. For this reason, the final assessment before leaving for Switzerland has to be done not more than three weeks before the journey. Since all the Alzheimer cases were at an early stage, no patient had lost capacity by that time, but two patients had already lost capacity when I first saw them and thus could not be considered for MAID. Here are three pseudonymized vignettes of patients who did retain capacity and had a medically assisted death in Switzerland.

Henry was in his early eighties and had been a senior figure in a major industry. A long-time supporter of voluntary euthanasia, he had developed early Alzheimer's disease and the busy social life that he and his wife used to enjoy was no longer possible, because he lost the thread of conversations. Retaining enough insight to realize that things would only get steadily worse and that he might soon not have the mental capacity to decide the manner of his death, he joined Dignitas—the oldest and largest of the Swiss organizations that can provide MAID to foreigners. When I visited him at his home for an assessment, he met me at the door and clearly understood why I had come and what we had to discuss. Testing showed definite short-term memory impairment, but he never lost track of the purpose of our conversation, remained clear and consistent in his desire to go to Switzerland and recognized the consequences of this decision. He was not depressed and not even gloomy. In short, he had mental capacity as legally defined. After a very good life, he wanted a good death and his decision clearly stemmed from existential views that he had held for many years. He also told me that he wanted his sizeable estate to go to his family and not to the staff and shareholders of care homes.

Jean had a well-documented history of bipolar (manic-depressive) illness since middle-life that was well-controlled on lithium. She was regularly reviewed by the local psychiatric service, and it was clear from her own account and the casenotes that her mood had been stable for several years since the last episode. Alzheimer dementia had been diagnosed a year earlier, and her desire for MAID was therefore not an impulsive or panicked reaction. Well into her 80 s, she too felt that she had had a good life and wanted to avoid a death that would be inconsistent with her concepts of dignity and autonomy and very distressing to her family. There were no subjective or objective features of depression, and she continued to practise her creative hobbies. Although she could have travelled to Switzerland within a few weeks, she delayed her journey for several months because she wanted to be present for the birth of another grandchild.

Charlotte also had Alzheimer's disease and was in some respects more severely affected but her story has some positive features. It shows that Alzheimer patients who are arranging for MAID when the time seems appropriate can still get pleasure and fulfilment out of life, but there were also some important diagnostic issues. Charlotte's dementia was quite slowly progressive, and I first saw her four years after the initial diagnosis. By that time, it had reached the point where an alarm system had been installed in her house because she occasionally got confused and wandered the streets. However, she was able to continue living on her own in a small country town, not far from her very supportive family, and had recently been re-elected to the chair of the local Women's Institute, whose members all knew her diagnosis. The tipping point for her would come when she could no longer live in her own home. That point was approaching but she wanted to hang on until it was very near. Charlotte knew more than most patients about the realities of cerebral impairment because when her husband was still in his sixties, he got very severe pneumonia and the resultant anoxia caused catastrophic brain damage. She had to share in the eventual decision to switch off his life support.

Apart from assessing whether her mental capacity was impaired by the dementia, I was also asked to decide whether or not she was "psychotic". That label had been attached to her following a couple of nocturnal episodes where she was not just confused but believed, during one of them, that some local youths had been banging on her front door. True or not (her daughter thought it was most likely a hallucination or the misinterpretation of an innocent noise) that belief developed into a fear that her daughter had been kidnapped and Charlotte went out to look for her. She was examined at home by a locum (i.e. temporary) psychiatrist who had never seen her before and, according to her daughter, stayed for barely ten minutes. He not only gave her a "psychotic" label but also prescribed an antipsychotic drug to go with it. This caused unpleasant side effects, so Charlotte stopped it after a week.

Delusions and hallucinations are, of course, common and almost defining features of psychoses but in Charlotte's case, they did not last for more than an hour or so. We do not generally think of that kind of short-lived problem as a psychosis, especially if the patient already has a diagnosis of dementia. "Psychosis" in this context implies a condition lasting for at least some days and often for weeks or months, during which time medication may be indicated. I thought it significant that all these isolated delusional episodes had all occurred in the middle of the night. Even young and healthy people sometimes wake in a confused state and hear imaginary sounds, and it must be a lot easier if the brain is beginning to fail. They are sufficiently common in normal life that there are technical but not necessarily pathological terms for hallucinations that occur when people are falling asleep (hypnogogic) or waking up (hypnopompic). They can be frightening but do not normally need any treatment. In any case, there were no signs of any psychotic thought processes when Charlotte came to see me. Indeed, it was not immediately obvious that she had any cognitive impairment, since she could follow a conversation and correctly recalled a holiday with her daughter two weeks previously. However, more precise tests of memory showed definite impairments, consistent with the very detailed assessment at her local hospital the previous year. She was not at all depressed and said "I want to make the most of everything I can while I can". Nevertheless, within less than three months, she decided to activate her long-planned journey to Switzerland, where she had a peaceful death in the presence of her close family. Charlotte took the risk of losing her mental capacity before deciding to go to Switzerland but in her case, it paid off. She was a clergyman's daughter. Her religious faith was probably the strongest of all the patients I have seen and she had a firm belief in an afterlife.

This is an appropriate point at which to discuss two of the main variables—religion and ethnicity—that determine individuals' premorbid attitudes and preferences in Western countries to the management of more advanced dementia, if they were to suffer from it. The first point to make is that in most of those countries, religious belief—in the sense of adhering to a particular faith and its holy texts, belief in a god or gods and the expectation of an afterlife—is a steadily declining characteristic and often now a minority taste, especially among indigenous (or White) citizens and the young. That is not the case in most developing countries and although India and some East Asian countries have long histories of tolerating belief systems that do not involve a deity, traditional concepts of religion still typify majority beliefs in sub-Saharan Africa, Latin America, the Caribbean and Oceania, and in the Islamic countries of North Africa, the Middle East and South-Central Asia. In many of those countries, publicly espousing atheism remains a crime and can still have fatal consequences. (In Britain, the last such execution took place in Edinburgh in 1697, when Scotland was still an independent country.) The USA is unusual among Western countries in the prominence of religion in public life, but the USA also demonstrates the persisting differences between competing versions of Christianity that appeared in Northern Europe during the reformation early in the sixteenth century, leading to the rise of Protestantism and the various non-conformist sects that remain an important feature of religion in the USA and some other Anglophone countries.

In all countries where surveys of attitudes to the resuscitation of patients with advanced dementia (or advanced and untreatable physical illness) have been carried out, clear and often marked differences emerge between those who identify strongly with a religion and those who identify less strongly or not at all. Differences are also found between Protestant Christians and Roman Catholic Christians; between "main-line" Protestants and more fundamentalist or "Evangelical" Protestants; and between White and Black citizens in the USA. In the main British survey, already referenced, there were significant differences between White British citizens and those classified as "Black and Ethnic Minority" British citizens. The precise ethnic breakdown was not given but in S-E England where the survey was carried out, the main ethnic minorities are Afro-Caribbeans—who are both more religious than White citizens and tend towards Evangelical branches of Christianity—and predominantly Sunni Muslims from South Asia. Predictably, those minorities were less likely to refuse resuscitation, but they were also more likely to request resuscitation in unpromising circumstances. Similar findings emerged in the Dutch study mentioned earlier, which included a third group of Protestant and mainly White Dutch citizens. In the study comparing British and US attitudes to end-of-life decisions in dementia that found "a … preference for 'measures to end my life peacefully' when decision-making capacity was compromised, previous exposure to dementia increased that preference but … being of 'black' race/ethnicity" reduced it. As yet, there have been few surveys specifically directed at Muslim citizens in European countries but those few show a widespread preference that "everything must be done", even when the likelihood of a functional outcome is low or negligible. In Belgium, where voluntary euthanasia is legal, recent surveys among its Muslim communities found that in both first- and second-generation residents, "an absolute rejection of every act that deliberately terminates life, based upon the unconditional belief in an afterlife and in God's sovereign power over life and death" was the norm (Ahaddour et al., 2018).

Religion and race are sensitive areas for some people and also too large to merit more than this brief discussion in a chapter devoted to more specific issues. However, as one of those numerous Western unbelievers, it seems paradoxical to me that when Christians and Muslims (the most studied groups) are on the brink of death, and even though both religions envisage an enjoyable afterlife—about which Islam gives the most confident, detailed and lyrical descriptions—it seems that the more faithful the adherents, the keener they are to keep their barely–functioning bodies and minds in this imperfect world rather than sampling the joys and perfections of the next, compared with those who have less or no faith. Equally paradoxical (and arguably of greater concern) is that they are also less likely than the average secular European resident (of whichever ethnic group) to allow their organs to be harvested after death for life-saving transplant surgery (Boulware et al., 2002).

It is often argued that once dementia advances beyond the early stages, when most patients are still aware of their declining cognitive abilities and of the progressive nature of the disease, patients will not suffer much because by that stage they will be unaware of the diagnosis and prognosis. In some cases, that description will be accurate. All clinicians will have seen demented patients who seem reasonably happy and unconcerned. However, if patients are to be fully informed about the varied manifestations of dementia—as many will want to be—then that means informing them of the high incidence of features that may be incompatible with happiness (or even indifference) and may cause their families even more distress than the progressive loss of personality, cognition, recognition and awareness in general. Dementia may exacerbate pre-existing mood disorders and other psychiatric conditions but progressive cerebral atrophy—like any other form of significant brain damage—may cause disorders of thought and behaviour that never existed before the diagnosis of dementia. Persistent delusions and hallucinations are common and often distressing. Antipsychotic drugs are of limited effectiveness and may add unpleasant side effects to the existing levels of distress. A recent review noted that: "… the manifestations of [Alzheimer Dementia (AD)] are not limited to cognitive symptoms; rather, they include a range of neuropsychiatric symptoms (NPS) of AD. The near universal prevalence of NPS in AD, combined with the serious and disabling effects they have on patients and caregivers, has focused significant recent attention on the fact that few effective and safe treatments exist" (Geda et al., 2013).

The loss of inhibitions that commonly follows frontal lobe damage frequently leads to inappropriate behaviour. People who had always been considerate of the feelings of others may become very insensitive. At its most embarrassing, that can lead to unwanted and unwelcome sexual advances by both men and women and what the literature calls "Inappropriate Sexual Behaviour" (ISB) is not rare. While "the most likely change in the sexual behaviour of a person with dementia is indifference", ISB becomes commoner in the moderate and severe stages and may include hyper-sexuality, changes in sexual preference, paedophilia, promiscuity and marital infidelity (Cipriani et al., 2016). Here is another pseudonymized case history.

Mark was a divorced, urbane man in his late 70s. When he developed early Alzheimer's, he moved to supported accommodation in a retirement home, where he liked to drink and socialize. One summer evening, he hammered on a neighbour's door, walked in and asked: "Are you the woman who invited me in for drinks this evening?" She said he was mistaken. He explained that someone had invited him and he could not remember who it was. After being sent away, he knocked on the doors of several other women living in the block. Ten minutes later, he was back at his original port of call, barging in without knocking and waving a half-full bottle of wine. "Are you sure you didn't invite me?" She asked him to describe the woman who issued the invitation—tall, short, fat, thin, dark, fair?—but he had no idea. Asked where he was at the time of the invitation, he replied that he was in the shower. "I had to wrap a towel round me". This second time, he was hard to dislodge but left, plainly annoyed, to continue his search for the invisible woman.

A few minutes later he was back again and the mood had changed. During the first visit, he had been full of charm; on the second he was growing more insistent; by the third he was showing annoyance bordering on aggression. "I know it was you and that you want to have a drink with me". He tried to sit down but was ushered out very firmly, and this time the door was locked behind him. Mark liked to display himself, naked, in front of his first-floor bedroom window, which gave the resident living opposite a shock. Who would want to be remembered for that sort of behaviour—or to expose their friends or families to it if they were being cared for in a multigenerational household? Within a few months, Mark's disinhibition became so unmanageable that he had to be moved to a specialist unit.

Although it is clear that a majority of citizens in many Western countries hope that death will free them from dementia sooner rather than later, not many actually seek MAID while they still retain capacity. This is true of MAID in general and even in countries and legislatures where MAID in various forms is legal, if only for terminal physical illnesses, the percentage of people who avail themselves of it remains in single figures. This probably reflects a widespread reluctance to accept the inevitability of death and to prepare for it. The people who request MAID remain overwhelmingly white, well-educated citizens with no history of significant psychiatric problems. Few of them are financially hard-pressed. Nearly all of them have spent some time when they were in good health thinking about death and how they might plan for it. Many are supporters or paid-up members of Right-to-Die societies. That in itself makes assessment much easier because when someone has clearly held and expressed views supportive of MAID for many years, it is very difficult to argue that an actual request for MAID is merely a panicked reaction to the diagnosis of a terminal or progressive disease that will fade with time and a course of antidepressants or psychotherapy.

Finally, since this is a book that is aimed at a medical and nursing readership, it is very relevant to note the comparatively high incidence of MAID among doctors facing terminal and progressive illness, including dementia, though in these cases, the medical assistance is often supplied by the doctors themselves, rather than by—or as well as—their medical colleagues and attendants. Doctors have both the pharmacological knowledge and the access to suitable medications that enable them to avoid unpleasant deaths and most experienced clinicians in Western countries will know from personal friendships or medical gossip about physicians who took matters into their own hands when faced with advanced cancer, Parkinsonism or motor-neurone disease—or dementia. A few years ago, a well-known professor of psychiatry told me, over lunch with his wife, that he had recently been diagnosed with early Alzheimer dementia and intended to take a lethal overdose before it got much beyond the early stage. In the event, he had a fatal heart attack while on holiday a few months later. The true cause of these medically assisted deaths hardly ever appears on death certificates or in obituaries because both families and medical attendants join in a benevolent conspiracy to conceal it. In many countries—including Britain—if death is expected within a month or so and the patient has been seen by a doctor within the two weeks before death, a death certificate can be issued without further examination of the body (unless the body is to be cremated) and an inquest will not be held if no suspicions of suicide emerge. Naturally, families want to minimize the risk of being accused of assisting a suicide, but even if some suspicions emerge, I have been told by more than one coroner that coroners are very reluctant to add to families' distress by publicly examining such matters. Like all doctors, I have seen some very unpleasant deaths and from early in my medical career, I was comforted by the knowledge that I would almost certainly be able to avoid that sort of unpleasantness myself. I also felt that it is unfair and hypocritical to deny that comfort to our patients. Now that I am retired and approaching the expected shelf life for my social class, sex and country, experience has only reinforced those medical school thoughts.

Some sections of this chapter are taken or adapted from my recent book—O, let me not get Alzheimer's, sweet heaven! Why many people prefer death or active deliverance to living with dementia. Foreword by Henry Marsh. Skyscraper Books. 2019.

References

Ahaddour, C., Van den Branden, S., & Broeckaert, B. (2018, January–March). God is the giver and taker of life: Muslim beliefs and attitudes regarding assisted suicide and euthanasia. *AJOB Empirical Bioethics, 9*(1), 1–11. https://doi.org/10.1080/23294515.2017.1420708. Epub 2018 January 29.

Boulware, L. E., Ratner, L. E., Sosa, J. A., Cooper, L. A., LaVeist, T. A., & Powe, N. R. (2002). Determinants of willingness to donate living related and cadaveric organs: Identifying opportunities for intervention. *Transplantation, 73*(10), 1683–1691.

Cantor, N. (2018, July–August). On avoiding deep dementia. Hastings Center Rep (pp. 15–24).

Cardona-Morrell, M., Kim, J., Turner, R. M., et al. (2016). Non-beneficial treatments in hospital at the end of life: A systematic review on extent of the problem. *International Journal for Quality in Health Care, 28*(4), 456–469. https://doi.org/10.1093/intqhc/mzw060

Chen, J. H., Lamberg, J. L., Chen, Y. C., Kiely, D. K., Page, J. H., Person, C. J., & Mitchell, S. L. (2006). Occurrence and treatment of suspected pneumonia in long-term care residents dying with advanced dementia. *Journal of the American Geriatrics Society, 54*(2), 290–295.

Cipriani, G., Ulivi, M., Danti, S., Lucetti, C., & Nuti, A. (2016). Sexual disinhibition and dementia. *Psychogeriatrics, 16*(2), 145–153. https://doi.org/10.1111/psyg.12143 Epub 2015 Jul 28.

Clarke, G., Fistein, E., Holland, A., Barclay, M., Theimann, P., & Barclay, S. (2017, April 5). Preferences for care towards the end of life when decision-making capacity may be impaired: A large scale cross–sectional survey of public attitudes in Great Britain and the United States. *PLoS One, 12*(4), e0172104. https://doi.org/10.1371/journal.pone.0172104.eCollection2017

Dehkhoda, A., Malpas, P., & Owens, R. G. (2019, March 7–9). Views on assisted dying for individuals with dementia: A netnographic approach. In *Paper presented at ICEL3 conference Ghent*.

Evans, N., Pasman, H. R., Deeg, D., & Onwuteaka-Philipsen, B. (2014, December). EURO IMPACT. Collaborators (17) How do general end-of-life treatment goals and values relate to specific treatment preferences? A population-based study. *Palliatic Medicine, 28*(10), 1206–12. https://doi.org/10.1177/0269216314540017. Epub 2014 June 18.

Geda, Y. E., Schneider, L. S., Gitlin, L. N., et al. (2013). Neuropsychiatric symptoms in Alzheimer's disease: Past progress and anticipation of the future. *Alzheimer's & Dementia, 9*(5), 602–608. https://doi.org/10.1016/j.jalz.2012.12.001

Gillick, M. R., & Volandes, A. E. (2008). The standard of caring: Why do we still use feeding tubes in patients with advanced dementia? *Journal of the American Medical Directors Association, 9*(5), 364–367. https://doi.org/10.1016/j.jamda.2008.03.011

Hendriks, S. A., Smalbrugge, M., Deliens, I., et al. (2017). End-of-life treatment decisions in nursing home residents dying with dementia in the Netherlands. *International Journal of Geriatric Psychiatry, 32*(12), e43–e49. https://doi.org/10.1002/gps.4650 Epub 2016 Dec 29.

Hinkka, H., Kosunen, E., Lammi, E. K., Metsänoja, R., Puustelli, A., & Kellokumpu-Lehtinen, P. (2002, May). Decision making in terminal care: A survey of Finnish doctors' treatment decisions in end-of-life scenarios involving a terminal cancer and a terminal dementia patient. *Palliative Medicine, 16*(3), 195–204.

Lamberg, J. L., Person, C. J., Kiely, D. K., & Mitchell, S. L. (2005). Decisions to hospitalize nursing home residents dying with advanced dementia. *Journal of the American Geriatrics Society, 53*(8), 1396–1401.

Macedo, F., Nunes, C., Ladeira, K., et al. (2018). Antimicrobial therapy in palliative care: An overview. *Supportive Care in Cancer*. https://doi.org/10.1007/s00520-018-4090-8

Mitchell, S. L., Kiely, D. K., & Hamel, M. B. (2004). Dying with advanced dementia in the nursing home. *Archives of Internal Medicine, 164*(3), 321–326.

Nakanishi, M., & Hattori, K. (2014). Percutaneous endoscopic gastrostomy (PEG) tubes are placed in elderly adults in Japan with advanced dementia regardless of expectation of improvement in quality of life. *The Journal of Nutrition, Health & Aging, 18*(5), 503–509. https://doi.org/10.1007/s12603-014-0011-9

Orgeta, V., Tabet, N., Nilforooshan, R., & Howard, R. (2017). Efficacy of antidepressants for depression in Alzheimer's disease: Systematic review and meta-analysis. *Journal of Alzheimer Disease, 58*, 725–733. https://doi.org/10.3233/JAD-161247

Piili, R. P., Metsänoja, R., Hinkka, H., Kellokumpu-Lehtinen, P., & Lehto, J. (2018, May 30). Changes in attitudes towards hastened death among Finnish physicians over the past sixteen years. *BMC Medical Ethics, 19*(1), 40. https://doi.org/10.1186/s12910-018-0290-5

van der Maaden, T., de Vet, H., Wilco, P., Achterberg, W., et al. (2016). Improving comfort in people with dementia and pneumonia: A cluster randomized trial. *BMC Medicine, 14*, 116. Published online 2016 August 11. https://doi.org/10.1186/s12916-016-0663

Van Wijmen, M., Pasman, H., Widdershoven, G., & Onwuteaka-Philipsen, B. (2014). Continuing or forgoing treatment at the end of life? Preferences of the general public and people with an advance directive. *Journal of Medical Ethics*. Published online September 2nd September 2014. https://doi.org/10.1136/medethics-2013-101544

Warnock, M. (2015). Easeful death for the very elderly. In C. Brewer & M. Irwin (Eds.), *I'll See Myself Out, Thank You. The arguments for rational suicide* (pp. 131–134). Newbold on Stour, Warwickshire. Skyscraper.

Williams, N., Dunford, C., Knowles, A., & Warner, J. (2007). Public attitudes to life–sustaining treatments and euthanasia in dementia. *International Journal of Geriatric Psychiatry, 22*(12), 1229–1234.

Chapter 21
Dementia in Under 65s—An Overview

Vidya Giri Shankar

Abstract Dementia is a broad category used to describe a neurodegenerative syndrome characterized by multidomain cognitive impairment that shows progressive decline and is sufficient to impact activities of daily living and occupation. When the dementia becomes clinically manifest before age 65, it is called early onset dementia (EOD), whereas if the symptoms arise onset before age 45, it is called young onset dementia (YOD). These terms are often used interchangeably. In the UK, approximately 850,000 people have a diagnosis of dementia, of which about 42,500 have YOD. 4.4% of people with Alzheimer's dementia are less than 65 years of age. Various epidemiological studies show that the prevalence rates of early onset dementia (EOD) in the 30–64 age group range between 54 and 260 cases per 100,000. The rise in numbers is an emerging concern and is attributed to medical advances leading to increased life expectancy. YODs have unique presentations, atypical features and an extensive list of differential diagnoses which often make for a challenging clinical situation. There are often non-cognitive symptoms such as apathy, personality changes and resistant depression that precede the cognitive symptoms of dementia leading to delayed diagnosis. Additionally, EOD patients are at a point in life in which they play an active role in society and often have young families or children. The disruption to these roles and responsibilities is, therefore, greater than in older people. They also have to deal with specific problems such as marital conflicts, setbacks in their profession and financial issues. This chapter aims to summarize the symptomatology, etiologies, diagnostic strategies and psychosocial challenges in dementia in under 65 year olds.

Keywords Young onset dementia · Early onset dementia · Presenile dementia · Neurodegenerative disease · Cognitive decline · Ageing · Onset · BPSD

V. Giri Shankar (✉)
Specialist Registrar in Older Adult and General Adult Psychiatry, South London and Maudsley NHS Foundation Trust Affiliated to the Institute of Psychiatry, Psychology and Neuroscience, King's College, London, UK

M. K. Shankardass (ed.), *Dementia Care*,
https://doi.org/10.1007/978-981-16-3864-0_21

Introduction

Dementia is a broad category of neurodegenerative illness defined by a clinical syndrome comprising of progressive multidomain cognitive impairment which is severe enough to interfere with the activities of daily living, social and occupational functioning. That is, there is an underlying disease of the brain.

Cognitive decline is the defining feature of dementia which comprises of deterioration in memory (amnesia) with one or more deficits from other cognitive domains which include difficulty with motor planning and execution (apraxia), inability to recognize or interpret auditory, visual and tactile stimuli, objects or people (agnosia), impairment of language function, affecting the ability to produce or comprehend speech, reading and writing (aphasia) and visuospatial deficits.

Another important clinical entity is mild cognitive impairment (MCI). Both dementia and MCI are characterized by a decline from previous cognitive level, but in MCI, the activities of daily living (ADLs) are well-preserved, e.g. ability to do household chores, personal care, managing personal financial affairs and ability to enjoy social activities. MCI may be the harbinger of dementia, and the clinical boundaries between the two may be blurred during the clinical journey.

In 2015, there were about 47 million people living with dementia in the world. This number is predicted to triple by 2050 as the world's demography is moving towards an increase in the ageing population like never before. The rate of growth is set to rise especially in the lower and middle income countries (LMIC). This is probably due to the increasing life expectancy. In 2015, the estimated global cost of dementia was US$818 billion, and this is expected to only increase in the coming years (Livingston et al., 2017).

There are different types of dementia and various ways to classify them. DSM 5 uses the term major neurocognitive disorder to describe dementia. Alzheimer's dementia (AD) is the most common form of dementia followed by vascular dementia (VD) and Lewy body dementia (LBD). Other types include fronto-temporal dementia (FTD), Parkinson's disease dementia (PDD), posterior cortical atrophy (PCA), dementias secondary to chronic alcohol use, traumatic brain injury, infections and lesions in the brain. Mixed dementias are those that are due to more than one cause. In this chapter, dementia would be the common term used to represent one or all of the above terms. Description of each of these dementia syndromes is beyond the scope of this chapter.

Dementia is stigmatized as an illness of the older persons. Definitely, age is the one of the greatest unmodifiable risk factors for developing dementia. However, there is a group of individuals in whom the illness develops in mid or late adulthood. Hence, one of the ways to classify dementias is based on the age of onset of the illness. Young onset dementia (YOD) is called so when the dementia syndrome affects people less than 45 years of age. Dementia in people between 45 and 65 years of age is termed as early onset dementia (EOD), and in those above 65 years of age it is called late onset dementia. This cut-off is an artificial and traditional one.

Early onset dementia (EOD) and young onset dementia (YOD) are often used interchangeably in literature. For the ease of understanding, in this chapter, we shall be using the term early onset dementia to describe both EOD and YOD. Presenile dementia is a synonym for this group of dementias.

Prevalence rates of early onset dementia (EOD) have been reported to range between 54 and 260 cases per 100,000 in the 30–64 age group (Vliet et al., 2010). A study conducted in the UK in 2003 found that from the age of 35 onwards, the prevalence of dementia approximately doubled with each 5 year increase in age. The study estimated that there are approximately 18,319 people with dementia under the age of 65 in the UK (Harvey, 2003).

EOD is a rapidly emerging global mental health problem. The progression of EOD was found to range between 8.3 and 22.8 new cases per 100.000 in those aged under 65 years (Vieira, 2013). It is an extremely debilitating experience for the patient and the carer. In this chapter, we aim to provide an overview of EOD to bring to light the challenges and nuances of an illness that is not only puzzling in presentation but often has unique challenges along the way. Dementia is a vast subject, and this chapter cannot do justice to the depth of the matter; however, it intends to help practically comprehend EOD and seeks to highlight key points of the illness. This chapter, if anything, hopes to encourage the reader to explore more on this subject.

Etiology of Early Onset Dementia

It is important to ascertain the aetiology of EOD as much as possible, in order to provide accurate medical management. While the underlying illness may not be cured, the presentation can be managed with psychotropic medication with the aim to preserve quality of life. EOD, if not managed well, can prove to be arduous and demanding on the caregiver and healthcare services.

Various illnesses have been implicated in EOD. Some of them include Alzheimer's dementia, vascular dementia, Lewy body dementia, fronto-temporal dementia, Parkinson's disease dementia, alcohol-related brain damage, traumatic brain injury and rarer illnesses like AIDS dementia complex, Wilson's disease, Huntington's disease, autoimmune encephalopathies, paraneoplastic encephalitis, multiple sclerosis, neurosyphilis, prion disease, down's syndrome and dialysis dementia to name a few. A majority of the cases are given a diagnosis of Alzheimer's pathology (Vieira, 2013).

Often EODs are found to have rapid progression, with more extensive brain damage, complications are more life-limiting (Manthorpe & Jill, 1997). They often have rarer causes than late onset and dementias and may have an additional neuropathology besides an Alzheimer's type of degeneration.

Risk Factors for Developing Dementia

The World Alzheimer Report estimates that 2–10% of all cases of dementia start before the age of 65 (Prince & Jackson, 2009). There are two types of risks that are associated with developing dementia. They are as follows:

1. *Unmodifiable risk factors*:

These are risk factors that cannot be changed, e.g. age, family history of dementia, down's syndrome, gender, pre-existing neurological disease, chronic and enduring mental illnesses (like chronic Schizophrenia) or a compromised central nervous system. Unmodifiable risk factors help in keeping the susceptible individuals under review and regular screening for early detection and prompt treatment. Unmodifiable risks also help in planning future care needs in anticipation.

2. *Modifiable risk factors*:

These are risks that can be identified and altered which can reduce the impact of the illness and economic costs. These factors help public health organizations for policymaking. Recognizing these risks help with planning and implementing population-wide primary preventive strategies. It also helps to design public awareness campaigns and allocate resources to the target groups.

Physical comorbidities play a major role in developing dementia. This includes chronic alcohol use, nutritional deficiency especially of Vitamin B12, folic acid, vitamin D, being overweight, poorly managed hypertension and diabetes, dyslipidaemias, lack of aerobic exercise, autoimmune conditions and depression.

Environmental factors are reported to contribute to the risk of developing dementia. Examples of these are as follows: air pollution (all types); exposure to aluminium; silicon; selenium; pesticides; and electric and magnetic fields (Killin et al., 2016). Statistically, across all age groups, neurodegeneration is the most substantial aetiology of dementia; however, reversible causes are relatively more often noted in the young (Rossor et al., 2010).

Many EODs are more likely to have genetic and/or metabolic causes, but the presence of other pre-existing physical and mental illness may make the individual vulnerable to manifest a late onset dementia at an earlier age (Masellis et al., 2013). This probably indicates that the brain in such people is undergoing an accelerated ageing process.

The presumed contribution from various causes with respect to age of onset can be broadly understood as follows based on literature review by Masellis et al:

Age 30–45 years: Strong genetic/metabolic causes

Age 45–65 years: Moderate genetic and mild/moderate ageing and comorbidities

Age 65–85 years: Weak genetic/strong ageing and comorbidities (Masellis et al., 2013).

Diagnosing of Early Onset Dementia—Approach to a Case

While the core symptoms of EOD are not different from that of late onset dementia, the index presentation may be atypical.

Dementia before the age of 40 is extremely rare. They may present with a dementia plus syndrome, wherein cognitive impairment occurs in the context of genetic and metabolic disorders which cause neurological problems.

Additional features include focal neurological deficits, pyramidal, extrapyramidal, cerebellar or peripheral nerve involvement which may hint at the presence of another genetic disorder (Sampson, 2004).

EOD requires a wider consideration of differential diagnosis compared with late onset dementia. As mentioned earlier, diagnostic delays are not uncommon, and the psychosocial needs of EOD require specialist services to manage. While creating a diagnostic formulation, the aetiologies have to be taken into consideration like family history, other physical comorbidities, such as cerebrovascular risk factors, that may influence the clinical presentation and age at onset (Masellis et al., 2013).

Since the list of differential diagnosis is long, a systematic and thorough investigation of all possible treatable causes must be identified and treated. Hence, the complete assessment for a dementia has various stages of assessments as described below.

Diagnosis of EOD often requires liaison between psychiatry and neurology along with input from the multidisciplinary team which includes neuropsychology and occupational therapy.

A detailed clinical history is invaluable in psychiatric practice in order to construct the timeline of events and understand the onset, duration and progress of the illness. With a good history, neurocognitive patterns can be mapped. It also identifies significant stressors, life events that may aggravate or relieve symptoms. A psychiatric interview also brings to light underlying psychiatric problems that may be causing or contributing to the presentation, e.g. depressive pseudo-dementia where depression may be causing the dementia-like picture, chronic schizophrenia in which cognitive decline is observed over the years. The other aspect of clinical interviewing is that it provides deep insights into the personality, expectations and life story of the patient. It helps with forming a therapeutic alliance which lays the foundation for all future work to be done with the patient and the family. The first consultation often sets the tone for future engagements.

Next most important aspect is a thorough physical examination, especially neurological examination to look for neurological deficits, movement disorders, bradykinesia and other forms of sensory and motor deficits. Neurological disease can increase the risk of developing dementia because it is, in essence, a disease of the brain.

One of the ways to understand the presentation is to base it on which neurotransmitter system is affected—

Dopaminergic system—motor and executive dysfunction

Serotonergic system—mood and hallucinations

Noradrenergic system—gait, mood and inattention.

Cholinergic system—hallucinations, inattention, impairment of semantic fluency, executive dysfunction and visuospatial impairment (Compta et al., 2011; Kehagia et al., 2010).

Another way to classify symptoms is based on functions of the various lobes of the brain. It is a simplified view, because many functions overlap and may not be strictly restricted to a particular lobe.

Physical health investigations are needed to rule out and treat potentially treatable causes of dementia. It ascertains the current level of physical fitness of the patient and forms a baseline for future reference. The dementia work-up includes blood tests like complete blood count, liver function test, renal function test, diabetes profile, thyroid profile, serum vitamin D levels, vitamin B12 levels, folic acid levels, urine examination and ECG.

Neuroimaging is another important investigation. First and foremost, it is used to rule out any sinister cause for the presentation and then to look for anatomical evidence of neurodegeneration. A normal neuroimaging does not automatically rule out a dementia. Ideally, a high-quality MRI will help look for any neurodegenerative changes, atrophy of the brain, ischemic changes, space occupying lesions or inflammation. When an MRI is not possible a well-reported, good-quality CT scan will provide reasonable information. A DAT scan may be required in cases with a suspicion of Lewy body dementia.

All patients with suspected EOD should have electroencephalography (EEG). This may assist diagnosis in the neurodegeneration (it is usually normal in frontotemporal dementia) and is particularly important in detecting unrecognized complex partial seizures that may produce an epileptic pseudo-dementia (Høgh et al., 2002). If seizures are well-managed, the dementia-like symptoms resolve.

In those under the age of 40 and in those with a family history of autosomal dominant or recessive disease, genetic counselling along with testing and specialized biochemical screening may be needed (Masellis et al., 2013).

Cognitive testing includes simple and quick screening tests like test your memory (TYM), Montreal cognitive assessment (MoCA), mini-mental state examination (MMSE) among others which are carried out by the GPs or at screening clinics. Further, more detailed formal cognitive testing includes Addenbrookes cognitive examination (ACE-R) and neuropsychological assessment which are done by specialist psychiatry services to identify the specific domains that are affected (e.g. attention, memory, language, verbal fluency and visuospatial). Neuropsychiatric/neuropsychological evaluation is also to establish the premorbid IQ of the patient. This facilitates the understanding of the degree of the cognitive decline and is a vital piece of investigation in very complex cases.

Assessment of functioning of the individual is conducted by occupational therapists which provides information on the ability of the individual's functional capacity around activities of daily living. This assessment also identifies risks in the home environment and the ability of the individual to mitigate those risks. Occupation therapy assessments also work as a baseline assessment to measure progression of the illness.

Recommendations regarding early onset dementia approved by the 4th Canadian Consensus Conference on the Diagnosis and Treatment of Dementia (CCCDTD) in 2012 include the following:

- All patients with EOD should be referred to a specialist (memory or movement disorders) clinic, preferably one with access to genetic counselling and testing as well as biochemical screening when appropriate.
- The investigation of genetic aetiologies should be guided by the detailed assessment and investigations (MRI, functional imaging, laboratory investigations, etc.) performed by the specialist neurologist in conjunction with the clinical geneticist, who will provide information to the patient and their family regarding implications of the potential findings.
- The cost of genetic counselling and testing should be covered by public funding.
- If age at onset is <40 years, then rare genetic metabolic disorders should always be investigated if the specific dementia diagnosis is not evident through other investigations.
- Physicians should be sensitive to the special issues associated with EOD, in particular with regard to loss of employment and access to support services appropriate for that age group.
- Considering the rarity of EOD, a national registry for interested at-risk individuals, mutation carriers and symptomatic patients will facilitate therapeutic research.
- This registry should be supported by public funding (Gauthier et al., 2012).

The above-mentioned investigations are available in most hospitals and are the most widely used. However, the more recent diagnostic advances that are available in only highly specialized settings like DAT scan, FDG-PET, amyloid PET, SPECT, functional MRIs and CSF proteomic studies. These facilities will pave way for more research into understanding the illness and therapeutics (Masellis et al., 2013).

With the help of the information gathered, a personalized medicine approach must be taken to manage modifiable risks, formulate a diagnosis and treatment plan. Practically, there are times when a specific diagnostic code based on ICD or DSM cannot be reached. In such a case, diagnostic and aetiologic formulation must be attempted. Along the course of the illness, if pathognomonic features of the illness begin to show, a formal diagnosis can be made.

Challenging Behaviour in EOD-Behavioural and Psychological Symptoms of Dementia (BPSD)

Although dementia is a progressive neurocognitive disorder, there are non-cognitive, i.e. behavioural and psychological symptoms that arise in all stages of the disease. It is also called neuropsychiatric symptoms of dementia. It is estimated to affect almost 90% of cases of dementia irrespective of the cause. It is independently associated with poorer outcomes (Cerejeira et al., 2012). BPSD is an umbrella term used to describe a range of psychiatric problems, abnormal behaviour and psychological responses seen in dementias arising out of any cause. BPSDs have a direct impact on the quality of life of the patient and carers. BPSDs increase rate of hospitalization, prolong hospital stay and increase healthcare cost. It is a major cause of carer stress and burnout along with suffering in the patient. The increased number of BPSD correlates negatively with survival rates over a 3-year period (Tun et al., 2007). Thus, BPSDs are a focus in management (Katona et al., 2007).

There is a delay in onset of BPSDs from the onset of dementia. This may indicate that BPSDs occur when there is extensive neurodegeneration indicating a more advanced disease setting. In typical cases of AD, the BPSDs are noted to occur after 45–50 months of diagnosis (Hart et al., 2003; Mirakhur et al., 2004).

Initially, there may be partial insight into these challenging behaviours, and it is followed by remorse and guilt. With the advancement of the illness, the insight wanes and duration of lucidity contract until it no longer occurs.

BPSDs are unpredictable and vary over the course of time. Many of these symptoms can be aggravated by pain (arthritic pain, toothache which is often missed, headaches), constipation, infections, dehydration, discomfort in the environment (like loud sounds, too much or too little light, temperature), sleep disturbances, dry itchy skin or sores.

BPSDs can be distressing and require crisis management. BPSDs can be broadly classed as follows:

Psychotic features include delusions and hallucinations, disorganized behaviour (e.g. smearing of faeces, tearing up items of clothing).

Delusional beliefs are strongly held, enduring and false ideas that cannot be challenged. It can be different on a case-to-case basis depending on the content of the delusion, systematization, complexity, associated distress and response to it. The delusions seen in dementias are typically simpler than those observed in non-dementia patients with psychosis. The usual content of delusional thoughts involves suspiciousness, abandonment and misidentification (Jeste et al., 2006).

Visual hallucinations are part of the cardinal symptoms of Lewy body dementia. They are recurrent and typically consist of well-formed images of animals or persons that the patient describes in detail (McKeith et al., 2005). Visual impairment and visual sensory deprivation may result in a condition called Charles Bonnet syndrome which mimics visual hallucinations of Lewy body dementia. Hence, vision testing is useful. Posterior cortical atrophy too may present with visual hallucinations and misperceptions.

These are particularly distressing BPSDs, and in some cases presence of hallucinations is an indicator of worsening of the dementia.

Mood features—depression, easy irritability, dysphoria, hypomania, euphoria or elation, sexual disinhibition, anxiety, apathy and aggression. These disturbances in emotional experiences are not what is classically seen in routine psychiatry practice. Since the ability is to identify and name the emotion is lost, apathy and depression may present similarly with loss of interest and motivation and general slowing. Mania and depression both can present as irritability.

Physical behaviour factors—apathy, wandering, insomnia, aggression, psychomotor agitation and restlessness, excessive vocalization (or shouting), aberrant motor behaviour and changes in appetite (Mirakhur et al., 2004).

Patients with EOD are generally physically stronger than the older age group; hence, when they become aggressive, it can result in significant harm to others and damage to property. Aggression is a particularly upsetting experience for the family and carers.

Some symptoms are shared across domains.

Understanding the domain under which these symptoms arise helps with decisions about what class of medication to use and what psychosocial interventions may be needed.

Management of Early Onset Dementia and BPSDs

EODs are unique in that they occupy several territories within health and social care. EODs are a fine example of multidisciplinary working and collaboration, an "all for one" approach. Management of EODs is complex, and input from a multidisciplinary team consisting of psychiatric doctors, nurses, psychologists, occupational therapists, social workers and support workers is vital. The multidisciplinary team works in close collaboration with the patient, family and the GP through the course of the illness. At each stage, risks are assessed as are the needs. Third sector organizations and local support groups also form an important part of the therapy network. Just as dementia affects every aspect of life, the same is reflected in the input from the treating team. The management is holistic and needs to address not only the medical needs but also the emotional, social and occupational requirements of the patient. The management can be divided into pharmacological and non-pharmacological interventions.

Pharmacological interventions:

The first and foremost step in management is the treatment of potentially reversible causes of the dementia and monitoring the patient for a relapse of the condition.

Any underlying anxiety, depression or other psychiatric illness (including alcohol use) must be addressed.

Cognitive enhancers such as donepezil, galantamine, rivastigmine and memantine remain the mainstay of dementia management. They are not disease modifying drugs but serve to slow the progression of the illness. These are currently licensed to be used in the UK. These medications increase the availability of the neurotransmitter acetyl choline, deficiency of which is implicated in the neuropathology of dementia, in the brain.

Depending on the type of BPSDs, medications are prescribed. Antidepressants are useful in patients with low mood, anxiety, impulsivity, irritability, sexual disinhibition and in some cases of aggression. A short course of benzodiazepines in low doses may be helpful in alleviating anxiety and treating insomnia. Antipsychotics are used to manage psychotic symptoms. They are cautiously prescribed due to their propensity for causing side effects especially in this patient population.

Cognitive enhancers are also shown to help with BPSDs. Prescribing for BPSDs may vary based on local guidelines, consensus and individual preferences. Details regarding the mechanism of action and dosing of these medications are beyond the scope of this chapter.

Non-pharmacological Interventions:

All behavioural issues need assessment. A behaviour chart consisting of antecedent incident, behaviour and consequence (ABC) chart can be useful in identifying patterns, triggers and timings of the challenging behaviour. It serves to document changes in presentation over time and objectively feedback regarding the effectiveness of interventions.

Another important aspect of management includes the approach taken towards the patient, i.e. speaking loudly, clearly, gently but assertively, speaking short easy sentences, frequent prompts and reminders, introduction of self at each interaction and to explain to them the purpose before doing anything for them (bathing, shaving, administering tablets).

Non-pharmacological interventions have been classified into the following categories (O'Neil et al., 2011):

(i) Cognitive/emotion-oriented interventions (reminiscence therapy, simulated presence therapy, validation therapy)
(ii) Sensory stimulation interventions (acupuncture, aromatherapy, light therapy, massage/touch, music therapy, Snoezelen multisensory stimulation, transcutaneous electrical nerve stimulation)
(iii) Behaviour management techniques; this includes distraction and meaningful engagement (e.g. gardening, group singing/dancing, therapy kitchen, playing games and puzzles, colouring books)
(iv) Other psychosocial interventions such as animal-assisted therapy and exercise.

Research is lacking in these areas, and the results are inconsistent (Kong et al., 2009).

Herbert Protocol:

The Herbert Protocol is a form that carers, family or friends of a vulnerable person can fill in and submit to the local police station. This is handy when the patient has a tendency of wandering away. It contains information to help the police if the person goes missing, such as

- Basic medical history and personal details
- Contact numbers
- Places most likely to visit/previously located
- A recent photograph.

Having a completed form saves the trouble of trying to recall the information during the stressful time of someone going missing. It also saves time for the police in gathering information, allowing the search to commence sooner (https://www.met.police.uk).

Occupational therapy can suggest environmental aids and adaptations to minimize risk and improve quality of life while still in the community. Speech therapists can help with the assessment of swallowing difficulties, speech difficulties and suggest alternative methods of communication.

The topics of end-of-life care, resuscitation and palliative pathway are sensitive conversations that need to be delicately handled and may call for increased support.

Living wills and Lasting Power of Attorney:

A systematic social evaluation needs to be done in order to help care plan for the future and to make advanced decisions regarding financial and legal matters. The patient must be supported in every way to be able to make the decision. Preferably, these assessments must be carried out as early as possible, while the patient still retains some capacity and insight. Support groups and dementia charities are an important resource.

Driving with Dementia:

All patients with an MCI or dementia diagnosis must inform the local licensing authority and the insurance company regarding the diagnosis. Driving is a motor activity that requires all the brain faculties. Having the diagnosis does not automatically mean the patient cannot drive, it means that there may be a driving and road safety assessment undertaken and restrictions may be put in place. This aspect is particularly dreadful for patients of EOD as they are often young and drive. The independence that comes with driving is threatened and the prospect of not being able to drive again can be an unwelcome decision. Addressing safety while driving is another vital question in history taking.

Differences Between Early Onset Dementia and Late Onset Dementia

Diagnostic and Therapeutic differences:

Late onset dementia (LOD) often overshadows the early onset dementia because of more awareness and the predominance of memory impairment in the former. It is not known whether EOD and LOD are variants of the same disease or if these are two distinct entities. Neuropathological findings are seemingly the same in both types, but they differ in their clinical presentation and course, considering the age of onset is an important determinant of the differences observed in the disease (Sá et al., 2012). Characteristics of early onset AD in comparison with late onset AD include a stronger genetic predisposition, faster deterioration, more frequent delay in diagnosis due to atypical presentations, higher prevalence of traumatic brain injury, fewer amnestic features, greater involvement of other cognitive domains on presentation and short survival time. There are also greater psychosocial difficulties. On neuroimaging, early onset AD more often spare the hippocampi and have greater tau burden in posterior neocortices. The neural networks involved in the two disease processes are also different (Mendez, 2019).

Challenges in the assessment and diagnosis of EOD are partly due to the fact that neurodegenerative dementias are less common and secondary dementias more common than in late onset dementia. Early symptoms are non-specific and insidious. They include depression, behavioural change, gradual change in personality, neurological disorders, systemic disorders and mild cognitive impairment (MCI) (Draper & Withall, 2016).

Owing to the age group affected by late onset dementias, there is a higher likelihood of associated sensory impairment (visual impairment, hearing impairment) and frailty which is not the case in the EOD cohort. They also have greater physical comorbidities and are often on multiple medications. They are more likely to develop delirium owing to the underlying physical health issues and polypharmacy.

While the medical management of early onset AD is similar to that of late onset AD, they differ in their psychosocial needs, patient profile and physical capabilities. EOD requires special emphasis on psychosocial support.

Psychological differences:

Coming to terms with the cognitive decline is psychologically distressing. Most EOD patients retain some insight into their deficits and have difficulty coping with it. A study by Lai et al. in dementia patients above the age of 50 years reported that although patients with prior dementia diagnosis had lower risk of suicidal behaviour, the risk increased with a recent dementia diagnosis and also an increased risk of new mental illness. Furthermore, the study revealed that mental health services are underutilized, thus indicating a need for multidisciplinary and interdepartmental collaboration of services in caring for dementia patients (Lai et al., 2019). Younger dementia patients are more physically able and may have more access to means for self-harm and suicide.

There may be gradual social alienation and increased dependency on family members leading to unmet needs in the families. They are at high risk of carer stress making them prone to physical and emotional burnout. Concerns about the heritability of dementia can be distressing to the family (Draper & Withall, 2016). The loss experienced in dementia is psychologically complex and incremental over time. It is as though the family experiences the grief of two deaths, the loss of the loved one before the physical death—a living bereavement. When the patient is young, this recession from the family is disturbing and ill-timed. It leaves young partners and children in ambiguity without the hope of reconciliation (Blandin, 2016). There is a change of role within the family, and it threatens the pre-existing family structure. The swiftly changing relationship landscape, loss of shared reality and disenfranchised grief interject life prematurely at a time when there are still more years ahead and dreams to realize.

Socio-occupational differences:

A diagnosis of EOD may have financial consequences such as need for workplace adjustments, reduced ability to work, early retirement, loss of income and increased spending on health care (Draper & Withall, 2016). There is a risk of financial exploitation and vulnerability. Management includes a greater focus on engagement in meaningful activity, environmental aids and adjustments to help with the reduced ability to work and assistance with financial and legal planning. The input for psychological services is invaluable in helping the patient cope with this loss of role and functional impairment.

Eventually, people with EOD are managed in care homes and institutions; these are primarily designed to cater to the geriatric population. This puts the patients with EOD in an unusual situation and presents with its own set of difficulties. There is now a move towards focussing on personhood and providing more personalized management strategies with more involvement of the family with an aim to prolong life in the community as much as possible and also taking this approach to the care homes when the time comes.

EODs are projected to rise in the years to come. It raises more questions than there are answers. Much remains to be understood about the illness and much more about diagnostic and treatment modalities. Since the past few years, this disease entity has been gradually capturing the attention of public health and researchers alike. This brings hope for better funding for research as well as public awareness and allocation of resources. EOD is a difficult diagnosis to deliver, more difficult to endure and most difficult to bear witness. An illness with no tomorrows means an illness in which new learning is impaired and one cannot remember recent event but only that of the distant past. It is like a tug of war with time where memory is the rope which silently but surely slips away.

References

Blandin, K. (2016, January 29). Dementia Grief—What Makes It Unique? Dementia.org. https://www.dementia.org/dementia-grief-characteristics

Cerejeira, J., Lagarto, L., & Mukaetova-Ladinska, E. B. (2012). Behavioral and psychological symptoms of dementia. *Frontiers in Neurology, 3*. https://doi.org/10.3389/fneur.2012.00073.

Compta, Y., Parkkinen, L., O'sullivan, S. S., Vandrovcova, J., Holton, J. L., Collins, C., Lashley, T. et al. (2011). Lewy- and Alzheimer-type pathologies in Parkinson's disease dementia: Which is more important? *Brain, 134*(5), 1493–1505. https://doi.org/10.1093/brain/awr031.

Draper, B., & Withall, A. (2016). Young onset dementia. *Internal Medicine Journal, 46*(7), 779–786. https://doi.org/10.1111/imj.13099

Gauthier, S., Patterson, C., Chertkow, H., Gordon, M., Herrmann, N., Rockwood, K., Rosa-Neto, P., & Soucy, J.-P. (2012). Recommendations of the 4th Canadian consensus conference on the diagnosis and treatment of dementia (CCCDTD4). *Canadian Geriatrics Journal, 15*(4). https://doi.org/10.5770/cgj.15.49.

Hart, D. J., Craig, D., Compton, S. A., Critchlow, S., Kerrigan, B. M., Mcilroy, S. P., & Passmore, A. P. (2003). A retrospective study of the behavioural and psychological symptoms of mid and late phase Alzheimer's disease. *International Journal of Geriatric Psychiatry, 18*(11), 1037–1042. https://doi.org/10.1002/gps.1013

Harvey, R. J. (2003). The prevalence and causes of dementia in people under the age of 65 years. *Journal of Neurology, Neurosurgery & Psychiatry, 74*(9), 1206–1209. https://doi.org/10.1136/jnnp.74.9.1206

Høgh, P., Smith, S. J., Scahill, R. I., Chan, D., Harvey, R. J., Fox, N. C., & Rossor, M. N. (2002). Epilepsy presenting as AD: Neuroimaging, electroclinical features, and response to treatment. *Neurology, 58*(2), 298–301. https://doi.org/10.1212/wnl.58.2.298

https://www.met.police.uk/notices/af/herbert-protocol/

Jeste, D. V., Meeks, T. W., Kim, D. S., & Zubenko, G. S. (2006). Research agenda for DSM-V: Diagnostic categories and criteria for neuropsychiatric syndromes in dementia. *Journal of Geriatric Psychiatry and Neurology, 19*(3), 160–171. https://doi.org/10.1177/0891988706291087

Katona, C., Livingston, G., Cooper, C., Ames, D., Brodaty, H., & Chiu, E. (2007). International psychogeriatric association consensus statement on defining and measuring treatment benefits in dementia. *International Psychogeriatrics, 19*(3), 345–354. https://doi.org/10.1017/s1041610207005145

Kehagia, A. A, Barker, R. A., & Robbins, T. W. (2010). Neuropsychological and clinical heterogeneity of cognitive impairment and dementia in patients with Parkinson's disease. *The Lancet Neurology, 9*(12), 1200–1213. https://doi.org/10.1016/s1474-4422(10)70212-x.

Killin, L. O. J., Starr, J. M., Shiue, I. J., & Russ, T. C. (2016). Environmental risk factors for dementia: A systematic review. *BMC Geriatrics, 16*(1). https://doi.org/10.1186/s12877-016-0342-y.

Kong, E.-H., Evans, L. K., & Guevara, J. P. (2009). Nonpharmacological intervention for agitation in dementia: A systematic review and meta-analysis. *Aging & Mental Health, 13*(4), 512–520. https://doi.org/10.1080/13607860902774394

Lai, A., Li, Y., & Byers, A. (2019). High risk of new psychiatric disorders and suicidal behavior in dementia. *Innovation in Aging, 3*, no. Supplement_1. https://doi.org/10.1093/geroni/igz038.1150.

Livingston, G., Sommerlad, A., Orgeta, V., Costafreda, S. G., Huntley, J., Ames, D., Ballard, C., Banerjee, S., Burns, A., Cohen-Mansfield, J., Cooper, C., Fox, N., Gitlin, L. N., Howard, R., Kales, H. C., Larson, E. B., Ritchie, K., Rockwood, K., Sampson, E. L., Mukadam, N., et al. (2017). Dementia prevention, intervention, and care. *The Lancet, 390*(10113), 2673–2734. https://doi.org/10.1016/s0140-6736(17)31363-6

Manthorpe, L., Tindall, & Jill. (1997). Early onset dementia: A case of Ill-timing? *Journal of Mental Health, 6*(3), 237–250. https://doi.org/10.1080/09638239718770.

Masellis, M., Sherborn, K., Neto, P., Sadovnick, D. A., Hsiung, G.-Y. R., Black, S. E., Prasad, S., Williams, M., & Gauthier, S. (2013). Early-onset dementias: Diagnostic and etiological considerations. *Alzheimer's Research & Therapy, 5*, no. Suppl 1. https://doi.org/10.1186/alzrt197.

McKeith, I. G., Dickson, D. W., Lowe, J., Emre, M., O'Brien, J. T., Feldman, H., Cummings, J., Duda, J. E., Lippa, C., Perry, E. K., Aarsland, D., Arai, H., Ballard, C. G., Boeve, B., Burn, D. J., Costa, D., Del Ser, T., Dubois, B., Galasko, D., Gauthier, S., Goetz, C. G., Gomez-Tortosa, E., Halliday, G., Hansen, L. A., Hardy, J., Iwatsubo, T., Kalaria, R. N., Kaufer, D., Kenny, R. A., Korczyn, A., Kosaka, K., Lee, V. M., Lees, A., Litvan, I., Londos, E., Lopez, O. L., Minoshima, S., Mizuno, Y., Molina, J. A., Mukaetova-Ladinska, E. B., Pasquier, F., Perry, R. H., Schulz, J. B., Trojanowski, J. Q., Yamada, M., and Consortium on DLB. (2005). Diagnosis and management of dementia with Lewy bodies: third report of the DLB Consortium. *Neurology 65*, 1992.

Mendez, M. F. (2019). Early-onset Alzheimer disease and its variants. *CONTINUUM: Lifelong Learning in Neurology, 25*(1), 34–51. https://doi.org/10.1212/con.0000000000000687.

Mirakhur, A., Craig, D., Hart, D. J., McIlroy, S. P., & Passmore, A. P. (2004). Behavioural and psychological syndromes in Alzheimer's disease. *International Journal of Geriatric Psychiatry, 19*(11), 1035–1039. https://doi.org/10.1002/gps.1203

O'Neil, M. E., Freeman, M., Christensen, V., Telerant, R., Addleman, A., & Kansagara, D. (2011). *A systematic evidence review of non-pharmacological interventions for behavioral symptoms of dementia*. Department of Veterans Affairs.

Prince, M., & Jackson, J. (2009). World Alzheimer's report 2009. Alzheimer's Disease International, London.

Rossor, M. N, Fox, N. C., Mummery, C. J., Schott, J. M., & Warren, J. D. (2010). The diagnosis of young-onset dementia. *The Lancet Neurology, 9*(8), 793–806. https://doi.org/10.1016/s1474-4422(10)70159-9.

Sá, F., Pinto, P., Cunha, C., Lemos, R., Letra, L., Simões, M., & Santana, I. (2012). Differences between early and late-onset Alzheimer's disease in neuropsychological tests. *Frontiers in Neurology, 3*. https://doi.org/10.3389/fneur.2012.00081.

Sampson, E. L. (2004). Young onset dementia. *Postgraduate Medical Journal, 80*(941), 125–139. https://doi.org/10.1136/pgmj.2003.011171

The Driving with dementia or Mild Cognitive Impairment Consensus Guidelines for Clinicians, UK (endorsed by Alzheimer's Society, Driving Mobility, Royal College of Psychiatrists, Royal College of General Practitioners, Memory Services National Accreditation Programme, Royal College of Occupational Therapists and the British Psychological Society).

Tun, S.-M., Murman, D. L., Long, H. L., Colenda, C. C., & Eye, A. V. (2007). Predictive validity of neuropsychiatric subgroups on nursing home placement and survival in patients with Alzheimer disease. *The American Journal of Geriatric Psychiatry, 15*(4), 314–327. https://doi.org/10.1097/01.jgp.0000239263.52621.97.

Vieira, R. T. (2013). Epidemiology of early-onset dementia: A review of the literature. *Clinical Practice & Epidemiology in Mental Health, 9*(1), 88–95. https://doi.org/10.2174/1745017901309010088

Vliet, D. V., Bakker, C., Raymond Koopmans, T. C. M., Vernooij-Dassen, M. J. F. J., Verhey, F. R. J., & De Vugt, M. E. (2010). Research protocol of the need YD-study (Needs in Young Onset Dementia): A prospective cohort study on the needs and course of early onset dementia. *BMC Geriatrics, 10*(1). https://doi.org/10.1186/1471-2318-10-13.

Chapter 22
Pharmacological Management of Dementia—An Overview

Reetika Dikshit, Sagar Karia, Abha Thakurdesai, Heena Merchant, and Avinash De Sousa

Abstract The medical management and pharmacotherapy of dementia are at interesting crossroads. There have been many recent developments with newer drugs being developed that would probably halt the progression of dementia. The following review looks at the various options we currently have in the pharmacotherapy of dementia and provides an overview of these drugs for the busy clinician. It also looks at the future and various newer drugs that would be developed and be available in the next decade. The future of the pharmacotherapy of dementia is promising with many new drugs in the pipeline.

Keywords Pharmacotherapy · Newer drugs · Dementia · Anticholinesterases · Medical management

Introduction

The pharmacological management of dementia is remains challenging since response to treatment and disease course varies from patient to patient. (Standridge, 2004) Four cholinesterase inhibitors and an N-methyl-D-aspartate (NMDA) glutamate receptor antagonist are approved by the US FDA for the treatment of Alzheimer's disease (AD). (Doody et al., 2001) The pharmacotherapy of dementia is clinically efficacious and associated with reduced caregiver burden. The pharmacotherapy of dementia aims at preventing brain degeneration, improving cognition, managing behavioural symptoms and improving the overall quality of life. However, the risk/benefit ratio for each patient should be critically evaluated, because pharmacological treatment has been associated with adverse events in older patients and caution must be exercised when using medications in patients with dementia. (Moretti et al., 2011) The following paper is an overview of the various pharmacotherapy option used in the management of dementia from a dementia perspective and does not look at

R. Dikshit · S. Karia · A. Thakurdesai · H. Merchant · A. De Sousa (✉)
Department of Psychiatry, Lokmanya Tilak Municipal Medical College, Mumbai, India

M. K. Shankardass (ed.), *Dementia Care*,
https://doi.org/10.1007/978-981-16-3864-0_22

the management of the behavioural and psychological symptoms associated with dementia. An overview of the current drugs available is provided along with an insight into the drugs that are in the pipeline for the future.

Cholinesterase Inhibitors

The nucleus basalis of Meynert in the basal forebrain is the major centre for cholinergic neurons which project to other areas of the cerebral cortex. (Kurosawa et al., 1989) These neurons have an important role in memory formation. It is suspected that degeneration of these neurons causes short-term memory impairment. (Whitehouse et al., 1982) Thus, boosting cholinergic neurotransmission here improves memory symptoms in Alzheimer's disease. When Acetylcholine is released by neurons, its action is terminated by either Acetylcholinesterase or Butyrylcholinesterase. (Francis et al., 1999) Acetyl cholinesterase is thought to be the main enzyme for inactivating Acetylcholine although Butyrylcholinesterase can also take on this activity if Acetylcholinesterase diffuses to nearby glia. Drugs which inhibit these enzymes lead to an increase in Acetylcholine levels in the brain. This enhances memory and can also slow the decline in functioning. (Collerton, 1986) There are four major cholinesterase inhibitors—Tacrine, Donepezil, Rivastigmine and Galantamine.

Tacrine

Tacrine is the first cholinesterase inhibitor approved by the FDA for use in Alzheimer's disease. It is a reversible cholinesterase inhibitor. Response to treatment is dose-dependent, and it often takes up to 12 weeks for results. Its biggest disadvantageous is a chance of hepatotoxicity which can even be fatal. When used, monitoring of liver function is essential. It has been removed from the market subsequently due to concerns about hepatic safety. (Eagger et al., 1991; Farlow et al., 1992)

Donepezil

Donepezil is one of the most commonly used medications for the treatment of Alzheimer's disease. It is a non-competitive inhibitor of acetylcholinesterase. It is appreciated by clinicians due to its once-a-day dosing and safety in terms of hepatic and renal functioning. It is available as tablets of 5, 10 and 23 mg and as a syrup. For mild–moderate cases, 5–10 mg is used and for severe cases, 10 and 23 mg are used. The tablet is to be given once a day, generally at night, so that the peaking of the plasma concentration occurs at night. Thus, even if mild adverse effects occur, they are unlikely to bother the patient. The once-a-day dosing is possible due to Donepezil's long half life of 70 h. It can be taken irrespective of food intake. We start with 5 mg, and after 4–6 weeks, increase the dose to 10 mg. After 3 months of being on 10 mg, the dose can be up titrated to 23 mg. It can be used along with other medications safely; however, one should be vigilant when using a CYP 3A4 or 2D6 inducer with it. (Birks & Harvey, 2006; Burns et al., 1999; Howard et al., 2012)

Rivastigmine

Rivastigmine is the only drug to be approved for both Alzheimer's and Parkinsonian dementia. It is a non-competitive, pseudoirreversible cholinesterase inhibitor due to its prolonged enzyme inhibition activity. It inhibits Acetylcholinesterase and butyrl-cholinesterase equally well. Rivastigmine tablets formulation should be consumed with food to minimize gastrointestinal adverse effects. The dose is 6–12 mg in 2 divided doses. Patients are started on 1.5 mg twice a day, and the dose is gradually increased every 2–4 weeks by 1.5 mg daily until the maximum daily dose of 12 mg/day is reached. Rivastigmine patches are also available. Rivastigmine is safe to use in hepatic and renal problems and has no interactions with CYP3A4 inhibitors or inducers. (Farlow et al., 2000; Potkin et al., 2001; Rösler et al., 1999)

Galantamine

Galantamine is extracted from daffodils of the plant *Galanthus Nivalis.* It is a competitive, reversible inhibitor that is more selective for Acetylcholinesterase. Galantamine also produces allosteric modulation of nicotinic receptors. Allosteric modulators upregulate or potentiate ion channel activity of nicotinic receptors in response to acetylcholine. Galantamine is the only drug of this class to have this action. The dose range is 16–32 mg/day given twice a day in a divided dose. It should be given with food. The initial dose is 8 mg/day, and after a minimum of 4 weeks, the dose can be increased. Galantamine is metabolized by hepatic CYP 2D6 and CYP 3A4 isozymes and patients should be monitored closely when known CYP 2D6 and 3A4 inhibitors are given along with galantamine. Galantamine is contraindicated in patients with severe renal and hepatic impairment. Even in moderate hepatic impairment, dose should not exceed 16 mg. (Coyle & Kershaw, 2001; Olin & Schneider, 2002; Wilcock et al., 2003)

Indications

Cholinesterase inhibitors are indicated for mild–moderate Alzheimer's disease. Donepezil is the only drug of this class approved for severe Alzheimer's dementia (Clegg et al., 2001). Rivastigmine is approved for use in mild–moderate dementia associated with Parkinson's disease. A few studies have shown efficacy in vascular and Lewy body dementia as well; however, due to their cardiovascular and cerebrovascular side effect profile, they have not received FDA approval for the same (Polinsky, 1998).

Caution is advised when patients are taking anticholinergic medications, as donepezil may interfere with their activity due to its mechanism of action. Donepezil may also lengthen the action of succinylcholine-type muscle relaxants during anaesthesia (Takeda et al., 2006).

Side Effects

Cholinesterase inhibitors are safe medications–common side effects are nausea, vomiting and diarrhoea which can sometimes lead to discontinuation. One should be careful in case of patients on high doses, those using NSAIDS and those with pre-existing gastric problems. These adverse effects occur more often in those patients with low body weight (Colombres et al., 2004). Other side effects are sleep disturbances, decreased appetite, fatigue and muscle cramps. Rivastigmine is also associated with tremors. Galantamine can also cause dizziness, headache and weight loss. Seizures, exacerbation of COPD and cerebrovascular accidents are known to occur; however, they are not very common (Eagger & Harvey, 1995). Due to their cholinergic action, cholinesterase inhibitors should be used with caution in patients with cardiac arrhythmias. Some patients using transdermal patch reported skin erythema. All cholinesterase inhibitors are Category C drug and should be avoided in pregnancy. Its excretion in breast milk is yet unknown (Greig et al., 2005).

NMDA Receptor Antagonists

Apart from cholinesterase inhibitors, Memantine, an NMDA-inhibitor is also used for the treatment of dementia. It is hypothesized that, in Alzheimer's disease, there is excess release of glutamate (Hynd et al., 2004). The neurofibrillary tangles and amyloid plaques that accumulate in Alzheimer's disease cause a steady leak of glutamate which affects glutamatergic neurotransmission and thus learning and memory. Memantine interferes with the abnormal glutamatergic neurotransmission and thus improves cognitive function and slows down deterioration (Francis, 2003).

Memantine is a selective uncompetitive inhibitor of the NMDA type of glutamate receptors. Its specific benefit is in blocking the chronic, low-level excitation brought about by glutamate in Alzheimer's disease (Tariot et al., 2004).

Memantine is approved for the treatment of moderate to severe Alzheimer's disease. It has also been tried in mild cases; however, the benefits have not been found to be very significant. The common side effects reported with memantine are constipation, dizziness, headache and confusion. Memantine is a renal excreted drug and must be avoided in renal failure. It must also be avoided in severe hepatic failure (Sonkusare et al., 2005).

Memantine must be used with caution with hydrochlorothiazide, triamterene, ranitidine, quinidine, and nicotine as they are renally secreted and can affect each others concentration. Memantine clearance is affected in alkaline urine caused by carbonic anhydrase inhibitors, sodium bicarbonate and medical conditions such as renal tubular acidosis. Memantine should be cautiously used in these situations (Rogawski & Wenk, 2003).

The recommended starting dose of Memantine is 5 mg daily. The recommended target dose is 20 mg/day. Drug is administered twice daily in separate doses with 5 mg increments weekly depending on tolerability. (Danysz & Parsons, 2003)

Donepezil–Memantine Combination

We have read about Donepezil and Memantine individually in the preceding sections. Donepezil, as we know is one of the best-tolerated cholinesterase inhibitors, and thus, the preferred drug in Alzheimer's dementia. It is approved for treatment in all severities of Alzheimer's disease. We also know that Memantine is used to treat moderate to severe Alzheimer's disease. (Atri et al., 2013)

As both Donepezil and Memantine have different mechanisms of action, it is often used in a fixed-dose combination to treat moderate to severe Alzheimer's disease. The pharmacokinetics of these drugs do not interfere with one another and their adverse effects are similar to what we see with Donepezil alone. To minimize the number of tablets the patients' takes, it is advantageous to use them in combination. It is also more economical to use the combination than to use the individual molecules separately. However, it is recommended to stabilize the patient on Donepezil and Memantine individually before using the combination. (Patel & Grossberg, 2011)

Nootropics

The term "Nootropic" was invented by Guirgea, who developed piracetam and was coined to describe a class of drugs that improve "higher cognitive function". There has been wide number of compounds that have been labelled as nootropic agents (Nicolaus, 1982). Here we are describing few of them playing important role in the treatment of Alzheimer's disease.

Piracetam

It was the first Nootropic drug is a cyclic derivative of the neurotransmitter γ-aminobutyric acid (GABA). Its role in dementia can be explained by its neuronal and vascular effects (Gouliaev & Senning, 1994).

Following are its neuronal effects helpful in cognition improvement:

1. Amyloid peptides have been known to cause lipid disorganization within the cell membranes. Piracetam significantly decreases the destabilizing effect of amyloid peptides. It also counteracts the negative curvature induced by fusogenic peptides and thereby decreases the likelihood of membrane fusion (Nicholson, 1990).
2. Piracetam increases the number of postsynaptic receptors and/or restore the function of these receptors of all the systems like cholinergic, serotoninergic, noradrenergic and glutamatergic. Effect on cholinergic and glutamatergic system is important in dementia. In animal models, it modifies hippocampal acetylcholine levels, increases the population of muscarinic cholinergic receptors in the frontal cortex of aged, normalizing the age-related elevated affinity of L-glutamate for the NMDA receptor, thereby restoring NMDA receptor function (Cohen & Mutter, 1993).
3. It promotes neuroplasticity when neural circuits are recoverable.
4. It also activates brain adenylate kinase and thereby increasing brain glucose utilization.

Its vascular effects helpful in dementia are as follows: (Waegemans et al., 2002).

1. It reduces adhesion of erythrocytes to endothelium and thereby improving their movement in circulation.
2. It stimulates prostacyclin synthesis which has vasodilator as well as platelet aggregation inhibitory properties.
3. It influences blood coagulation by reducing levels of fibrinogen and von Willebrand factor in a dose-dependent manner by up to 40%.

Thus, it improves microcirculation as a result of combined effects on erythrocytes, blood vessels and blood coagulation.

Piracetam has good oral bioavailability close to 100% and is rapidly absorbed.

Side effects observed are mild like anxiety, insomnia, irritability, headache, agitation, nervousness, tremor and hyperkinesia (Tariska & Paksy, 2000).

Curcumin

Curcumin (derived from rhizome of *Curcuma longa*—Haldi), the source of the spice Turmeric, is used in curries and other spicy dishes from India, Asia and the Middle East. Apart from usage in food, it has pain-relieving, anti-inflammatory and anticancer properties too. Studies have shown benefits of curcumin in Alzheimer's disease (AD) too (Mishra & Palanivelu, 2008).

Various mechanisms by which it plays beneficial role in dementia are as follows:

1. *Effects on Macrophages*: It may help macrophages to clear the amyloid plaques found in AD. It has antiproliferative action on microglia, which is known to secrete some cytokines that exacerbate A-beta pathology. It inhibits neuroglial proliferation dose dependently. The overall effect is decreased astrocyte proliferation, improved myelogenesis and increased activity and differentiation of oligodendrocytes (Gupta et al., 2013).
2. *Anti-inflammatory action*: Curcumin inhibits Aβ-induced expression of Egr-1 protein and Egr-1 DNA-binding activity in THP-1 monocytic cells, thereby reducing inflammation. The chemotaxis of monocytes occurring in response to chemokines from activated microglia and astrocytes in the brain is also decreased by curcumin. Exposure to curcumin impairs production of pro-inflammatory cytokines like IL-1,IL-6, TNF-α (Shishodia et al., 2005).
3. *Antioxidant effects*: It has strong antioxidant action as it inhibits the formation and propagation of free radicals. It decreases the low-density lipoprotein oxidation and the free radicals that cause the deterioration of neurons in AD. Administration of curcumin significantly reduces lipid peroxidation and lipofuscin accumulation that is normally increased with ageing. It is known to increase the activities of superoxide dismutase, sodium–potassium ATPase that decreases with ageing. Curcumin protects brain mitochondria against various oxidative stress. It is also a potent inducer of hemoxygenase, a protein providing efficient cytoprotection against various oxidative stress (Jayaprakasha et al., 2006).
4. *Effect on Beta Amyloid Plaques*: Due to its lipophilic nature, it crosses the blood–brain barrier and binds to plaques. It is better A-beta 40 aggregation inhibitor and also destabilizes the A-beta polymer (Lim et al., 2001).
5. *Metal Chelation Effect*: By interaction with heavy metals like cadmium and lead, it prevents neurotoxicity caused by these metals. Curcumin suppresses inflammatory damage by preventing metal induction of NF-kappa (Yang et al., 2005).
6. *Cholesterol-Lowering Effect*: It exerts beneficial effects in AD by inhibiting cholesterol formation and decreasing serum peroxides (Soudamini et al., 1992).

Common side effects are gastrointestinal upset, chest tightness, skin rashes, swollen skin are said to occur with high dose. Curcumin has poor oral bioavailability due to low absorption and rapid hepatic and intestinal metabolization. New formulations are being developed like co-administering with piperidine (piperidine may block enteric and hepatic glucuronidation, thereby resulting in higher curcumin plasma and tissue levels), nanoparticles, micelles and liposomes (Baum et al., 2008).

Ginkgo Biloba

Active components of Ginkgo biloba include flavonoids, terpenoids and terpene lactones (ginkgolides and bilobalide). Ginkgolides and bilobalide, which are unique to Ginkgo biloba. A well-defined extract, EGb 761, is produced from the ground up leaves which contains 24% w/w (weight of active compound/total weight of extract) flavone glycosides and 6% w/w terpene lactones. (Birks & Evans, 2009).

Ginkgo biloba has been used for variety of conditions like asthma, bronchitis, heart dysfunction. In neurology, it has been prescribed extensively for cerebral insufficiency, a diagnosis that can cover a range of conditions including memory and concentration problems, confusion, depression, anxiety, dizziness, tinnitus and headache. (DeKosky et al., 2008)

Its role in dementia can be explained by combination of effects like increasing blood supply by dilating blood vessels, reducing blood viscosity, by modification of neurotransmitter systems and by reducing the density of oxygen free radicals. Recently, an in vitro study has shown that ginkgo extract has an anti-amyloid aggregation effect too. (Sierpina et al., 2003)

Studies have shown that ginkgo extract improves cognition, activities of daily living, behavioural symptoms and overall quality of life in AD patients. The daily dose is 120–240 mg daily. The main side effects noted are mild gastrointestinal symptoms, headache, dizziness or allergic skin reactions. (Ernst & Pittler, 1999)

Huperzine A

Huperzine A (Hup-A) is a novel *Lycopodium* alkaloid isolated from Chinese folk medicine *Huperzia serrata* (*Qian Ceng Ta*) and has been used in the treatment of contusions, strains, swelling and schizophrenia in China. Hup-A is known to improve cognitive functions, learning abilities and memory in the elderly. (Ma et al., 2007)

Its usefulness in AD can be explained by the following actions:

1. *Acetylcholinesterase inhibitor*: It is a selective, reversible and more potent inhibitor of AChE than donepezil, rivastigmine and galantamine in vivo. Also Hup-A possesses a longer half life ($t1/2 = 35$ min) for the AChE/Hup-A complex, and a slower rate of dissociation. (Zhang et al., 2008)
2. *Reduction in Glutamate induced toxicity*: It is proposed that Hup-A may reduce Ca2+ influx initiated by the binding of glutamate to its receptor, or may directly interfere with influx of Ca2+ through the associated ion channel. Thereby improving survival of neurons. (Ho et al., 2011)
3. *Neuroprotective Action*: It is known to protect neurons from toxins induced cell death. Soluble secretory Amyloid Precursor Protein (sAPP) exhibits neuroprotective effects like promotion of cell proliferation and neurite outgrowth and prevention of intracellular calcium accumulation, thereby delaying AD

progression. Hup-A is known to stimulate release of sAPP. (Perry & Howes, 2011)

4. *Antioxidative Effect*: Pre-incubation of cells with Hup-A prior to Aβ exposure was found to enhance cell survival and activities of antioxidant enzymes including GSH-Px, SOD and CAT and decrease MDA production. (Mantle et al., 2000)
5. *Anti-apoptosis Effect*: Hup-A significantly improves Aβ induced apoptotic changes like DNA laddering, cell shrinkage, generation of nuclear apoptotic bodies. It also attenuates the increase of Caspase-3 activity induced by Aβ in cultured primary cortical neurons. (Xu et al., 2012)
6. *Improving mitochondrial function*: It improves mitochondrial function in Aβ exposed cells by reducing levels of reactive oxygen species and increasing activities of key components of the respiratory chain and TCA cycle. (Singh et al., 2011)
7. *Anti-inflammatory action*: Hup-A can significantly restore regional cerebral blood flow, reduce infarct area, decrease over-expression of proinflammatory factors including TNF-α and IL-1β and suppress activation of glial cells in ischemic penumbra. (Wang et al., 2010)

Hup-A has advantages of longevity, good blood–brain barrier penetrating ability, oral bioavailability, as well as good safety profile. Daily dosing is 200 mcg in divided doses. Side effects include mild nausea, vomiting and dizziness. (Jesky & Hailong, 2011).

Piribedil

It has been proved that age-related dopamine decline leads to cognitive impairment of elderly. Brain imaging technique study has demonstrated an association between age-related decrease in dopamine D2 receptors and impairment on neurocognitive tasks involving the frontal lobe (such as executive function and response inhibition) by a mechanism that may involve reduced frontal and cingulate metabolism. Patients with mild cognitive impairment had improvement in global cognitive function when treated with the dopamine receptor agonist piribedil. Piribedil, a D2/D3 receptor agonist, has been shown to improve reaction time, immediate and delayed free recall and dual coding tasks. Dose of 50 mg per day is advisable. But with increase in age there might be further decline in dopamine receptors thereby warranting higher doses. (Nagaraja & Jayashree, 2001; Peretti et al., 2004)

Pyritinol

Also known as Pyrithioxine and Pyridoxine disulphide. Mechanism of action is enhanced blood glucose metabolism, it is known to increase glucose uptake and utilization, tending to increase transport across blood–brain barrier. It improves oxygen consumption, nucleic acid metabolism and cortical acetylcholine release. Due to improvement in blood flow and cell membrane structure and functioning improving after damage, pyritinol helps in protecting nerve cells. Studies have shown that pyritinol improves EEG findings in dementia patients by decreasing slowing and improving fast alpha and beta activities, thereby improving vigilance of the patients clinically. The dosage is 200 mg three times a day. (Fischhof et al., 1992; Heiss et al., 1994).

Targeting Amyloid as a Future Disease—Modifying Treatment of Alzheimer's Disease

With increasing numbers of asymptomatic amyloidosis (stage I presymptomatic Alzheimer's disease) or cognitive changes with early neurodegeneration (stage II Alzheimer's disease, MCI), it is becoming even more urgent to intervene in this disorder at an earlier stage when brain changes are present without any overt cognitive decline, or certainly before dementia sets in. (Salomone et al., 2012) The newer drugs are as follows.

Secretase Modulators

These are drugs that are inhibitors of β-secretase, γ-secretase or α-secretase stimulators–

1. **Beta secretase inhibitors**: Inhibitors of the β-secretase enzyme have been difficult to synthesize, but compounds such as SCH1381252, CTS21666, and others are moving forward in clinical development, and their results are eagerly awaited because of their theoretical promise as a mechanism of preventing β-amyloid formation. The rationale of beta—site APP cleaving enzyme [BACE] inhibition is that it represents a way of interfering upstream in the amyloid cascade. The finding that BACE1 knockout mice have abolished Aβ production without any clinical phenotype made BACE1 inhibitors an attractive therapeutic strategy. β-secretase inhibitors have been developed to reduce brain Aβ concentrations in Alzheimer's disease transgenic mice. The orally active BACE1 inhibitor, BI 1,181,181, is the first generation of BACE1 inhibitor failed in Phase I trials because of low oral bioavailability and low blood–brain barrier penetration Subsequent, second-generation BACE1 inhibitors, RG7129, LY2811376 and

LY2886721 also failed in clinical trials because of liver toxicity. However, third-generation BACE1 inhibitors including E2609, AZD329, CNP520, JNJ-54861911 have shown satisfactory pharmacokinetics and encouraging clinical data in ongoing studies. Beta secretase inhibitors are Tramprosate, colostrinin and sylloinositol. (Ghosh et al., 2012; Nunan & Small, 2000)

2. **Gamma secretase inhibitors**: The enzyme γ-secretase is targeted to reduce Aβ production in a dose-dependent manner; however, this has not translated into a clinical benefit yet. The enzyme γ-secretase cleaves embedded peptides, which in some cases leads to release of toxic peptides (particularly Aβ42). Inhibition of this enzyme could therefore prevent the formation of toxic peptides. Semagacestat (LY450139) was terminated from clinical trials for safety reasons as it impaired cognition and function *more* than placebo did, and also increased the occurrence of skin cancer. Avagacestat was discontinued in a Phase II clinical trial after causing serious adverse events such as cerebral microbleeds, dose-dependent glycosuria and non-melanoma skin cancer. The orally available, small molecule, selective γ-secretase modulator EVP-0962 reduces $A\beta_{1\text{-}42}$ production by shifting the APP cleavage towards the production of shorter and less toxic Aβ peptides, without affecting Notch cleavage. Although EVP-0962 showed promise in transgenic Alzheimer's models, reducing $A\beta_{1\text{-}42}$ peptide levels, decreasing amyloid plaque build-up, reversing behavioural deficits and reducing brain inflammation associated with AD, EVP-0962 was discontinued in Phase II clinical trials in the USA. NIC5-15, pinitol, is a naturally occurring cyclic sugar alcohol that modulates γ- secretase to reduce Aβ production. Gamma secretase inhibitors are semagacestat, tarenflurbil and avagacestat. (Panza et al., 2010; Tamayev & D'Adamio, 2012)
3. **Alpha secretase stimulators**: Drugs that stimulate α-secretase can shift APP processing towards the non-amyloidogenic pathway, thus reducing Aβ production. Bryostatin, a protein kinase C activator currently tested in clinical trials as an anticancer drug, substantially enhances α-secretase processing of APP and reduces brain Aβ42 concentrations in Alzheimer's disease transgenic mice. (Small, 2005) (Fig. 22.1).

Anti Tau Drugs

Drug candidates that reduce tau phosphorylation by inhibiting tau kinases, such as CDK5 and GSK-3β, are in the preclinical phase. However, since tau phosphorylation is regulated by the balance between multiple kinases and phosphates, inhibition of a single kinase might be insufficient to normalize tau phosphorylation. (Yamada & Nabeshima, 2000).

Enhancement of microtubule stabilization by tau stabilizers: Epothilone D is a small molecule microtubule stabilizer; TPI 287 is a tubulin-binding and microtubule-stabilizing drug in Phase I trials in patients with mild-to-moderate AD Tau stabilizers

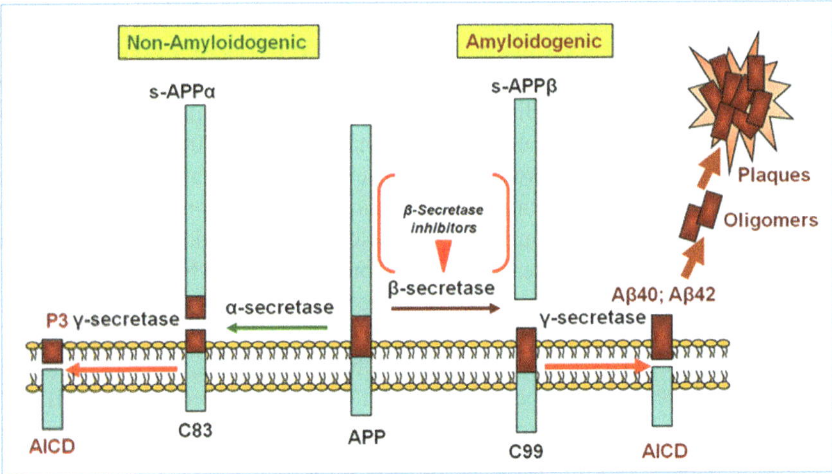

Fig. 22.1 Amyloid and tau hypothesis in Alzheimer's disease

fail to reach the clinic, due to toxic side effects (paclitaxel) or have been discontinued or are in Phase I clinical development. (Spillantini & Goedert, 1998).

Prevention of tau aggregation by tau aggregation inhibitors: Prevention of aggregation regardless of phosphorylation or other tau modification is another therapeutic approach. Derivatives of methylene blue have been shown to disrupt the aggregation of tau, such as Rember™ (a first-generation tau protein aggregation inhibitor), which showed some improvement in AD-related symptoms but failed in Phase II because of emergent side effects including diarrhoea, urinary urgency, painful urination, dizziness and falls. TRx0237 (LMTM) is a second-generation tau protein aggregation inhibitor currently in three Phase III trials. (Heutink, 2000).

Enhancement of phosphorylated—tau clearance by active immunotherapy: High-affinity antibodies against phosphorylated-tau from active immunization of phosphorylated-tau are one approach, such as AADvac-1, which contains synthetic tau peptides, and also ACI-35, which contains phosphorylated S396 and S404 tau fragments; AADvac-1 is in trial. AADvac1 contains a synthetic protein that corresponds to a naturally occurring truncated and misfolded tau protein coupled to keyhole limpet hemocyanin and aluminium hydroxide adjuvant. ACI-35 is a liposomal-based 16-amino acid tetrapalmitoylated phosphor-tau peptide (a synthetic peptide of human protein tau sequence 393–408 of the longest tau isoform and is phosphorylated at S396 and S404). (Morgan, 2011).

Conclusions

This paper has given an overview of the various pharmacological agents available in the management of dementia. There are many agents in the pipeline, and the future of dementia pharmacology is at interesting crossroads. It is only time that will tell us how the newer agents shall help us prevent and manage dementia better in the years to come.

References

Atri, A., Molinuevo, J. L., Lemming, O., Wirth, Y., Pulte, I., & Wilkinson, D. (2013). Memantine in patients with Alzheimer's disease receiving donepezil: New analyses of efficacy and safety for combination therapy. *Alzheimer's Research Theraphy, 5*(1), 6.

Baum, L., Lam, C. W., Cheung, S. K., Kwok, T., Lui, V., Tsoh, J., Lam, L., Leung, V., Hui, E., Ng, C., & Woo, J. (2008). Six-month randomized, placebo-controlled, double-blind, pilot clinical trial of curcumin in patients with Alzheimer disease. *Journal of Clinical Psychopharmacology, 28*(1), 110–113.

Birks, J., Evans, J. G. (2009). Ginkgo biloba for cognitive impairment and dementia. *Cochrane Database Syst Rev*, 1. (CD003120)

Birks, J., Harvey, R. J. (2006). Donepezil for dementia due to Alzheimer's disease. *Cochrane Database Systematic Reviews*, 1. (CD001190)

Burns, A., Rossor, M., Hecker, J., Gauthier, S., Petit, H., Möller, H. J., Rogers, S. L., & Friedhoff, L. T. (1999). The effects of donepezil in Alzheimer's disease—Results from a multinational trial. *Dementia and Geriatric Cognitive Disorders, 10*(3), 237–244.

Clegg, A., Bryant, J., Nicholson, T., McIntyre, L., De Broe, S., Gerard, K., & Waugh, N. (2001). Clinicial and cost-effectiveness of donepezil, rivastigmine and galantamine for Alzheimer's disease: A rapid and systematic review. *Health Technology Assessment, 5*(1), 1–37.

Cohen, S. A., & Mutter, W. E. (1993). Effects of piracetam on N-methyl-D-aspartate receptor properties in the aged mouse brain. *Pharmacology, 47*(4), 217–222.

Collerton, D. (1986). Cholinergic function and intellectual decline in Alzheimer's disease. *Neuroscience, 19*(1), 1–28.

Colombres, M., Sagal, J. P., & Inestrosa, N. C. (2004). An overview of the current and novel drugs for Alzheimer's disease with particular reference to anti-cholinesterase compounds. *Current Pharmaceutical Design, 10*(25), 3121–3130.

Coyle, J., & Kershaw, P. (2001). Galantamine, a cholinesterase inhibitor that allosterically modulates nicotinic receptors: Effects on the course of Alzheimer's disease. *Biological Psychiatry, 49*(3), 289–299.

Danysz, W., & Parsons, C. G. (2003). The NMDA receptor antagonist memantine as a symptomatological and neuroprotective treatment for Alzheimer's disease: Preclinical evidence. *International Journal of Geriatric Psychiatry, 18*(S1), S23-32.

DeKosky, S. T., Williamson, J. D., Fitzpatrick, A. L., Kronmal, R. A., Ives, D. G., Saxton, J. A., Lopez, O. L., Burke, G., Carlson, M. C., Fried, L. P., & Kuller, L. H. (2008). Ginkgo biloba for prevention of dementia: A randomized controlled trial. *JAMA, 300*(19), 2253–2262.

Doody, R. S., Stevens, J. C., Beck, C., Dubinsky, R. M., Kaye, J. A., Gwyther, L. M., Mohs, R. C., Thal, L. J., Whitehouse, P. J., DeKosky, S. T., & Cummings, J. L. (2001). Practice parameter: Management of dementia (an evidence-based review) Report of the quality standards subcommittee of the American academy of neurology. *Neurology, 56*(9), 1154–1166.

Eagger, S. A., & Harvey, R. J. (1995). Tacrine and other anticholinesterase drugs in dementia. *Current Opinion in Psychiatry, 8*(4), 264–267.

Eagger, S. A., Levy, R., & Sahakian, B. J. (1991). Tacrine in Alzheimer's disease. *Lancet, 337*(8748), 989–992.
Ernst, E., & Pittler, M. H. (1999). Ginkgo biloba for dementia. *Clinical Drug Investigation, 17*(4), 301–308.
Farlow, M., Gracon, S. I., Hershey, L. A., Lewis, K. W., Sadowsky, C. H., Dolan-, J., Asher, S. W., Beaver, C., Hamilton, D., Bergman, S. M., & Roger, L. F. (1992). A controlled trial of tacrine in Alzheimer's disease. *JAMA, 268*(18), 2523–2529.
Farlow, M., Anand, R., Messina, J., Jr., Hartman, R., & Veach, J. (2000). A 52-week study of the efficacy of rivastigmine in patients with mild to moderately severe Alzheimer's disease. *European Neurology, 44*(4), 236–241.
Fischhof, P. K., Saletu, B., Rüther, E., Litschauer, G., Möslinger-Gehmayr, R., & Herrmann, W. M. (1992). Therapeutic efficacy of pyritinol in patients with senile dementia of the Alzheimer type (SDAT) and multi-infarct dementia (MID). *Neuropsychobiology, 26*(1–2), 65–70.
Francis, P. T. (2003). Glutamatergic systems in Alzheimer's disease. *International Journal of Geriatric Psychiatry, 18*(S1), S15-21.
Francis, P. T., Palmer, A. M., Snape, M., & Wilcock, G. K. (1999). The cholinergic hypothesis of Alzheimer's disease: A review of progress. *Journal of Neurology, Neurosurgery and Psychiatry, 66*(2), 137–147.
Ghosh, A. K., Brindisi, M., & Tang, J. (2012). Developing β-secretase inhibitors for treatment of Alzheimer's disease. *Journal of Neurochemistry, 120*, 71–83.
Gouliaev, A. H., & Senning, A. (1994). Piracetam and other structurally related nootropics. *Brain Research Reviews, 19*(2), 180–222.
Greig, N. H., Ruckle, J., Comer, P., Brownell, L., Holloway, H. W., Flanagan, J., Douglas, R., Canfield, C. J., & Burford, R. G. (2005). Anticholinesterase and pharmacokinetic profile of phenserine in healthy elderly human subjects. *Current Alzheimer Research, 2*(4), 483–492.
Gupta, S. C., Kismali, G., & Aggarwal, B. B. (2013). Curcumin, a component of turmeric: From farm to pharmacy. *BioFactors, 39*(1), 2–13.
Heiss, W. D., Kessler, J., Mielke, R., Szelies, B., & Herholz, K. (1994). Long-term effects of phosphatidylserine, pyritinol, and cognitive training in Alzheimer's disease. *Dementia and Geriatric Cognitive Disorders, 5*(2), 88–98.
Heutink, P. (2000). Untangling tau-related dementia. *Human Molecular Genetics, 9*(6), 979–986.
Ho, Y. S., So, K. F., & Chang, R. C. (2011). Drug discovery from Chinese medicine against neurodegeneration in Alzheimer's and vascular dementia. *Chinese Medicine, 6*(1), 15.
Howard, R., McShane, R., Lindesay, J., Ritchie, C., Baldwin, A., Barber, R., Burns, A., Dening, T., Findlay, D., Holmes, C., & Hughes, A. (2012). Donepezil and memantine for moderate-to-severe Alzheimer's disease. *New England Journal of Medicine, 366*(10), 893–903.
Hynd, M. R., Scott, H. L., & Dodd, P. R. (2004). Glutamate-mediated excitotoxicity and neurodegeneration in Alzheimer's disease. *Neurochemistry International, 45*(5), 583–595.
Jayaprakasha, G. K., Rao, L. J., & Sakariah, K. K. (2006). Antioxidant activities of curcumin, demethoxycurcumin and bisdemethoxycurcumin. *Food Chemistry, 98*(4), 720–724.
Jesky, R., & Hailong, C. (2011). Are herbal compounds the next frontier for alleviating learning and memory impairments? An integrative look at memory, dementia and the promising therapeutics of traditional Chinese medicines. *Phytotherapy Research, 25*(8), 1105–1118.
Kurosawa, M., Sato, A., & Sato, Y. (1989). Stimulation of the nucleus basalis of Meynert increases acetylcholine release in the cerebral cortex in rats. *Neuroscience Letters, 98*(1), 45–50.
Lim, G. P., Chu, T., Yang, F., Beech, W., Frautschy, S. A., & Cole, G. M. (2001). The curry spice curcumin reduces oxidative damage and amyloid pathology in an Alzheimer transgenic mouse. *Journal of Neuroscience, 21*(21), 8370–8377.
Ma, X., Tan, C., Zhu, D., Gang, D. R., & Xiao, P. (2007). Huperzine A from Huperzia species—An ethnopharmacolgical review. *Journal of Ethnopharmacology, 113*(1), 15–34.
Mantle, D., Pickering, A. T., & Perry, E. K. (2000). Medicinal plant extracts for the treatment of dementia. *CNS Drugs, 13*(3), 201–213.

Mishra, S., & Palanivelu, K. (2008). The effect of curcumin (turmeric) on Alzheimer's disease: An overview. *Annals of Indian Academy of Neurology, 11*(1), 13–19.
Moretti, A., Gorini, A., F Villa, R. (2011). Pharmacotherapy and prevention of vascular dementia. *CNS Neurological Disorders Drug Targets*, 10(3), 370–390.
Morgan, D. (2011). Immunotherapy for Alzheimer's disease. *Journal of Internal Medicine, 269*(1), 54–63.
Nagaraja, D., & Jayashree, S. (2001). Randomized study of the dopamine receptor agonist piribedil in the treatment of mild cognitive impairment. *American Journal of Psychiatry, 158*(9), 1517–1519.
Nicholson, C. D. (1990). Pharmacology of nootropics and metabolically active compounds in relation to their use in dementia. *Psychopharmacology (berl), 101*(2), 147–159.
Nicolaus, B. J. (1982). Chemistry and pharmacology of nootropics. *Drug Development Research, 2*(5), 463–474.
Nunan, J., & Small, D. H. (2000). Regulation of APP cleavage by α-, β-and γ-secretases. *FEBS Letters, 483*(1), 6–10.
Olin, JT., Schneider, L. (2002). Galantamine for alzheimer's disease. *Cochrane Database Systematic Reviews*, 3. (CD001747)
Panza, F., Frisardi, V., Imbimbo, B. P., Capurso, C., Logroscino, G., Sancarlo, D., Seripa, D., Vendemiale, G., Pilotto, A., & Solfrizzi, V. (2010). γ-Secretase inhibitors for the treatment of Alzheimer's disease: The current state. *CNS Neuroscience & Therapeutics, 16*(5), 272–284.
Patel, L., & Grossberg, G. T. (2011). Combination therapy for Alzheimer's disease. *Drugs and Aging, 28*(7), 539–546.
Peretti, C. S., Gierski, F., & Harrois, S. (2004). Cognitive skill learning in healthy older adults after 2 months of double-blind treatment with piribedil. *Psychopharmacology (berl), 176*(2), 176–182.
Perry, E., & Howes, M. J. (2011). Medicinal plants and dementia therapy: Herbal hopes for brain aging ? *CNS Neuroscience & Therapeutics, 17*(6), 683–698.
Polinsky, R. J. (1998). Clinical pharmacology of rivastigmine: A new-generation acetylcholinesterase inhibitor for the treatment of Alzheimer's disease. *Clinical Therapeutics, 20*(4), 634–647.
Potkin, S. G., Anand, R., Fleming, K., Alva, G., Keator, D., Carreon, D., Messina, J., Wu, J. C., Hartman, R., & Fallon, J. H. (2001). Brain metabolic and clinical effects of rivastigmine in Alzheimer's disease. *International Journal of Neuropsychopharmacology, 4*(3), 223–230.
Rogawski, M. A., & Wenk, G. L. (2003). The neuropharmacological basis for the use of memantine in the treatment of Alzheimer's disease. *CNS Drug Reviews, 9*(3), 275–308.
Rösler, M., Bayer, T., Anand, R., Cicin-Sain, A., Gauthier, S., Agid, Y., Dal-, P., Stähelin, H. B., Hartman, R., & Gharabawi, M. (1999). Efficacy and safety of rivastigmine in patients with Alzheimer's disease: International randomised controlled trial commentary: Another piece of the Alzheimer's jigsaw. *BMJ, 318*(7184), 633–640.
Salomone, S., Caraci, F., Leggio, G. M., Fedotova, J., & Drago, F. (2012). New pharmacological strategies for treatment of Alzheimer's disease: Focus on disease modifying drugs. *British Journal of Clinical Pharmacology, 73*(4), 504–517.
Shishodia, S., Sethi, G., & Aggarwal, B. B. (2005). Curcumin: Getting back to the roots. *Annals. New York Academy of Sciences, 1056*(1), 206–217.
Sierpina, V. S., Wollschlaeger, B., & Blumenthal, M. (2003). Ginkgo biloba. *American Family Physician, 68*(5), 923–926.
Singh, N., Pandey, B. R., & Verma, P. (2011). An overview of phytotherapeutic approach in prevention and treatment of Alzheimer's syndrome & dementia. *International Journal of Pharmaceutical Sciences and Drug Research, 3*(3), 162–172.
Small, D. H. (2005). Acetylcholinesterase inhibitors for the treatment of dementia in Alzheimer's disease: Do we need new inhibitors ? *Expert Opinion on Emerging Drugs, 10*(4), 817–825.
Sonkusare, S. K., Kaul, C. L., & Ramarao, P. (2005). Dementia of Alzheimer's disease and other neurodegenerative disorders—Memantine, a new hope. *Pharmacological Research, 51*(1), 1–7.

Soudamini, K. K., Unnikrishnan, M. C., Soni, K. B., & Kuttan, R. (1992). Inhibition of lipid peroxidation and cholesterol levels in mice by curcumin. *Indian Journal of Physiology and Pharmacology, 36*, 239–242.

Spillantini, M. G., & Goedert, M. (1998). Tau protein pathology in neurodegenerative diseases. *Trends in Neurosciences, 21*(10), 428–433.

Standridge, J. B. (2004). Pharmacotherapeutic approaches to the treatment of Alzheimer's disease. *Clinical Therapeutics, 26*(5), 615–630.

Takeda, A., Loveman, E., Clegg, A., Kirby, J., Picot, J., Payne, E., & Green, C. (2006). A systematic review of the clinical effectiveness of donepezil, rivastigmine and galantamine on cognition, quality of life and adverse events in Alzheimer's disease. *International Journal of Geriatric Psychiatry, 21*(1), 17–28.

Tamayev, R., & D'Adamio, L. (2012). Inhibition of γ-secretase worsens memory deficits in a genetically congruous mouse model of Danish dementia. *Molecular Neurodegeneration, 7*(1), 19.

Tariot, P. N., Farlow, M. R., Grossberg, G. T., Graham, S. M., McDonald, S., Gergel, I. (2004). Memantine Study Group. Memantine treatment in patients with moderate to severe Alzheimer disease already receiving donepezil: a randomized controlled trial. *JAMA, 291*(3), 317–324.

Tariska, P., & Paksy, A. (2000). Cognitive enhancement effect of piracetam in patients with mild cognitive impairment and dementia. *Orvosi Hetilap, 141*(22), 1189–1193.

Waegemans, T., Wilsher, C. R., Danniau, A., Ferris, S. H., Kurz, A., & Winblad, B. (2002). Clinical efficacy of piracetam in cognitive impairment: A meta-analysis. *Dementia and Geriatric Cognitive Disorders, 13*(4), 217–224.

Wang, Y., Huang, L. Q., Tang, X. C., & Zhang, H. Y. (2010). Retrospect and prospect of active principles from Chinese herbs in the treatment of dementia. *Acta Pharmacologica Sinica, 31*(6), 649–664.

Whitehouse, P. J., Price, D. L., Struble, R. G., Clark, A. W., Coyle, J. T., & Delon, M. R. (1982). Alzheimer's disease and senile dementia: Loss of neurons in the basal forebrain. *Science, 215*(4537), 1237–1239.

Wilcock, G., Howe, I., Coles, H., Lilienfeld, S., Truyen, L., Zhu, Y., Bullock, R., & Kershaw, P. (2003). A long-term comparison of galantamine and donepezil in the treatment of Alzheimer's disease. *Drugs and Aging, 20*(10), 777–789.

Xu, Z. Q., Liang, X. M., Zhang, Y. F., Zhu, C. X., & Jiang, X. J. (2012). Treatment with Huperzine A improves cognition in vascular dementia patients. *Cell Biochemistry and Biophysics, 62*(1), 55–58.

Yamada, K., & Nabeshima, T. (2000). Animal models of Alzheimer's disease and evaluation of anti-dementia drugs. *Pharmacology Therapeutics, 88*(2), 93–113.

Yang, F., Lim, G. P., Begum, A. N., Ubeda, O. J., Simmons, M. R., Ambegaokar, S. S., Chen, P. P., Kayed, R., Glabe, C. G., Frautschy, S. A., & Cole, G. M. (2005). Curcumin inhibits formation of amyloid β oligomers and fibrils, binds plaques, and reduces amyloid in vivo. *Journal of Biological Chemistry, 280*(7), 5892–5901.

Zhang, H. Y., Zheng, C. Y., Yan, H., Wang, Z. F., Tang, L. L., Gao, X., & Tang, X. C. (2008). Potential therapeutic targets of huperzine A for Alzheimer's disease and vascular dementia. *Chemico-Biological Interactions, 175*(1–3), 396–402.

The manufacturer's authorised representative in the EU is Springer Nature Customer Service Centre GmbH, Europaplatz 3, 69115 Heidelberg, Germany. If you have any concerns regarding our products, please contact ProductSafety@springernature.com

Printed and bound by CPI Group (UK) Ltd, Croydon, CR0 4YY
30/09/2026
02225295-0012